Statistical Procedures for Machine and Process Qualification

Statistical Procedures for Machine and Process Qualification

Edgar Dietrich
Alfred Schulze

ASQ Quality Press
Milwaukee, Wisconsin

Statistical Procedures for Machine and Process Qualification
Edgar Dietrich
Alfred Schulze

Library of Congress Cataloging-in-Publication Data
Dietrich, Edgar, 1951–
[Statistische Verfahren zur qualifikation von Messmittein, Maschinen und Prozessen. English]
Statistical procedures for machine and process qualificaion / Edgar Dietrich, Alfred Schulze.
p. cm.
Includes bibliographical references.
ISBN 0-87389-447-2 (alk. paper)
1. Process control—Statistical methods. I. Schulze, Alfred, 1952– II. Title.
TS156.8.D52 1999
670.42′75—dc21

10 9 8 7 6 5 4 3 2 1

ISBN 0-87389-447-2

Acquisitions Editor: Ken Zielske
Project Editor: Annemieke Koudstaal
Production Coordinator: Shawn Dohogne

ASQ Mission: The American Society for Quality advances individual and organizational performance excellence worldwide by providing opportunities for learning, quality improvement, and knowledge exchange.

Attention: Bookstores, Wholesalers, Schools and Corporations: ASQ Quality Press books, videotapes, audiotapes, and software are available at quantity discounts with bulk purchases for business, educational, or instructional use. For information, please contact ASQ Quality Press at 800-248-1946, or write to ASQ Quality Press, P.O. Box 3005, Milwaukee, WI 53201-3005.

To place orders or to request a free copy of the ASQ Quality Press Publications Catalog, including ASQ membership information, call 800-248-1946. Visit our web site at http://www.asq.org.

Printed in the United States of America

Printed on acid-free paper

American Society for Quality

Quality Press
611 East Wisconsin Avenue
Milwaukee, Wisconsin 53202
Call toll free 800-248-1946
http://www.asq.org

Preface

This book describes the statistical methods and procedures required for machine and process qualification. The procedures are presented by way of a review. For all considerations, the requirements of the DIN ISO norm, as well as company and association guidelines, are to the fore. Contrary to the usual statistical textbooks, where statistical procedures are often treated without any practical relation, efforts are made here to present rules and handling procedures to the practical person. The case examples included originate from questions arising daily in the industry.

In the course of a great number of seminars and training courses as well as presentations of papers (all based on traditional training material), the authors were able to establish the fact that the knowledge contained in this material is sufficient for basic comprehension of the statistical procedures but does not answer any practice-related questions. Without the use of computer systems, the procedures presented are applicable only to ideal cases, which rarely occur in practice, in a realistic period of time. Handling of the procedures is usually emphasized without giving details on their significance and targets and the related application possibilities. Also, the derivation of the procedures takes an unnecessary amount of time. The graphical displays are used insufficiently because of the handling. It is impossible to poll the information required for decision-making from the process data arising continuously at different locations.

The authors in this book take a different approach in connection with a software program. They present the procedures, describe their application possibilities, show interactions, and give interpretations of the results. They treat practical cases and evaluation criteria. Special effort is made to present the matter to the reader in a transparent way by using informative graphical displays.

Company standards are taken into account also. Nomenclature is based on definitions taken from norms or set by the DGQ (German Society for Quality). The statistical procedures must be regarded as tools (Tool Box) for collecting information. The reader will learn

- Which tools exist
- Where they may be used
- How to handle them
- How to present and interpret the results
- Which quality criteria may be used for evaluation of the results

The reader finds his daily tasks mirrored in a sufficient way. Thus, the book may also be used as a *reference book* to freshen up on one or the other subject. Basic knowledge of statistics assumed, most examples may be calculated using the appropriate tables and with the help of pocket calculators. Depending on the example, this may be rather cumbersome. The examples may be calculated using a computer and the software package qs-STAT® from Q-DAS® GmbH, Birkenau. The Demo version is sufficient for this purpose. You may order it directly from: Q-DAS® GmbH, Balzenbacher Str. 57, D-69488 Birkenau, Telephone: ++49-6201/3941-0, Fax: ++49-6201/3941-24, according to the currently valid price list, or from the American branch: Q-DAS Inc., 2582 Product Drive, Rochester Hills, MI 48309, USA, Fax: ++1-248-650-6226.

The authors would like to express their thanks to the numerous decision-makers from industry and institutions for participating in many discussions on this subject. The experiences gained there form the basis for this book. Our special thanks go to Mrs. Mesad, Q-DAS® GmbH, for preparation of text and graphics.

Birkenau, September 1998
Edgar Dietrich and Alfred Schulze

Table of Contents

Figures

Tables

1 Introduction

1.1 The Use of Statistical Methods

Since the early 1970s, when Statistical Process Control (SPC) was "rediscovered" in the Western industrialized world, there has been continual discussion as to the "pros and cons" of SPC.

As a positive result of this discussion, it is now recognized, on the basis of facts and experience, that, correctly and properly applied, statistical process control is a highly valuable tool in the field of industrial manufacture and production. A modern, forward-looking company will therefore make use of SPC in order to benefit from the resultant improvement in the quality of its products. Today's increasing competition seems certain to ensure that all companies will need the competitive advantages that SPC brings in order to survive in tomorrow's marketplace.

Hence no company can dispense with the logical and correct use of SPC as the driving force behind improvements in quality. Considered in this manner, SPC is more than a control chart or an index of capability. It is a system that, based on process data, describes a manufacturing process, including the environmental factors affecting it, in model form (see Figure 1.1–1). It provides important information for decision making. Only from this viewpoint is it economical and logical to make use of SPC.

While the benefits of SPC are no longer open to doubt, as they were in the early 1980s, it still has not been possible to arrive at uncontested standard procedures for control charting. Also, it has not yet been possible to set out a universally accepted standard on how to determine capability indices. There is certainly a need for action in this area, and a substantial step in the direction of improvement of product quality will be possible when standardization of procedures throughout industry leads to the use of a standard language. One important initial step in this direction is the VDA Paper 4, which appeared earlier this year. This VDA Paper is concerned with quality improvement and all tools and methods that can be used in industry to improve product quality to the benefit of the customer. Specifically, Chapter 10, "Process Evaluation," was taken into account for this book.

Quality tools may be divided into two categories. The first group, to which SPC belongs, is concerned with monitoring and improving ongoing production. This group is also known as "on-line quality tools."

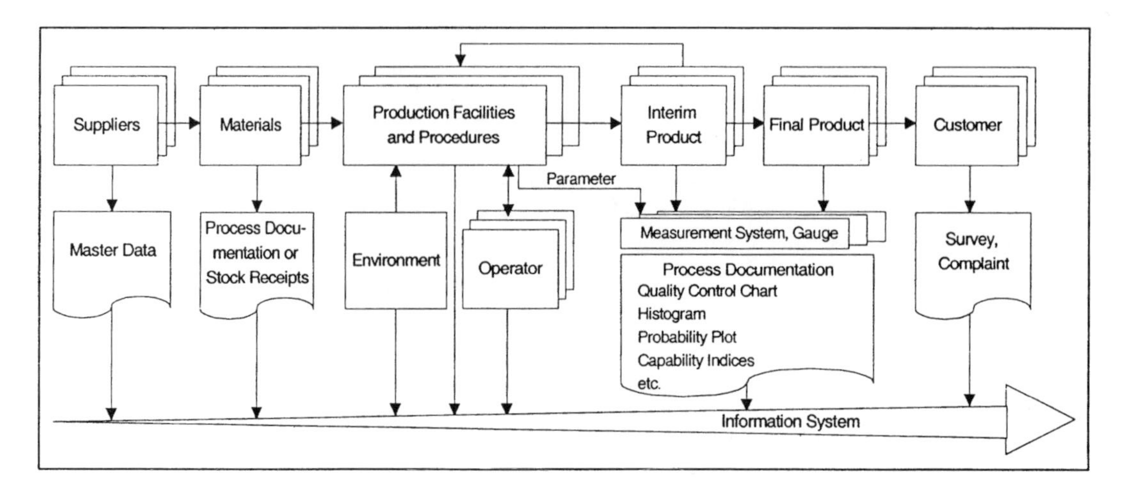

Figure 1.1–1. SPC as a process information system.

The second group, which includes among others the whole subject of experimentation (DOE, Design of Experiments), is concerned with the statistical analysis of test runs, with the object of obtaining crucial information about product quality while still in the early development stage in order to exert a positive influence on product quality right from the very beginning of the product cycle. This group of quality tools is also known as "off-line quality tools." Although it is not the main content of this book, we shall at this point refer to the whole subject of experimentation as a very important off-line quality tool.

The procedures laid down by Shainin and Taguchi are particularly highly regarded as methods for experimentation. The carrying out of such experiments and the statistically based analyses of their results make it possible to detect quality risks at a very early planning stage and with relatively small expenditure, and to take suitable action to prevent the appearance of defects and quality problems in subsequent production.

The basic procedure for determining the quality capability of a manufacturing process is described in a very detailed and easy-to-follow manner in this book. Based on Volume 4 of the VDA Paper, this book deals with the procedures for control charting and the determination of capability indices in great detail and so can also be used as an instruction guide for process capability studies while providing a wealth of supplementary general explanations.

The procedures described should assist the user to accept SPC as an easily understandable tool and to use it to improve quality. In this connection, the different procedures for determining capability indices are of particular importance and are explained, commented upon, and compared with one another; and a standard procedure for assessing process capability is suggested. It is hoped that this method, which at present forms the basis for a suggested ISO standard, will soon be adopted by industry as a standard procedure. Such a step would represent a substantial simplification for the whole of the supplier industry, which would then finally be in a position to apply a single standard procedure for all clients.

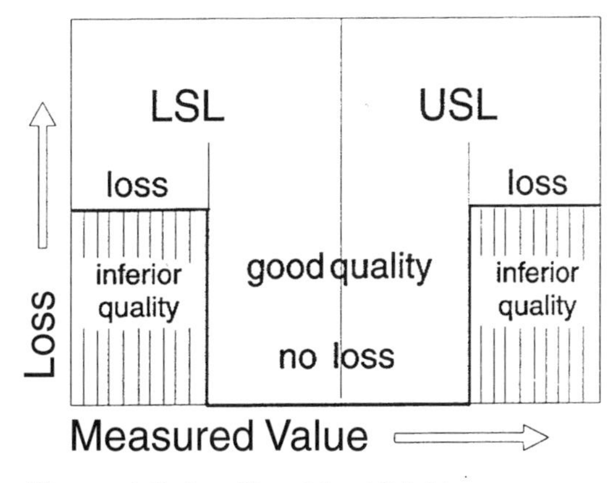

Figure 1.2–1. Good-bad thinking.

1.2 SPC as the Basis of "Never-Ending Improvement": The Principle of Continuous Quality Improvement

The original definition of quality as "keeping within the specified limits" offers no incentive to the never-ending improvement of product quality. As long as the products lie within the tolerances given in the drawings, nobody will, on the basis of this theory, feel the need to seek any improvement in the process concerned.

All those taking part in a process must, however, using technically feasible and economically justifiable measures, make an effort to produce a product with the least possible variation about the desired target value. The need for such action arises from the continually increasing desire of the customer for high-quality products. The readiness to act accordingly and to commit oneself to making the product as good as possible must be firmly fixed in the thoughts and in the quality understanding of all those taking part in the manufacture of a product.

The following explanation illustrates the situation and underscores the need to apply the principle of "never-ending improvement" in a modern, quality-conscious company.

First, let us describe the original view of quality in the form of a model (see Figure 1.2–1). All products that were within tolerance were described as "good." No distinction was made between a part having its actual value in the middle of the tolerance range and a part that lay just inside the tolerance limit. According to this definition, both parts were regarded as good. Considering the parts' intended use, this decision is, however, extremely questionable.

If we compare a part just outside tolerance with a part just inside, it is difficult to understand why one part is suitable without restriction for its intended purpose and the other part is classified as completely useless. These two parts, with only a negligible difference in their actual values, cannot exhibit such great differences with regard to their capacity for use as the conventional point of view implies.

To obtain an accurate picture of the quality performance of a given product, a model should be used that gives a better description of the quality loss caused by deviations from the nominal value. The model of the loss function is ideal for

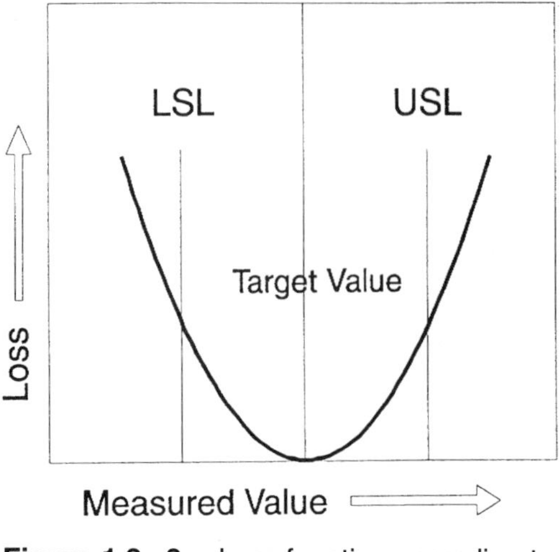

Figure 1.2.–2. Loss function according to Taguchi.

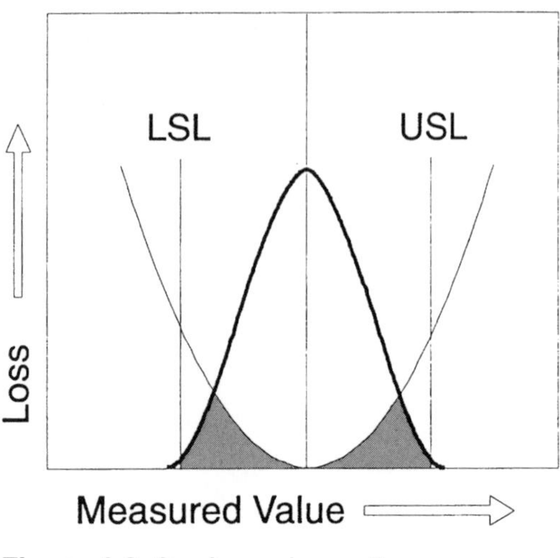

Figure 1.2–3. Loss depending on process location and spread.

description of this situation (see Figure 1.2–2). To understand the model, it is particularly important to realize that monetary loss and quality impairment are considered over the whole area of the product's practical use. Taguchi has very aptly described the loss incurred in this area as a loss for society and states that one should do everything possible to incorporate into the model all factors that could lead to such a loss. According to his model, the loss incurred is a continuous variable having a minimum at the target value. This target value is a value dependent on the function of the product, which corresponds to the optimum expectation of the customer and, given correct planning and design, corresponds to the nominal value laid down in the specification. The more a product or a product characteristic deviates from its target value, the greater the loss caused by this deviation.

In Taguchi's model, this relationship is illustrated by means of a parabola that reaches its minimum at the target value. This model clearly shows the importance of achieving the target value in order to minimize loss. If we superpose a normal distribution curve on the loss function (see Figure 1.2–3), it is immediately seen that a process producing normally distributed parts will cause a much smaller loss than a process with a rectangular distribution that just manages to squeeze within the predetermined limits of the specification. If we make further use of this model, we can see that a reduction in spread will also cause a reduction in loss. Furthermore, we see the importance not only of taking into account the spread of the process, but also of taking a similar interest in the mean value (i.e., centering the process on the target value), even if the process spread only uses up a small part of the overall tolerance. This model representation does, of course, sometimes provoke discussions as to the extent to which the illustrations used here provide an exact description and are fully valid for all production processes, but the really important point is to recognize that thought and action oriented exclusively to compliance with specification will never be consistent with the goal of continuous quality improvement.

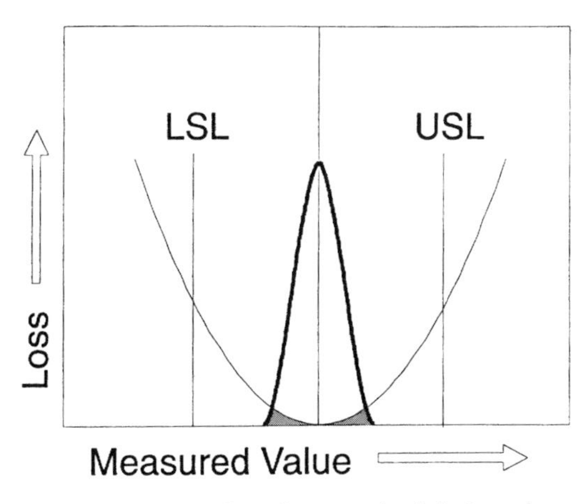

Figure 1.2–4. Small spread minimizes loss.

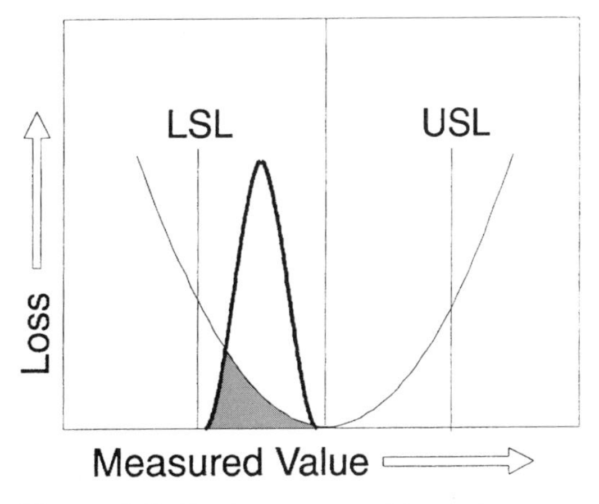

Figure 1.2–5. Loss caused by a change in process location.

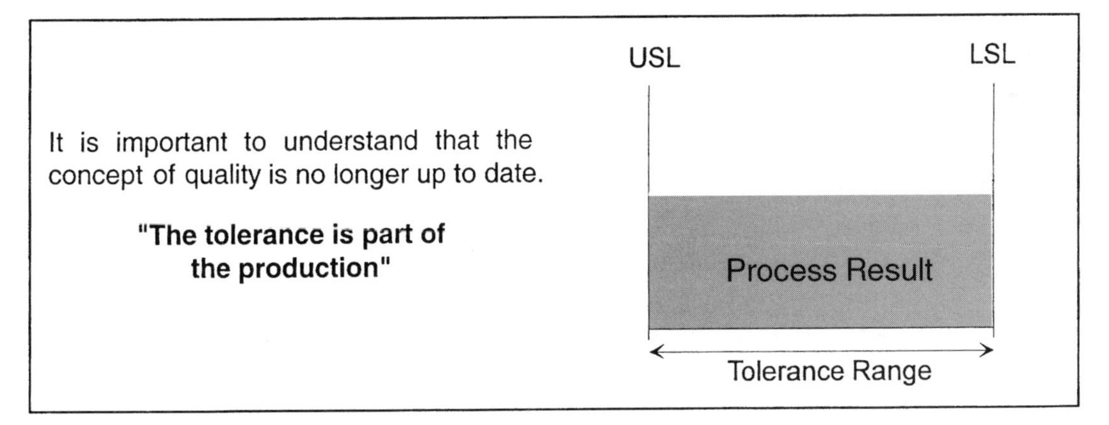

Figure 1.2–6. Use of the entire tolerance range.

Based on these considerations, the methods used in manufacturing, both in process design and actual process operation, should be reviewed.

To survive in today's intense industrial competition, it is absolutely necessary to meet the customer's wishes for optimum quality. It is not sufficient to offer products produced in accordance with the overall tolerance. All those involved in a process must do what they can to center the process on its target value (nominal) (see Figure 1.2–5). At the same time, efforts should be made to keep the spread as low as possible (see Figure 1.2–4) and continually reduce it further, as far as is economically feasible.

Figures 1.2–6 and 1.2–7 show clearly that a process that uses up the total specified tolerance in the manufacture of a product provides lower quality than a process controlled by the principle of operation centered on the target value.

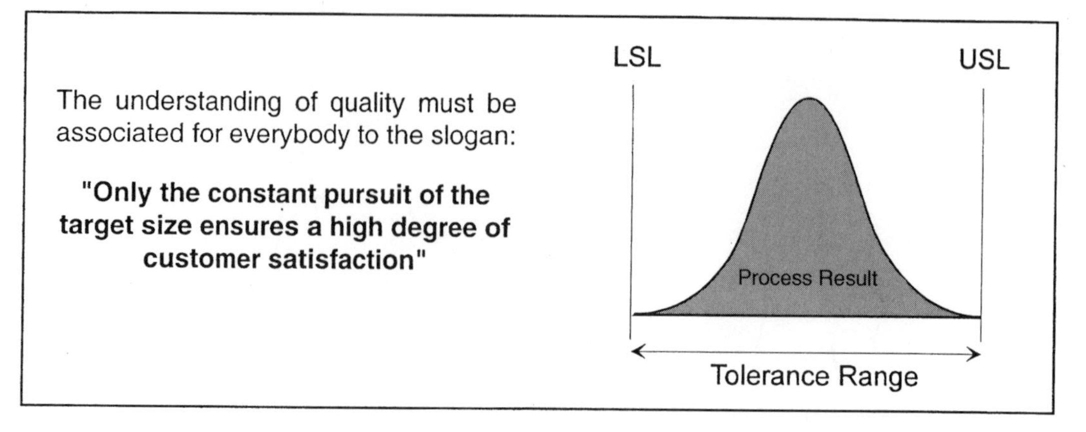

Figure 1.2–7. Target-value-oriented use of the tolerance range.

1.3 Different Types of Process Models

To obtain a comprehensive and correct machine or process assessment, it is necessary to describe the situation accurately by a model, using statistical methods. The better this modeling, the better and more accurate the results and the more valuable the information so obtained. How can suitable models be found?

For this purpose we can, for a given production process or machine, measure either process parameters or certain characteristics of the finished products. These data indicate—on the assumption of a suitable measurement procedure and sufficiently large samples—the performance of the machine or process. For a deterministic description, a function $f(x_i)$ would have to be found that clearly defines the relationship between the parameter x_i and the relevant value of the function. In general, this is not possible due to the complex technical relationships. Hence the process is simulated in the form of a mathematical model. The parameters of this model are derived from the measured data by means of statistical procedures. All assessments and capability studies are based on this model. The accuracy of the results is therefore linked directly to the accuracy of the theoretical model.

Experience has shown that real-life processes can be represented by three different process models. In this book, these will be named

- Process model A (Figure 1.3–1)
- Process model B (Figure 1.3–3)
- Process model C (Figure 1.3–4)

The models shown are represented in an idealized form. In reality, there are no clear boundaries between these model scenarios. Hence it remains for the observer to find the best model in any given case on the basis of the procedures indicated. Comparative representations in particular will facilitate assessment and help avoid erroneous interpretations.

The aim of the assessment is, with the aid of statistical procedures,

- To find the appropriate model
- To assess the goodness of fit of the model
- To calculate statistics based on the model
- To interpret the results

For this purpose there are purely numerical procedures and also graphical methods available. Both are treated in this book.

Process model A is characterized by the following features:

- The distribution of the characteristic values at time (t) is a normal distribution with standard deviation σ (t) and mean value μ (t).
- The standard deviation σ (t) of the process is constant.
- The mean value μ (t) of the process is constant.

For this process, $\sigma_{tot} = \hat{\sigma}$, i.e., the overall standard deviation is identical to the short-term spread of the process. The process model described here represents an ideal case (classical Shewhart model), which is rarely found in practice.

Normal distribution is usually not applicable in the case of certain characteristic types like shape and location as well as unilateral physically limited characteristics. For these, process model A1 may be allocated (see Figure 1.3–2).

Process model A1 is characterized by the following features:

- The momentary distribution of the characteristic values at time (t) is no normal distribution.
- Process location, variation, skewness, and kurtosis are constant over time.

Process model B is characterized by the following features:

- The distribution of the characteristic values at time (t) is a normal distribution with standard deviation σ (t) and mean value μ (t).
- The standard deviation σ (t) of the process is constant.
- The mean values μ (t) of the process are normally distributed about a mean that remains constant over a long period.

Process model C is characterized by the following features:

- The distribution of the characteristic values at time (t) is a normal distribution with standard deviation σ (t) and mean value μ (t).
- The standard deviation σ (t) of the process is constant.
- The mean values μ (t) change in accordance with known laws.

In practice, process behavior frequently does not correspond to the idealized form presented here. This makes mixed forms possible that lead to

- Process model D (Figure 1.3–5)
- Process model E (Figure 1.3–6)
- Process model F (Figure 1.3–7)
- Process model G (Figure 1.3–8)

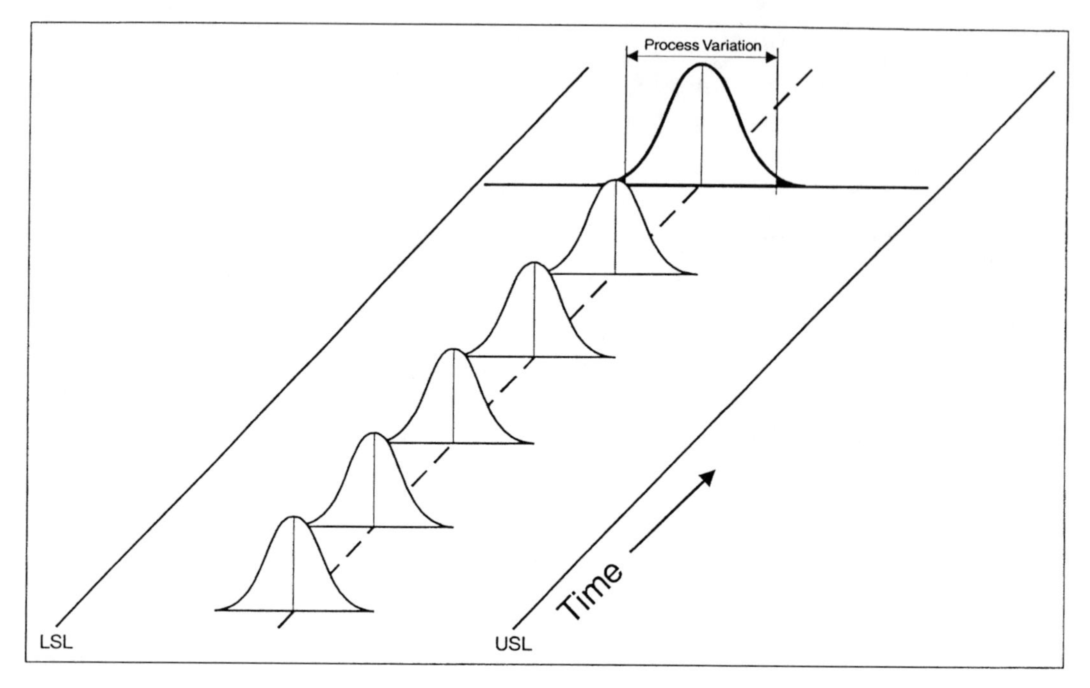

Figure 1.3–1. Process model A.

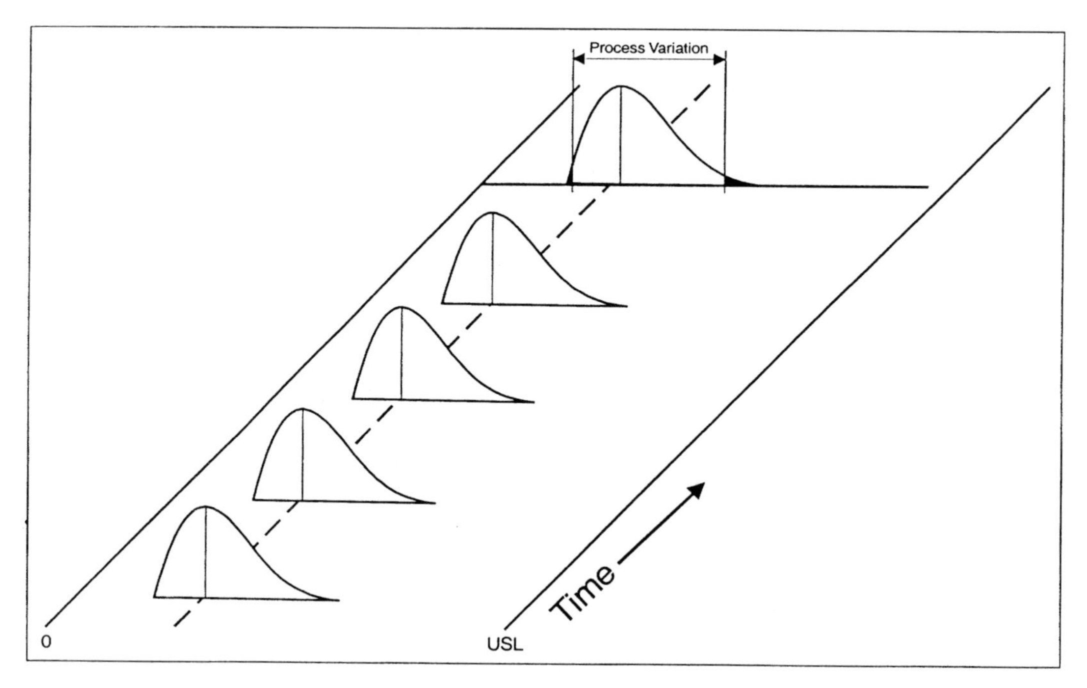

Figure 1.3–2. Process model A1.

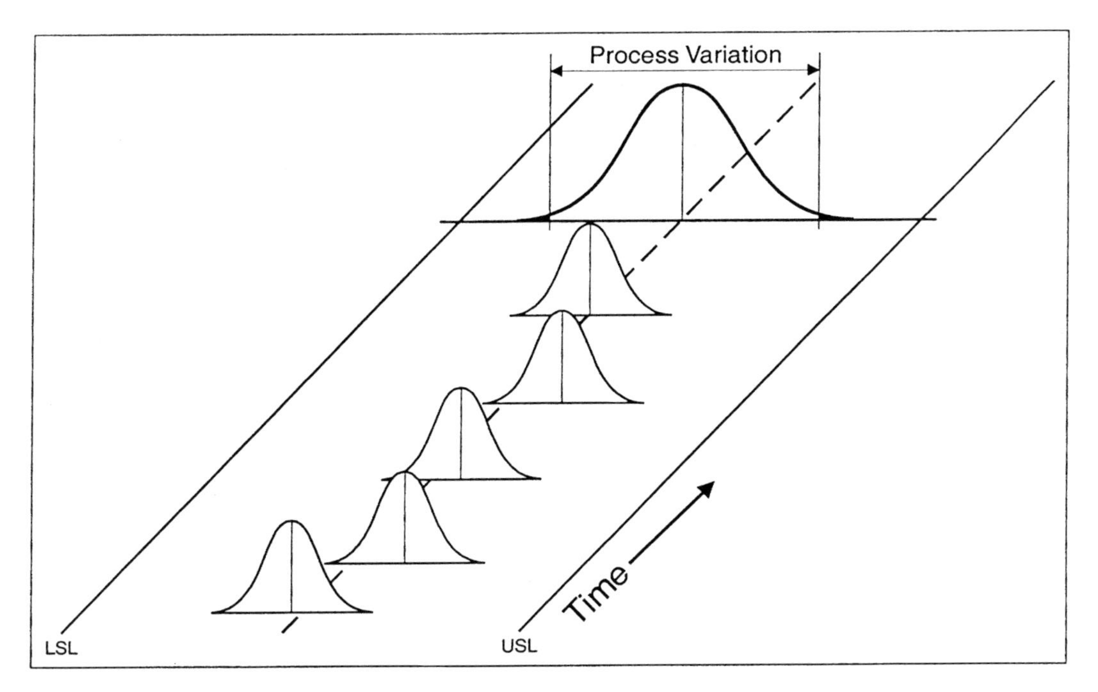

Figure 1.3–3. Process model B.

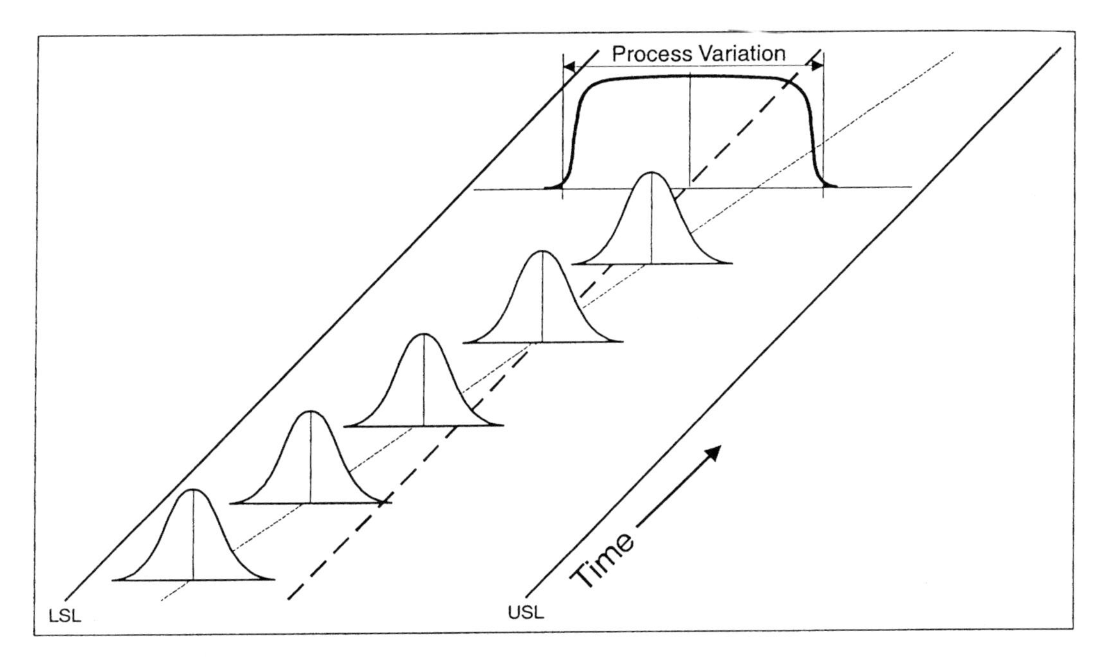

Figure 1.3–4. Process model C.

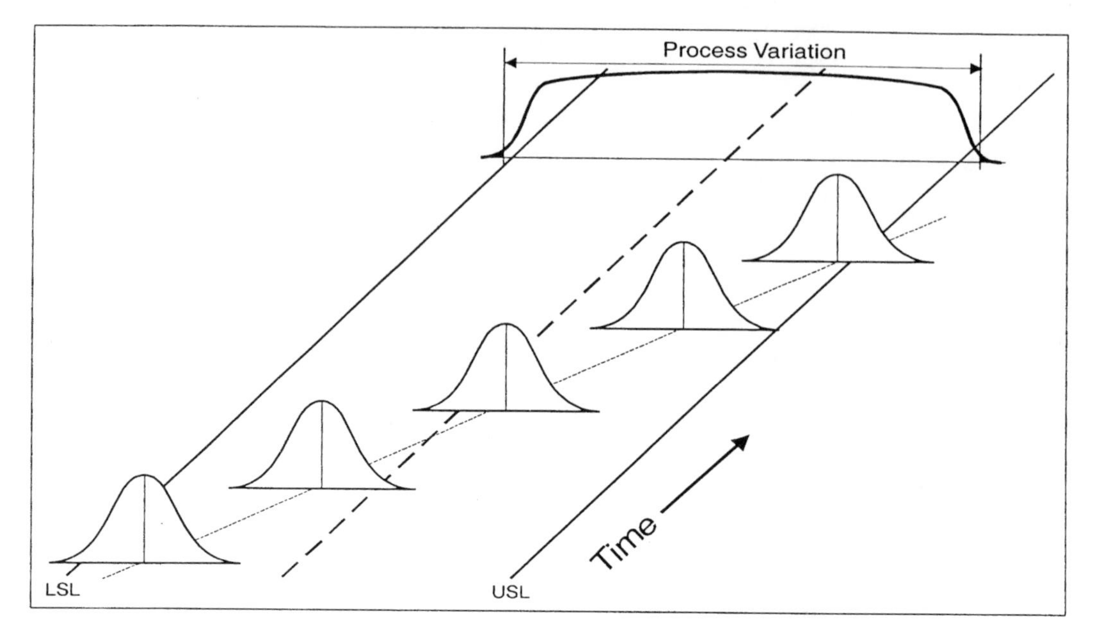

Figure 1.3–5. Process model D.

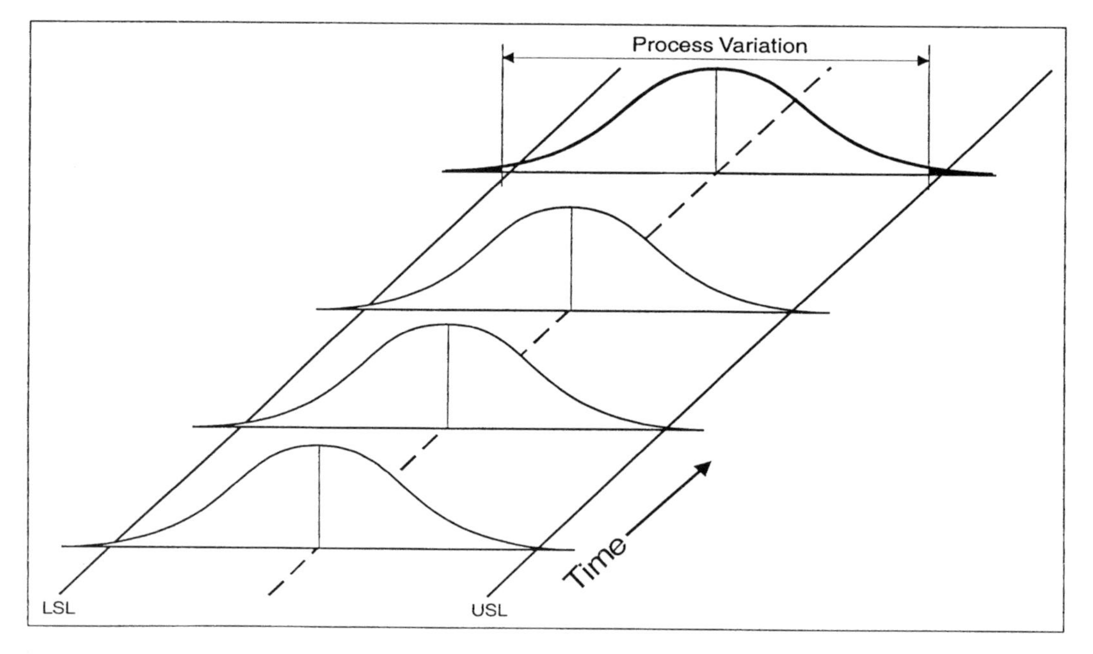

Figure 1.3–6. Process model E.

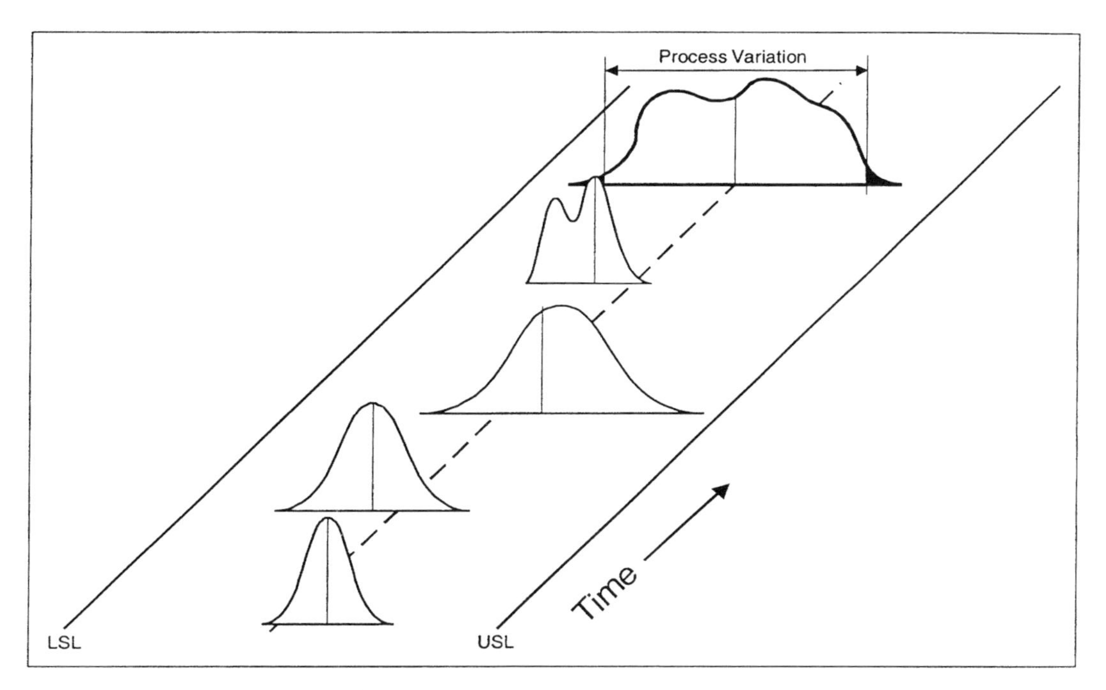

Figure 1.3–7. Process model F.

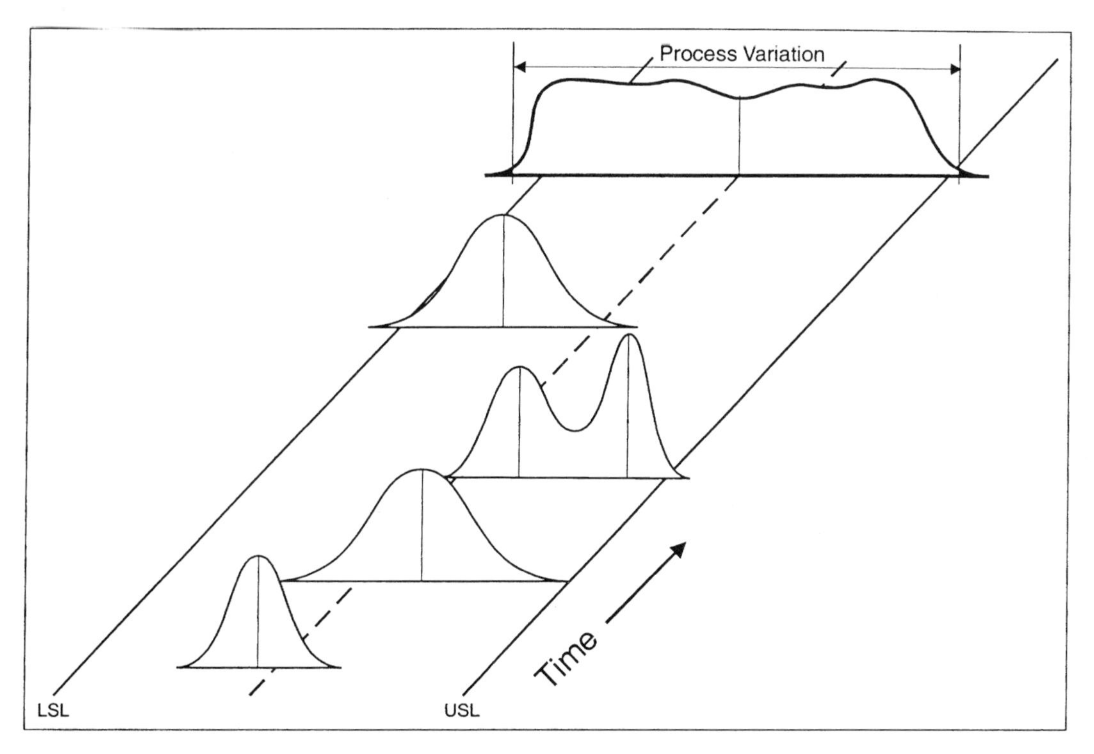

Figure 1.3–8. Process model G.

Process model D, in most cases, matches the behavior of process model C. The disproportionately large trend of the averages over time increases process variation in excess of the allowed limits. Model C may be attained through corrective procedures.

If this process behavior is caused by tool wear, then quality capability may be achieved by well-timed and proper adjustment.

Process model E, in most cases, matches process model A. The random variation interference leads to an improper process variation. This interference must be reduced with model A as objective.

Process model F does not show a steady variation σ (t) nor a steady average μ (t). Process variation does not exceed the given limits, so this process may be considered capable. As long as this situation is given, this process model may be treated like process model C or B.

Process model G does not show a steady variation σ (t) nor a steady average μ (t). Process variation exceeds the allowed limits. Process location and variation must be improved. If process variation reaches permitted values and, thus, quality capability, this process model may be treated like process model B or C.

Notes

According to these arguments, all process types (models D to G) may be traced back to the idealized models A, B, or C.

Initial results of substantial examinations carried out as part of an unpublished study of real processes show that mainly process models B, C or D, F, and G are applicable. Only a small portion may be approximated with a normal distribution (type A or B, E). Process type A1 is very rare and applies only with small subgroup sizes.

The study is based on about 1000 data sets from various companies and process types. Data sets ranged from less than 100 values to more than 200,000 values. The data sets were selected and evaluated according to equal criteria.

This study may be regarded as unique (at least in German-speaking territory) and clearly shows that the theory prepared by Mr. Shewhart in the 30s (process model A) is not suitable in practice for a realistic process description. All the results based on this model therefore must be wrong.

1.4 Stages of Assessment

The application of statistical methods and the evaluation of results depend on the stage at which machine or process assessment is to be performed. The main differences are between an analysis

- On purchase
- At introduction at the supplier's or on location
- Before start of mass production
- After start of mass production

Figure 1.4–1 shows typical applications and defines the individual stages. This definition will be referred to many times in the following chapters.

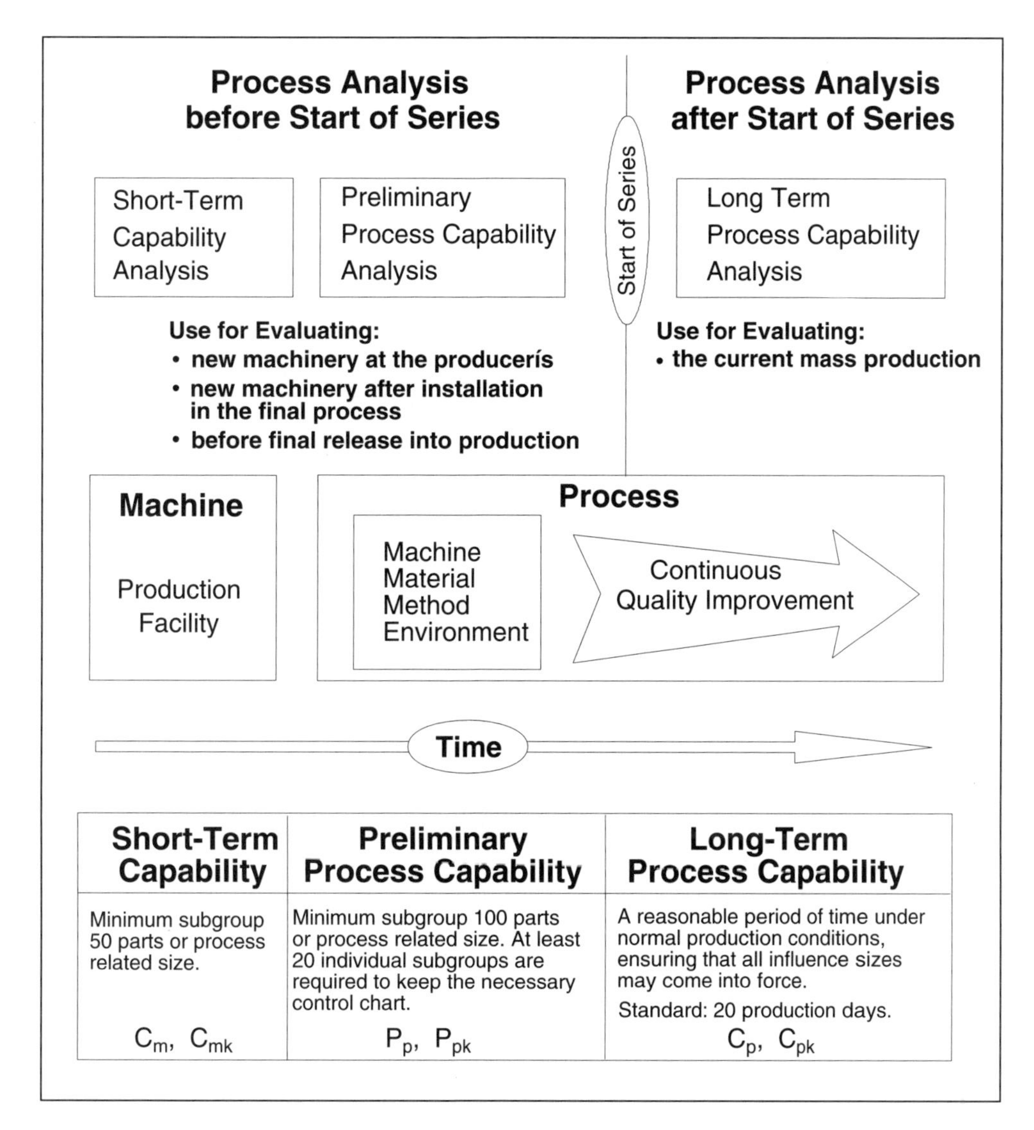

Figure 1.4–1. Process capability analysis.

Machine acceptance based on a short-term capability study (Figure 1.4–2) is applicable only if it is possible to study the machine exclusively; i.e., other influence factors—like material, man, environment, etc.—may be regarded as stable during evaluation.

Machine and process evaluation is based on characteristic values collected using measurement systems. For this reason, the capability of the measurement system must be established prior to every study (see Chapter 8: Measurement System Capability). Using defective or incorrect measurement systems will result in values that only insufficiently mirror the behavior of the facility or the process. This necessarily leads to an illegitimate evaluation.

The target of the initial system qualification and ongoing process control is

- To prove the capability in principle
- To establish process qualification
- To recognize significant changes in the running process

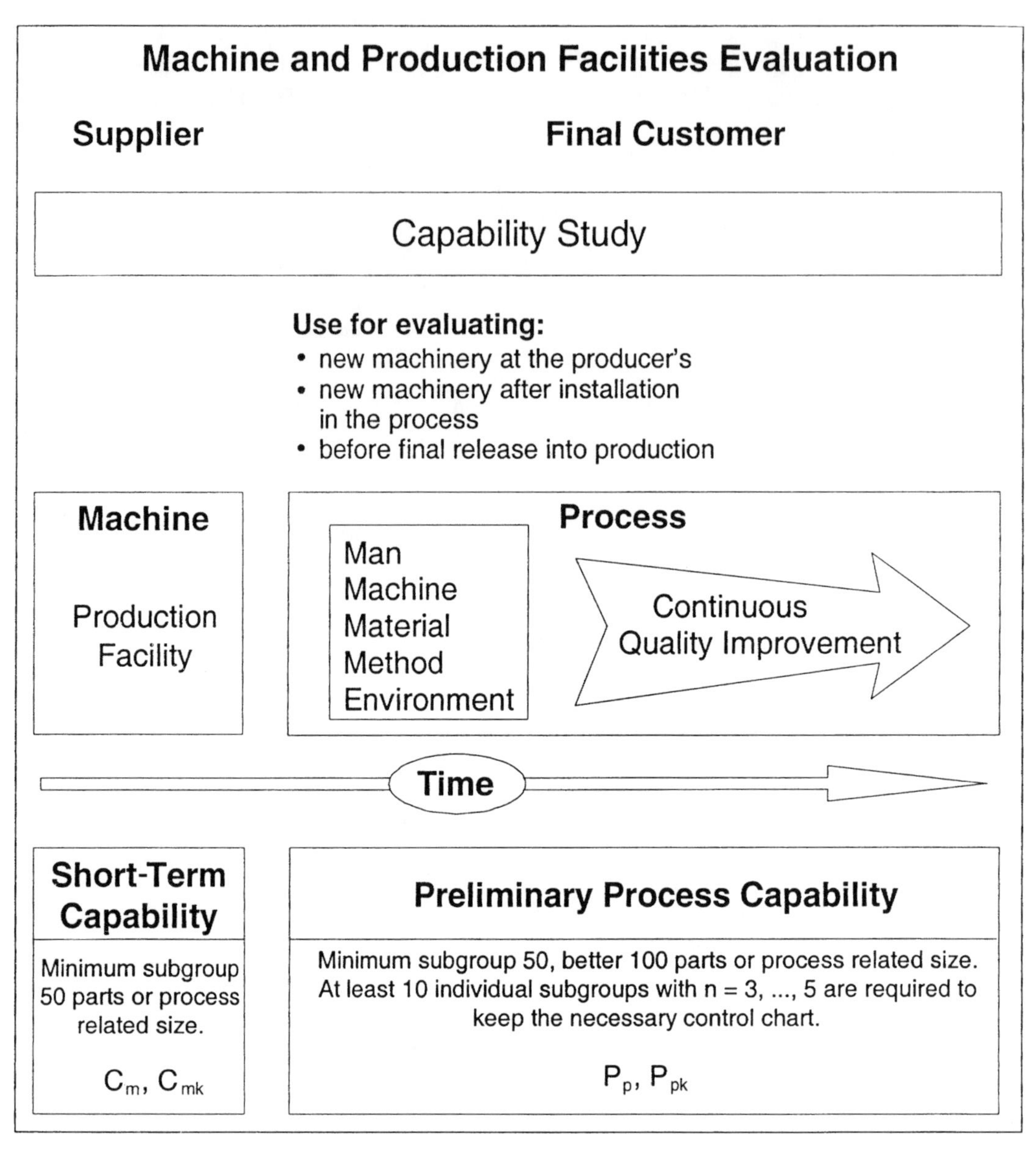

Figure 1.4–2. Acceptance of machines and production facilities.

In time, an information system is built up from which quality data may be polled for improvements and product or process optimization.

The initial qualification of a system and process evaluation before the start of a series is an especially unique procedure carried out for a given reason. This procedure is also called Machine Capability Study or Process Capability Study. It is often carried out off-line. For this purpose, parts are produced and measured and the resulting data is analyzed according to the given "Rules of the Game." This book describes the procedures in common practice today as well as the utilized methods. For reason of comparability and reproducibility at a later time, it is important to use the same procedure for data evaluation every time. The results are compared to the acceptance criteria. If insufficient results were obtained, improvement measures must be carried out or other equipment must be used. If the process is regarded as capable, then the results will be taken on as standards (i.e., suitable quality control charts) for later process control.

If machinery, facilities, and processes are regarded as capable and thus usable, it must be established whether the systems undergo significant changes during use as opposed to the results. This control usually is carried out in the running production. Parts are taken from the process as samples, then measured, and the measuring results are controlled according to the requirements of SPC (Statistical Process Control) with the help of quality control charting. If these results show deviations from the standards, improvement measures must be taken for the production facility, and already produced parts must be controlled. Because process control is carried out on the production line, it must be ensured that the required quality information may easily be gained from the process. Today, this is realized through automatic measurement systems and transfer of the results to a computer system. The measurement values are to be displayed on-line and evaluated according to the required statistical regulations. Deviations must be displayed to the user in an easily understandable way together with possible required corrective measures and reactions. Acceptance of this method is higher the less this procedure interferes with production and if process behavior is displayed to the user in an understandable and transparent way. This system must be a help to the user and interfere only a little or not at all with his usual work flow. The collected values must be analyzed at regular intervals to evaluate long-term process behavior.

1.5 Structure of the Book

Chapter 1 is concerned with the question "Why should statistical methods be used?" It gives the reasons that statistical methods are applied today in a wide range of manufacturing processes and illustrates different process models.

Chapters 2 and 3 describe basic statistical methods. Chapter 2 gives an introduction to engineering statistics; describes discrete, continuous, and parametric distributions; and gives an overview of the most important statistical concepts. In addition to graphical methods, numerical test procedures are discussed.

Quality control charting (Chapter 3) is at the heart of the statistical assessment of machines, processes, and products. Different types of control charts are discussed, both for discrete and continuous data. Shewhart control charts (conventional, with moving statistics and extended limits), acceptance charts, Pearson charts, Pre-Control, and Cusum charts are discussed. In addition, operating characteristic (OC) curves for assessment of the sensitivity of a given control chart method are described.

Chapters 4, 5, and 6—Assessment Criteria for Selection of a Distribution Model, Procedures for Process Assessment, and Process Capability Indices—build on the two preceding chapters. The methods described are used to show how suitable distribution and process models can be found. Assessment criteria are discussed that help to evaluate the quality of a model approximation. Based on these models, various procedures for calculating quality indices (capability indices) are discussed.

Chapter 7 deals with the development of an "assessment system" for tracking the assessment status of production equipment and processes. Economical use of SPC requires the application of the described methods.

Any statistical analysis is based on data collected by means of a checking or measuring system. Chapter 8 addresses the question of whether the measurement system used is suitable for the given process situation. In addition to general rationales and procedures, company standards for determining gauge capability indices are discussed.

Chapters 9 and 10 are devoted to the most important formulas and tables. One chapter deals with the subject of "analysis of variance" (ANOVA).

The test samples for assessment of SPC systems used by Ford Motor Company reflect practical applications and indicate the associated calculation procedures for control charting and capability assessment.

2
Introduction to Engineering Statistics

2.1 Overview

By the term statistics, we understand methods for obtaining, collecting, classifying, and evaluating observational data to reach reliable decisions based on these data. The economical aspect of statistical methods is assuming increasing importance.

Since it is frequently not possible (e.g., due to the time required or to destructive testing) to check each individual item in a population for a certain characteristic, investigations are carried out based on random samples. The important point is that then inferences are drawn from the sample data about the whole population.

A simple transfer of sample results to the whole population is, however, not possible. But, given certain conditions, it is permissible to draw conclusions about the whole population from the sample statistics.

To reach a sound understanding of the statistical procedures for machine and process assessment, a fundamental acquaintance with the following subject areas is required:

- Basic statistical concepts
- Probability distributions
- Statistical parameters
- Numerical test procedures
- Transformation of measurement data
- Distribution models and their assessment
- Stability criteria
- Operating characteristics
- Control charting methodology
- Analysis of variance (ANOVA)
- Capability studies and capability indices
- Regression and correlation
- Suitability of measurement procedures

These subjects are discussed in the following chapters.

If the use of these methods is to be an economic proposition, they must be carried out as far as possible automatically. As far as the interpretation of data is concerned, it is of prime importance that the original data are reliable and that they are interpreted correctly. The statistical procedures used are correct in themselves. The only question is whether they can or cannot be applied to any given situation. Apart from the procedures themselves, it is therefore necessary to consider what criteria can be used to establish their suitability in a given situation.

2.1.1 Basic Model of Inferential Statistics

While *descriptive statistics* suffices for the investigation and description of the population as far as possible, *inferential statistics* investigates only a part of the population, a *sample,* which is representative of the whole. In other words, inferences are drawn from the sample about the population as a whole (see Figure 2.1–1). In taking the sample, it must be ensured that each item of the population has the same chance of being included in the sample. The sample may then be regarded as representative of the population.

Inferential statistical methods are necessary whenever results cannot be reproduced as frequently as desired or to the desired degree of precision. The causes of this

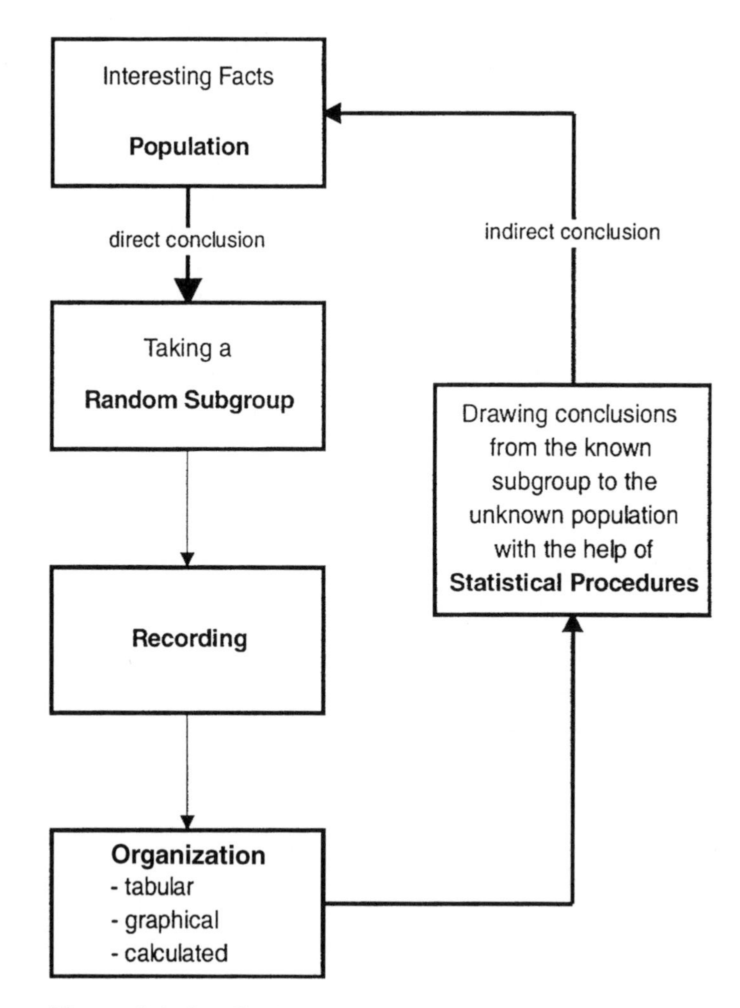

Figure 2.1–1. Basic model of inferential statistics.

lack of reproducibility lie in uncontrolled and uncontrollable influences. These influences lead to variation in the measured values of the characteristic concerned. Since, due to this variation, individual values are hardly ever reproducible, it is impossible to draw certain and unambiguous conclusions. Only with a correspondingly large sample is it possible to make a statement with a high probability of correctness.

When drawing conclusions from a sample about the population, one says that one is *making an inference,* and it is necessary to establish a *confidence interval* (see Section 2.1.4) for each estimated parameter. A typical application of this type of approach is the calculation of capability indices. The calculated result is based on a random subgroup. Another subgroup necessarily will lead to different results. This means that the *true value* of the result may be slightly larger or smaller. It is possible to estimate, by way of the confidence region, in which region the value lies with what probability. This range would have to be indicated for quality evaluation of every statistical value, especially for capability indices. One must be aware of this situation to avoid unnecessary discussion in practice.

When conclusions are drawn from a known population—or from a population that is assumed to be known—about the likely behavior of samples, one is *making a prediction.* In this context, it is important to determine *random dispersion intervals* (see Section 2.1.5). This type of approach forms the basis of control charting.

2.1.2 Classification of Characteristics

A manufactured product (part, element, sample item, etc.) is assessed according to its *properties.* Products may differ in their properties due to differing realizations of

- Quantitative and
- Qualitative

characteristics [11].* One piece is, in general, characterized by a number of characteristics (see Figure 2.1–2). We use the term *observation* to describe the result of measuring or checking a given characteristics, e.g., age or sex. The values of measurable characteristics are known as *measurement data,* those of countable characteristics as *count data.*

Continuous Characteristics

Observation of these characteristics produces measurement values. If we place these on a scale, each selected point on the scale can be occupied by a measured value. Examples include

- The age of a person
- Diameter of turned parts
- Monthly production costs

Discrete Characteristics

These produce whole number observed values. No value is possible between successive numbers of a scale. Examples include

- Number of nonconforming items in a sample
- Number of operational failures
- Number of parts manufactured per hour

**Bracketed numbers refer to the Bibliography located on page 380.*

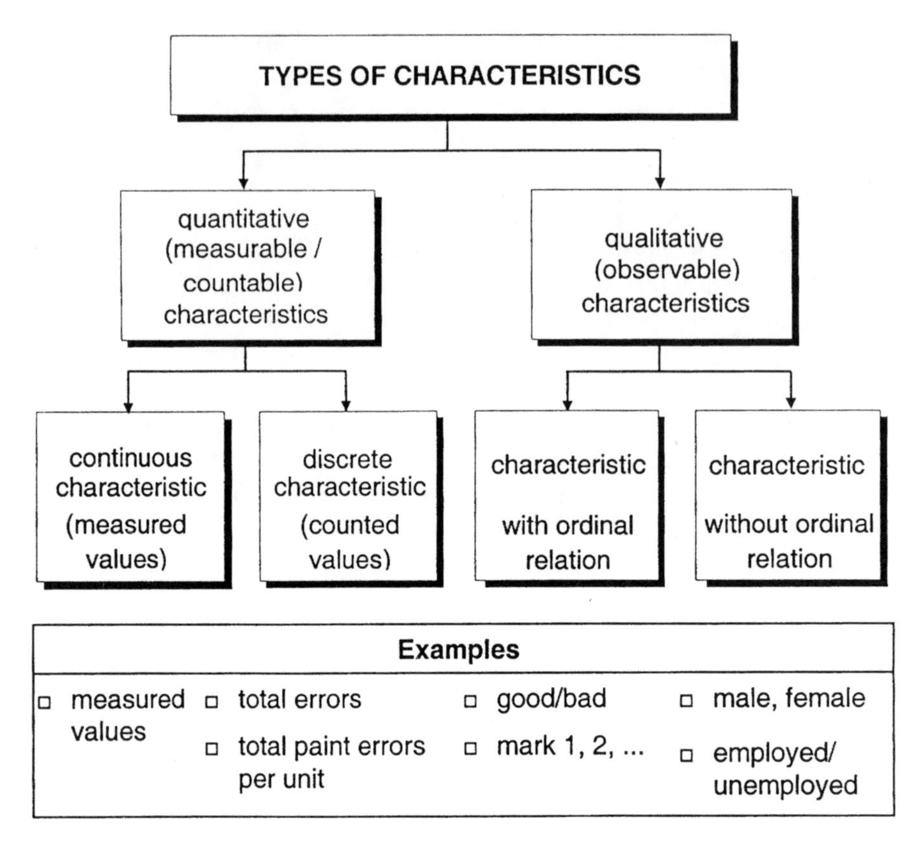

Figure 2.1–2. Schematic illustration of various types of characteristics.

Binary Characteristics

These occur when the observation of a characteristic gives a result that may be allocated by way of a binary scale (ordinal relation in the meaning of a sequence). Intervals may be selected at random. A typical example is an inspection using a go/no-go gauge. The inspection can only lead to the results *pass* or *fail.* At best, *fail* may be subdivided into *scrap* or *rework.* Further examples include

- Salary groups: 1, 2, . . ., 8
- Performance: Unsatisfactory/satisfactory/good/outstanding

Nominal Characteristics

These occur when no ordinal relation exists for the description of the possible outcomes for the characteristic value. The result of the observation of the characteristic is to be allocated to a nominal scale. The categories are without order or rank. For this reason, order of values may be selected at random. Examples include

- Sex: Male, female
- Suitability: Suitable/unsuitable
- Employment status: Employed/unemployed
- Type of costs: Material/machining/administrative/other costs

To maintain the principle of *never-ending improvement* mentioned in Chapter 1, Section 1.2, assessment should be effected by observation of *continuous characteristics* (measurement data) wherever possible. By using modern measurement procedures, measurement can in most cases take place without the involvement of an operator, and the results can be transferred for further processing directly into a computer system. The degree of automation is thereby increased and the likelihood of recording errors minimized.

2.1.3 Classification of Distributions

The aim of sample analysis is to draw an inference about the population in question by generalization from the *observed* behavior of the relevant characteristic. However, this generalization is possible only when certain plausible assumptions are made about the behavior of the characteristic and certain facts may be assumed true.

This leads to the use of various distributions, which either

- Describe as *probability distributions* (distributions of random variables) the population from which the observed values originate or
- Are applied as *sampling distributions* (probability distributions of sample parameters) to obtain statistical results such as random dispersion or confidence intervals, test procedures, etc.

According to the possible values of product characteristics, the probability distributions may be divided into

Discrete distributions

- Hypergeometric distribution
- Binomial distribution
- Poisson distribution

Continuous distributions

- Normal distribution
- Logarithmic normal distribution (lognormal distribution)
- Weibull distribution
- Rayleigh distribution
- Folded normal distribution
- Pearson function
- Johnson transformation
- Mixed distribution

The appropriate probability function is written as g(x) and its distribution function as G(x). This means that for every x a probability g(x) may be indicated for its occurrence.

Sampling distributions are divided into

- Normal distribution
- t-distribution
- χ^2-distribution
- F-distribution

Figure 2.1–3 shows a schematic view of the most important probability distributions. Under certain conditions, some distributions merge with others. The individual distributions are described in detail in the following chapters.

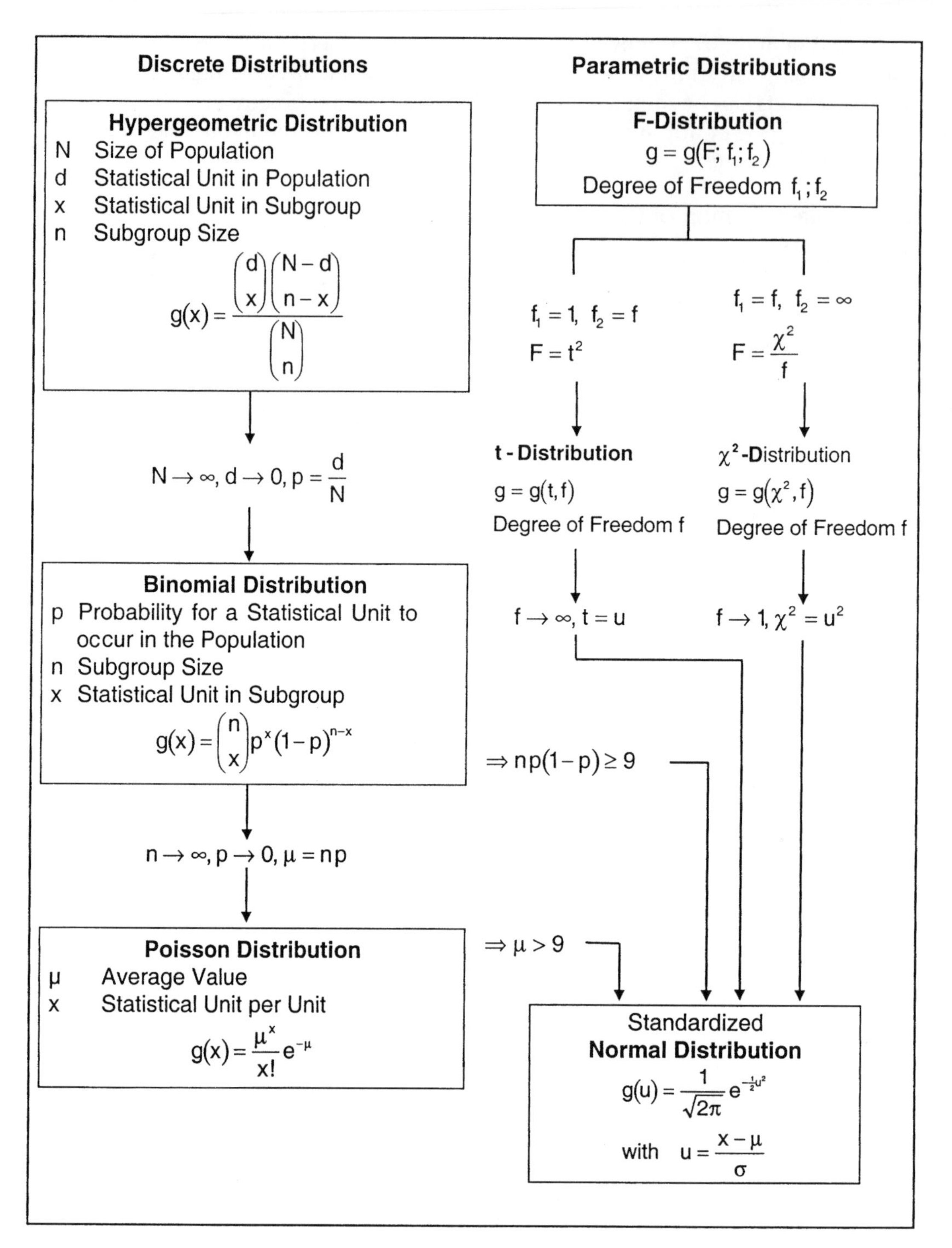

Figure 2.1–3. Correlation and transition of the various distributions.

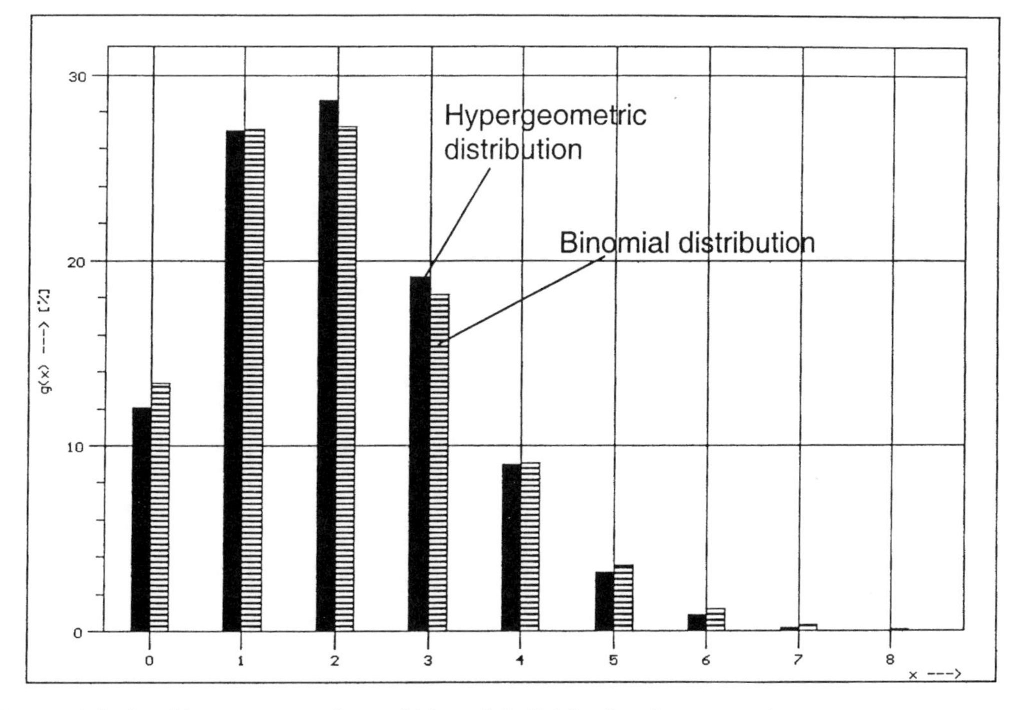

Figure 2.1–4. Hypergeometric and binomial distribution in comparison with N = 3000
n = 200
d = 20

Figures 2.1–4 and Figure 2.1–5 show, for discrete distributions, how the hypergeometric distribution merges into the binomial distribution and the binomial distribution into the Poisson distribution. In both cases, it will be seen that there is good agreement under the specified conditions.

The main application of the statistical procedures described here ends with the model approximation of practical conditions. This means that real conditions are approximated through a theoretical model. Typical process models were presented in Chapter 1, Section 1.3. The model approximation is also carried out with the help of the probability distributions presented in this chapter. The measurement values that represent the real condition are used to estimate the parameters of the respective distribution. As a rule, the distribution for a certain characteristic type is determined from the parameters, and conformity between the found distribution and the set of data on hand are compared using quality criteria. Possible evaluation criteria are presented in Chapter 4.

The results of a process evaluation depend largely on the accuracy of the measurement procedure (see Chapter 8, Measurement System Capability) and the correct model description. Too little store was set on both points in the past because of missing possibilities. The consequence was usually insufficient process evaluations, making the user unsure and bringing the procedures described here into disrepute.

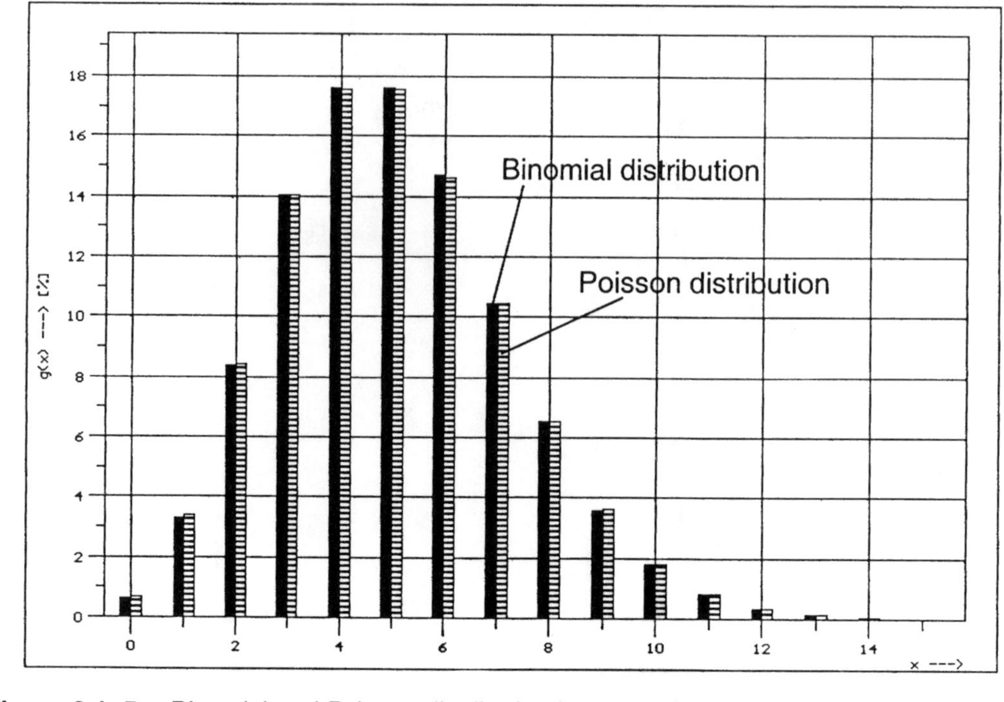

Figure 2.1–5. Binomial and Poisson distribution in comparison with n = 500 p = 1%

2.1.4 Definition of the Confidence Interval

If the samples taken are used to draw a conclusion about the population, this is known as a *statistical inference.* The object is to make use of the sample statistics (e.g., sample mean $\bar{x}$, sample standard deviation s, etc.) to make a statement about the parameters of the population (e.g., population mean μ, standard deviation σ, etc.). The true value (parameter) for the population will lie within the confidence interval (see Figure 2.1–6) with a probability of $1 - \alpha$ (α = probability of error). $1 - \alpha$ is also known as the confidence level. Typical values for $1 - \alpha$ are 95 percent, 99 percent, or 99.9 percent. Of course, the confidence interval is also dependent on the sample size. Table 2.1–1 shows, for different capability indices, the confidence intervals based on varying sample sizes. Figure 2.1–7 illustrates the point for a capability index of 1.33 for the population.

Given knowledge of the confidence interval, the sample results can be used to draw useful conclusions about the population. Depending on the problem at hand, it is possible to determine a single-sided or double-sided confidence interval. The formulas for the calculations are given in Section 2.4.

Example

It is good practice to give a confidence interval for the capability index. If a capability study leads to an index of 1.29 and the minimum specified for the index is 1.33—a customary requirement—then the capability index is too small. If we take

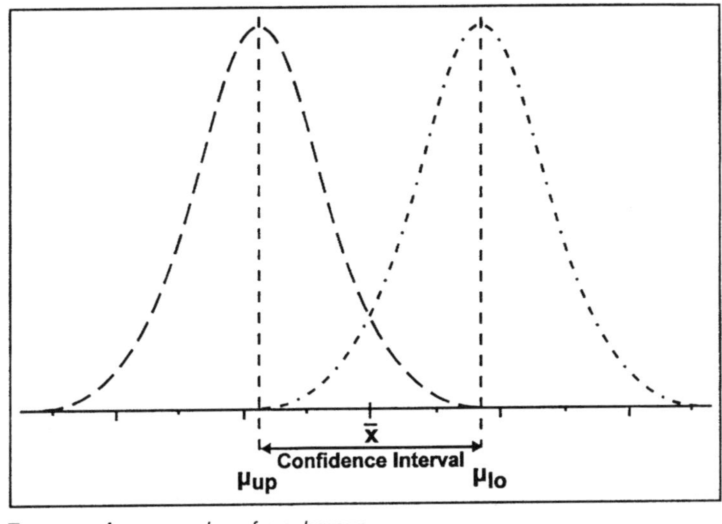

$\overline{x}$ *Average value of a subgroup*
μ_{up}, μ_{lo} *Confidence interval for the average of the population*

Figure 2.1–6. Confidence interval.

Table 2.1–1. Capability indices with confidence interval 1 – α = 95%.

n	C-value 1.00	1.33	1.67	2.00
10	0.548–1.454	0.729–1.934	0.915–2.428	1.096–2.908
15	0.634–1.366	0.843–1.817	1.059–2.281	1.268–2.732
20	0.685–1.315	0.911–1.749	1.143–2.196	1.369–2.630
25	0.719–1.281	0.956–1.703	1.200–2.139	1.438–2.561
30	0.744–1.256	0.989–1.670	1.242–2.097	1.488–2.511
40	0.779–1.221	1.036–1.624	1.301–2.039	1.558–2.442
50	0.802–1.197	1.067–1.592	1.340–1.999	1.605–2.394
60	0.820–1.180	1.090–1.569	1.369–1.970	1.640–2.360
70	0.833–1.166	1.108–1.551	1.392–1.948	1.667–2.333
80	0.844–1.155	1.123–1.537	1.410–1.930	1.689–2.311
90	0.853–1.147	1.135–1.525	1.425–1.915	1.706–2.293
100	0.861–1.139	1.145–1.515	1.438–1.902	1.722–2.278
250	0.912–1.088	1.213–1.447	1.523–1.816	1.824–2.175
500	0.938–1.062	1.247–1.412	1.566–1.774	1.876–2.124
1000	0.956–1.044	1.272–1.388	1.597–1.743	1.912–2.088

into account that the result is based on a sample size of n = 100 and that the true value will, with a 95 percent probability, lie approximately between the confidence limits of 1.145 and 1.515 (see Table 2.1–1), the statement is immediately put into proportion. There is, therefore, no real basis for the frequently encountered discussions as to whether capability indices slightly below the limits can still be regarded as *OK*.

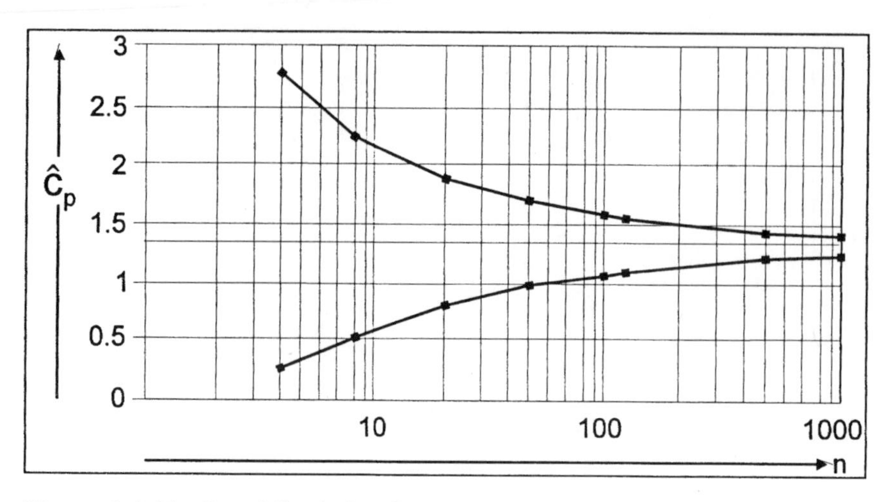

Figure 2.1–7 Capability index C = 1.33 with $1 - \alpha = 99\%$.

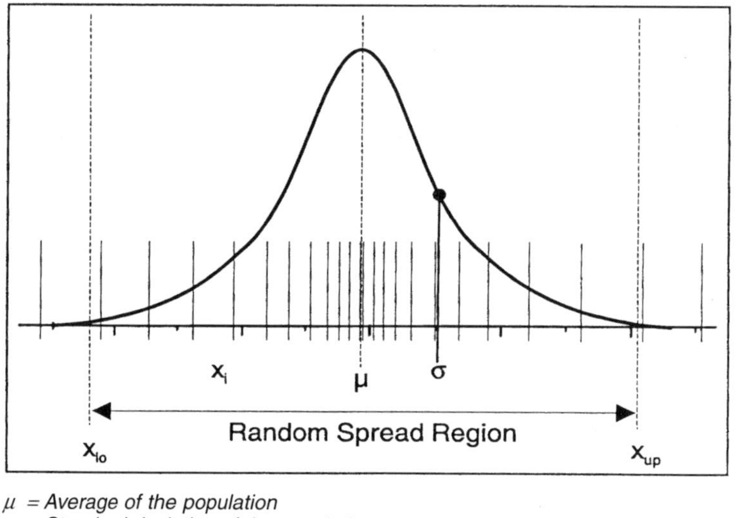

μ = Average of the population
σ = Standard deviation of the population
$\bar{x}_i$ = Average of a subgroup

Figure 2.1–8. Random spread region.

2.1.5 Definition of the Random Dispersion Interval

We speak of a *prediction* when conclusions are drawn from the known population about the sampling results that are likely to occur. In other words, an attempt is made to give an interval within which the results of a sample will lie within a given probability (see Figure 2.1–8). This prestated probability is denoted by $P = 1 - \alpha$. Typical values for this probability are 99 percent and 99.73 percent.

Depending on the problem at hand, we may work with single-sided or double-sided random dispersion intervals. (As previously noted, the formulas for calculation are given in Section 2.4.)

Example

A large-scale preliminary production run for plastic parts led to a 0.5 percent proportion of nonconforming items. During this run, there was evidence that the production process remained constant. Hence, this value was regarded as a parameter of the population.

To spot potential changes in the production process, a sample of n units is taken each day. This leads to the following question: How many faulty items may there be in the sample before we have to draw the conclusion that the process has changed?

The technique of quality control charting forms an important sphere of application of the random dispersion interval. For example, if the characteristics of the sample lie outside the control limits (calculated from the population), this is assumed to be due to an undesirable change in the process.

2.1.6 The Purpose of Probability Functions

The initial purpose of probability functions is to describe an actual set of facts in model form with the aid of a distribution function. Which function is chosen depends partly on the type of characteristic (see Section 2.1.2) and partly on the process model concerned (see Chapter 1, Section 1.3). The better it succeeds in mathematically reproducing the true facts, the better the results and the more reliable the statements based on them. Chapter 4 deals in more detail with the assessment criteria for selection of distribution models.

If a distribution model g(x) or G(x) has been determined, then it is possible to determine

- From g(x), the probability of occurrence P for a given value x
- From the parameters of the population, the random dispersion intervals
- From sample statistics, the confidence intervals

Section 2.2 shows basic distribution models and discusses their application in accordance with these objectives. Table 2.1–2 shows the essential points. In Figure 2.2–2, typical questions regarding the probability calculation were compiled, not depending on the type of distribution.

2.2 Probability Distributions

2.2.1 Probability Distributions for Discrete Data

Discrete probability distributions define for each frequency of occurrence of a particular event (e.g., the number of nonconformities in a sample) the corresponding probability (i.e., the probability that exactly this number of nonconformities is

Table 2.1–2. Determination of random spread and confidence regions.

Characteristic	**Distribution model** g(x) or G(x)	**Random spread region** Typical values for $1 - \alpha = 95\%, 99\%$, or 99.73%	**Confidence interval** Typical values for $1 - \alpha = 95\%, 99\%$, or 99.9%
Discrete			
Total or share of defective units	Binomial distribution	$p, 1 - \alpha, n \rightarrow x_{lo} \dots x_{up}$	$\hat{p}, 1 - \alpha, n \rightarrow p_{lo} \dots p_{up} \dots$
Discrete			
Total errors per unit	Poisson distribution	$\mu, 1 - \alpha, \rightarrow x_{lo} \dots x_{up}$	$x, 1 - \alpha, \rightarrow \mu_{lo} \dots \mu_{up}$
Continuous			
Series of measured values	Normal distribution	$\mu, \sigma, 1 - \alpha, \rightarrow \bar{x}_{lo} \dots \bar{x}_{up}$ $s_{lo} \dots s_{up}$	$\bar{x}, s, 1 - \alpha, \rightarrow \mu_{lo} \dots \mu_{up}$ $\sigma_{lo} \dots \sigma_{up}$

P = Probability from G(x)
α = Error probability
x = Number of faulty items or number of faults per unit (sample)
p = Number of faulty items (population)
μ = Average of population
$\bar{x}$ = Average of subgroup
σ = Standard deviation of population
s = Standard deviation of sample
n = Sample size

A statistic marked with the circumflex ^ (read as hat*) is to be regarded as an estimator for the relevant parameter.*

observed). The formulas and limiting conditions are given in the following paragraphs. Examples are then given to explain the facts. The results are given both numerically and graphically in the form of a bar chart.

2.2.1.1 Hypergeometric Distribution

The hypergeometric distribution makes it possible to treat the general problem of *the number of nonconforming units in a sample.* Here it should be noted that the sample taken is not replaced after removal and assessment. Regard is thus paid to the change in the population as a result of removal of the sample.

The probability that—given a batch (population) of N items, which includes d nonconforming items—a sample of n items will include *exactly x nonconforming items* can be calculated from the following formula:

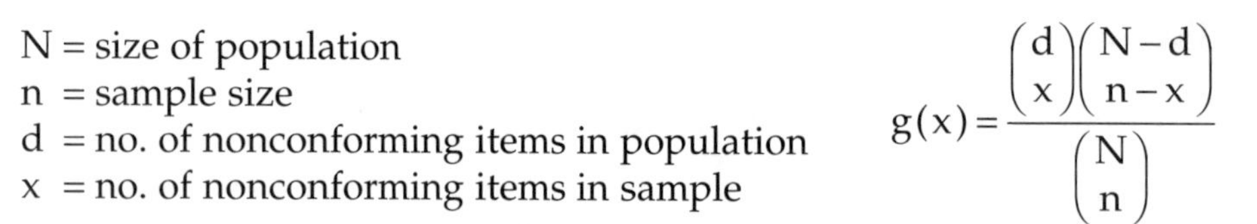

N = size of population
n = sample size
d = no. of nonconforming items in population
x = no. of nonconforming items in sample

$$g(x) = \frac{\binom{d}{x}\binom{N-d}{n-x}}{\binom{N}{n}}$$

If we wish to find for all values of x (x = 0, 1, 2, . . .) the probability that *exactly x nonconforming units* will be found in the sample, the distribution function is given as a function of the parameters d, N, and n. Figure 2.2–1 shows changes in the shape of the distribution.

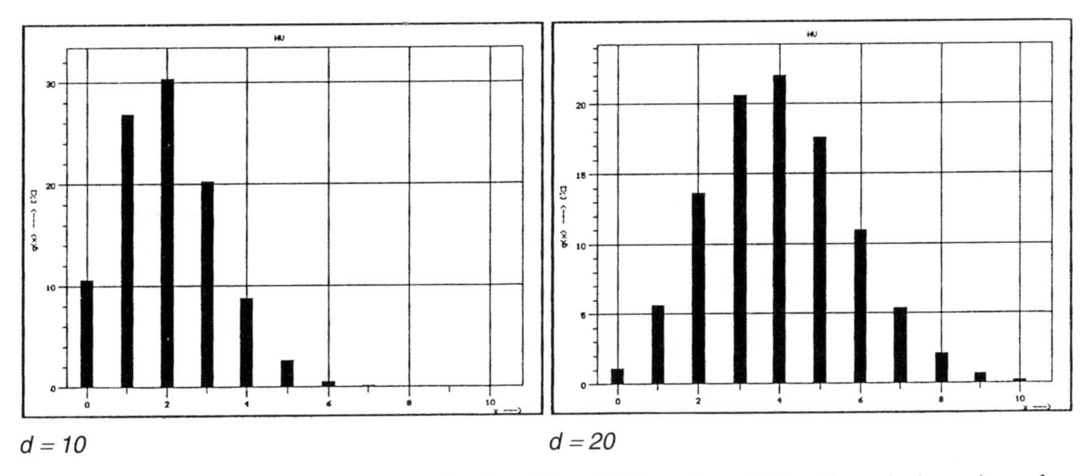

d = 10 *d = 20*

Figure 2.2–1. Hypergeometric distribution (N = 1000 and n = 250 with varied number of errors in the population).

A question that often arises is: What is the likelihood of finding *up to x nonconforming items* in a sample? This probability can be found using the formula

$$G(x) = \sum_{i=0}^{x} g(i)$$

Typical questions (Figure 2.2–2) are

- The probability of finding *exactly x* nonconforming items:

  g (x) or
  G (0) for x = 0
  G (x) – G (x – 1) for x ≧ 1

- The probability of finding *up to x* or *at most x* nonconforming items:

  G (x)

- The probability of finding *at least x* nonconforming items:

  1 for x = 0
  1 – G (x – 1) for x ≧ 1

- The probability of finding *more than x* nonconforming items:

  1 – G (x)

- The probability of finding *fewer than x* nonconforming items:

  G (x – 1) for ≧ 1

These questions arise for all discrete distributions. Figure 2.2–2 illustrates the questions for x = 3 nonconforming items.

Examples

Batch size N	Sample size n	No. of nonconforming items d
1000	200	100

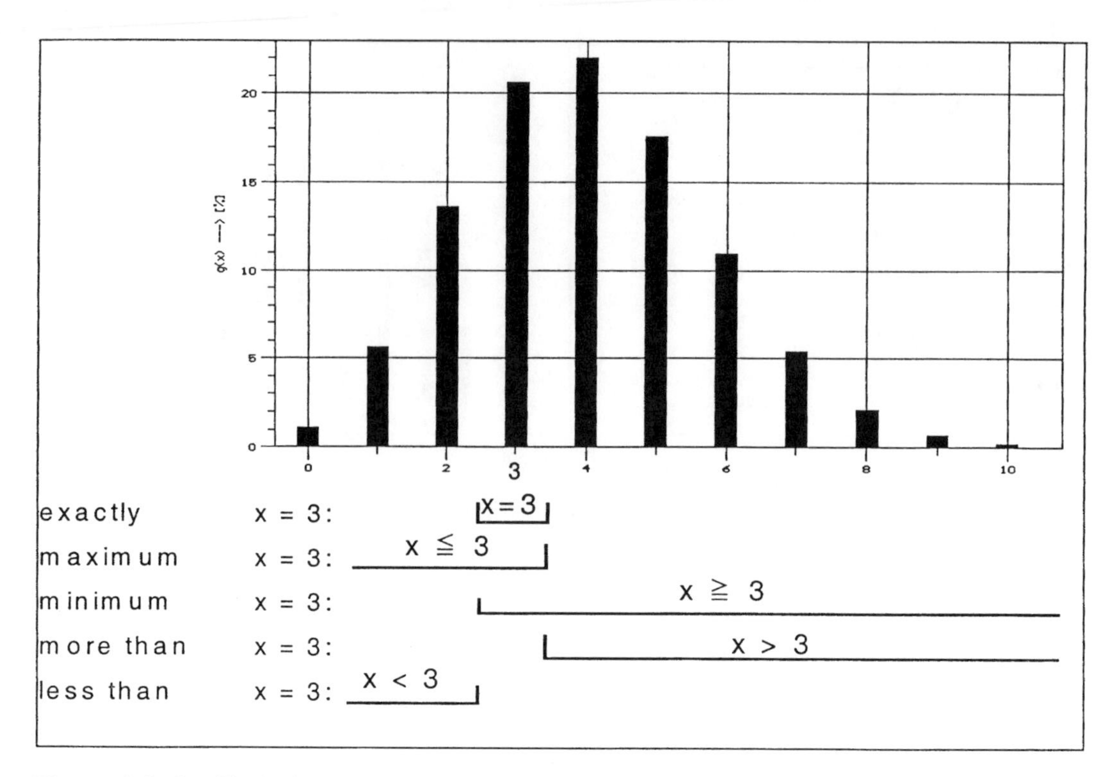

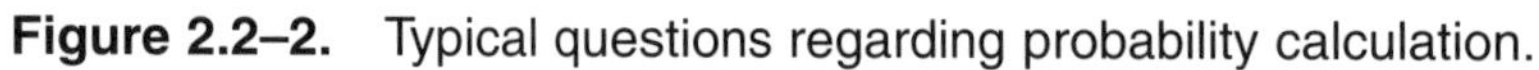

Figure 2.2–2. Typical questions regarding probability calculation.

What is the likelihood of finding at least x = 15 and at most x = 20 nonconforming items in the sample?

HYPERGEOMETRIC DISTRIBUTION **Individual or cumulative probability**		
Population size	N :	1000
Subgroup size	n :	200
Number of defective units (in population)	d :	100
Number of defective units (in subgroup min.)	x_1 :	15
Number of defective units (in subgroup max.)	x_2 :	20
Cumulative probability G = 49.0792%		

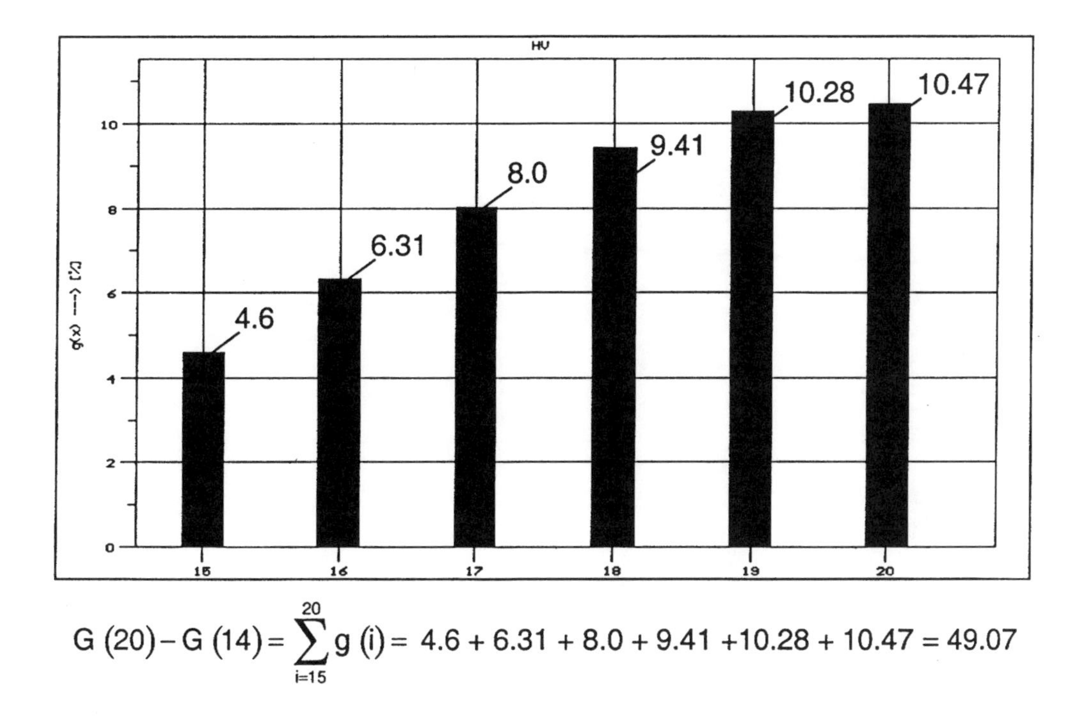

$$G(20) - G(14) = \sum_{i=15}^{20} g(i) = 4.6 + 6.31 + 8.0 + 9.41 + 10.28 + 10.47 = 49.07$$

Batch size N	Sample size n	No. of nonconforming items d
500	100	10

What is the likelihood of finding up to eight nonconforming items in the sample?

HYPERGEOMETRIC DISTRIBUTION Individual or cumulative probability		
Population size	N :	500
Subgroup size	n :	100
Number of defective units (in population)	d :	10
Number of defective units (in subgroup min.)	x_1 :	0
Number of defective units (in subgroup max.)	x_2 :	8

Cumulative probability
G = 99.9997%

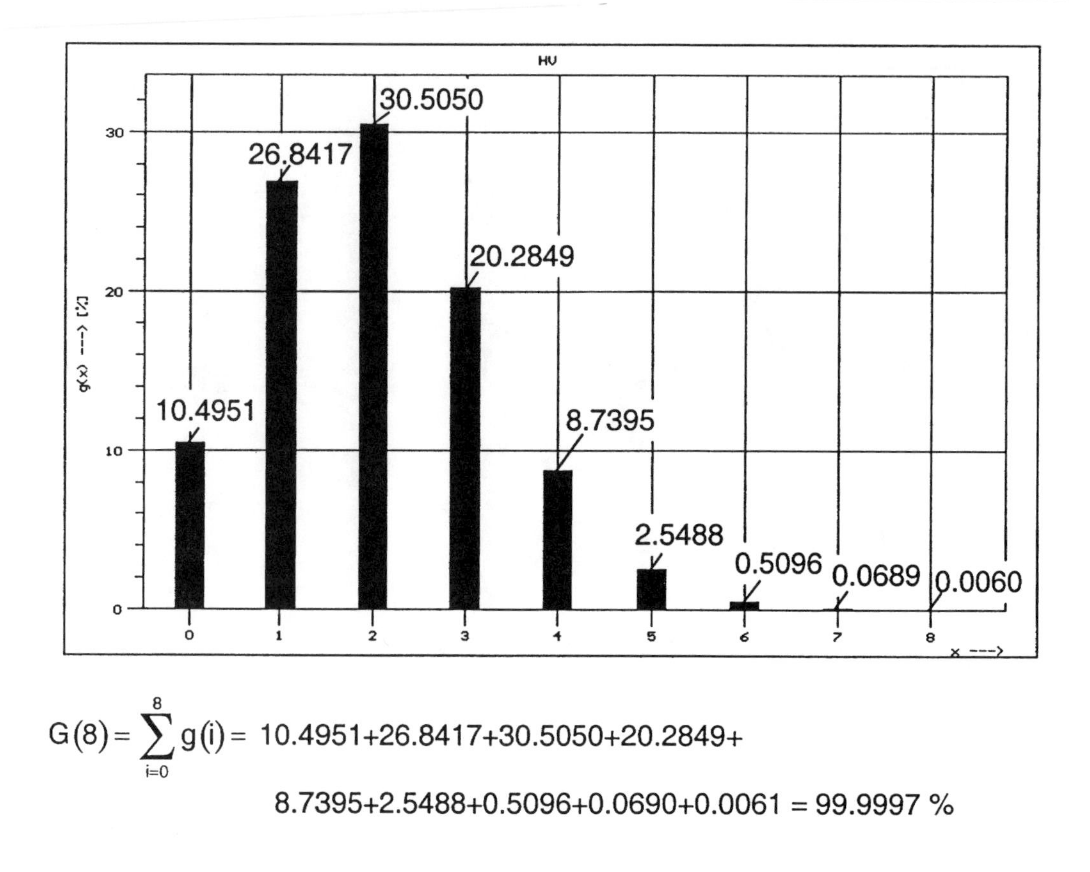

$$G(8)=\sum_{i=0}^{8} g(i)= 10.4951+26.8417+30.5050+20.2849+ 8.7395+2.5488+0.5096+0.0690+0.0061 = 99.9997\ \%$$

2.2.1.2 Binomial Distribution

Like the hypergeometric distribution, the binomial distribution is used for evaluations of *nonconforming items.* For the sake of simplicity, the change in the population caused by removal of the sample is frequently ignored. The effect of removal of the sample on the population is negligible provided that

$$n < \frac{N}{10} \quad \text{where} \quad N = \text{size of population}, \quad n = \text{sample size}$$

The probability of finding x nonconforming items in a sample of n items is given by the binomial distribution as

$$g(x) = \binom{n}{x} \cdot p^{x} (1-p)^{n-x}$$

where p = probability that an item is defective
1 − p = probability that an item is not defective

The distribution function is obtained from

$$G(x;\, n,\, p) = \sum_{i=1}^{x} g(i;\, n,\, p) \quad \text{where } i = 0,\ 1,\ 2,\ \ldots$$

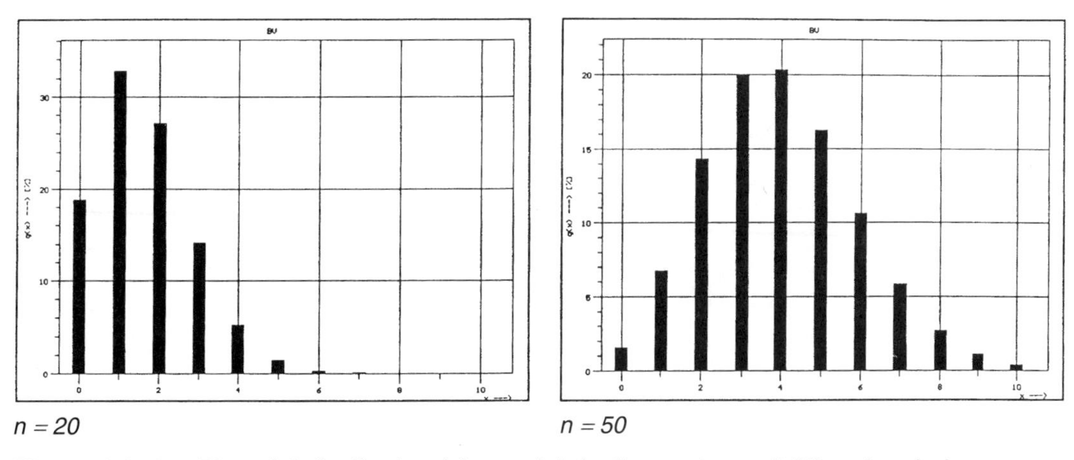

Figure 2.2–3. Binomial distribution (share of defective parts p = 0.08 and varied subgroup size).

Further we have

$$G(x;\, n,\, p) = 1 - G(n - x - 1;\, n, 1 - p)$$

Since in general the proportion p of nonconforming items in the population is unknown, it is estimated from a *large preproduction run:*

$$p \leftarrow \hat{p} = \frac{x_1 + x_2 + x_3 + \ldots + x_k}{n_1 + n_2 + n_3 + \ldots + n_k} = \frac{\sum_{i=1}^{k} x_1}{\sum_{i=1}^{k} n_i}$$

where x_i = number of nonconforming items in sample no. i
n_i = size of sample no. i
k = number of samples

Hence $\hat{p}$ gives an estimate for the parameter p of the population. In this case, too, the shape of the distribution (Figure 2.2–3) is dependent on the parameters n and p.

In practice, the requirement that $n < N/10$ is frequently satisfied. In this case, the hypergeometric distribution may be replaced by the binomial distribution (Figure 2.1–4). This function enables us to answer the same questions as those listed in the section on the hypergeometric distribution.

Subject to the condition $n \cdot p(1 - p) \geq 9$, the binomial distribution may be approximated by the normal distribution. Then,

the mean $\mu = p \cdot n$

the standard deviation $\sigma = \sqrt{\dfrac{p \cdot (1 - p)}{n}}$

Subject to the condition $n \to \infty$ and $p \to 0$, the binomial distribution may be replaced by the Poisson distribution (Figure 2.1–5) with the mean $\mu \approx n \cdot p$.

In addition to the question of individual and cumulative probabilities, we can also obtain, depending on the task at hand (inference or prediction),

Table 2.2–1. Formulas for random spread and confidence regions in binomial distribution.

	Random spread region $x_{lo} \le x \le x_{up}$		Confidence interval $p_{lo} \le p \le p_{up}$		
Single-sided upper	$G(x; n, p)$	$\ge 1 - \alpha$	$G(x; n, p_{up})$	$= \alpha$	for $x < n$
			$0 \le p \le 1 - \sqrt[n]{\alpha}$		for $x = 0$
Single-sided lower	$G(x - 1; n, p)$	$\le \alpha$	$G(x - 1; n, p_{lo})$	$= 1 - \alpha$	for $x \ge 1$
			$\sqrt[n]{\alpha} \le p \le 1$		for $x = n$
Double-sided	$G(x - 1; n, p)$	$\le \alpha/2$	$G(x - 1; n, p_{lo})$	$= 1 - \alpha/2$	for $x \ge n$
	$G(x; n, p)$	$\ge 1 - \alpha/2$	$G(x; n, p_{up})$	$= \alpha/2$	for $x \le n$
			$0 \le p \le 1 - \sqrt[n]{\alpha/2}$		for $x = 0$
			$\sqrt[n]{\alpha/2} \le p \le 1$		for $x = 0$

x = number of defective units
p = number of defective units
α = level of significance
n = subgroup size
G(x) = distribution function

- Random dispersion intervals (Section 2.1–5)
- Confidence intervals (Section 2.1–4)

as functions of an error probability α. Typical values are $\alpha = 0.1$ percent, 1 percent, and 5 percent, which corresponds to $P = 99.9$ percent, 99 percent, and 95 percent. Possible types of intervals are: single-sided with upper limit, single-sided with lower limit, and double-sided. The formulas for determination of these intervals are summarized in Table 2.2–1. Figure 2.2–4 shows the probability function for $p = 0.08$ and $n = 100$ with the random dispersion intervals 0.95 and 0.99 corresponding respectively to $\alpha = 0.05$ and 0.01.

The *random dispersion intervals* are given with a *nominal* probability $P = 1 - \alpha$. For the double-sided interval, the largest possible end is cut off from each side of the distribution that satisfies the condition $\leqq \alpha/2$ or $\ge 1 - \alpha/2$.

In accordance with Figure 2.2–4, we have the double-sided random dispersion intervals

P	**Double-sided**
95%	$2 \le x \le 14$
99%	$1 \le x \le 16$

Likewise, single-sided random dispersion intervals may be determined:

P	**Single-sided with upper limit**	**Single-sided with lower limit**
95%	$0 \le x \le 13$	$4 \le x \le 100$
99%	$0 \le x \le 15$	$2 \le x \le 100$

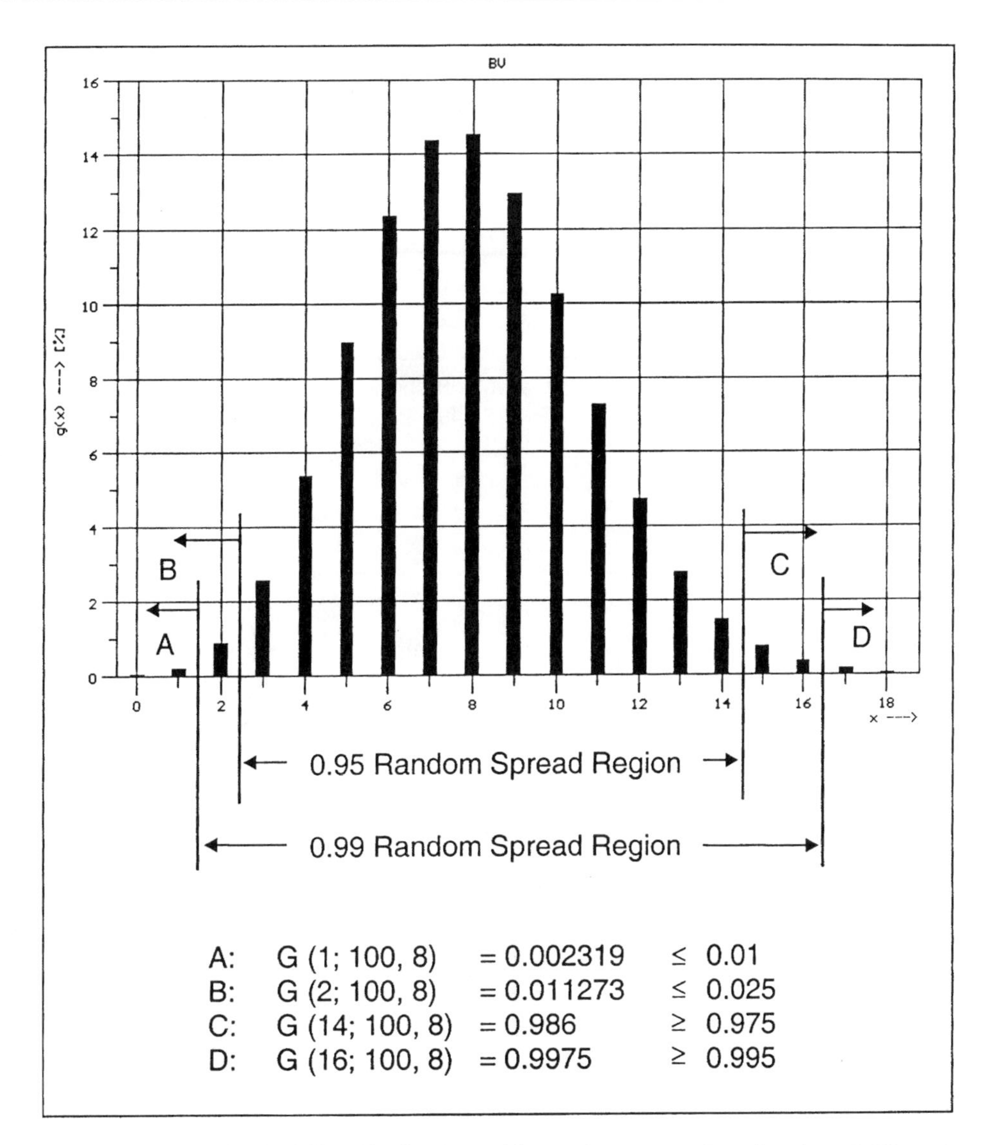

Figure 2.2–4. Probability distribution with random spread region ($p = 0.08$ and $n = 100$).

Examples

1. *Cumulative Probability.* It is known that in mass production of a certain article the whole production contains a proportion of $p = 1.5$ percent of nonconforming items.

What is the likelihood of finding at most two faulty items in a sample of size $n = 200$?

BINOMIAL DISTRIBUTION **Individual or cumulative probability**		
Subgroup size	n :	200
Number of defective units (in population)	p in % :	1.5
Number of defective units (in subgroup min.)	x_1 :	0
Number of defective units (in subgroup max.)	x_2 :	2

Cumulative probability
G = 42.1496%

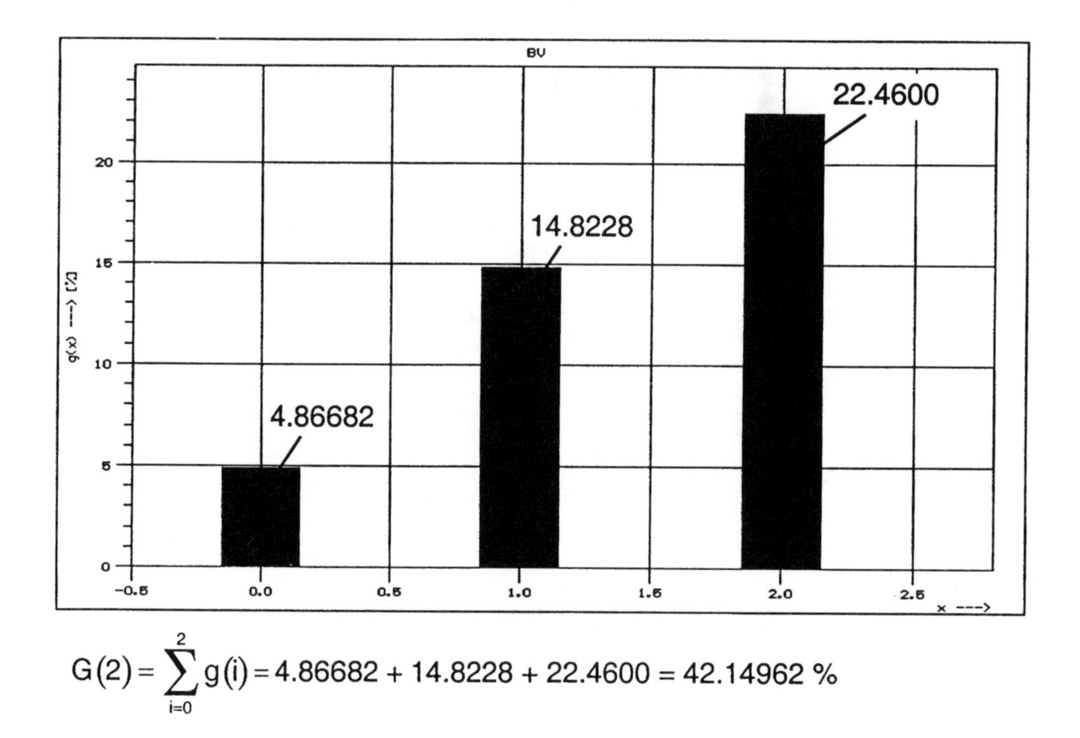

$$G(2) = \sum_{i=0}^{2} g(i) = 4.86682 + 14.8228 + 22.4600 = 42.14962\ \%$$

It is known that in mass production of a certain article the whole production contains a proportion of p = 0.5 percent of nonconforming items.

What is the likelihood of finding exactly three faulty items in a sample of size n = 100?

BINOMIAL DISTRIBUTION
Individual or cumuiative probability

Subgroup size	n :	500
Number of defective units (in population)	p in % :	0.5
Number of defective units (in subgroup min.)	x_1 :	3
Number of defective units (in subgroup max.)	x_2 :	3

Individual probability
g = 21.4353%

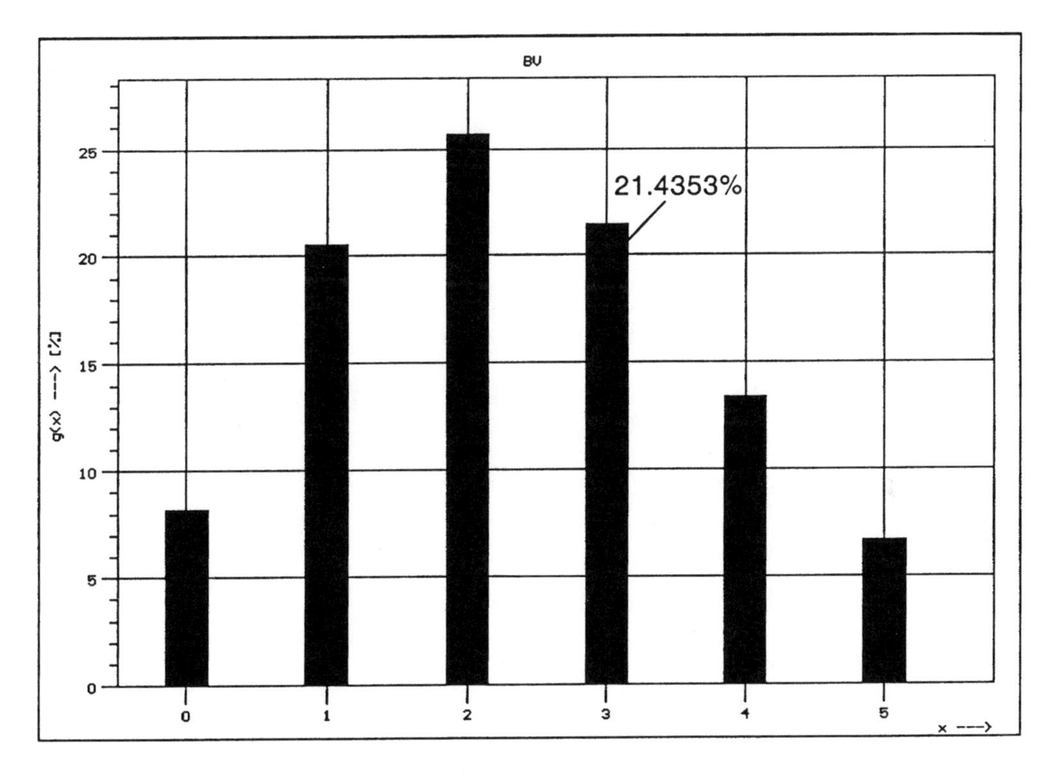

g(3) = 21.435 %

2. *Confidence Interval (Single-Sided with Lower Limit).* Ten faulty parts were found in a sample of size n = 315. Given the requirement of a confidence level of 95 percent $\rightarrow \alpha = 0.05$, what minimum proportion of nonconforming items is to be expected in the population?

BINOMIAL DISTRIBUTION Confidence limits		
Subgroup size	n :	315
Number of defective units (in subgroup)	x :	10
Confidence level	1 – α in % :	95
Single-sided lower		
Confidence limits for the error share of the population p (lo) = 1.73099%		

3. *Random Dispersion Limits (Double-Sided).* A sample of size n = 200 is taken from a population with a binomial distribution with p = 0.05 (5 percent). Find the random dispersion interval (based on an error probability of 5 percent).

BINOMIAL DISTRIBUTION Random spread limits		
Subgroup size	n :	200
Number of defective units (in population)	p in % :	5.0
Confidence level	1 – α in % :	95
Double-sided		
Random spread limits		
Upper limit	x(up) = 16	
Share	> x(up) = 2.37995%	
Lower limit	x(lo) = 4	
Share	< x(lo) = 0.90484%	

2.2.1.3 *Poisson Distribution*

A further quantitative characteristic is the *number of defects per unit,* e.g., the number of scratches on a surface, the insulation defects per roll of cable, etc. In other words, we are not asking whether items are defective or nonconforming, but are instead concerned with the question of whether *a given number of defects per unit* is or is not exceeded. The reference *unit* needs to be clearly determined. This may be one production item or a definite quantity:

- Scratches per automobile hood
- Knotholes in a board
- Insulation faults per 25m length of wire
- Knots per 100m

In this case, the mathematical basis for calculation of the individual and cumulative probabilities is the *Poisson distribution* with the distribution function

$$G(x) = \sum_{i=1}^{x} g(i) \quad \text{where } i = 0, 1, 2, \ldots$$

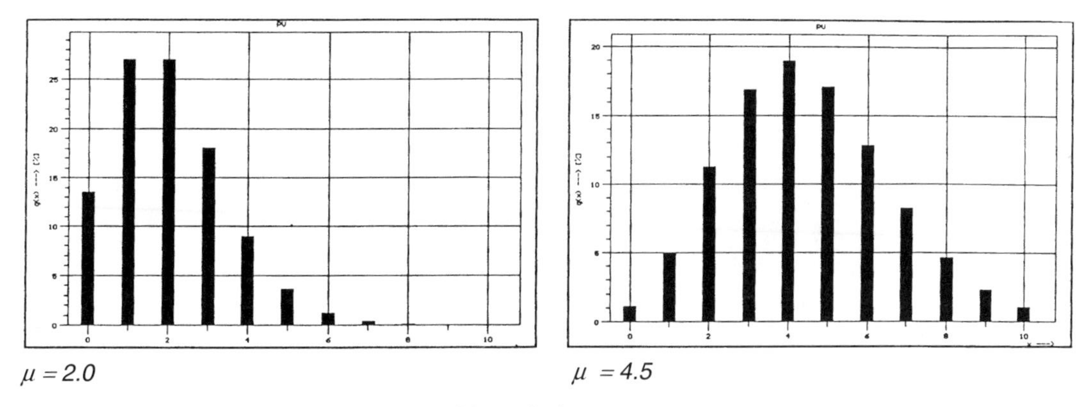

$\mu = 2.0$ $\mu = 4.5$

Figure 2.2–5. Poisson distribution with varied μ.

and the probability function g(x):

$$g(x) = \frac{\mu^x}{x!} e^{-\mu} \quad \text{where } x = \text{number of defects per unit}$$

$$x! = 1 \cdot 2 \cdot 3 \cdot \ldots \cdot x$$

The estimate of the mean μ is based on sample data:

$$\mu \leftarrow \hat{\mu} = \frac{x_1 + x_2 + x_3 + \ldots + x_k}{n_1 + n_2 + n_3 + \ldots n_k} = \frac{\sum_{i=1}^{k} x_i}{\sum_{i=1}^{k} n_i}$$

where x_i = number of defects per unit in sample no. i
n_i = size of sample no. i
k = number of samples

By means of the Poisson distribution, we can now calculate the probability of finding 0, 1, 2, . . . defects in a sample. From this we can calculate the probability of finding *up to x defects.*

The shape of the distribution is determined by the parameter μ. Figure 2.2–5 shows two different shapes of the distribution.

The average number of defects per unit is frequently given; e.g., one states that a wire 100m long has on average 0.8 insulation faults. This information is then to be transferred to the actual length (e.g., 250m). In this case, we have

$$\mu = \frac{250\,\text{m}}{100\,\text{m}} \times 0.8 = 2$$

For $\mu \geq 9$, the Poisson distribution may be approximated by the normal distribution with

- Mean μ and
- Standard deviation $\sqrt{\mu}$

As with the binomial distribution, random dispersion intervals and confidence intervals can be found. These are summarized in Table 2.2–2.

Examples

1. *Cumulative Probability* We have a Poisson distribution with a mean value of $\mu = 12$. What is the likelihood of finding between $x = 5$ and $x = 18$ defects in a sample?

Table 2.2–2. Formulas for random spread and confidence regions in poisson distribution.

	Random spread region $x_{lo} \leq x \leq x_{up}$		Confidence interval $\mu_{lo} \leq \mu \leq \mu_{up}$		
Single-sided upper	$G(x; \mu)$	$\geq 1 - \alpha$	$G(x; \mu_{up})$	$= \alpha$	
Single-sided lower	$G(x - 1; \mu)$	$\leq \alpha$	$G(x - 1; \mu_{lo})$	$= 1 - \alpha$	for $x \geq 1$
Double-sided	$G(x - 1; \mu)$	$\leq \alpha/2$	$G(x - 1; \mu_{lo})$	$= 1 - \alpha/2$	for $x \geq n$
	$G(x; \mu)$	$\geq 1 - \alpha/2$	$G(x; \mu_{up})$	$= \alpha/2$	

x = number of defects per unit *n = subgroup size*
α = error probability, *G(x) = distribution function*

POISSON DISTRIBUTION Individual or cumulative probability		
Population average	μ :	12
Number of defects (in subgroup min.)	x_1 :	5
Number of defects (in subgroup max.)	x_2 :	18

Cumulative probability
G = 95.4983%

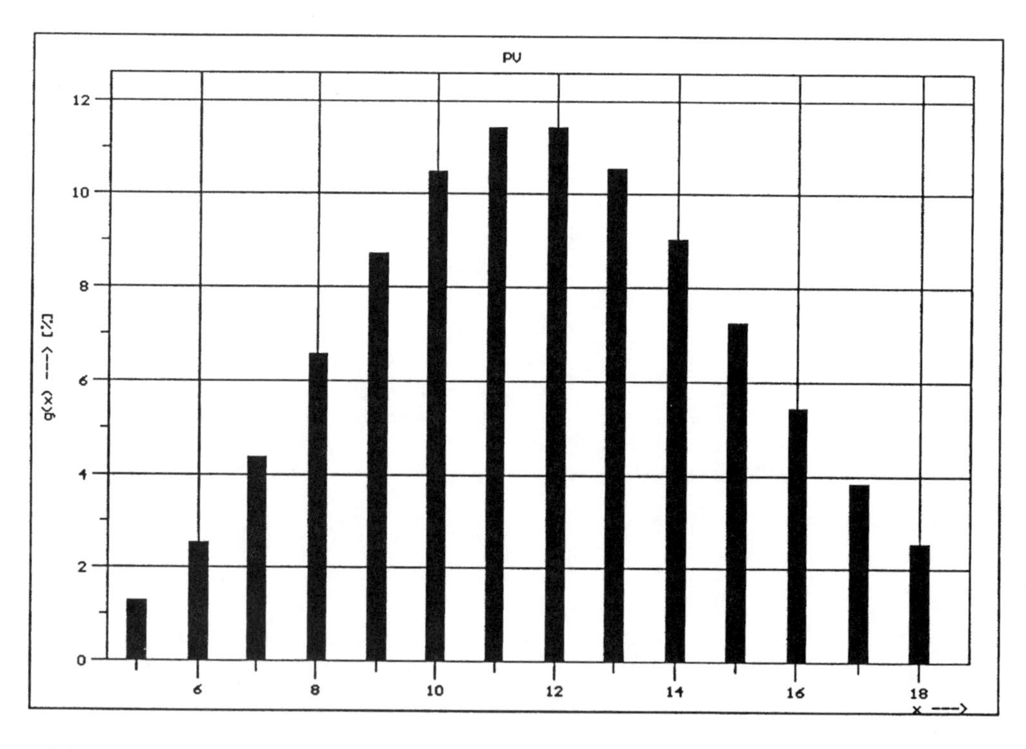

G(18) – G(4) = 96.2584% – 0.76004% = 95.4983 %

We have a Poisson distribution with a mean value of $\mu = 25$. What is the likelihood of finding $x = 33$ defects in a sample?

POISSON DISTRIBUTION Individual or cumulative probability		
Population average	μ :	25.0
Number of defects (in subgroup min.)	x_1 :	33
Number of defects (in subgroup max.)	x_2 :	33
	Individual probability g = 2.16757%	

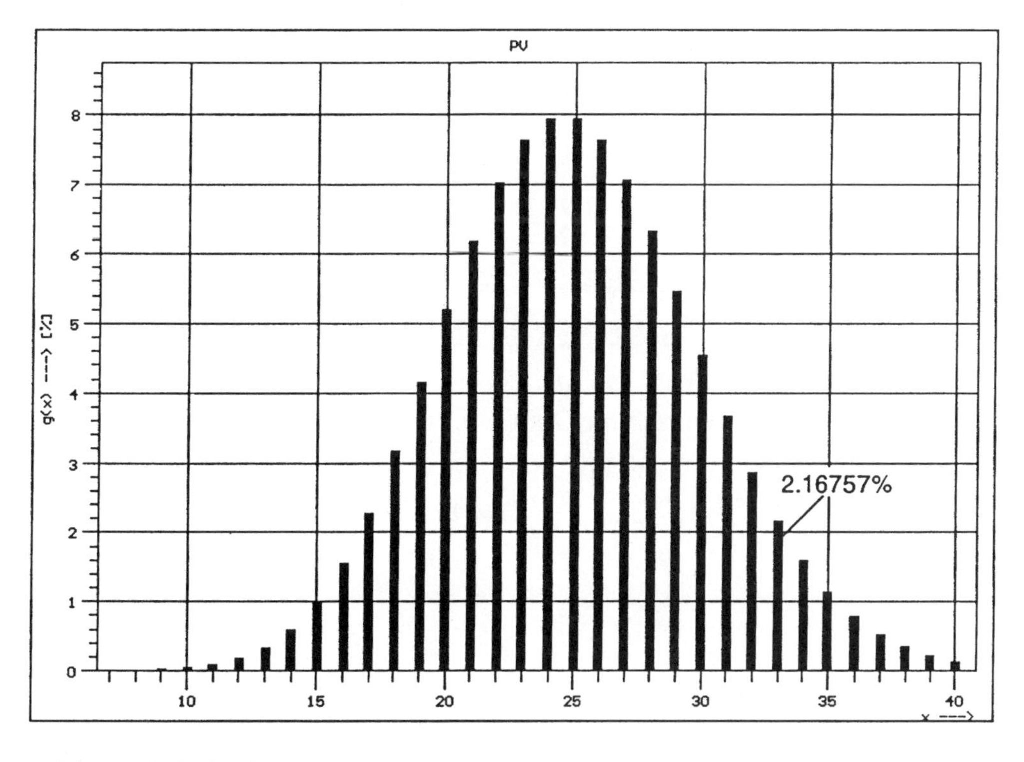

g(x) = 2.16757 %

2. *Confidence Interval (Single-Sided with Lower Limit).* In twisting thread in a twisting machine, six breaks are observed in eight hours. What is the smallest number of breaks to be expected with $1 - \alpha = 0.95$?

POISSON DISTRIBUTION Confidence limits		
Number of defects (in subgroup)	x :	6
Confidence level Single-sided lower	1 – α in % :	95

Confidence limit for the share of defects in the population
μ' (lo) = 2.613010%

3. *Random Dispersion Limits (Double-Sided).* A sample is taken from a population with a mean of $\mu = 6.8\%$ and a Poisson distribution. What are the smallest and largest numbers of defects to be expected in the sample? The probability of error should be $\alpha = 0.05$ (5 percent).

POISSON DISTRIBUTION Random spread limits		
Population average	μ :	6.8
Confidence level Double-sided	1 – α in % :	95

Random spread limits

Upper limit	x(up) =	12
Share	> x(up) =	2.20968%
Lower limit	x(lo) =	2
Share	< x(lo) =	0.86874%

2.2.2 Probability Distributions for Continuous Data

The normal distribution or Gaussian distribution is frequently used as a model distribution for characteristics that generate continuous data. There are many reasons for this:

- Many characteristics encountered in experiments and in real-life observations are normally distributed. Specifically, it has been found that the sum of a number of independent random variables will, regardless of the variables' individual distribution patterns, tend to be normally distributed with the approximation becoming closer as the number of variables increases.
- Even characteristics that are not normally distributed may frequently be approximated by the normal distribution. The assumption that the distribution is normal leads in many cases to results that make sense and are of practical use.
- Characteristics that are not normally distributed may in certain cases be transformed such that the transformed variable is normally distributed.
- Some more complicated distributions may be replaced by the normal distribution in limiting cases.
- Ease of use of the distribution as a mathematical model.

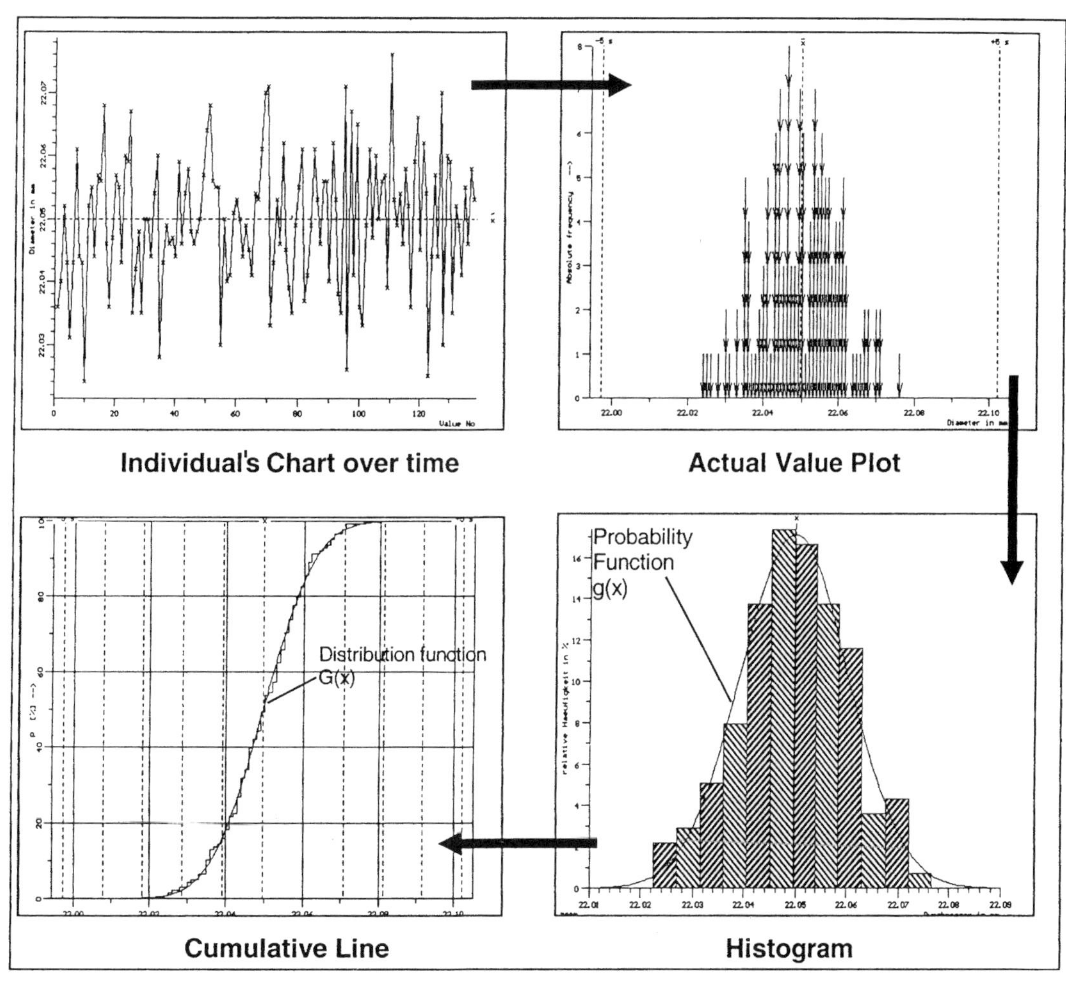

Figure 2.2–6. Buildup of the distribution function.

2.2.2.1 Normal Distribution

If we take an infinite number of samples from a normally distributed population, the frequency plot or histogram (Figure 2.2–6) will merge with the smooth bell-shaped distribution curve of the normal distribution.

The distribution function G(x) is arrived at on the following basis:

Run Chart of Individual Values

In the run chart, the original observations are plotted in their original time sequence. A constant scale is used for the x axis, and the observed values are numbered.

Frequency Plot

If the sequence of observation is disregarded, the values may be represented in the form of a frequency plot. This shows the frequencies of individual measurement values and thus provides an answer to the question: How frequently does value x appear?

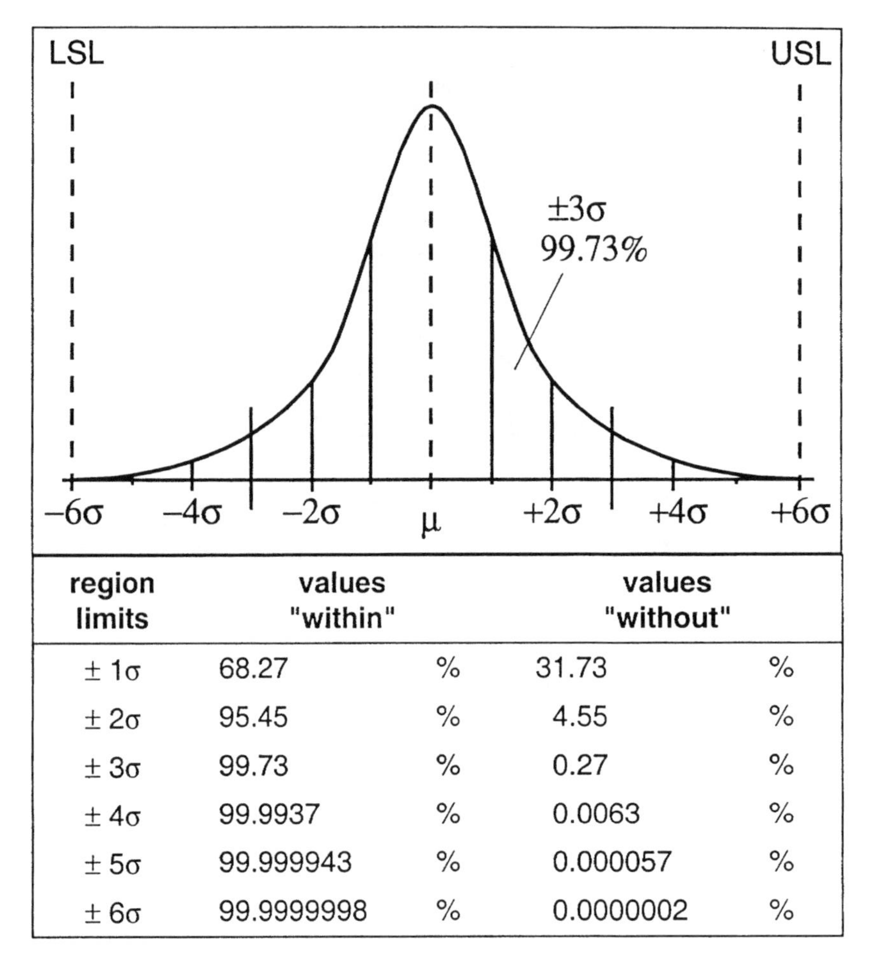

region limits	values "within"		values "without"	
± 1σ	68.27	%	31.73	%
± 2σ	95.45	%	4.55	%
± 3σ	99.73	%	0.27	%
± 4σ	99.9937	%	0.0063	%
± 5σ	99.999943	%	0.000057	%
± 6σ	99.9999998	%	0.0000002	%

Figure 2.2–10. Normal distribution with σ intervals.

Figure 2.2–10 shows the normal distribution with typical σ intervals and gives percentages inside and outside the intervals.

2.2.2.2 Mathematical Description of the Normal Distribution

The probability density function g(x; μ, σ) of the normal distribution is given by

$$g(x) = \frac{1}{\sigma \times \sqrt{2\pi}} \times e^{-\frac{1}{2}\left(\frac{x-\mu}{\sigma}\right)^2}$$

As this formula shows, the characteristics of a normal distribution are defined by the parameters μ and σ.

Standardized Normal Distribution

For practical calculations, the standardized distribution function of the normal distribution with mean $\mu = 0$ and variance $\sigma^2 = 1$ is used; its values are obtainable from tables. The actual values need to be transformed for this. The distance from the mean value is given as a multiple of the standard deviation and is denoted by μ:

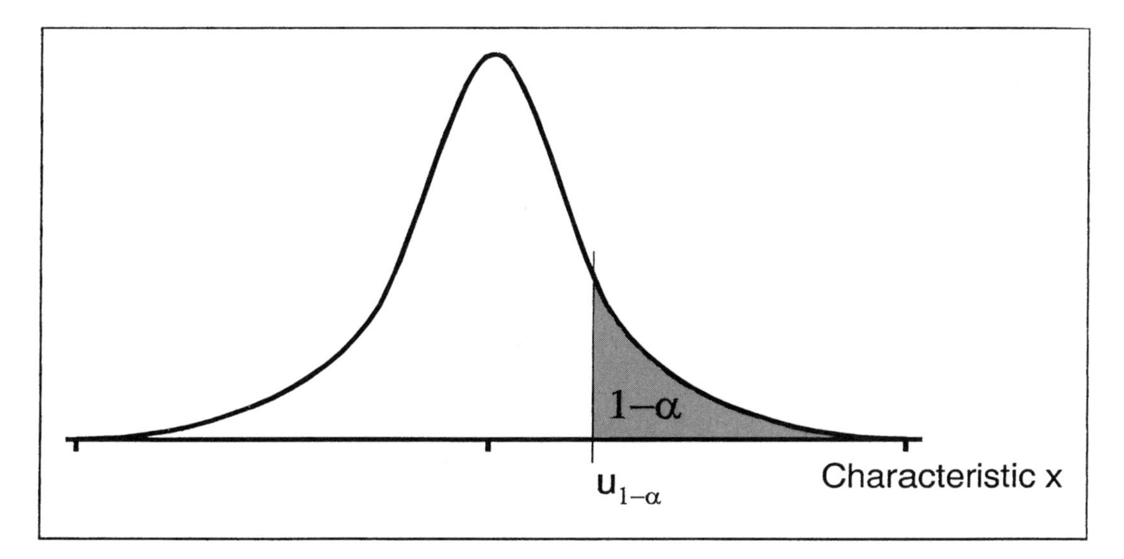

Figure 2.2–11. Lower limit.

$$u = \frac{x - \mu}{\sigma}$$

The formula for the probability density function then becomes

$$g(u) = \frac{1}{\sqrt{2\pi}} e^{-\frac{1}{2}u^2}$$

The area under the curve, i.e., the total probability between u = − ∞ and u = + ∞, is 1, or 100 percent, and is described mathematically by the integral of the probability density function:

$$G(u) = \frac{1}{\sqrt{2\pi}} \int_{-\infty}^{+\infty} e^{-\frac{1}{2}u^2} du = 1$$

The following tables show, for commonly occurring values, the size of the proportion 1 − α of the total area that is below a *single-sided upper limit* (Figure 2.2–11). This means that one table is used to read the percentage proportion, and the other is used for the u value of the given percentage proportion.

Single-sided upper limit $u_{1-\alpha}$	Proportion *below* $1-\alpha$ in %
0	50
1	84.13
2	97.725
3	99.8650
4	99.996833

Proportion *below* $1-\alpha$ in %	Single-sided upper limit $u_{1-\alpha}$
90	1.282
95	1.645
99	2.326
99.9	3.090

The true values are obtained from the formula $x_{upper} = \mu + u_{1-\mu} \cdot \sigma$.

Examples

If the parameters μ and σ of the normal distribution are known, then about 84 percent of the values of the whole population will not exceed $x_{upper} = \mu + \sigma$.

If *single-sided lower limits* are required, these are equal to the negative of the upper limits, due to the symmetry of the normal distribution (Figure 2.2–11).

$$u_{\alpha} = -u_{1-\alpha}$$

In addition to single-sided units, *double-sided symmetrical limits* for the normal distribution (Figure 2.2–12) are frequently required.

The following tables show the size of the proportion 1 – α of the total area that is within the double-sided limits, along with the converse. This means that one table is used to read the percentage proportion, and the other is used for the u value of the given percentage proportion.

Double-sided limit $u_{1-\alpha/2}$	Proportion *between* $1 - \alpha/2$ in %
0	0
1	68.27
2	95.45
3	99.73
4	99.9937

Proportion *between* $1 - \alpha/2$ in %	Double-sided limit $u_{1-\alpha/2}$
90	1.654
95	1.960
99	2.576
99.73	3.0
99.9	3.291

The true values are obtained from the formulas:

$$x_{upper} = \mu + u_{1-\alpha/2} \times \sigma$$
$$x_{lower} = \mu - u_{1-\alpha/2} \times \sigma$$

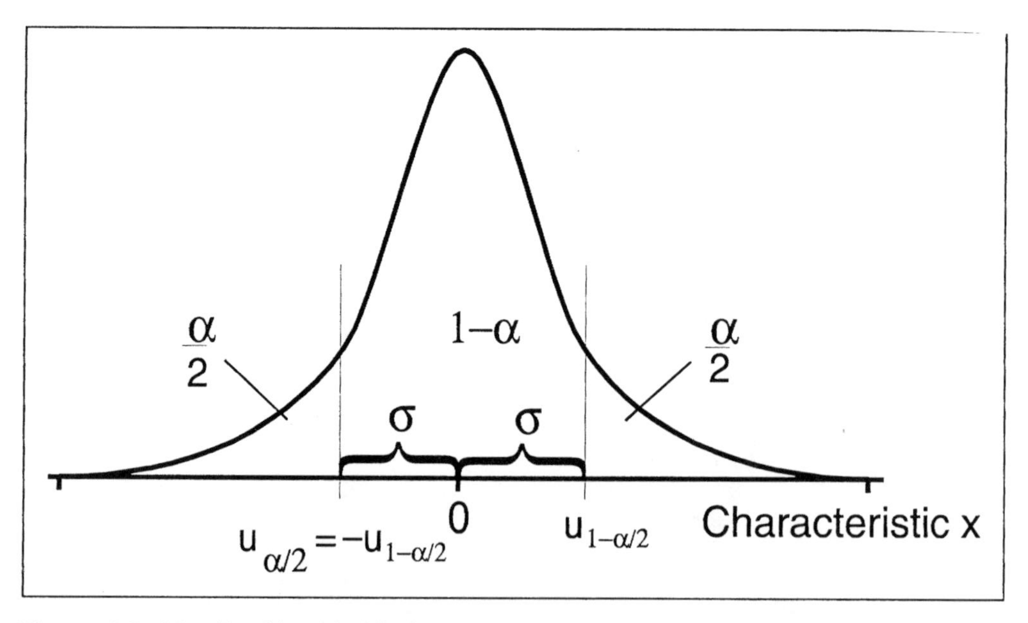

Figure 2.2–12. Double-sided limit.

Examples

1. *Determination of the Cumulative Probability.* The relevant cumulative probability is to be determined from a value of u and the type of limit (single-sided upper/lower or double-sided). In the standardized normal distribution ($\mu = 0$, $\sigma = 1$), the value of u corresponds to the standard deviation.

CUMULATIVE PROBABILITY Normal distribution			
		u values	$1 - \alpha$ in %
Single-sided lower	u (α)	3	0.135%
Single-sided upper	u ($1 - \alpha$)	2	97.725%
Double-sided	($1 - \alpha/2$)	1	68.2689%

A normally distributed population has parameters $\mu = 301$ and $\sigma = 10$. What percentage of the values are outside specification if the upper specification limit is 325 and the lower specification limit 275?

Solution

We first calculate the u values by transformation. The appropriate p values can then be determined.

$$u_{up} = \frac{USL - \mu}{\sigma} = \frac{325 - 301}{10} = 2.4 \rightarrow p_{upper} = 0.82\%$$

$$u_{lo} = \frac{LSL - \mu}{\sigma} = \frac{275 - 301}{10} = -2.6 \rightarrow p_{lower} = 0.47\%$$

$$p_{tot} = p_{upper} + p_{lower} = 1.29\%$$

2. *Determination of Critical Values.* Limits (single-sided lower, single-sided upper, and double-sided) are to be determined for a given cumulative probability.

NORMAL DISTRIBUTION Critical values (u values)			
		$1 - \alpha$ in % :	u values
Single-sided lower	u (α)	95%	−1.64485
Single-sided upper	u ($1 - \alpha$)	99%	2.3263
Double-sided	u ($1 - \alpha/2$)	99.73%	3.0

2.3 Other Continuous Distributions

In contrast to the distributions discussed up to now, in which we were concerned with the probability distributions of discrete or continuous characteristics, we shall now discuss *sampling distributions,* i.e., distributions of sample statistics (means, variances, etc.). If, for example, we consider the mean values of a number of samples from a normally distributed population, these are again normally distributed. In this case, the normal distribution would be a sampling distribution. The ranges of application of these distributions are indicated in the following chapters.

2.3.1 The t-Distribution

In addition to the limit values or percentage points of the normal distribution, statistical procedures are frequently used in which limit values of the t-distribution are applied. The formation of the t-distribution can be described by the following idealized experiment:

A large (theoretically infinite) number of samples is taken from a normally distributed population having parameters σ and μ, these samples being of size n, and $\overline{x}$ and s are calculated for each sample. The statistic t

$$t = \frac{\overline{x} - \mu}{s/\sqrt{n}}$$

then conforms to a distribution known as the t-distribution.

The t-distribution (Figure 2.3–1) is sometimes called *Student's t-distribution* after W. S. Gossett, who studied it and published papers on it under the pen name *Student*. The t-distribution is also a symmetrical distribution; hence, the methods concerning single-sided or double-sided limits are analogous to those used for the normal distribution. For the t-distribution, there is an additional parameter f (number of degrees of freedom). The number of degrees of freedom is given by $f = n - 1$ (n = sample size).

For large values of n, the t-distribution approximates to the normal distribution and merges with it in the limiting case.

Practical Application

The t-distribution is used in the calculation of a *confidence interval for the population mean* where σ is unknown.

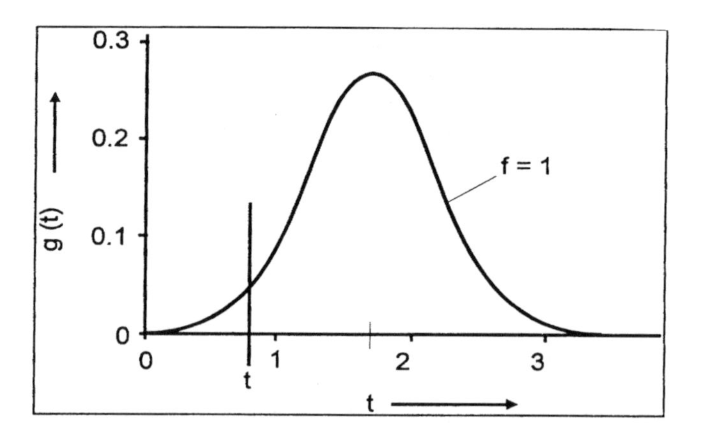

Figure 2.3–1. t-distribution.

Example

For $1 - \alpha = 0.95$ ($\alpha = 0.05$) and $f = 30$, the following t values are to be calculated:

1. Single-sided lower t limit
2. Double-sided (upper and lower) t limits

t-DISTRIBUTION Critical values			
Degree of freedom		f :	30
Confidence level		$1 - \alpha$ in % :	95
Critical values for the t-distribution			
Single-sided lower	$t(f; \alpha)$	=	−1.6973
Double-sided upper	$t(f; 1 - \alpha/2)$	=	2.04228
Double-sided lower	$t(f; \alpha/2)$	=	−2.04228

2.3.2 χ^2 Distribution

Whereas the mean values of samples taken from a normally distributed population are themselves normally distributed, the sample variances follow the χ^2 distribution. The χ^2 distribution is in consequence an asymmetric distribution (variances have a zero limit). Hence, lower and upper limits are not obtained by reversal of sign of the opposite limits. They must be separately determined. Further, regard must be paid to the number of degrees of freedom ($f = n - 1$).

The following idealized experiment illustrates the formation of the χ^2 distribution:

If from a normally distributed population of variance σ^2 a large (theoretically infinite) number of samples of size n are taken and if the standard deviation s of each sample is calculated, then the statistic

$$\chi^2 = f\frac{s^2}{\sigma^2} \quad \text{where } f = n - 1$$

follows a distribution called the chi-squared or χ^2 distribution. (Chi is the name of the Greek letter χ and is pronounced ki, *to rhyme with* high.*)*

Practical Application

The χ^2 distribution (Figure 2.3–2) is applied in reaching conclusions from a sample variance s^2 about the *variance* σ^2 of the whole population, i.e., in calculation of the corresponding *confidence interval.*

It is also used in connection with the question as to what range the variance of a sample will lie in with a given probability when the corresponding value of the whole population is known, i.e., for calculating a *random dispersion interval.*

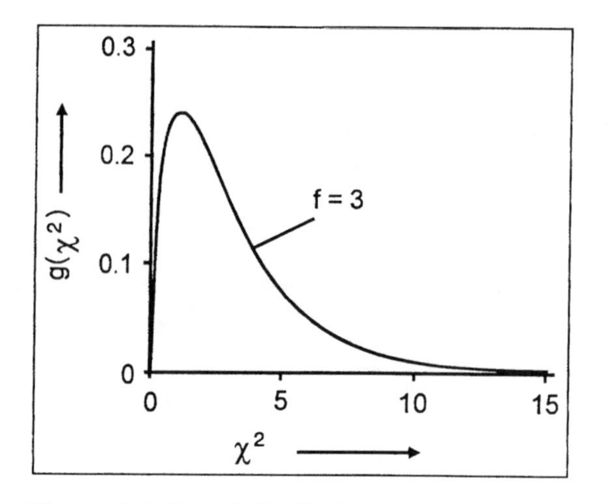

Figure 2.3–2. χ^2 distribution.

Example

For f = 50 and $1 - \alpha = 0.95$ ($\alpha = 0.05$), the following χ^2 values are to be calculated:

1. Single-sided lower χ^2 limit
2. Double-sided (upper and lower) χ^2 limit

χ^2-DISTRIBUTION Critical values		
Degree of freedom	f :	50
Confidence level	$1 - \alpha$ in % :	95
Critical values for the χ^2-distribution		
Single-sided lower	χ^2 (f; α) =	34.7643
Double-sided upper	χ^2 (f; $1 - \alpha/2$) =	71.4202
Double-sided lower	χ^2 (f; $\alpha/2$) =	32.3574

2.3.3 The F-Distribution

The F-distribution (Figure 2.3–3) is based on the following idealized experiment:

From a normally distributed population of known variance σ^2, two samples of sizes n_1 and n_2 are taken. If we repeatedly take pairs of samples, size n_1 and n_2, the ratio between the variances of the two samples

$$F = \frac{s_1^2}{s_2^2}$$

will follow a distribution called the F-distribution.

Since s_1 and s_2 may assume minimum values of zero, this distribution is asymmetric and can assume values between zero and infinity.

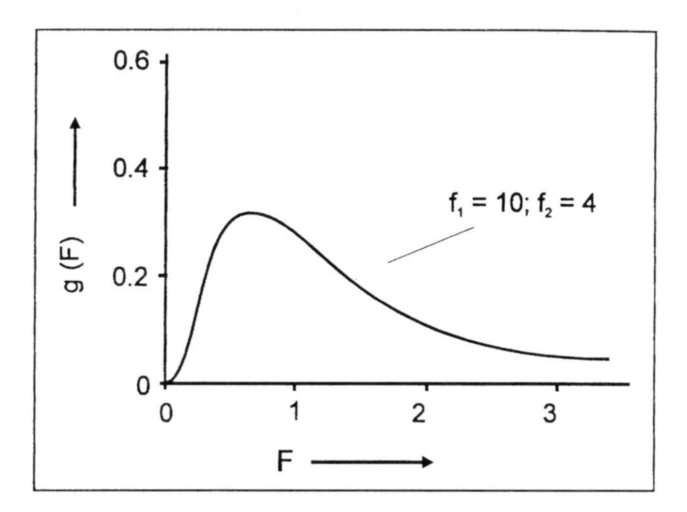

Figure 2.3–3. F-distribution.

Practical Application

This distribution is used, for example, to compare the variances of two populations on the basis of sampling results.

Example

For $f_1 = 79$ and $f_2 = 44$, $1 - \alpha = 0.95$ ($\alpha = 0.05$), the following F values are to be calculated:

1. Single-sided lower F limit
2. Double-sided (upper and lower) F limits

F-DISTRIBUTION Critical values			
Degree of freedom	f_1 :		79
Degree of freedom	f_2 :		44
Confidence level	$1 - \alpha$ in % :		95
Critical values of the F-distribution			
Single-sided lower	$F(f_1; f_2; \alpha)$	=	0.65342
Double-sided upper	$F(f_1; f_2; 1 - \alpha/2)$	=	1.72817
Double-sided lower	$F(f_1; f_2; \alpha/2)$	=	0.60220

2.4 Random Dispersion and Confidence Intervals

At this point, we summarize the necessary formulas for determination of random dispersion and confidence intervals for means, variances, and proportions exceeding in the case of a normal distribution.

Random spread region	Average $\bar{x}$
Single-sided lower	$\mu - \frac{u_{1-\alpha}\sigma}{\sqrt{n}} \le \bar{x} < \infty$
Single-sided upper	$-\infty < \bar{x} \le \mu + \frac{u_{1-\alpha}\sigma}{\sqrt{n}}$
Double-sided lower/upper	$\mu - \frac{u_{1-\alpha/2}\sigma}{\sqrt{n}} \le \bar{x} \le \mu + \frac{u_{1-\alpha/2}\sigma}{\sqrt{n}}$

Random spread region	For variance s^2 with degree of freedom $f = n - 1$
Single-sided lower	$\frac{\chi^2_{f;\alpha}}{f}\sigma^2 \le s^2 < \infty$
Single-sided upper	$0 \le s^2 \le \frac{\chi^2_{f;1-\alpha}}{f}\sigma^2$
Double-sided lower/upper	$\frac{\chi^2_{f;\alpha/2}}{f}\sigma^2 \le s^2 \le \frac{\chi^2_{f;1-\alpha/2}}{f}\sigma^2$

Confidence region	For average μ with known variance
Single-sided lower	$\bar{x} - \frac{u_{1-\alpha}\sigma}{\sqrt{n}} \le \mu < \infty$
Single-sided upper	$-\infty < \mu \le \bar{x} + \frac{u_{1-\alpha}\sigma}{\sqrt{n}}$
Double-sided lower/upper	$\bar{x} - \frac{u_{1-\alpha/2}\sigma}{\sqrt{n}} \le \mu \le \bar{x} + \frac{u_{1-\alpha/2}\sigma}{\sqrt{n}}$

Confidence region	For average μ with unknown Variance
Single-sided lower	$\bar{x} - \frac{t_{f;1-\alpha}s}{\sqrt{n}} \le \mu < \infty$
Single-sided upper	$-\infty < \mu \le \bar{x} + \frac{t_{f;1-\alpha}s}{\sqrt{n}}$
Double-sided lower/upper	$\bar{x} - \frac{t_{f;1-\alpha/2}s}{\sqrt{n}} \le \mu \le \bar{x} + \frac{t_{f;1-\alpha/2}s}{\sqrt{n}}$

Confidence region	For variance σ^2 with degree of freedom f = n – 1
Single-sided lower	$\frac{f}{\chi^2_{f;1-\alpha}} s^2 \le \sigma^2 < \infty$
Single-sided upper	$0 < \sigma^2 \le \frac{f}{\chi^2_{f;\alpha}} s^2$
Double-sided lower/upper	$\frac{f}{\chi^2_{f;1-\alpha/2}} s^2 \le \sigma^2 \le \frac{f}{\chi^2_{f;\alpha/2}} s$

For small samples with n > 30, the confidence interval for the estimated proportion exceeding may be determined from the Durrant nomogram [12]. For larger samples, confidence intervals are calculated from the following formulas.

Confidence region	For excess proportion
Single-sided lower	$u_{1-p_{lo}} = u_{1-p'} - u_{1-\alpha} \sqrt{\frac{1}{n} + \frac{u_{1-p'}}{2n-2}}$
Single-sided upper	$u_{1-p_{up}} = u_{1-p'} + u_{1-\alpha} \sqrt{\frac{1}{n} + \frac{u_{1-p'}}{2n-2}}$
Double-sided lower/upper	$u_{1-p^{up}_{lo}} = u_{1-p'} \pm u_{1-\alpha/2} \sqrt{\frac{1}{n} + \frac{u_{1-p'}}{2n-2}}$

Examples: Double-Sided Confidence Interval for Mean

A sample of size n = 200 from a normally distributed population has a mean value of $\bar{x} = 20.5$ and a standard deviation of s = 0.13. We wish to find the double-sided confidence interval for the mean, with a confidence level of $1 - \alpha = 0.95$.

NORMAL DISTRIBUTION Confidence region average value		
Subgroup average	$\bar{x}$:	20.50000
Subgroup standard deviation	s :	0.130000
Subgroup size	n :	200
Confidence level	$1 - \alpha$ in % :	95
Double-sided		
Confidence limits $20.4819 \le \mu \le 20.5181$		

Double-Sided Confidence Interval for Standard Deviation

A sample of size n = 100 taken from a normally distributed population was found to have a variance of $s^2 = 0.86$. We wish to find the double-sided confidence interval for the standard deviation, with a confidence level of $1 - \alpha = 0.95$.

NORMAL DISTRIBUTION Confidence region standard deviation		
Subgroup average	$\bar{x}$:	---
Subgroup standard deviation	s :	0.927362 $= \sqrt{0.86}$
Subgroup size	n :	100
Confidence level	$1 - \alpha$ in % :	95
Double-sided		
Confidence limits $0.81423 \leq \sigma \leq 1.07729$		

Random Dispersion Limits (Double-Sided)

For a lot of turned pieces, it is claimed that the mean μ is 100.00 mm and the standard deviation σ is 0.1 mm. If the statement is correct, the mean of a sample of size n = 25 must, with a 1 percent probability of error, lie within what interval?

NORMAL DISTRIBUTION Random spread region		
Population average	μ :	100.0000
Population standard deviation	σ :	0.1
Subgroup size	n :	25
Confidence level	$1 - \alpha$ in % :	99
Double-sided		
Random spread limits $99.9485 \leq \bar{x} \leq 100.052$		

Random Dispersion Limits (Single-Sided, Upper)

According to manufacturer's claims, $\mu = 12.00$ mm and $\sigma = 0.1$ mm. Assuming an error probability of 1 percent, how large may the standard deviation of a sample of 25 values be before we disbelieve the manufacturer?

NORMAL DISTRIBUTION Random spread region		
Population average	μ :	12.0000
Population standard deviation	σ :	0.1
Subgroup size	n :	25
Confidence level	$1 - \alpha$ in % :	99
Single-sided upper		
Random spread limits $0 \leq s \leq 0.13382$		

2.5 Determination of Sample Statistics

Before conclusions can be drawn from a sample about the population, the sample data must be structured either in tabular or in graphic form.

In a statistical investigation, the observed data are usually recorded in the order in which they occurred. The data thus recorded are known as the raw data.

Table 2.5–1 shows a series of measurements consisting of 138 values.

This *graveyard of numbers* is not very illuminating, and it conveys very little message. It is more useful to calculate statistics from the data that describe the entire set of measurements taken.

There follows a list of useful statistics in this context, including some commonly used abbreviations and symbols for them:

- Size of the series of measurements n
- Arithmetic means $\bar{x}$
- Median $\tilde{x}$
- Sample variance s^2, sample standard deviation s
- Range R
- Smallest and greatest values x_{min}, x_{max}
- Proportion exceeding α
- Kurtosis g_2
- Skewness g_1
- Capability indices C_m, C_{mk}, P_p, P_{pk}, C_p, C_{pk}
- Quantiles, percentiles
- Coefficient of variation coefficient
- Confidence intervals for means, variances, capability indices, and proportions exceeding

These statistics give, among other information, answers to the following questions:

- Where is the mean of the sample located (measure of location)?
- How widely are the values of the sample dispersed (measure of dispersion)?
- What percentage of the values may be expected to lie outside specification?
- Is the model of a normal distribution suitable as a basis for determining capability?
- Is the production device or process capable?

Table 2.5–1. Example for the determination of statistical values.

1	2	3	4	5
22.036	22.052	22.055	22.050	22.059
22.040	22.055	22.043	22.044	22.046
22.052	22.044	22.060	22.054	22.054
22.043	22.057	22.059	22.060	22.058
22.031	22.056	22.067	22.028	22.048
22.043	22.068	22.035	22.043	22.046
22.061	22.046	22.042	22.049	22.048
22.044	22.036	22.048	22.046	22.050
22.043	22.047	22.035	22.047	22.057
22.024	22.057	22.050	22.044	22.064
6	**7**	**8**	**9**	**10**
22.068	22.050	22.033	22.061	22.062
22.056	22.044	22.043	22.037	22.053
22.055	22.049	22.053	22.041	22.038
22.055	22.045	22.046	22.049	22.035
22.030	22.041	22.062	22.061	22.071
22.020	22.054	22.045	22.053	22.026
22.030	22.053	22.039	22.046	22.067
22.041	22.061	22.035	22.056	22.041
22.021	22.070	22.049	22.056	22.065
22.043	22.071	22.055	22.040	22.036
11	**12**	**13**	**14**	
22.033	22.053	22.062	22.035	
22.049	22.049	22.054	22.052	
22.061	22.054	22.025	22.049	
22.047	22.046	22.044	22.041	
22.060	22.058	22.057	22.055	
22.050	22.052	22.044	22.046	
22.056	22.036	22.070	22.058	
22.057	22.059	22.030	22.053	
22.039	22.066	22.060		
22.076	22.045	22.059		

The statistics may be determined by means of a pocket calculator or using graphical methods (e.g., probability paper). The use of computers is of very great assistance in this process. This is true in particular

- For large quantities of data
- For complex statistics
- For correct assessment of the available facts
- For automatic monitoring of the process using statistical parameters
- When comparing different methods

2.5.1 Important Sample Statistics

Arithmetic Mean

The most important measure of location is the arithmetic mean $\bar{x}$ (short: mean) of the data.

It is given by

$$\bar{x} = \frac{\text{sum of sample values}}{\text{number of sample values}}$$

$$\bar{x} = \frac{1}{n}\sum_{i=1}^{n} x_i \qquad \begin{aligned} &n = \text{sample size} \\ &i = 1, 2, \ldots, n \\ &x_i = \text{characteristic} \end{aligned}$$

Geometric Mean

The geometric mean is calculated as follows:

$$\bar{x}_G = \sqrt[n]{x_1 \cdot x_2 \cdots x_n}$$

Alternatively, it can be calculated using the logarithmic approach:

$$\log \bar{x}_G = \frac{1}{n}\sum_{i=1}^{n} \log x_i$$

Median (Central Value)

The median value $\tilde{x}$ is that value of the sample that divides the measured values, once these have been sorted according to size, into two equal parts, such that the same number of observed values lies above it as lies below it.

For an even number of observed values, the median is calculated as the mean of the two values lying in the middle of the sorted values.

$$\tilde{x} = x_{(m)} \qquad \text{for } n = 2m - 1 \quad (n \rightarrow \text{odd})$$

$$\tilde{x} = \frac{1}{2}\left[x_{(m)} + x_{(m+1)}\right] \qquad \text{for } n = 2m \quad (n \rightarrow \text{even})$$

Extreme values (outliers) have less effect on the median than on the arithmetic mean $\bar{x}$.

Variance and Standard Deviation

The *sample variance* is a measure of the spread or dispersion of the measured values. Its symbol is s^2, and it is defined by the following formula:

$$s^2 = \frac{1}{n-1}\sum_{i=1}^{n}(x_i - \bar{x})^2 = \frac{1}{n-1}\left(\sum_{i=1}^{n} x_i^2 - \frac{1}{n}\left(\sum_{i=1}^{n} x_i\right)^2\right) \qquad \begin{aligned} &n = \text{sample size} \\ &i = 1, 2, 3, \ldots, n \end{aligned}$$

The positive square root of the sample variance is denoted as the *sample standard deviation:*

$$s = +\sqrt{s^2}$$

The factor n – 1 is known as the number of *degrees of freedom*. This has the following basis: If, for example, the average (note the presence of the sample average in the preceding formula) of three measured values is known, then if two values are freely chosen, the third is automatically given from a knowledge of the average. For n measured values whose average is known, only n – 1 values may be freely chosen. For $n \to \infty$ (large sample sizes), this difference becomes insignificant.

The calculation of mean and standard deviation may also be performed on the basis of classed values. This results, of course, in some loss of information.

Range

The *range R* of a sample is, like the sample variance, a measure of dispersion. But, in contrast to the sample variance, the range is very easy to determine. It is simply the difference between the largest and smallest values of the sample:

$$R = x_{max} - x_{min}$$
$$x_{max} = \text{largest value}$$
$$x_{min} = \text{smallest value}$$

The range is, of course, highly affected by extreme values (e.g., outliers).

Coefficient of Variation Coefficient

The coefficient of variation V is used for purposes of comparison and is determined as the ratio between the standard deviation and the mean.

$$V = \frac{s}{|\bar{x}|} \quad \text{for } \bar{x} \neq 0$$

The relative coefficient of variation V_r is calculated as follows:

$$V_r = \frac{s}{\bar{x}\sqrt{n}} 100 \quad (\text{in } \%)$$

Skewness, Asymmetry

A unimodal (single-peaked) distribution may be skewed to the right (positively skewed) or skewed to the left (negatively skewed). The statistic g_1 is a measure that gives information about the direction and magnitude of the skewness.

$$g_1 = m_3/m_2^{3/2} \qquad m_2 = \frac{1}{n}\sum_{i=1}^{n}(x_i - \bar{x})^2 \qquad m_3 = \frac{1}{n}\sum_{i=1}^{n}(x_i - \bar{x})^3 \qquad \text{where } i = 1, 2, 3, \ldots, n$$

If $g_1 = 0$, the distribution is symmetrical. The more clearly the value is negative, the more the distribution is skewed to the left (Figure 2.5–1a). The more clearly the value is positive, the more the distribution is skewed to the right (Figure 2.5–1b).

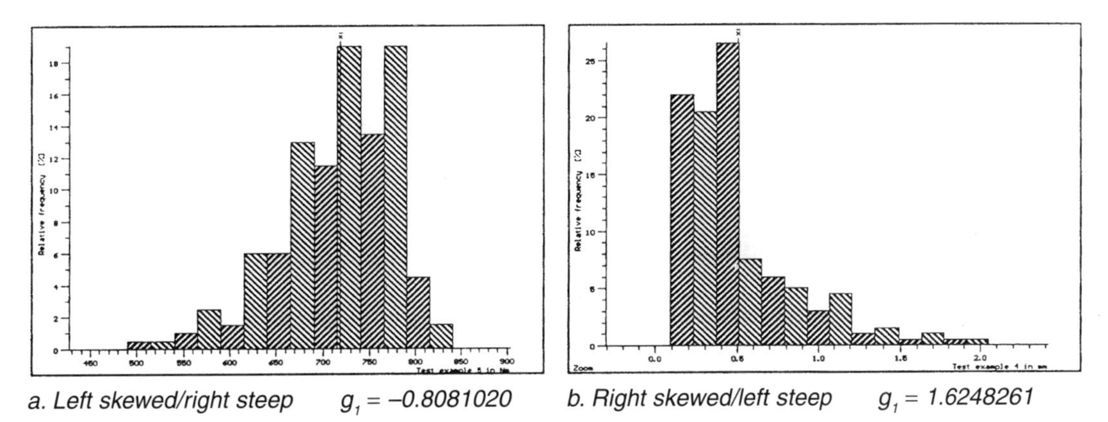

a. Left skewed/right steep $g_1 = -0.8081020$ *b. Right skewed/left steep* $g_1 = 1.6248261$

Figure 2.5–1. Skewness depending on the type of distribution.

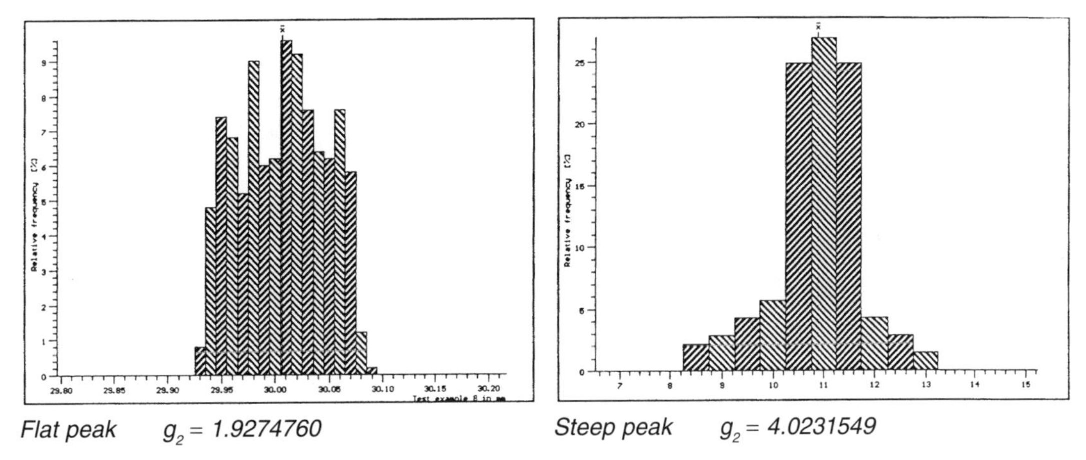

Flat peak $g_2 = 1.9274760$ *Steep peak* $g_2 = 4.0231549$

Figure 2.5–2. Kurtosis depending on the type of distribution.

Excess, Skewness (Kurtosis)

Excess is a measure that shows, for a unimodal distribution, whether the highest point of the distribution curve is higher than that of a normal curve with the same standard deviation. In other words, it tells us whether the distribution has a steep peak or a flat peek. The theoretical kurtosis value for a normal distribution is 0. If $g_2 > 0$, the absolute maximum of the distribution curve is greater than for the equivalent normal distribution; if $g_2 < 0$, it is smaller. Frequently, the term *kurtosis* is used. Excess g_2 equals kurtosis b_2 minus 3.

$$\text{Kurtosis:} \quad b_2 = m_4/m_2^2 \quad m_2 = \frac{1}{n}\sum_{i=1}^{n}(x_i - \bar{x})^2$$

where i = 1, 2, 3, ...

$$\text{Excess:} \quad g_2 = b_2 - 3 \quad m_4 = \frac{1}{n}\sum_{i=1}^{n}(x_i - \bar{x})^4$$

The larger the kurtosis value, the steeper the sides of the peak, and vice versa. (See Figure 2.5–2.)

Note: If the values for skewness and kurtosis differ substantially from 0 (or 3), then this is a hint that the population is not normally distributed. For this purpose, these parameters are included in the numerical test procedures.

Proportion Exceeding and Quantiles (Percentiles)

The proportion exceeding states what percentage of the measured values can be expected to lie below or above a given reference value (see also Section 2.2.2.2). If this reference value is equal to a specification limit, then the result gives the proportion nonconforming above or below the specification limit.

To enable better comparisons of results, a value may be determined below or above which a selected percentage of values are to be found, e.g., the value below which lie 5 percent (and above which lie 95 percent) of the values. Such values are called *quantiles* or *percentiles.*

Capability Indices (Normally Distributed Data)

In a short-term capability study (see Figure 1.4–1), the capability index is often referred to as a machine capability index. This is found from

$$c_m = \frac{USL - LSL}{6s}$$

USL = upper specification limit
LSL = lower specification limit
s = standard deviation

This value, sometimes also referred to as potential, does not take into account the location of the process data in relation to the specification limits.

Hence, critical values for the upper and lower specification limits have been defined for LSL and for USL:

$$c_{lo} = \frac{\bar{x} - LSL}{3s} \qquad c_{up} = \frac{USL - \bar{x}}{3s}$$

where

$\bar{x}$ = mean of the measurement values
s = standard deviation of the measurement values

The smaller of the two values is taken as the critical one and abbreviated to C_{mk}:

$$C_{mk} = \min\{c_{lo}; c_{up}\}$$

Both of these values must not fall below a predefined lower limit. This limit differs from company to company.

This formula leads to reliable capability indices (see Figure 2.5–3) only when the measurement data are normally distributed. Otherwise, a suitable mathematical description of the measurement data must be found. Calculation of the capability indexes is then carried out by means of other formulas. These are dealt with in Chapter 6.

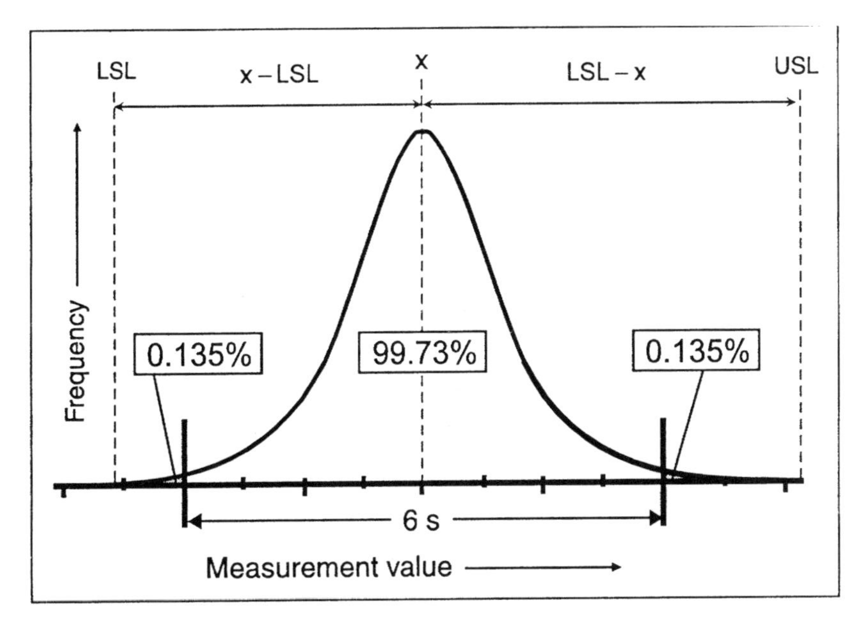

Figure 2.5–3. Capability index definition.

Example: Determination of Capability Index

Measurement of the external diameter (specification requirement *60 ± 0.1 mm) of n = 50 parts* has resulted in a set of readings with mean $\bar{x} = 59.95$ *mm* and standard deviation $s = 0.01$ *mm.*

From *60 ± 0.1 mm* we have:

lower specification limit $LSL = 59.9$ *mm*
upper specification limit $USL = 60.1$ *mm*

$$\mathbf{C_m} = \frac{USL - LSL}{6 \cdot s} = \frac{60.1 - 59.9}{6 \cdot 0.01} = \mathbf{3.33}$$

$$C_{lo} = \frac{\bar{x} - LSL}{3 \cdot s} = \frac{59.95 - 59.9}{3 \cdot 0.01} = 2.67$$

$$C_{up} = \frac{USL - \bar{x}}{3 \cdot s} = \frac{60.1 - 59.95}{3 \cdot 0.01} = 4.00$$

$$\mathbf{C_{mk}} = \min\{C_{lo}; C_{up}\} = \min\{2.67; 4.00\} = \mathbf{2.67}$$

Figure 2.5–4 explains the course of the calculation.

Figures 2.5–5 and 2.5–6 show process distributions with differing location and dispersion measures and the resulting effects on the capability indices.

The four typical cases are summarized in Figure 2.5–7.

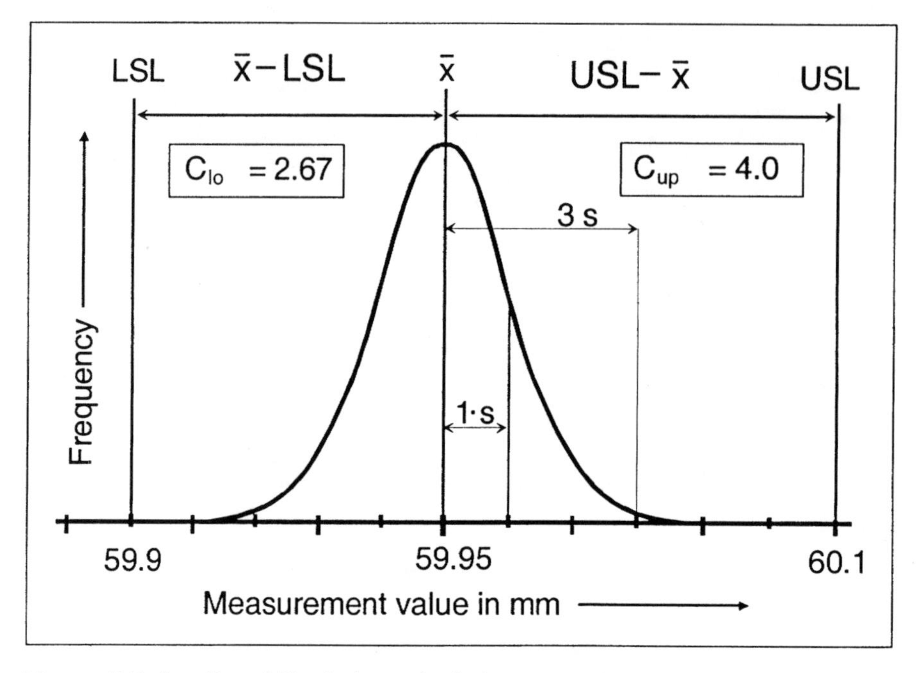

Figure 2.5–4. Capability index calculation example.

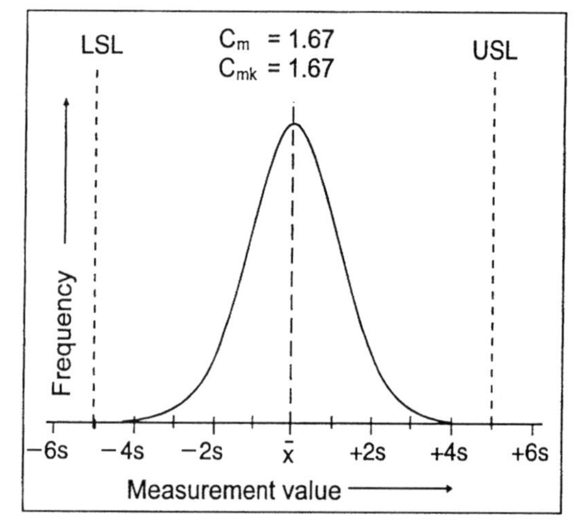

Figure 2.5–5. Centered process.

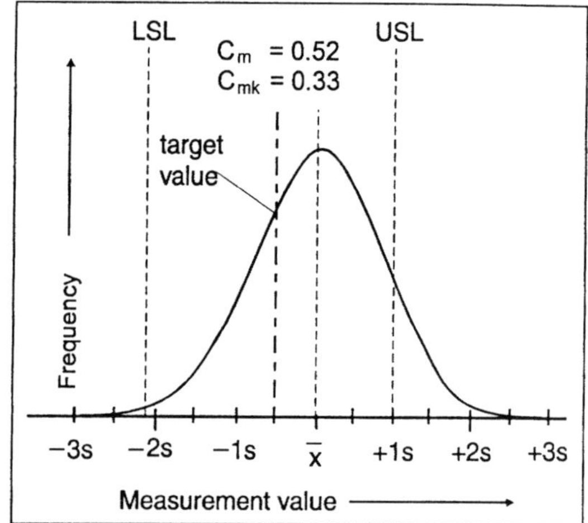

Figure 2.5–6. Noncentered process with exceeding spread.

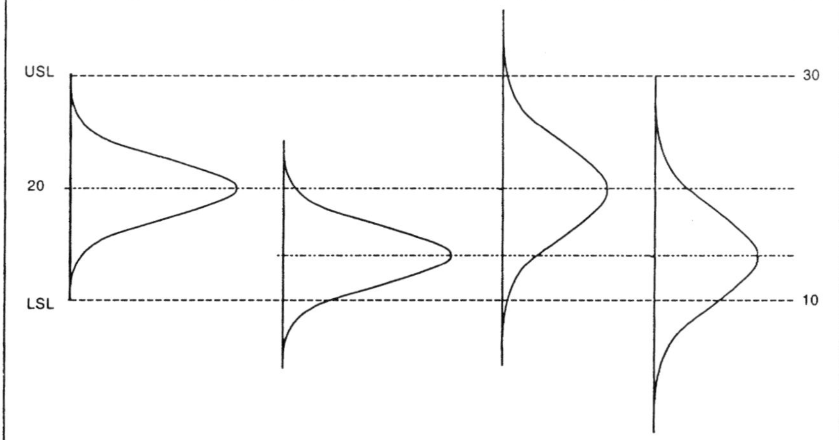

Spread	OK	OK	n.OK	n.OK
s	2	2	5	5
Location	OK	n.OK	OK	n.OK
$\bar{x}$	20	14	20	14
Capability	OK	n.OK	n.OK	n.OK
C_m	**1.67**	1.67	0.67	0.67
C_{mk}	**1.67**	0.67	0.67	0.33

OK = in order
n.OK = not in order

Figure 2.5–7. Capability indices depending on process location and spread.

2.5.2 Example of Sample Statistics Calculation

Based on the individual values of Table 2.5–1 and on a condition of the specification that the upper and lower limits are 22.1 and 22.0 mm respectively, we have the following results:

Subgroup size	n(tot)	:	138
Model distribution		:	Normal distribution
Arithmetic average	$\bar{x}$	:	22.04909
Confidence interval	$\mu_{lo}..\mu_{up}$	:	22.0472167 .. 22.0509717
Median value (central value)	$\tilde{x}$	:	22.049
Minimal value	x_{min}	:	22.020
Maximal value	x_{max}	:	22.076
Variance	s^2	:	0.0001244071
Standard deviation	s	:	0.01115379
Confidence interval	$\sigma_{lo}..\sigma_{up}$	:	0.009974974 ... 0.01265107
Range	R	:	0.056
Skewness	g_1	:	–0.2147893
Kurtosis	b_2	:	2.8431154
Variation coefficient	v	:	0.05058618%
Quantile		:	22.03075
Regression coefficient	r(tot)	:	0.99768844

Continued on next page

Subgroup size	n(tot)	:	138
Model distribution		:	Normal distribution
Total no. of values	> USL	:	0 (0.00000%)
Expected proportion	> USL	:	0.00025%
Confidence interval		:	0.00001% .. 0.00319%
Total no. of values	< LSL	:	0 (0.00000%)
Expected proportion	< LSL	:	0.00054%
Confidence interval		:	0.00004% .. 0.00580%
Capability index	C_m	:	1.49
Confidence interval		:	1.32 .. 1.67
Crit. capability index	C_{mk}	:	1.47
Confidence interval		:	1.28 .. 1.65
Capability index → USL	C_{up}	:	1.52
Capability index → USL	C_{lo}	:	1.47

The confidence regions are based on a confidence level of 95%.

2.6 Construction of Meaningful Graphs

2.6.1 Representation of Individual Values

Run Chart

The simplest, but also most informative, representation is a run chart of the individual values. In this graph, the y axis represents the relevant measurement scale, with the x axis representing time. Points are plotted in their original time sequence and at regular horizontal intervals. The x axis is labeled with a consecutive numbering for the measurement values, with date and time indications, or with lot numbers. The nature of the task at hand determines what type of labeling should be used in any given situation. For a more complete description of the process, it is desirable to relate individual values to events such as tool change, new batch of material, special environmental conditions, and/or other changes in the machine and/or process.

The run chart of the individual values (Figure 2.6–1) provides information about important aspects of process behavior. Values outside specification, periodicity, and outliers can, in most cases, easily be recognized.

In order to monitor prescribed sampling frequencies, the values may be shown in the form of an x(t) plot (Figure 2.6–2). In this type of graph, the values are not plotted at regular horizontal intervals, but are plotted instead against real time. This type of representation is particularly useful for detecting changes in process behavior after major interruptions.

In order to be able to recognize correlations between characteristics and/or changes in the process over time, run charts of different characteristics should be studied together and compared (Figure 2.6–3). It can also be useful to superimpose two or more chart traces on a single chart (Figure 2.6–4).

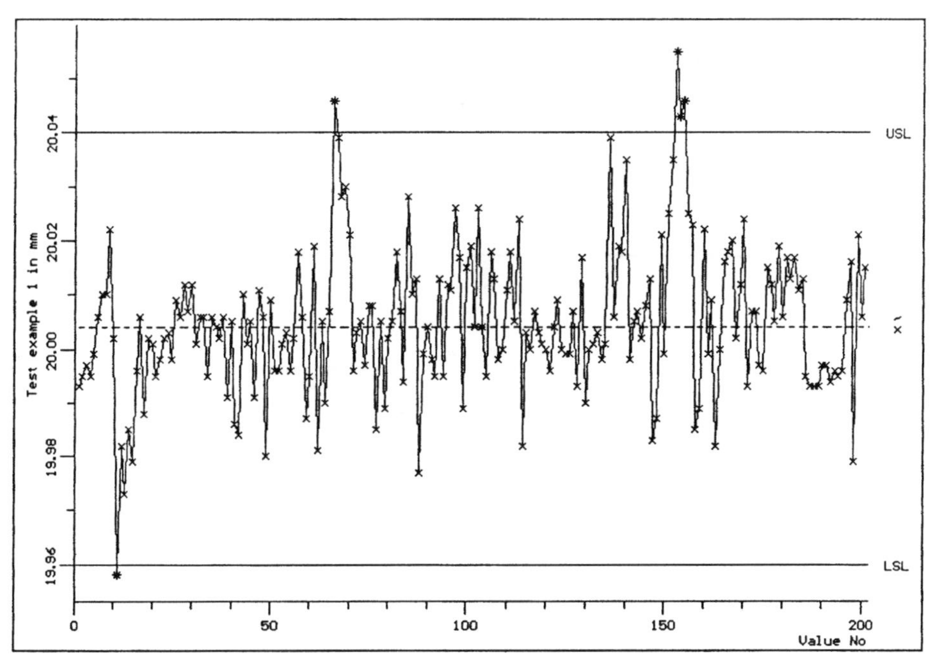

LSL = Lower Specification Limit
USL = Upper Specification Limit

Figure 2.6–1. Actual value chart.

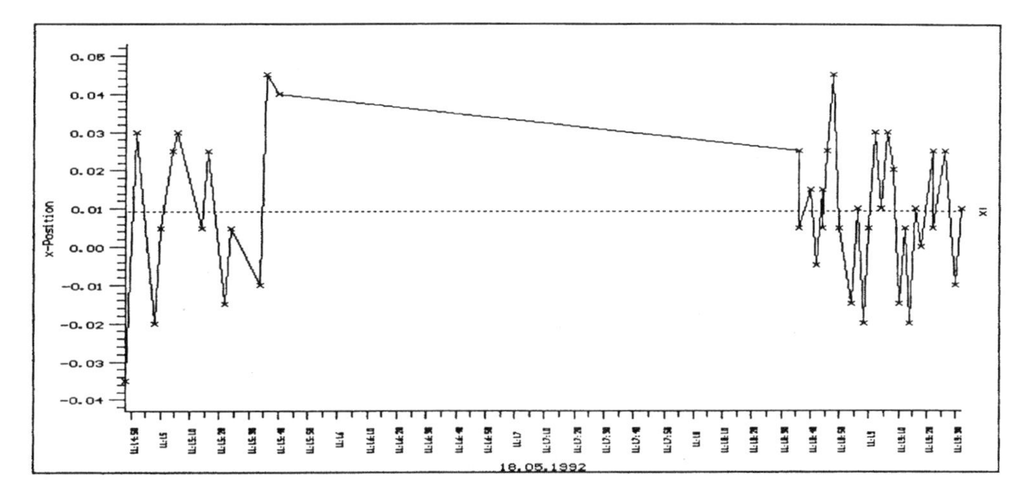

Figure 2.6–2. x(t) plot: Real-time display of the values.

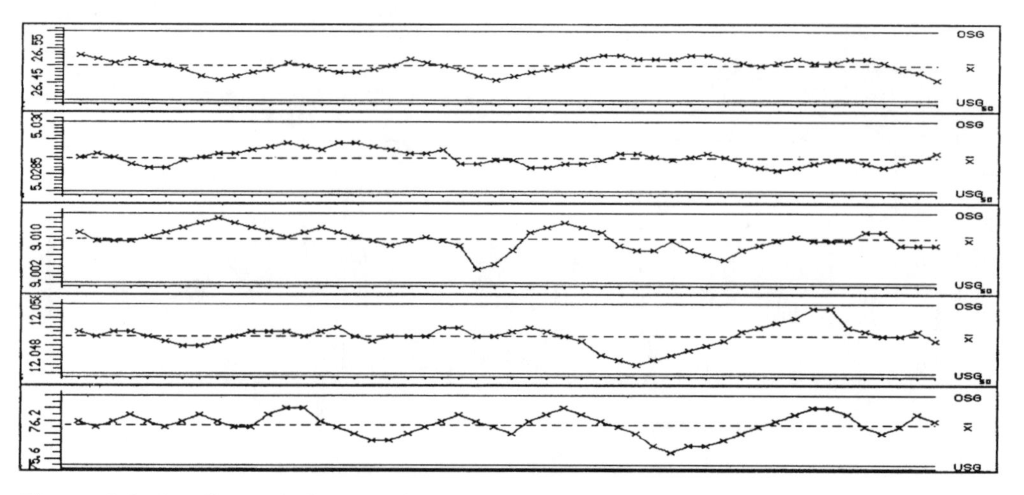

Figure 2.6–3. Several characteristics in comparison.

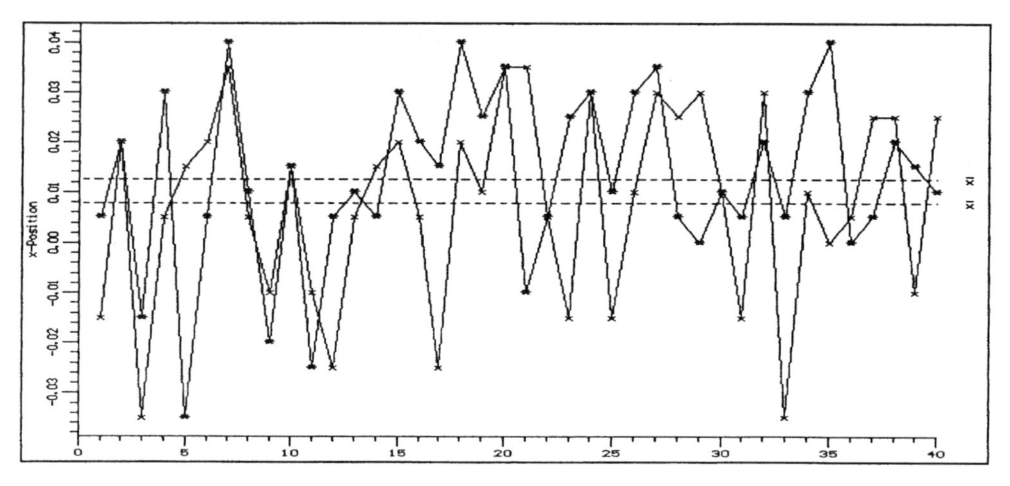

Figure 2.6–4. Two characteristics overlaid.

2.6.2 Actual Value Plot

The actual value plot (Figure 2.6–5) is obtained from the run chart of individual values by indicating the frequency of each value. The x axis represents the measurement scale, and the y axis is used to represent frequency. Due to the time-consuming task of plotting the values manually, this form of presentation of the values is of little practical importance when more than 50 values are concerned. But there is no problem in presenting individual values in this form if a computer is available. In contrast to the histogram, no information is lost by subdividing the values into classes. The resolution of the measurement system is apparent in the plot and can be visually compared to the spread of the data and/or the tolerance interval (if shown). It is thus possible to arrive at a rough assessment of the measurement system's suitability.

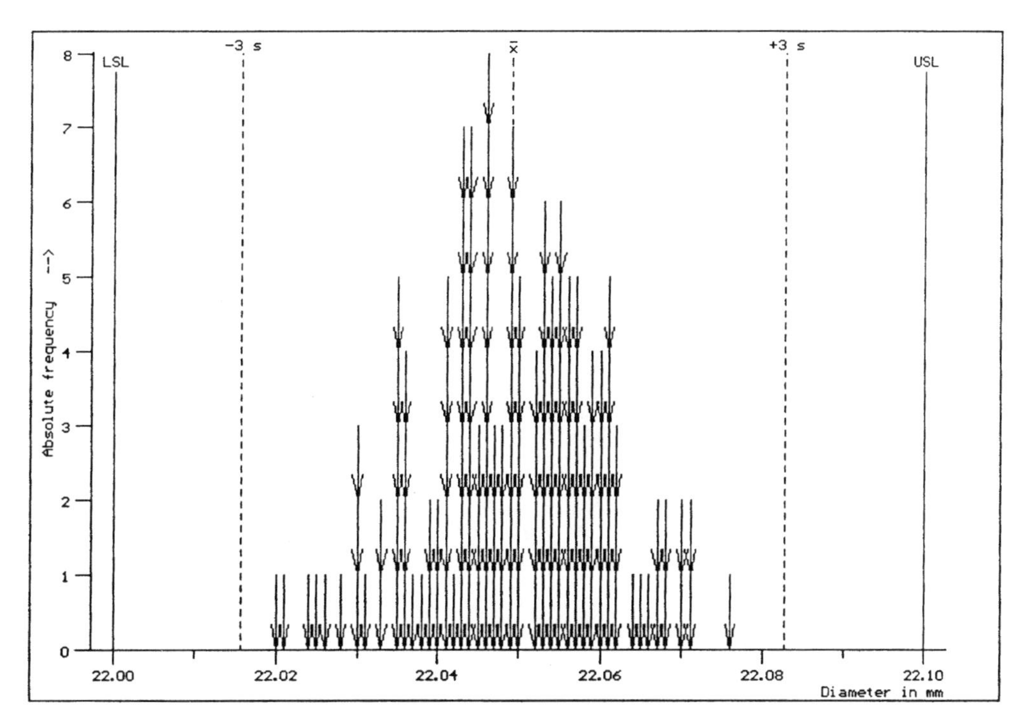

Figure 2.6–5. Actual value plot.

Figure 2.6–5 shows, in addition to the actual data, the mean, the ±3s interval, and the specification limits. The latter makes sense only in case of a normally distributed measurement series. If another distribution model is the basis for the set of data, then the upper percentile (99.865) and the lower percentile (0.135) must be plotted (see Chapter 6).

2.6.3 Histogram

A histogram is based on the same principles as the actual value plot, but has an x axis that is divided into classes (Figure 2.6–6). The number of values within each class is known as the *class frequency.* This may be shown, against the y axis, either in absolute figures or relative to sample size (Figure 2.6–7).

The most difficult part of generating a histogram consists of finding suitable class boundaries. A consistent procedure is required so that subsequent analyses remain comparable. In particular, it should be noted that the result of the χ^2 test (see Section 2.8.4) is dependent on the classification. If the number of measured values n ≥ 25, several measurement results may be combined to form a single class. A small amount of information is lost in this procedure, but the clarity of the graph is improved, especially for large sample sizes.

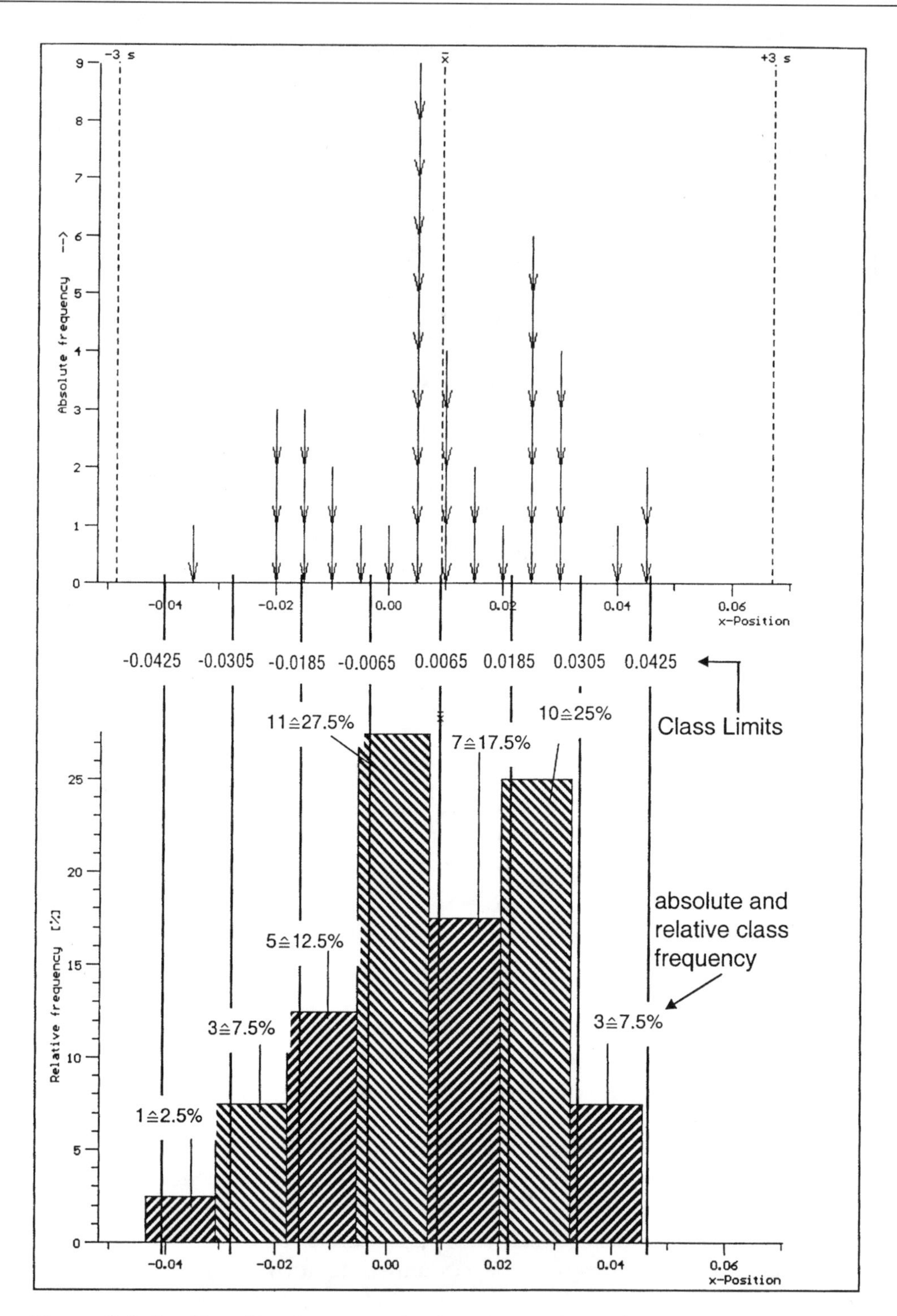

Figure 2.6–6. Transition from actual value chart to histogram.

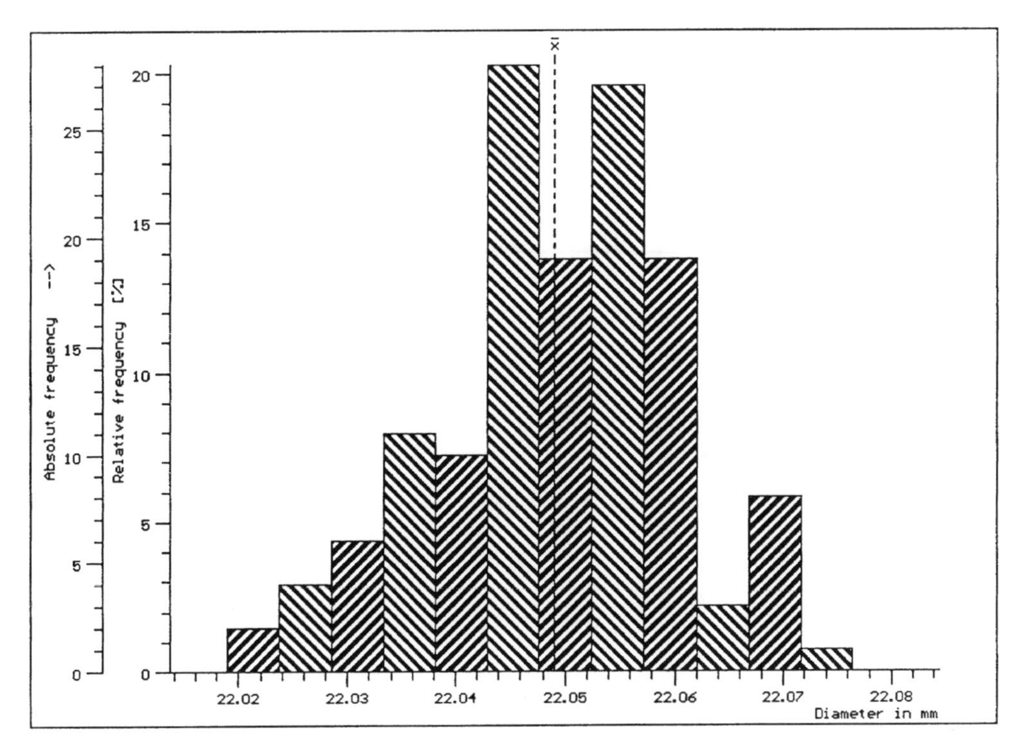

Figure 2.6–7. Histogram with classification from Table 2.6–1.

Option 1 (DIN Standard)

The German standard DIN 55302 contains the following recommendations regarding the number of classes:

Subgroup size n	Classes total k
up to 50	no classification
up to 100	min. 10
up to 1000	min. 13
up to 10000	min. 16

Option 2

The following procedure should be followed:

1. Select a consistent class width (class interval) throughout the whole range of measurement. The final nonzero digit (or decimal) of the class width should be one, two, or five.
2. The class boundaries should be the simplest possible numbers and coincide with rounding limits within the measurement range.
3. To ensure a simple unambiguous allocation of the values to the class, the class boundaries are to be stated to one more decimal place than the data values.

The number of classes may be calculated from the rule of thumb $k \approx \sqrt{n}$, where n is the sample size. The class boundaries should be the simplest possible numbers. For $n \geq 400$, the number of classes remains constant at 20.

Class width ω is then

$$\omega = \frac{x_{max} - x_{min}}{k}$$

where x_{max} = highest value and x_{min} = lowest value.

Option 3

The CNOMO norm [59] suggests the following procedure. The number of classes (rounded up to the next higher integer) is obtained through

$$k = 1 + \frac{10\log(N)}{3} \qquad \text{where } N = \text{subgroup size}$$

Class width ω of a subgroup N is calculated from

$$\omega = \frac{R}{k} \qquad \text{where } R = x_{max} - x_{min}$$

The theoretical class interval is rounded up to the next higher integer value compatible with measurement resolution.

Figure 2.6–6 shows the transition from actual value plot to histogram, giving class boundaries and frequencies (absolute and relative) within each class. The selected class width was 0.012. The readings themselves went to three decimal points. Hence, the class boundaries were extended to a precision of 0.0005.

Examples

The classification of the readings shown in Table 2.5–1 can, for example, be carried out as shown in Table 2.6–1. The corresponding histogram is shown in Figure 2.6–7.

Table 2.6–1. Classification with absolute frequency.

Class no.	Lower class limit	Upper class limit	Absolute frequency	Relative frequency in %
1	22.0189	22.0237	2	1.45
2	22.0237	22.0285	4	2.9
3	22.0285	22.0333	6	4.35
4	22.0333	22.0381	11	7.97
5	22.0381	22.0429	10	7.25
6	22.0429	22.0477	28	20.29
7	22.0477	22.0525	19	13.77
8	22.0525	22.0573	27	19.57
9	22.0573	22.0621	19	13.77
10	22.0621	22.0669	3	2.17
11	22.0669	22.0717	8	5.80
12	22.0717	22.0765	1	0.72

Table 2.6–2 gives another classification.

A comparison of Figure 2.6–7 and Figure 2.6–8 illustrates the considerable effect of the classification process on the shape of the histogram.

Table 2.6–2. Classification with absolute frequency.

Class no.	Lower class limit	Upper class limit	Absolute frequency	Relative frequency in %
1	22.0195	22.0263	5	3.62
2	22.0263	22.0332	7	5.07
3	22.0332	22.0400	15	10.87
4	22.0400	22.0468	31	22.46
5	22.0468	22.0537	28	20.29
6	22.0537	22.0605	32	23.19
7	22.0605	22.0673	13	9.42
8	22.0673	22.0742	6	4.35
9	22.0742	22.0810	1	0.72

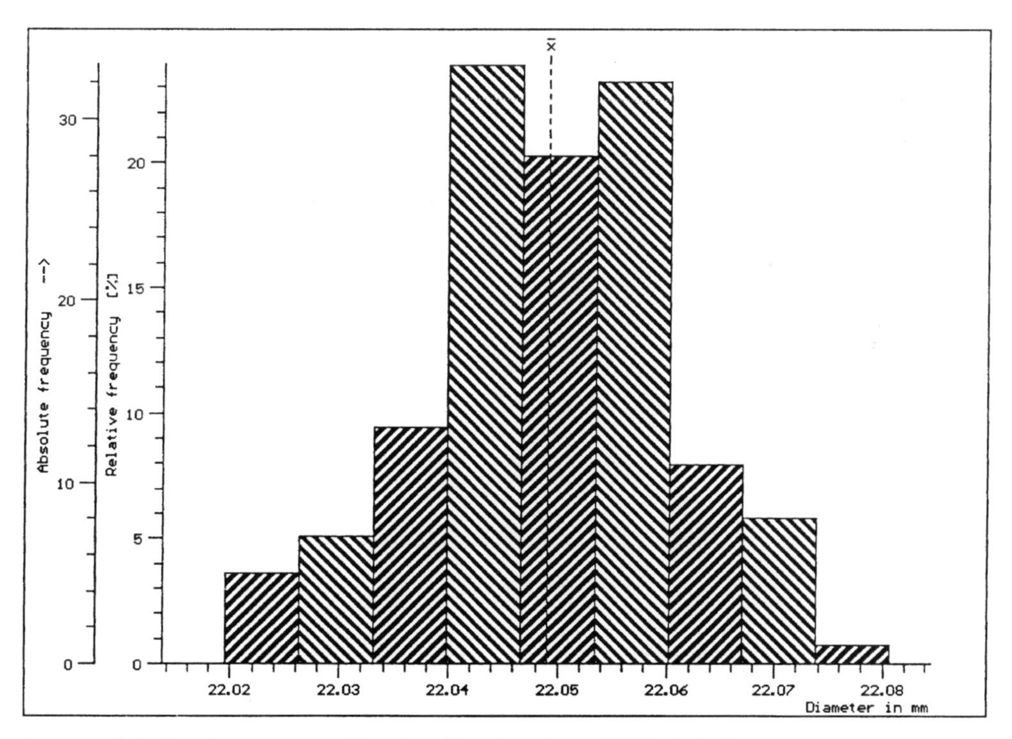

Figure 2.6–8. Histogram with classification from Table 2.6–2.

2.6.4 Cumulative Frequency Curve

By accumulation of the relative frequencies in the histogram, we can obtain a stepped graph. This is approximated by the distribution function G(x) (Figure 2.6–9).

The y axis has a linear scale. Hence, each value has the same weight. This representation makes it possible to estimate the agreement between the cumulative function and the actual values. If this agreement is good, the validity of the assumed distribution model is confirmed (see also Section 2.6.5). The resolution is very poor at the end of the graph, where the function approaches either 0 percent or 100 percent. In these regions, the values of G(x) either cannot be estimated or can be only approximated.

This form of representation is particularly unsuitable for manual evaluation, since plotting the function G(x) correctly by hand is a very difficult and time-consuming process. With the objective of easier handling and better estimation of G(x) in the outermost regions, the tool known as *probability paper* was developed (see Section 2.6.5).

2.6.5 Normal Probability Paper

Probability plots made on probability paper are closely related to the cumulative frequency curve. The only difference is that, by means of a transformation, the cumulative frequency curve is made to *unbend* on probability paper to yield a *straight-line probability plot.* By spreading out the values in the end regions (less than 0.1 and greater than 0.9), the estimation of proportions exceeding a given value in these regions can be carried out much more accurately than from the cumulative curve.

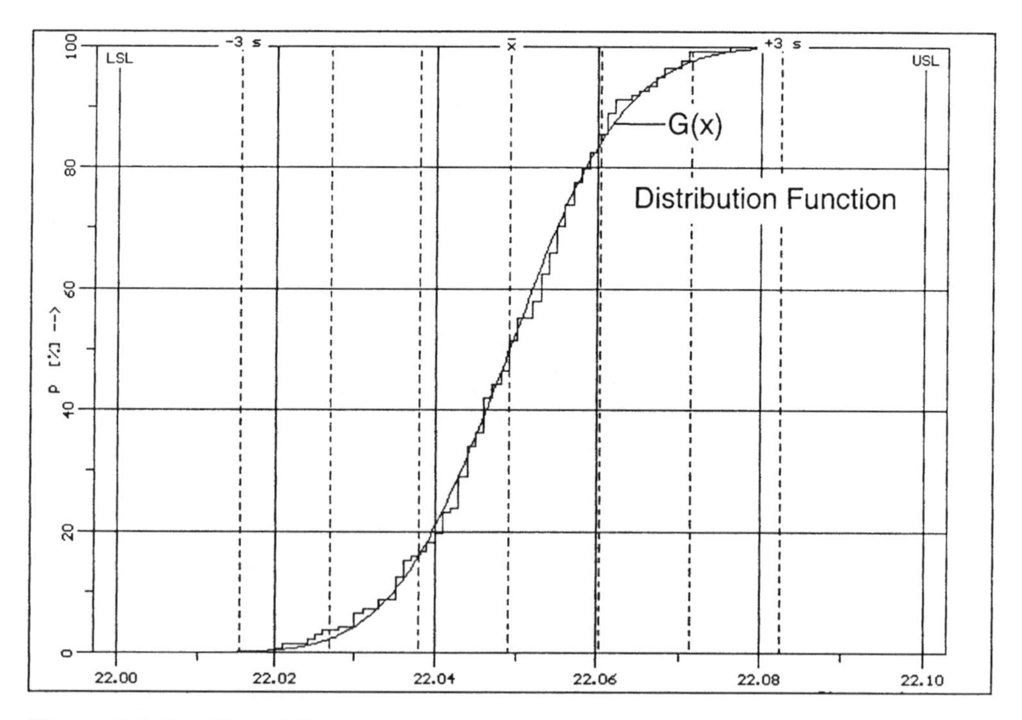

Figure 2.6–9. Cumulative curve.

Statistical Basis of Probability Paper

On probability paper, the standardized variable

$$u = \frac{x - \mu}{\sigma}$$

of the probability function of the normal distribution is plotted against the variable x. If x is normally distributed, the plot forms a straight line. Figure 2.6–10 shows the percentage values of the distribution function G(u) in addition to the u values.

Use of Probability Paper

To construct the straight-line probability plot, the individual values or class frequencies (depending on sample size) of a frequency diagram are plotted on the paper. In a computer-based evaluation, it is definitely preferable to plot the individual values. If the plot points lie approximately in a straight line, it can be assumed that the population is normally distributed.

From this line, we can read off the mean at 50 percent and the standard deviation, which is half the interval between 16 percent and 84 percent. Hence, process location and process dispersion may be determined on the basis of the straight-line probability plot. In addition, if the upper and lower specification limits are shown on the graph, the proportion exceeding may be estimated at the intersections of the specification limits with the straight line (see Figure 2.6–11).

For manual construction of the plot based on more than 50 values, it may be preferable to use grouped data rather than the individual values.

In order to determine whether the values entered confirm the assumption of a normal distribution, confidence limits based on a given level of confidence (e.g., 95 percent or 99 percent) may be marked on the graph (Figure 2.6–12).

Each type of distribution has a different scale for the ordinates (and sometimes for the abscissas as well) and, hence, its own probability paper.

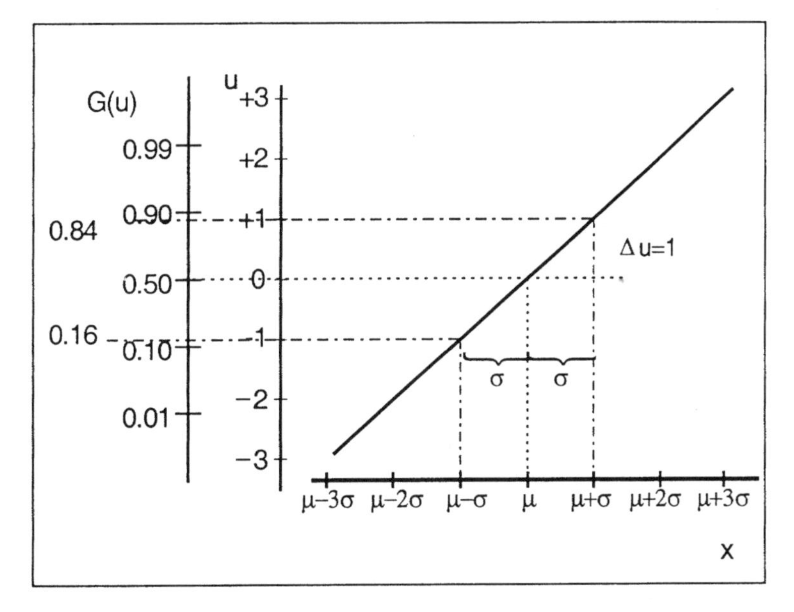

Figure 2.6–10. Probability plot (schematic).

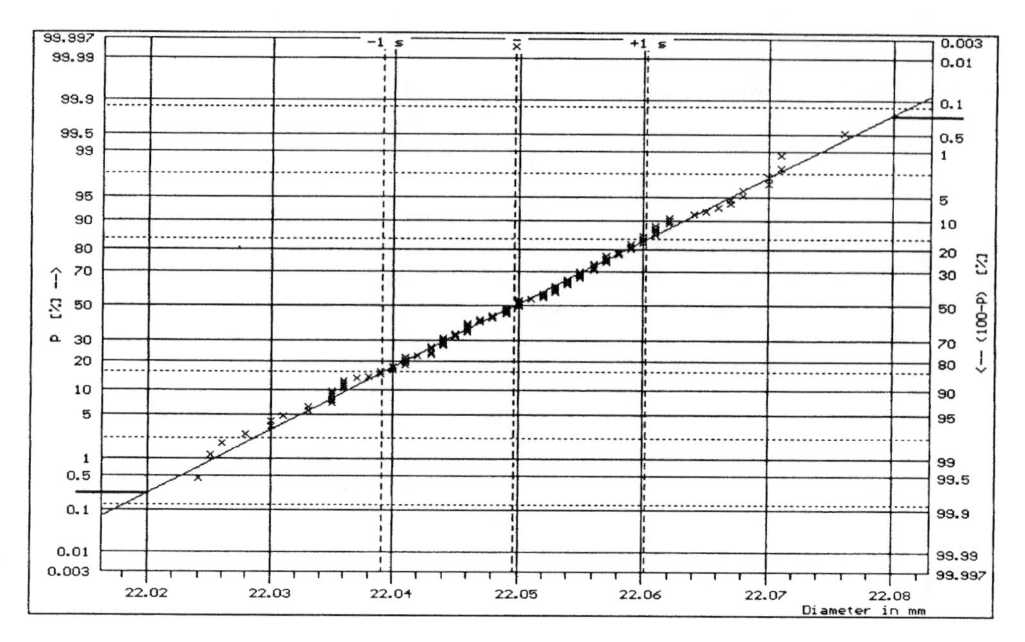

Figure 2.6–11. Probability plot with individuals.

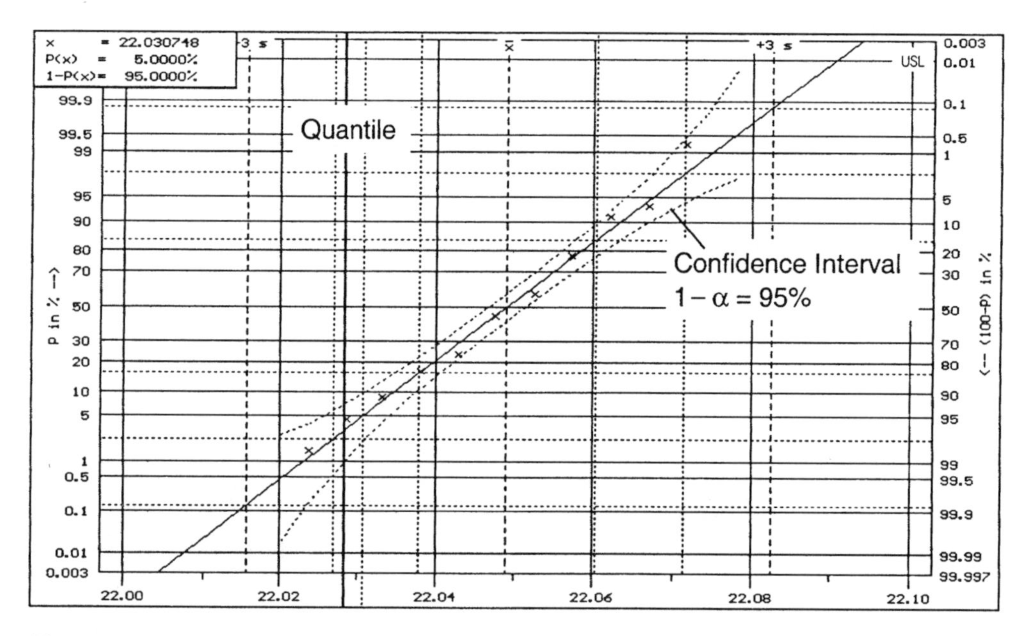

Figure 2.6–12. Probability plot with confidence interval and quantile.

2.6.6 Representations of Pairs of Values

x/y Plot

One frequently encounters characteristics whose measurement does not generate a single value, but instead a pair of values. Positional tolerances (e.g., for centers of bores) are a typical example. The pair of values comprises an x value and a y value. Individual pairs may be represented in Cartesian coordinates in terms of an x/y plot (Figure 2.6–13). The specification limits may be shown on the graph in the form of a rectangle. In contrast with the one-dimensional treatment, the desired distribution of values forms, depending on the scales of the axes, a circle or an ellipse in a plane. If a normal distribution is assumed for the values, probability ellipses may be drawn. These show what percentage of the population will lie within a designated range.

Polar Representation

If pairs of values are measured in terms of angle and distance from a reference point, it is advisable to plot them in a polar diagram (Figure 2.6–14). Specification limits may also be indicated in this type of diagram.

Representation of Two-Dimensional Distributions

Frequencies may be determined for pairs of values in a manner comparable to the determination for one-dimensional characteristics. In this case, these may be described by means of the two-dimensional normal distribution (Figure 2.6–15).

Based on this distribution, capability indices may be calculated for paired value characteristics.

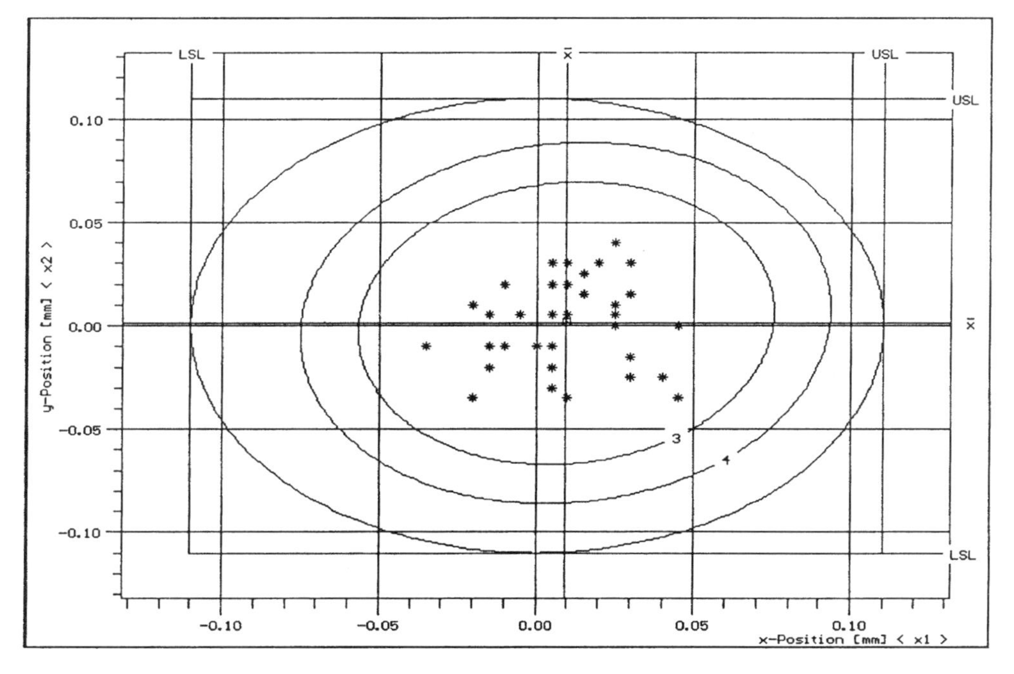

Figure 2.6–13. x/y plot.

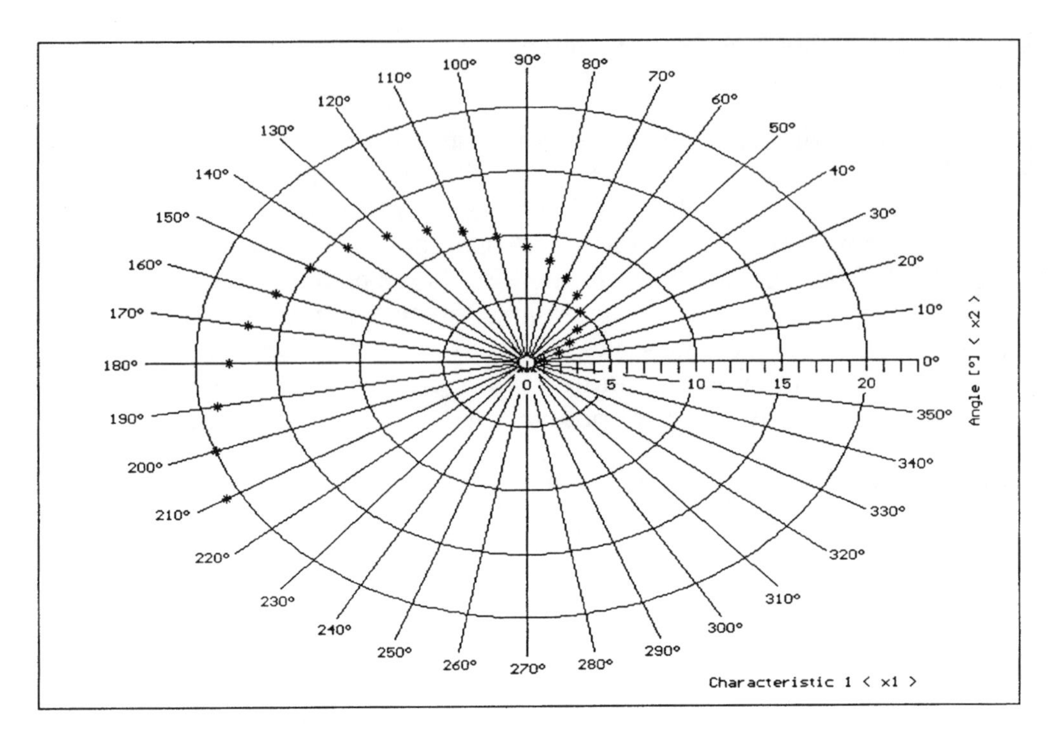

Figure 2.6–14. Polar display.

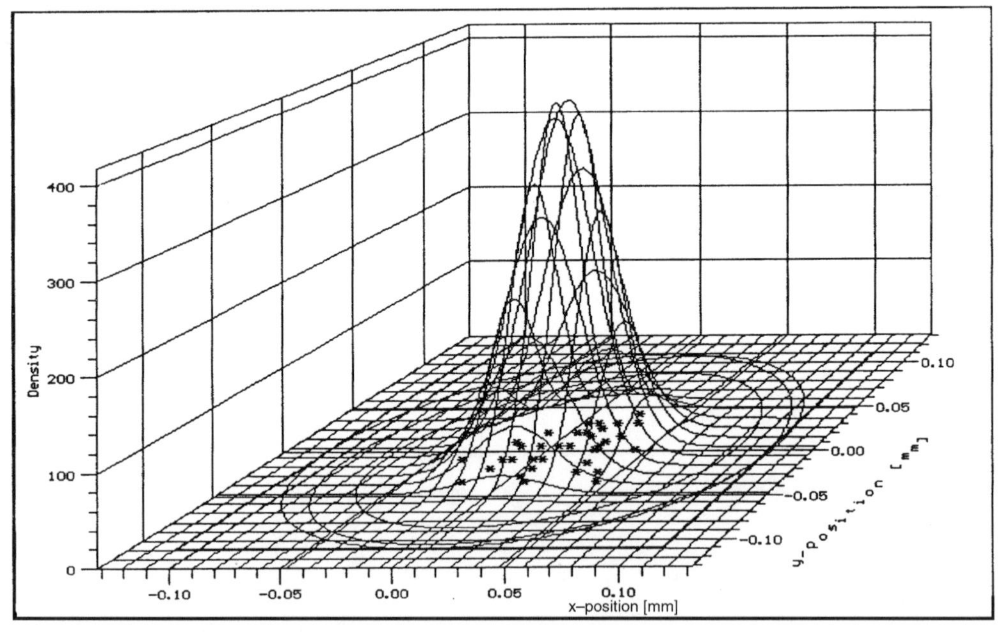

Figure 2.6–15. Two-dimensional normal distribution.

2.6.7 Representation of Sample Statistics

If samples are taken from a running process, the mean and variance may be calculated from the individual samples. In an assessment of the whole population, based on these samples, the question frequently arises whether

- The means and variances are randomly distributed
- The means themselves are normally distributed
- The underlying process variation as reflected in the sample variances may be assumed to be constant
- The variances conform to the χ^2 distribution

One of the most important means to display sample statistics is the control chart. Control charts will be discussed in Chapter 3. There are, however, also other means of representing sample statistics.

Moving Averages and Variances

The manner in which the averages and variances vary over time may be represented in the same way as the variation of individual values. It is also possible to superimpose *moving averages* on the individual values (Figures 2.6–16 and 2.6–17). Moving averages are determined by finding the mean of a group of values extending several samples to the left and to the right of the value in question. This leads to a smoothing of the curve; i.e., random influences are filtered out.

In the calculation of means, a distinction is made between *weighted* and *unweighted* calculations. In an unweighted summarizing of three values into one average, the first mean is calculated from the first, second, and third individual values, the second mean from the second, third, and fourth, etc.

In a weighted calculation, the mean may, for example, be calculated from seven values, as per the following formula:

$$\bar{x}_i = \frac{2x_{i-3} + 3x_{i-2} + 6x_{i-1} + 7x_i + 6x_{i+1} + 3x_{i+2} + 2x_{i+3}}{21} \qquad i = 4, 5, 6, \ldots$$

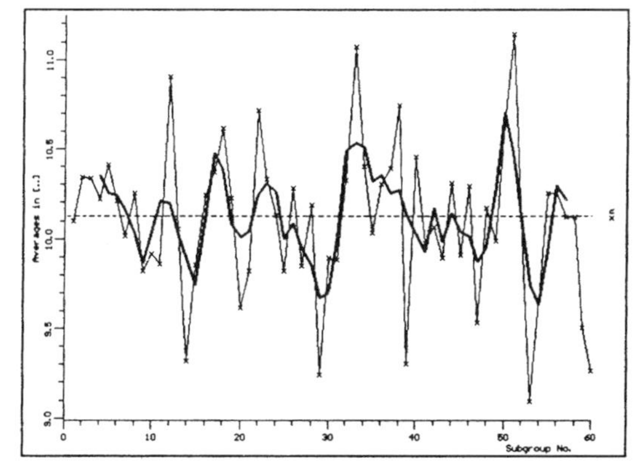

Figure 2.6–16 Averages chart with moving averages.

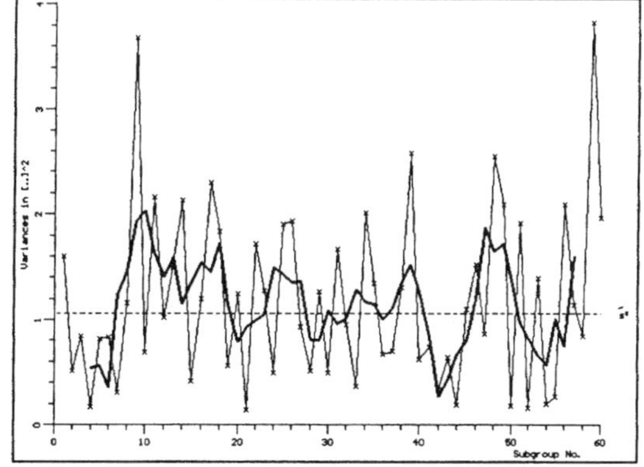

Figure 2.6–17. Variances chart with moving averages.

Probability Paper for Averages

Whether the means of a number of samples are normally distributed may be determined, as for the individual values, by plotting the means on probability paper (Figure 2.6–18). This allows the detection of fluctuations in process location (shifts in the process average). Hence, this probability plot has a similar function to that of the average chart.

χ^2 Paper for Variances

By plotting the variances on χ^2 paper (Figure 2.6–19), it may be determined whether the variances conform to the χ^2 distribution. This paper is interpreted in the same manner as normal probability paper. Disturbances in the dispersion of the process may be detected. Hence, this graph has a function similar to that of the standard deviation chart.

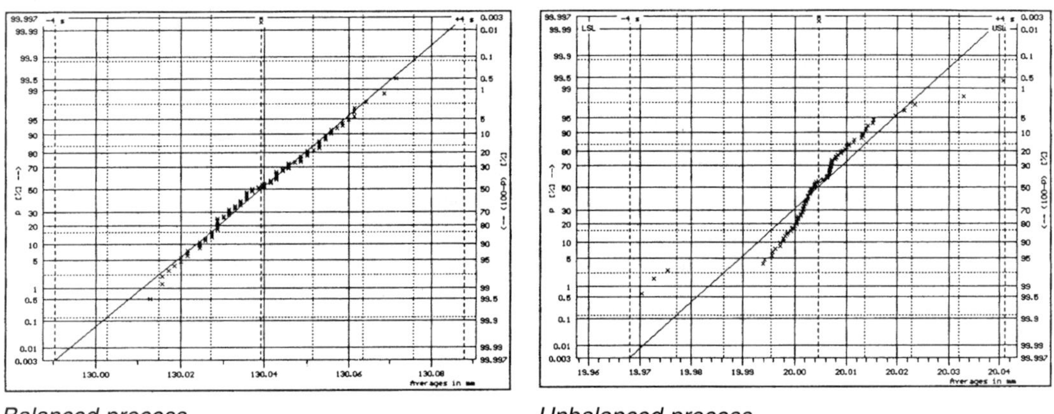

Balanced process *Unbalanced process*

Figure 2.6–18. Display of the averages in a probability plot.

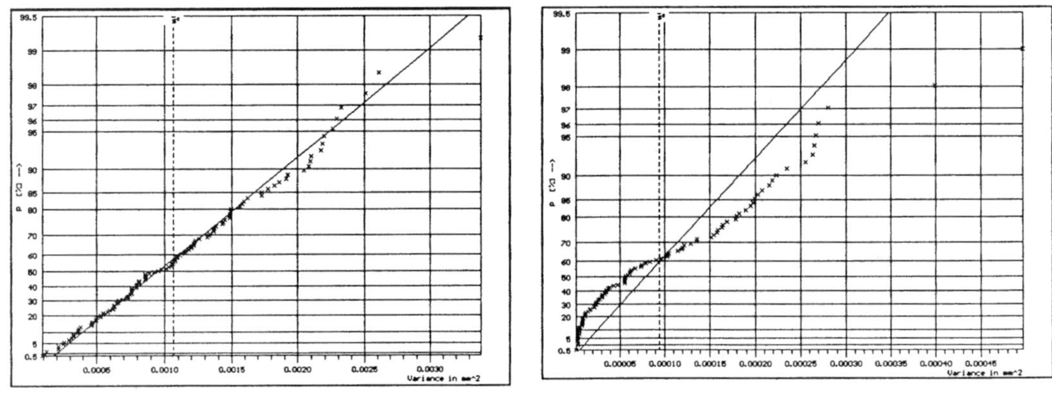

Balanced process *Unbalanced process*

Figure 2.6–19. Display of the variances in a χ^2 plot.

2.6.8 Pareto Analysis

The Italian economist Vilfredo Pareto (1848–1923) discovered over 100 years ago that in many cases about 20 percent of the causes are responsible for about 80 percent of the effects. This is known as *Pareto's law* or the *80/20 rule.* Pareto analyses (sometimes also called ABC analyses or fault frequency analyses) are performed with the aid of special diagrams to separate the *vital few* from the *trivial many* and thus enable the establishment of priorities for quality improvement actions.

First, changes in process parameters and ambient conditions are recorded as events on a check sheet or stored in a computer with suitable input means. For subsequent evaluation, these events must be cataloged and provided with an unambiguous key. In general, it is possible to link a number of events to a single recorded value or sample. Figure 2.6–20 shows a run chart of individual values with a number of allocated events.

In the present example, the events

- Slider closed
- Direction lever incorrect
- Magnet defective

were allocated to value no. 3.

If, over a given time interval, there are values with events linked to them, the individual and percentage frequencies of the events are determined and plotted in the form of a bar chart.

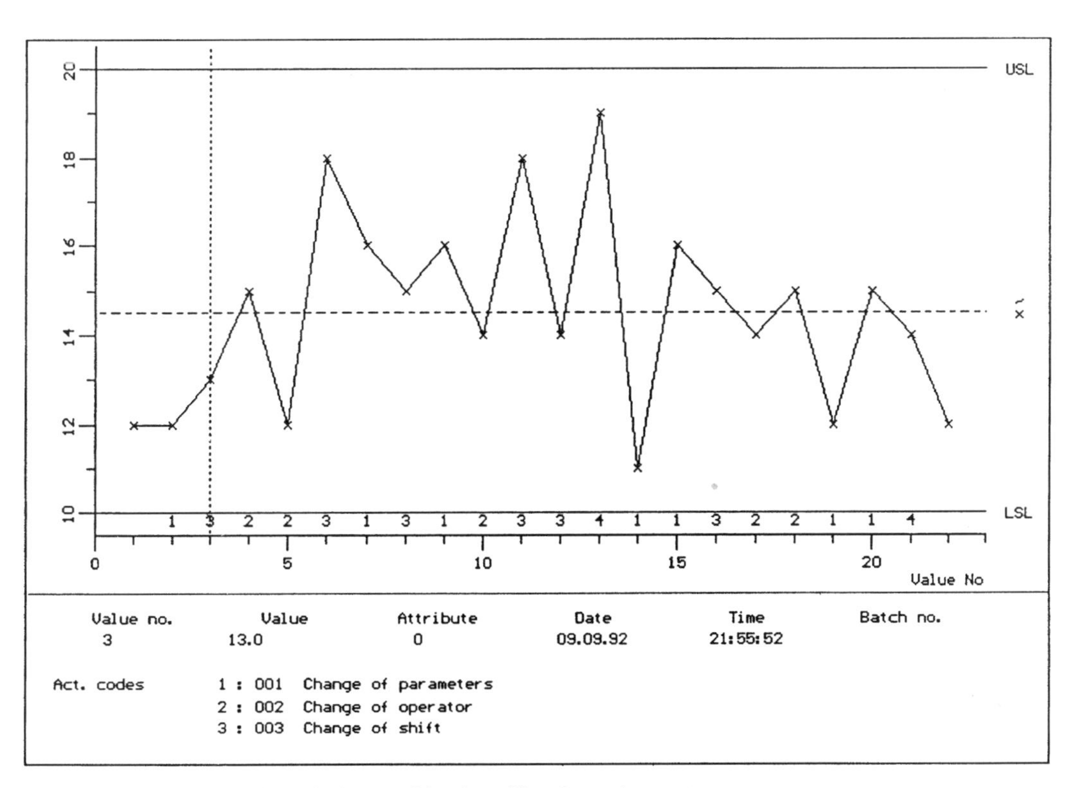

Figure 2.6–20. Individual chart with classification of events.

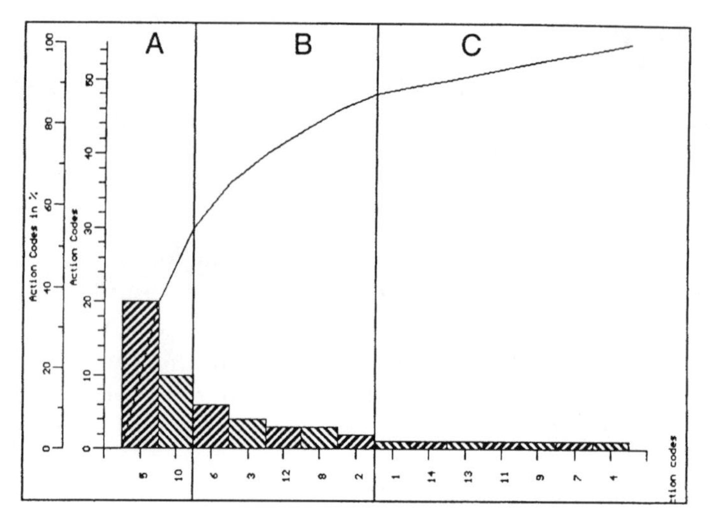

Figure 2.6–21. Pareto diagram.

As Figure 2.6–21 shows, the Pareto analysis rank orders the events. Plotting cumulative percentages yields a typical *Pareto curve* (also called *Lorenz line*). The part of the curve with the steepest slope corresponds to region A. This is followed by region B with a lesser slope. The part where the curve levels out corresponds to region C.

Region A

Here, two of the categories (14 percent of the total number of categories), i.e., the slider and the direction lever, between them cause 54.54 percent of the total rework time and thus represent the *vital few.* Process actions to prevent defects and reduce rework should focus on these two points.

Region B

The next five categories (36 percent) are responsible for a further 32.72 percent of the total rework time. The A and B categories together account for 87.26 percent of the rework.

Region C

The remaining seven categories (50 percent) cause only 12.74 percent of the rework time and thus, in comparison, constitute the *trivial many.*

Thus, Pareto analysis provides a simple but very versatile method for determining and displaying priority items such as

- Faulty components
- Frequent process changes
- Differing environmental conditions
- Etc.

The effects of the important items are worthy of investigation, and, where appropriate, this should lead to the establishment of rules and operational instructions. Figure 2.6–22 shows other ways of plotting Pareto diagrams.

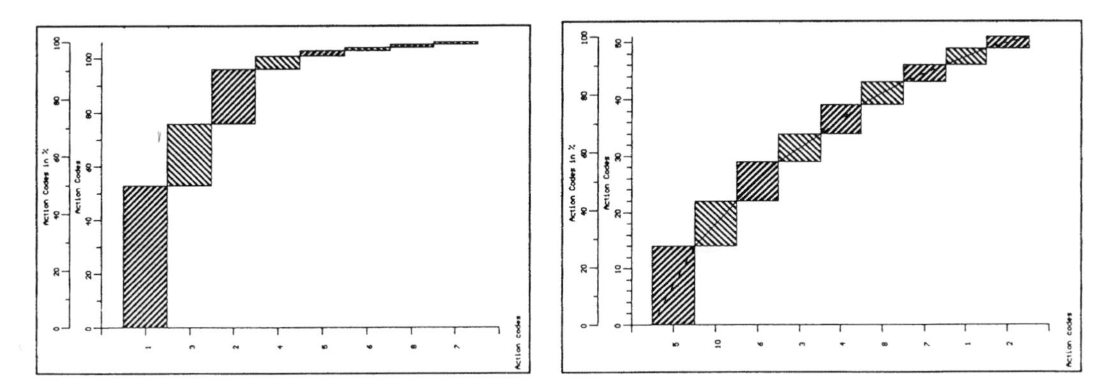

Figure 2.6–22. Other possible ways to display the Pareto diagram.

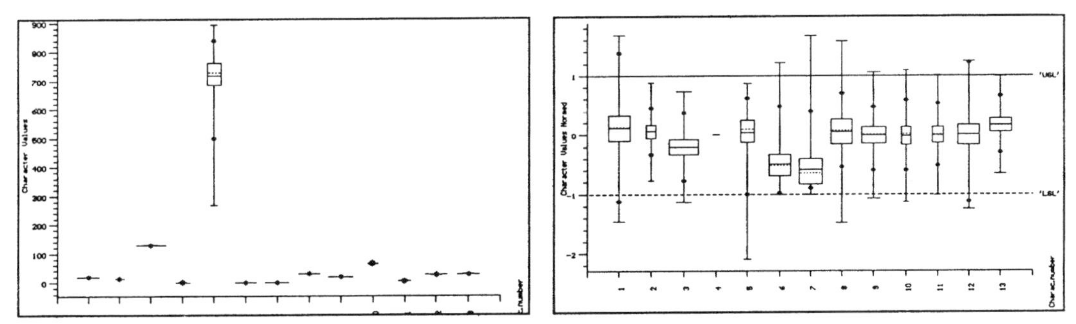

Not standardized without specification

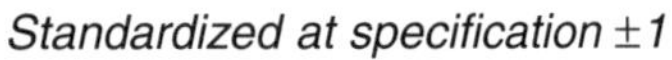

Standardized at specification ±1

Figure 2.6–23. Box plot.

2.6.9 Box Plots

A clear representation of the most important statistics is of vital importance for the comparative analysis of product or process data. The box plot (Figure 2.6–23) meets these requirements.

The height of the box shows the region in which 50 percent of the values lie, based on an assumed distribution model. In addition, the mean and/or median may be shown, with confidence intervals for 95 percent, 99 percent, and 99.9 percent. The width of a box depends on the sample size.

★ Largest or smallest value.

The size of the interval indicates the range within which a certain percentage of the values lies. A typical range is the 99.73 percent range (corresponding to ±3s for a normal distribution). If the distribution is not normal, this range may not be centered on the box.

In order to make the representation independent of the value ranges of the characteristics, it is useful to standardize the values. Then the specification limits become +1 and –1. By these means, statistical parameters such as process location and dispersion may be compared. If a specification limit is missing, the normed display of the characteristic is not possible.

If the value ranges of the characteristics are not too different, it is possible to dispense with standardization (see Figure 2.6–24).

2.6.10 Capability Indices: An Overview

An important result of any process or machine assessment is the process or machine capability as described by the potential and critical capability indices. For comparison of a number of characteristics or of results taken over different time intervals, a bar chart of the capability indices, as shown in Figure 2.6–25, is recommended.

Notes

1. In the case of characteristics with unilateral tolerances, only the critical capability index is shown (a potential index cannot be calculated due to there being only one specification limit).

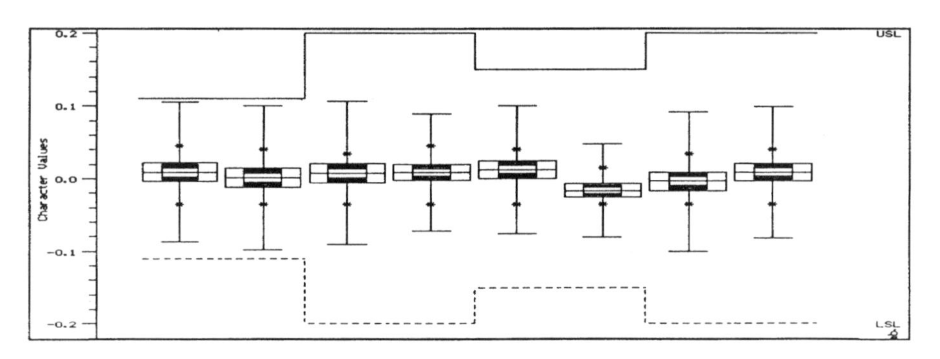

Figure 2.6–24. Nonstandardized box plot.

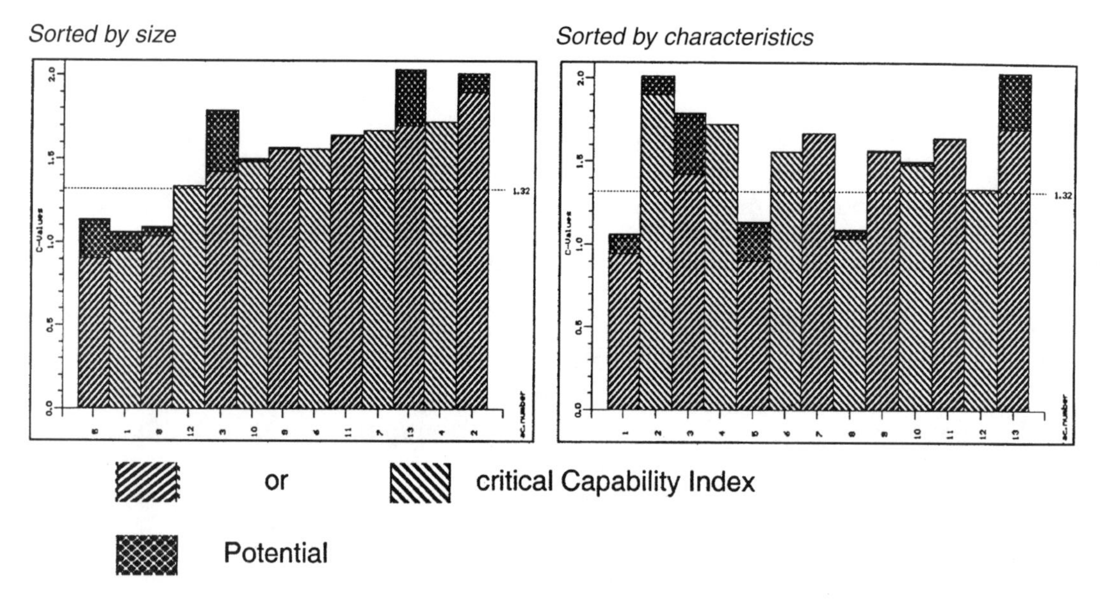

Figure 2.6–25. Critical capability indices.

2. If, however, there is a natural limit for a characteristic with a unilateral tolerance (e.g., for zero-limited characteristics, the lower specification limit is *zero*), the potential may be calculated (see Chapter 6, Section 6.8). In this case, the potential may be smaller than the critical capability index.

2.7 Other Unimodel Probability Distributions

Not all sets of measurements encountered in practice may be described by a normal distribution. In order to deal properly with these cases, other models are required:

Transformation of sets of measurements

Distribution models

- Logarithmic normal (*lognormal*) distribution
- Weibull distribution
- Rayleigh distribution
- Folded normal distribution

Pearson functions

Johnson transformations

Mixed distribution

Deviations from the normal distribution may result from the technical features of the process at hand. Process models B and C (see Chapter 1, Section 1.3) reflect this fact, which may arise from trends, fixed tooling, differing production periods, different machines, fluctuations in material, and similar causes. Here we are, in general, concerned with a mixture of different distributions with differing parameters (mixed distribution). This type of distribution may take many shapes.

In addition, there are characteristics that we would not expect to be normally distributed, due to technical reasons. Table 2.7–1 shows some typical characteristics and the types of distribution that can be expected for them. Among these are measurements of characteristics such as out-of-roundness and positional measurements that have a *natural zero limit.* On the assumption that the production process is not disturbed and that a homogeneous sample is taken, the expected distribution for these characteristics will be unimodal and skewed to the right. In these cases, it would be inappropriate to apply a normal distribution model. Figure 2.7–1 illustrates this point.

In the case of characteristics such as torque, coating thickness, or coating hardness, there may not be a zero limit in practice, but, nevertheless, there may be deviations from the normal distribution. This situation arises in the case of undisturbed processes when only a minimum or maximum value is specified. Typical examples are *minimum torque, maximum hardness, minimum coating thickness,* etc. Hence, attempts will be made to control the process with respect to those limits, and this may alter the shape of the distribution.

What type of distribution will be observed, therefore, depends on the physical properties of the characteristic. This is valid only for an undisturbed process and a sufficiently large number of values. Typical relationships are shown in Table 2.7–1 (see also [5]).

Table 2.7–1. Characteristic and evaluation method.

Characteristic		Evaluation Method	Characteristic		Evaluation Method
Shape tolerances			**Position tolerances**		
Symbol*	tolerated characteristic		Symbol[1]	tolerated characteristic	
⏤	Straightness	B1	∥	Parallelism	B1
⏥	Flatness	B1	⊥	Rectangularity	B1
○	Roundness	B1	∠	Gradient	B1
⌭	Cylinder shape	B1	⌖	Position	B1
⌒	Outline	B1	◎	Coaxiality	B2
⌓	Surface	B1	⌯	Symmetry	B1
			↗ ⌰	Radial run-out	B1/B2
				Axial run-out	B1

Others	Evaluation Method
Roughness	B1
Embalance	B2
Torque	N
Measure of length	N

N = Normal Distribution

B1 = Folded Normal Distribution

B2 = Rayleigh Distribution

**Symbols acc. to DIN ISO 1101, Shape and Position Tolerance.*

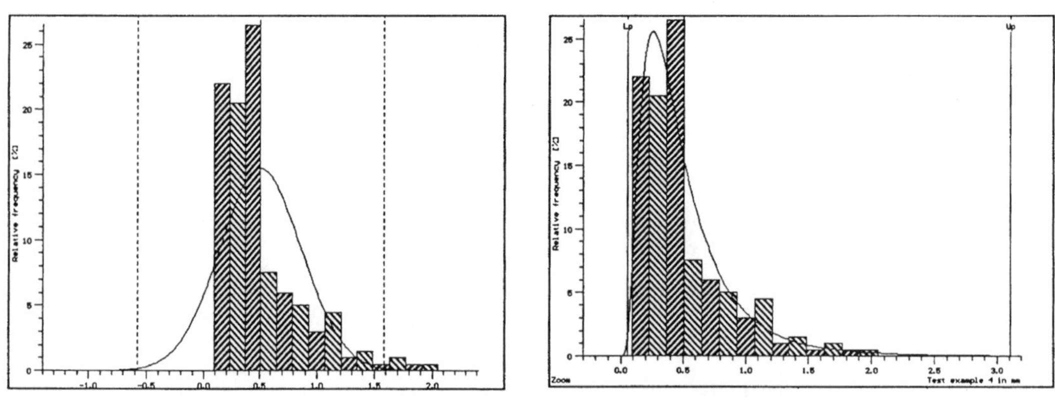

Normal distribution *Logarithmic normal distribution*

Figure 2.7–1. Histogram with probability function overlaid.

The following chapters will deal with the possibilities for describing a set of measurements in model form. This can well lead to ambiguous cases; i.e., a set of measurements may be described similarly well by several distribution models, depending on the circumstances. In order to be able to make the best choice, assessment criteria have been laid down that make it possible to make an objective statement as to the suitability of a given distribution model. These criteria are explained in Chapter 4.

2.7.1 Transformation

A model may also be generated with the aid of a transformation. In this case, the values are transformed by means of a suitable function so that the transformed values conform to one of the previously named distribution models (usually the normal distribution). Statistics are calculated on the basis of the model assumed. The specification limits must then also be transformed. Typical transformations are root transformations and logarithmic transformations with shift (log(x – a)) and/or with reflection (log(a – x)) (see Figure 2.7–2.)

This procedure is really only appropriate for computer-aided evaluation, since trying out various transformations to arrive at a suitable choice is a very time-consuming process if performed manually.

Example: Logarithmic Transformation

Figure 2.7–2 explains the procedure using the example of a logarithmic transformation ln(a + bx).

A logarithmic transformation will only make sense if the original distribution is skewed to the right, with the lowest values lying relatively close to zero. For a skew distribution of values located in an interval relatively far removed from zero, the logarithmic transformation will not change the distribution shape to a meaningful degree. The object of the transformation is to reposition the values such that the *new,* transformed values are approximately normally distributed. Chapter 4 deals with the selection and assessment of this transformation in more detail.

The set of values in the top left diagram in Figure 2.7–2 shows a right-skewed distribution, where, in addition, the distribution should be moved left (toward zero) using an (x – a) transformation. The distribution shown top right is skewed to the left; these values should be mirror reflected using (a – x). The signs of a and b are thus determined by what kind of shift is desirable and whether mirror reflection is required. Finally, the logarithms of the values are taken.

By taking logarithms of the values in the center illustration, we obtain the bottom right illustration. The transformed values now have an *approximately normal distribution.*

All transformations have the effect of altering the scale of the x axis in some way. This is usually very difficult for the inexperienced observer to understand. If the scale of the original values is to be retained, it is necessary to select an appropriate model distribution. In the present case, this is the logarithmic normal (lognormal) distribution. The illustration on the next page on the left shows the same values as the illustration in the center, but with a superimposed lognormal distribution curve.

The two options are illustrated on probability paper in Figure 2.7–3. The left-hand picture shows the original values (not transformed) plotted on lognormal probability paper. The right-hand picture shows the log-transformed individual values plotted on normal probability paper.

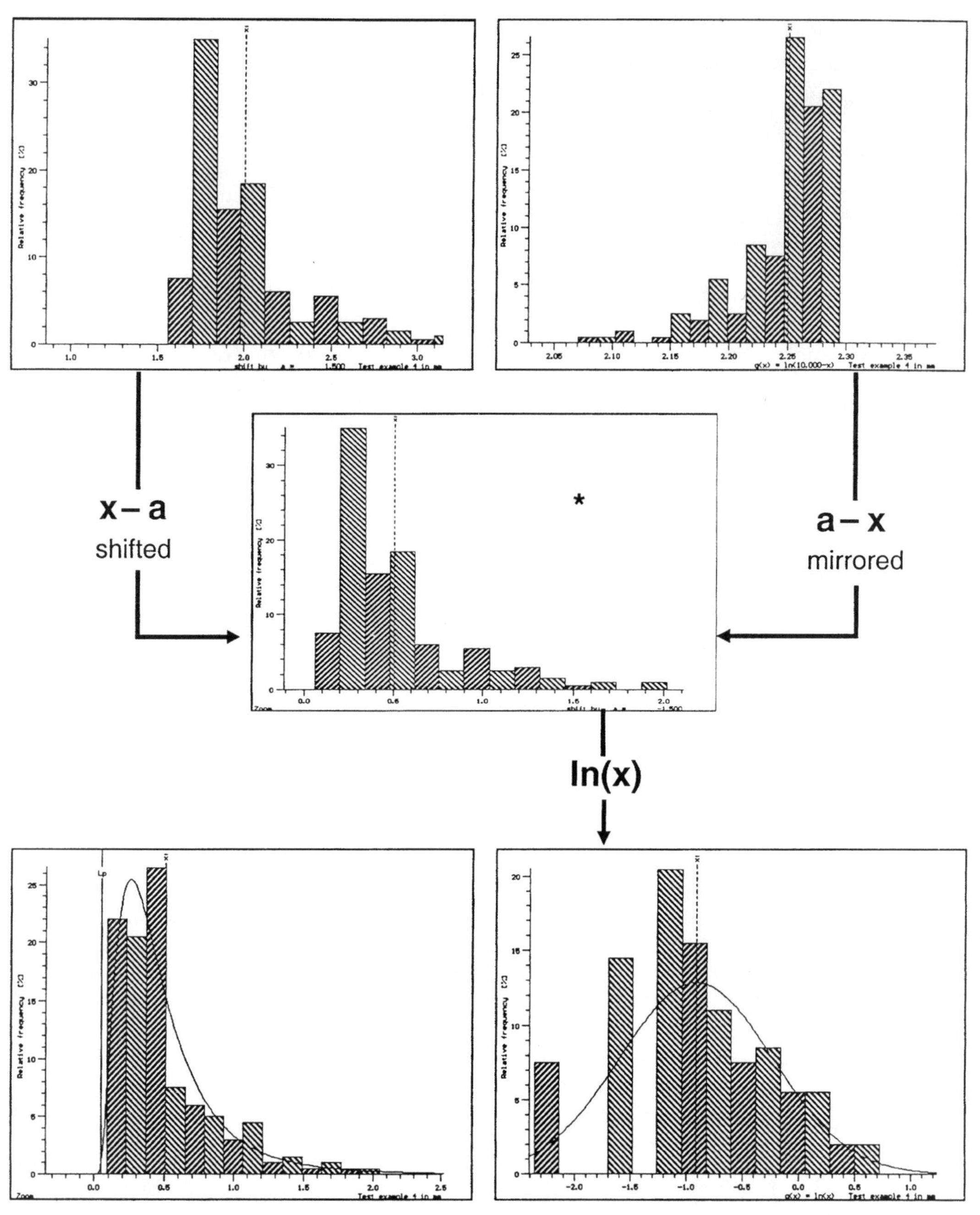

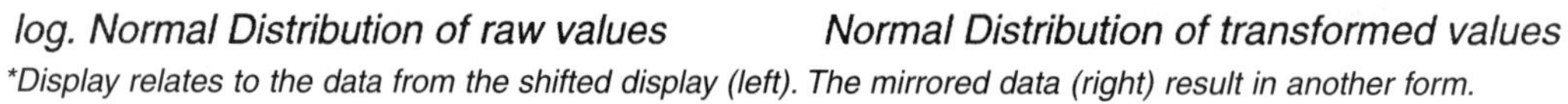
log. Normal Distribution of raw values *Normal Distribution of transformed values*

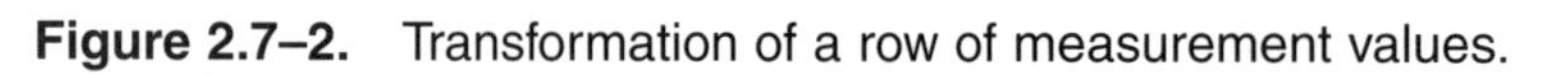
**Display relates to the data from the shifted display (left). The mirrored data (right) result in another form.*

Figure 2.7–2. Transformation of a row of measurement values.

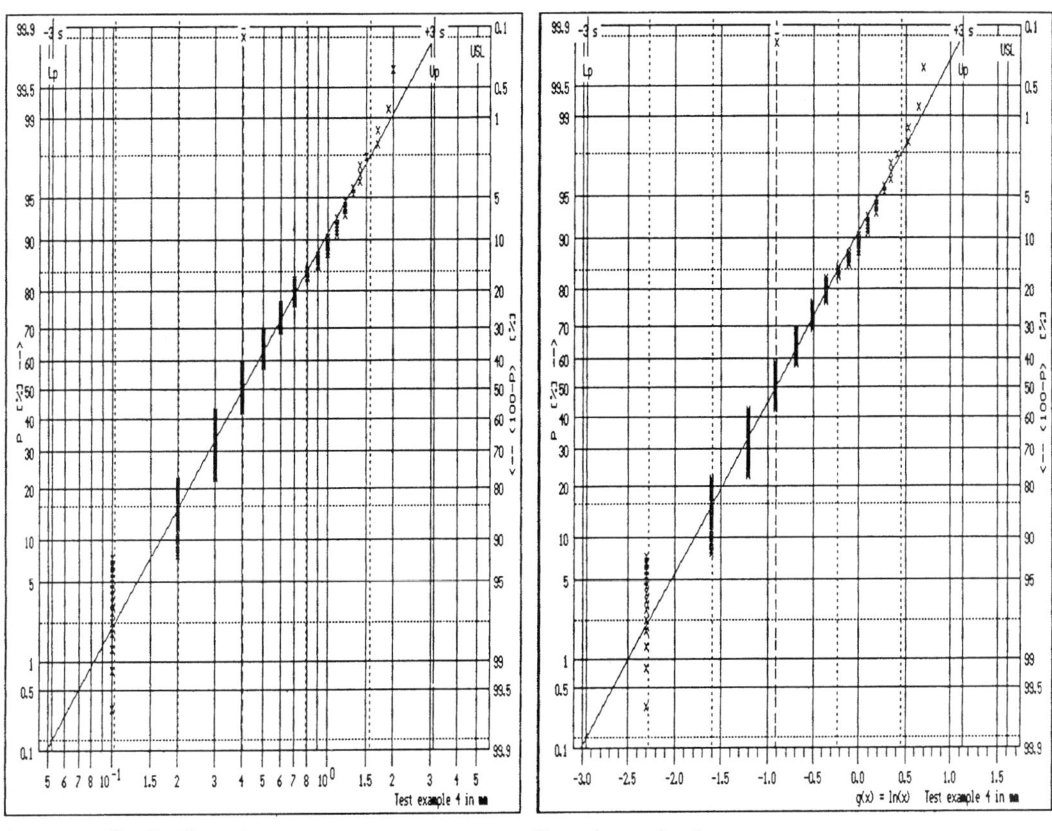

Lognormally distributed *Transformed values*

Figure 2.7–3. Probability plots.

When transforming the values to make them conform to the normal distribution, the specification limits must also be transformed in the same manner. Following this, it is possible, for example, to calculate capability indices as described in Section 2.7. If, on the grounds of a poor critical capability index, the result requires a correction of process location, the required shift in the process average must be determined by retransformation. For this reason, the indirect method of describing a set of values by means of a transformation is time-consuming and involves practical difficulties for the user. Today, in the age of computers, it seems to make more sense to use a suitable model distribution in the first place, rather than applying a transformation to make the data conform to a different distribution model. Subsequent chapters will deal with the various possibilities.

2.7.2 The Logarithmic Normal Distribution

The logarithmic normal distribution (frequently known as lognormal distribution) has today lost much of its importance as a model for describing sets of data, since, as is seen from Table 2.7–1, there are no standard correspondences between this distribution model and common technical characteristics. For manual evaluations, the logarithmic normal distribution was and still is used on account of its simplicity of application (see Section 2.7.1).

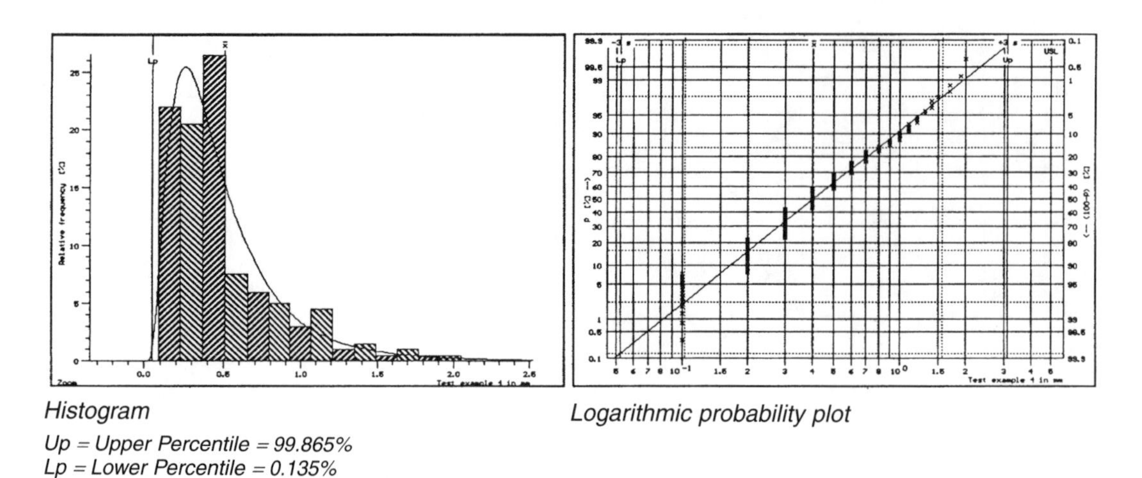

Histogram

Up = Upper Percentile = 99.865%
Lp = Lower Percentile = 0.135%

Logarithmic probability plot

Figure 2.7–4. Logarithmic normal distribution.

This distribution model may be used for a first approximation when a characteristic is limited on one side. Since a set of measurements can be shifted up or down and/or mirror reflected, the lognormal distribution may be used for both right-skewed and left-skewed distributions. In other words, this model may be used to describe distributions with a maximum or a minimum limit. Typical examples are: shape and positional measurements such as out-of-roundness, coaxiality, flatness, and, in some circumstances, also surface characteristics and coating thickness or hardness values.

In order to establish whether a given set of measurements is lognormally distributed, the values are plotted on lognormal probability paper. If the values form a straight-line probability plot, the distribution model is confirmed.

Probability function: $$g(x) = \frac{1}{\sqrt{2\pi}\,\sigma}\frac{1}{x-a}\exp\left\{-\frac{1}{2}\left(\frac{\ln(x-a)-\mu}{\sigma}\right)^2\right\}$$

$$a < x < \infty$$

Distribution function: $$G(x) = \int_a^x g(t)dt$$

Figure 2.7–4 shows the histogram of a set of readings with overlaid lognormal curve, along with the values plotted on lognormal probability paper.

Based on this distribution model, 0.135 percent of the points lie respectively below LP and above UP (lower and upper percentiles). This means that 99.73 percent of the values lie within the LP-UP interval. This corresponds to the normal distribution's $\pm 3\sigma$ interval.

2.7.3 Folded Normal Distribution

The folded normal distribution is a normal distribution that has been folded at a selected point $\leq \mu$. This folding adds the values left of the folding point to those to the right. The shape of the distribution curve is thus changed from *a* to the new shape *b* (see Figure 2.7–5).

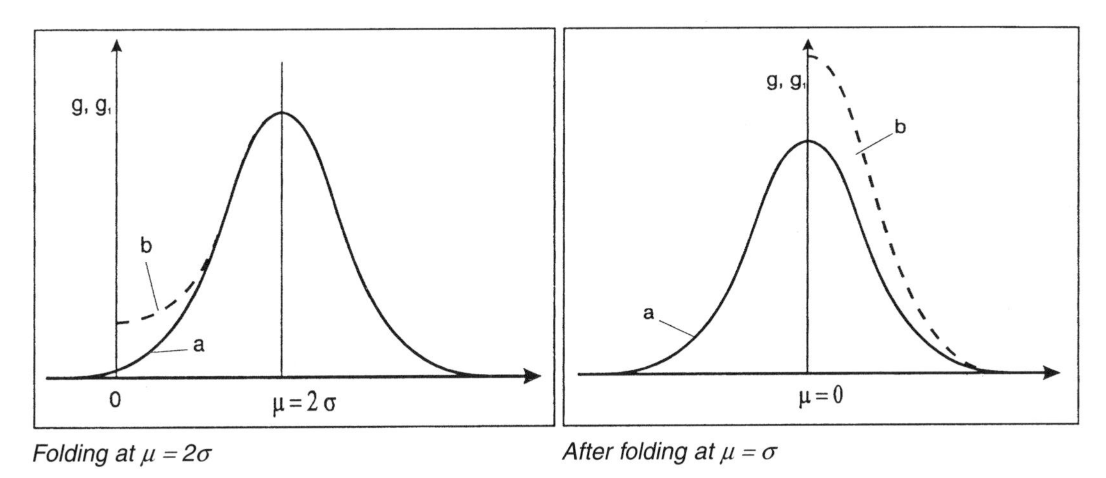

Folding at $\mu = 2\sigma$

After folding at $\mu = \sigma$

Figure 2.7–5. Folding of the normal distribution.

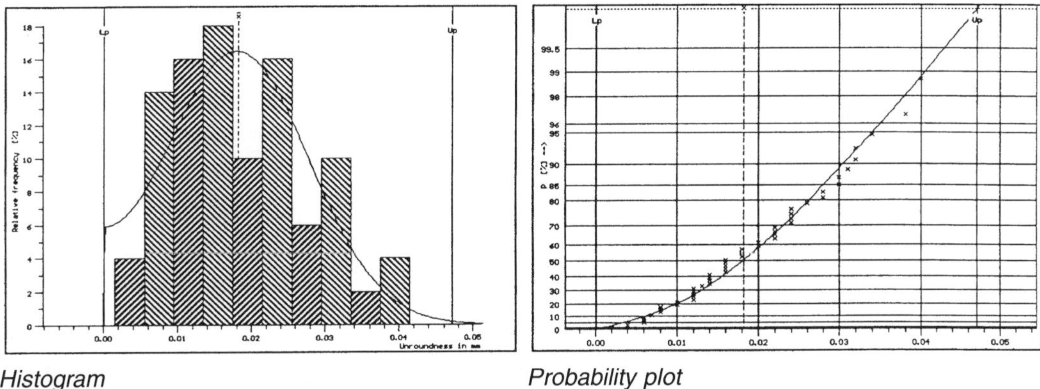

Histogram

Probability plot

Figure 2.7–6. Folded normal distribution.

Probability function:

$$g(x) = \frac{1}{\sigma\sqrt{2\pi}}\left[e^{\frac{1}{2}\frac{(x-\mu)^2}{\sigma^2}} + e^{\frac{1}{2}\frac{(x+\mu)^2}{\sigma^2}}\right] \text{ for } x \geq 0$$

Distribution function:

$$G(x) = \frac{1}{\sigma\sqrt{2\pi}}\int_0^x\left[e^{\frac{1}{2}\frac{(t-\mu)^2}{\sigma^2}} + e^{\frac{1}{2}\frac{(t+\mu)^2}{\sigma^2}}\right] dt$$

If plotted on probability paper, the values do not form a straight line, but a curve (see Figure 2.7–6). The curvature depends on the position of the folding point. The figure also shows the frequency distribution with the superimposed distribution function.

A special case of the folded normal distribution arises if the folding point is equal to the mean of the normal distribution. In addition, the values may be subject to an offset, e.g., due to incorrect calibration of the measuring system. In that case, the distribution is displaced by an offset factor a (Figure 2.7–7).

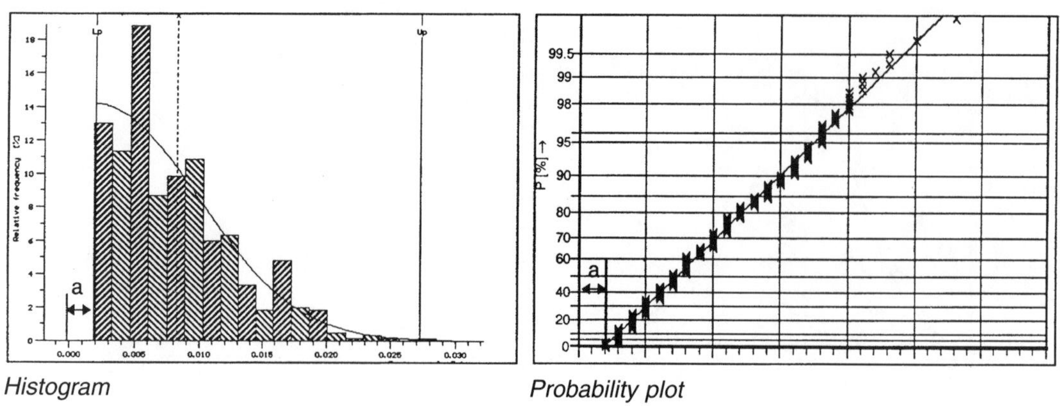

Histogram *Probability plot*

a = shift factor
Lp = Lower percentile *Up = Upper percentile*

Figure 2.7–7. Folded normal distribution: Folded at the mean value.

Probability function: $$g(x) = \frac{2}{\sqrt{2\pi}\sigma} \exp\left\{-\frac{1}{2}\left(\frac{|x-\mu|}{\sigma}\right)^2\right\}; \quad 0 \le |x-a| < \infty$$

Distribution function: $$G(x) = \int_0^{|x-a|} g(t)dt$$ a = zero draft

Examples

Play in steering (target value = 0)	Deviation in degrees ignoring whether the play is to the right (plus) or to the left (minus)
Shot weight	Deviation in grams ignoring whether the part exceeds the target value (e.g., 500 g) or falls short (absolute deviation)

2.7.4 Rayleigh Distribution

The Rayleigh distribution is a two-dimensional distribution that finds its application where the characteristic has two components and the dispersion of the individual components may be considered equal. This state of affairs is shown in Figure 2.7–8.

The two-dimensional normal distribution is described by the following probability function:

$$g_{x,y} = \frac{1}{2\pi\sigma^2} e^{\frac{x^2+y^2}{2\sigma^2}}$$

Due to its rotational symmetry, a description using polar coordinates is appropriate. Substituting $r^2 = x^2 + y^2$ and integrating from 0 to 2π gives the following probability function:

$$g_r = \frac{r}{\sigma^2} e^{-\frac{r^2}{2\sigma^2}}$$

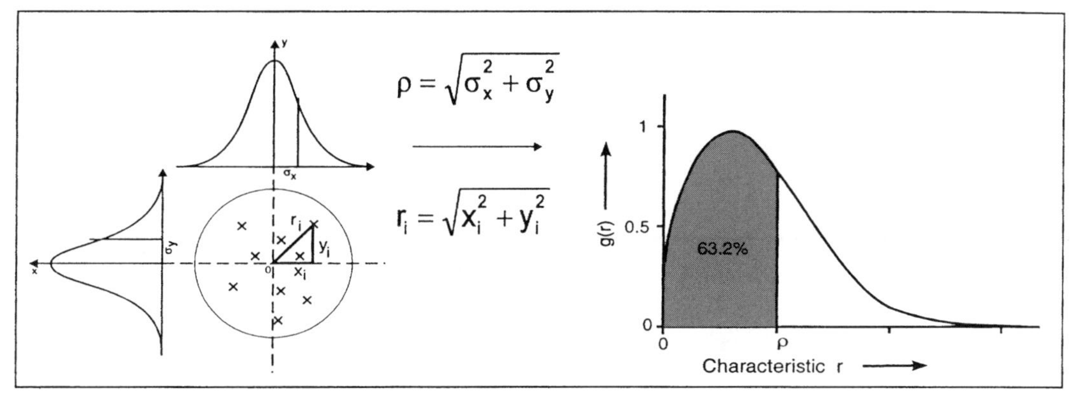

σ_x, σ_y = *Standard deviation of the individual components*

Figure 2.7–8. Origin of the Rayleigh distribution.

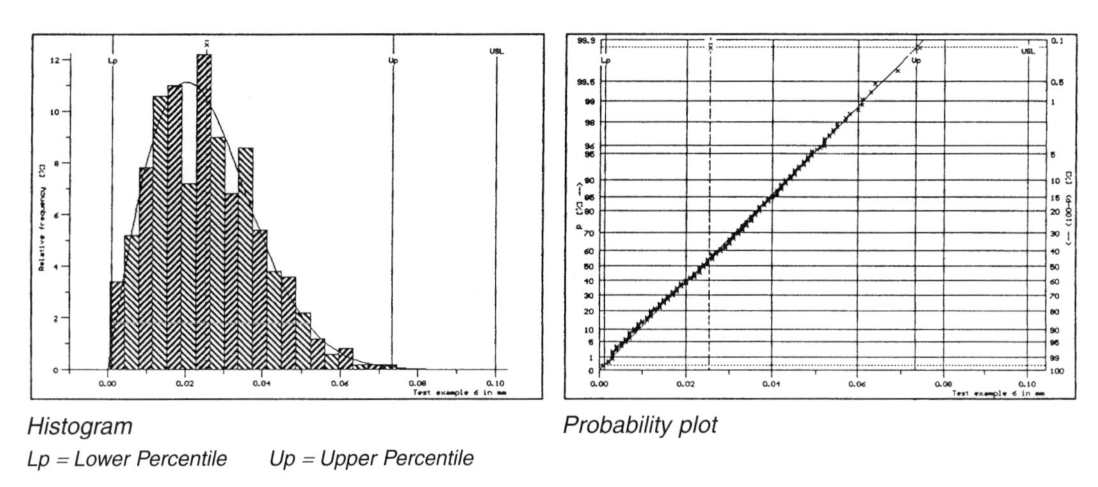

Histogram

Probability plot

Lp = Lower Percentile *Up = Upper Percentile*

Figure 2.7–9. Rayleigh distribution.

Figure 2.7–9 shows the histogram of a set of values with the distribution function superimposed, as well as the straight-line probability plot on Rayleigh paper.

Like the folded normal distribution (see Section 2.7.3), the Rayleigh distribution may also be shifted away from zero by an offset factor *a*. As in the folded normal distribution shown in Figure 2.7–7, this leads to a displacement both in the histogram and on probability paper.

If the Rayleigh distribution is folded at any selected point (see folded normal distribution, Section 2.7.3), we have

Probability function: $$g(x_B) = \frac{x_B}{2\pi\sigma^2} e^{\frac{1}{2\sigma^2}(a_2 + x_B^2)} \int_0^{2\pi} e^{\frac{a x_B}{\sigma^2}\cos\alpha}\, dx$$

Distribution function: $$G(x_B) = \frac{1}{2\pi\sigma^2} \int_0^{x_B} x e^{-\frac{1}{2\sigma^2}(a^2 + x^2)} \left(\int_0^{2\pi} e^{\frac{ax}{\sigma^2}\cos\alpha}\, d\alpha \right) dx$$

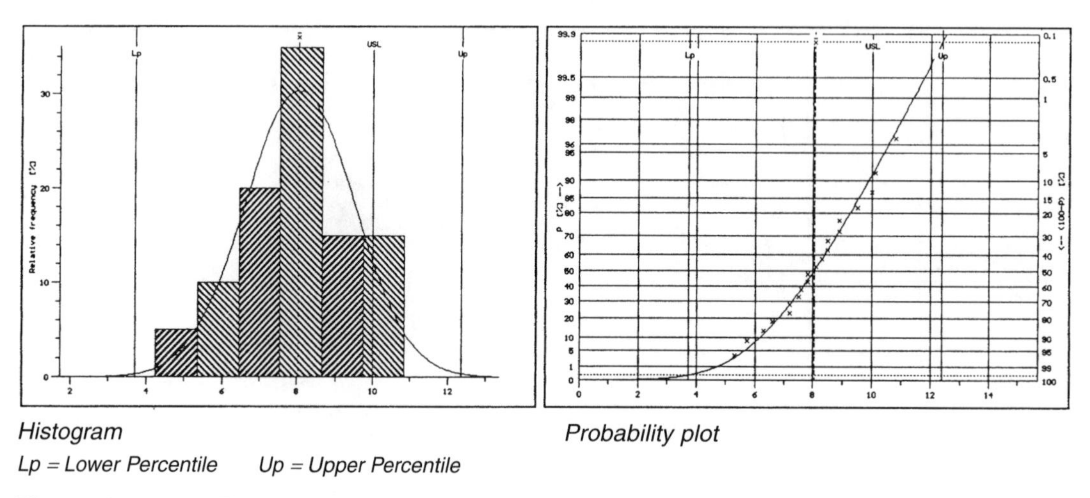

Histogram

Probability plot

Lp = Lower Percentile *Up = Upper Percentile*

Figure 2.7–10. Folded Rayleigh distribution.

Figure 2.7–10 shows the histogram of a set of values with the *folded Rayleigh distribution* superimposed and the values plotted on Rayleigh paper. The curvature of the probability curve depends on the folding point.

2.7.5 Weibull Distribution

The Weibull distribution is a very versatile model distribution of wide application, which, due to its mathematical properties, may be adapted to *any* form of frequency distribution. Its main use is in reliability and life-testing problems.

Probability function: $$g(x) = \frac{\beta}{\alpha}\left(\frac{x-a}{\alpha}\right)^{\beta-1} \exp\left\{-\left(\frac{x-a}{\alpha}\right)^{\beta}\right\}$$

Distribution function: $$G(x) = 1 - \exp\left\{-\left(\frac{x-a}{\alpha}\right)^{\beta}\right\}$$

The shape of the Weibull distribution thus depends on three parameters, α, β, and a, which are interpreted below in terms of reliability or life-testing analyses:

- Scale parameter = typical service life α
- Shape parameter = failure slope β
- Location parameter = failure-free time a

The distribution models described up to now may be regarded either as special cases of the Weibull distribution or may be approximated by the Weibull distribution. The distinguishing feature is the value of the shape parameter β.

$\beta = 1 \Rightarrow$ exponential distribution (special case for $\beta = 1$)
$1.5 \le \beta \le 3 \Rightarrow$ lognormal distribution
$\beta = 2 \Rightarrow$ Rayleigh distribution
$3.1 \le \beta \le 3.6 \Rightarrow$ lognormal distribution
$3.6 \Rightarrow$ normal distribution

Figure 2.7–11 shows the histogram for a set of measurements with the Weibull distribution superimposed, along with the values plotted on Weibull paper.

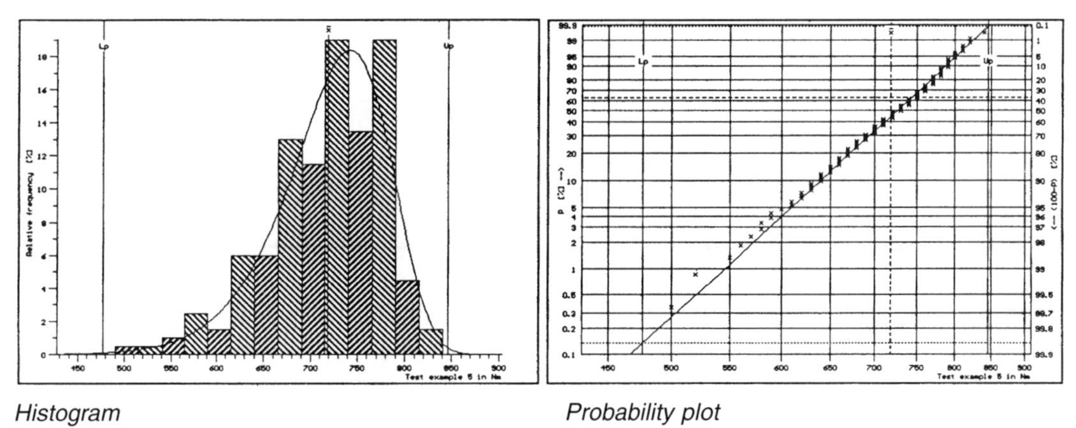

Histogram

Probability plot

Lp = Lower Percentile *Up = Upper Percentile*

Figure 2.7–11. Weibull distribution.

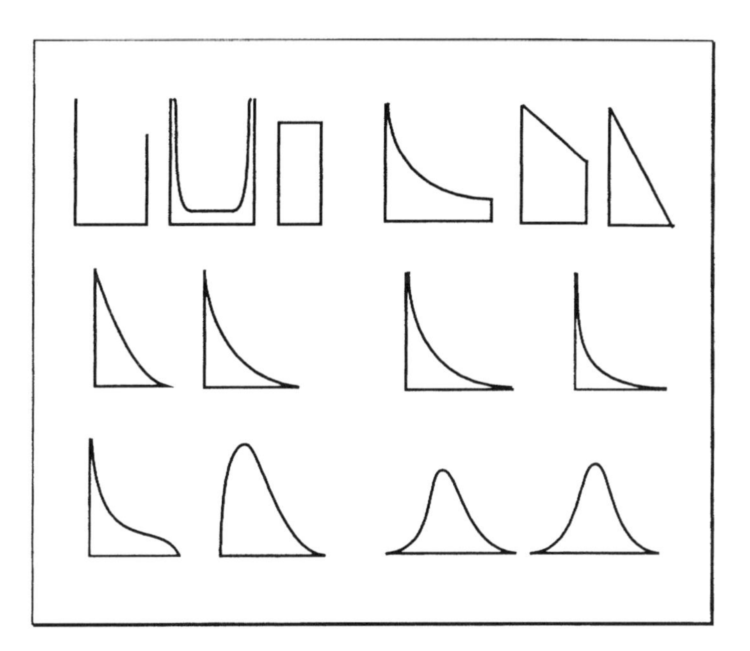

Figure 2.7–12. Different forms of the Pearson functions.

2.7.6 Pearson Functions

Pearson functions provide a further possibility of modeling unimodal distributions. They comprise a family of 14 distributions (see Figure 2.7–12). Selection of the appropriate distribution is governed by two sample statistics, skewness and kurtosis.

The Pearson function so obtained may be plotted on probability paper in order to evaluate the model. As there is no *Pearson* probability paper, normal probability paper is used.

In this case, the customary *straight-line plot* in fact becomes a curve depending on the distribution. The interpretation with regard to the validity of the distribution model (agreement between values and curve) is identical with that for other models.

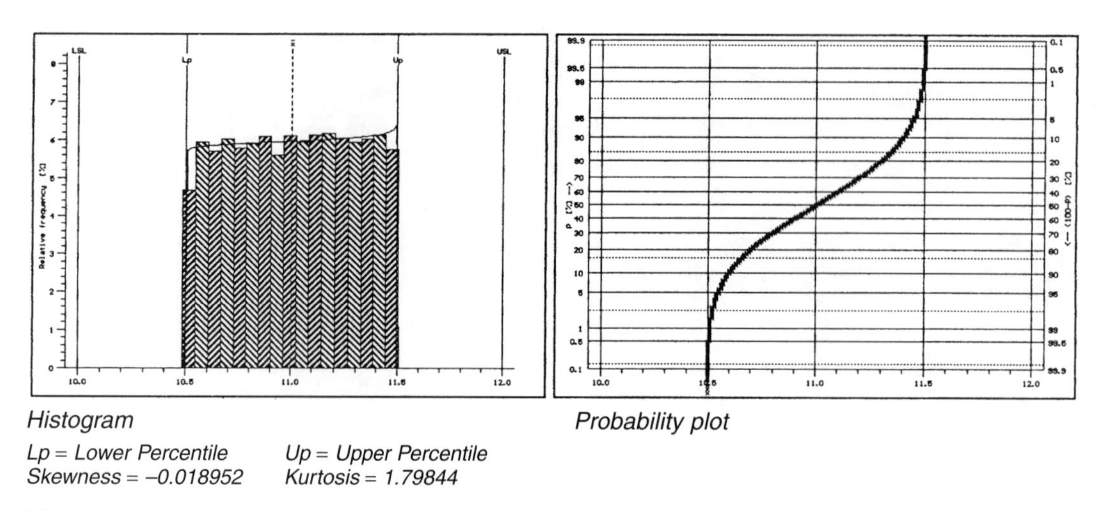

Histogram

Lp = Lower Percentile
Skewness = –0.018952

Probability plot

Up = Upper Percentile
Kurtosis = 1.79844

Figure 2.7–13. Model description of the rectangular distribution with Pearson function.

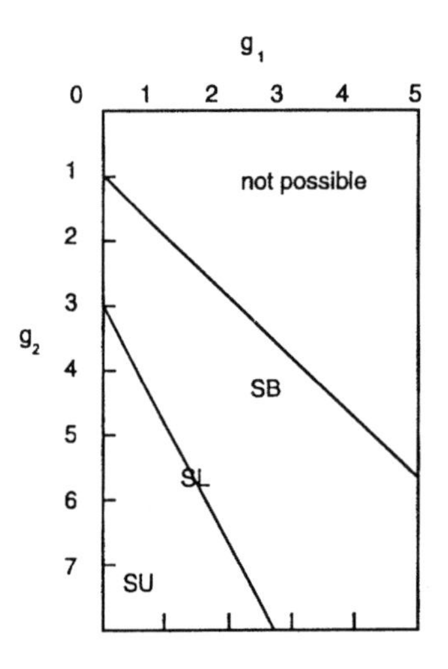

Figure 2.7–14. Types of the Johnson Systems in dependence of g_1 and g_2.

Figure 2.7–13 shows the histogram of a set of readings with the Pearson function superimposed, and the values plotted on probability paper. The values lie on the curve based on the Pearson function.

2.7.7 Johnson Transformations

Based on the possibilities of transforming values, the American mathematician Johnson developed a transformation system by means of which all the important types of continuous distribution may be converted into the normal distribution. The selection of the transformations is based on the characteristics of skewness (g_1) and kurtosis (g_2).

Viewed purely theoretically, there is an infinite number of pairs of values of these two characteristics. They occupy a plane that Johnson divides into three regions. These regions are shown in Figure 2.7–14.

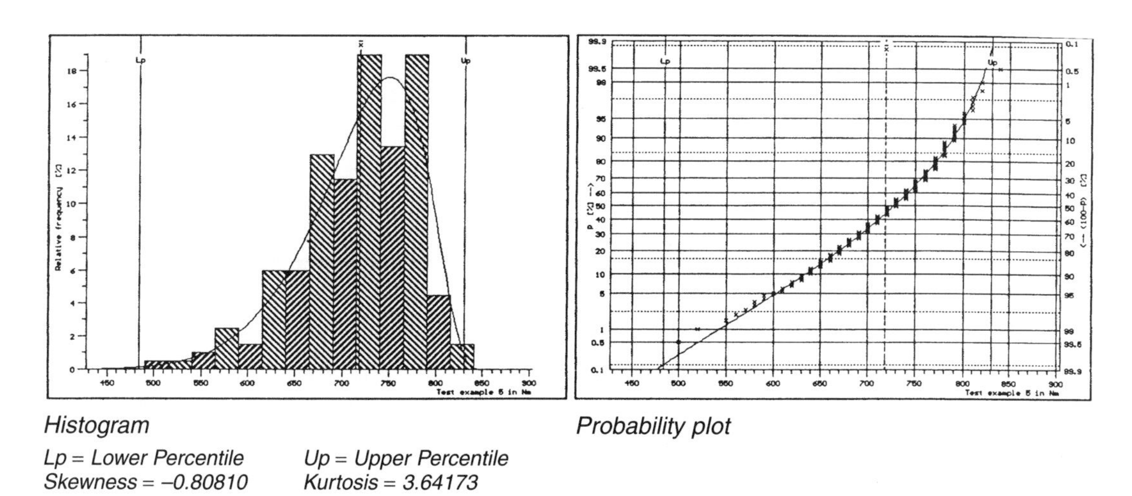

Histogram

Probability plot

Lp = Lower Percentile *Up = Upper Percentile*
Skewness = –0.80810 *Kurtosis = 3.64173*

Figure 2.7–15. Johnson transformation type SB.

$$\mu_k = \frac{\sum_{i=1}^{n}(x_i - \overline{x})^k}{n}$$

$$g_1 = \frac{\mu_3^3}{\mu_2^2} \qquad x_i = \text{individual value}$$

$$\overline{x} = \text{mean}$$

$$n = \text{sample}$$

$$g_2 = \frac{\mu_4}{\mu_2^2} - 3 \qquad u_k = \text{moment number k}$$

There is a region that cannot be mapped mathematically and is not observed in practical applications. The remaining area is divided into two regions. One area contains combinations of skewness and kurtosis that may be described by a system of equations whose stock of values is limited on two sides. Johnson calls this the System Bounded or SB. This area merges into a region where it is appropriate to use an unbounded type of equation, which Johnson calls the System Unbounded or SU.

Between *limited* on both sides and *unlimited* on both sides, there is a transition type, a system of equations limited on one side. This system involves a three-parameter logarithmic transformation and is abbreviated SL.

The fact that there is a blurred boundary between these three systems of equations leads to the recognition that one and the same set of data may be represented sufficiently well by a number of equation systems. Which system is accepted depends greatly on the user's assessment of the boundary conditions.

Figure 2.7–15 shows a histogram of a set of values with the distribution function superimposed, based on a Johnson transformation, along with the values plotted on probability paper. The values lie on the probability curve given by the Johnson transformation.

The set of values used in this example is identical to that used in the Weibull distribution example (see Section 2.7.5). This illustrates again that there may be several possible descriptive models that lead to identical or similar results.

2.7.8 Mixed Distribution

The mixed distribution is based on the method of moments. It is possible to characterize distributions extensively through the moments.

$$EX^k = \left(\frac{1}{n} \sum_{i=1}^{n} x_i \right)^k \qquad \begin{array}{l} i = 1, 2, ..., n \\ E = \text{expected value} \end{array}$$

or the central moments

$$E(X - EX)^k = \frac{1}{n} \sum_{i=n}^{n} (x_i - \bar{x})^k \qquad k = k^{th} \text{ moment}$$

Thus, it is possible to describe the *form* of the data set on hand using the estimator of the moments.

As an example, the skewness derived from the 3rd central moment (see Section 2.5.1) or the kurtosis derived from the 4th central moment result in a measure for the deviation of the data from the normal distribution. If higher moments are evaluated in addition, then it is possible to describe the data more exactly. In case of the mixed distribution, the first 27 moments (k = 27) are regarded, A higher number of moments is possible, though will increase only calculation trouble without greatly improving accuracy of the model adaptation. With the help of a polynomial interpolation of the values derived from the moment, overlaid normal distributions are found that describe the existing data set sufficiently well.

The example in Figure 2.7–16 shows the *conformity* of the data and the distribution model described with the help of the mixed distribution in the histogram as well as in the probability plot. Data were created through simulation. The proportion on the left was derived from a population with the average $\mu_1 = 2.5$ and a standard deviation of $\sigma_1 = 0.1$. The right proportion is derived from a population $\mu_2 = 3.5$ and a standard deviation of $\sigma_2 = 0.15$. The initial subgroup amounts to 30 percent of all the data. In all, 1,500 values were simulated.

Such a set of data cannot be described using the classical distribution models like normal distribution, lognormal distribution, folded normal distribution, Rayleigh

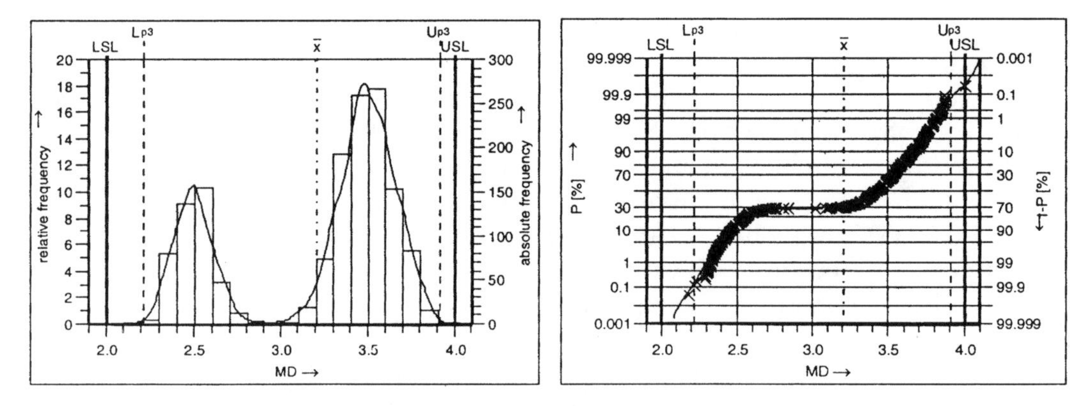

Figure 2.7–16. Data set with overlaid mixed distribution.

distribution, or Weibull distribution. Even distribution models like the Pearson function or the Johnson transformation do not lead to sensible results, as these models are suited mainly for unilateral distributions.

2.7.9 Two-Dimensional Normal Distribution

A two-dimensional distribution occurs, for example, in the case of positional tolerances. The density function is given by

$$g_{(x,y)} = \frac{1}{2\pi \times \sigma_x \times \sigma_y \times \sqrt{1-\rho^2}} \quad \text{EXP}\left[-\frac{1}{2(1-\rho^2)}(u^2 - 2\rho uv + v^2)\right]$$

where

$$u = \frac{x-\mu_x}{\sigma_x}; \quad v = \frac{y-\mu_y}{\sigma_y}; \quad -\infty < u, v < +\infty$$

$$\rho = \sqrt{\sigma_x^2 + \sigma_y^2}$$

and the parameters are

$\mu_x \rightarrow$ mean x co-ordinate of point

$\mu_y \rightarrow$ mean y co-ordinate of point

$\sigma_x^2 \rightarrow$ variance of x

$\sigma_y^2 \rightarrow$ variance of y

$\rho \rightarrow$ correlation coefficient of x and y

Figure 2.7–17 shows bore centers with their tolerance ellipse and the superimposed probability density function. The tolerances for the x and y (or z) coordinates result in an elliptical tolerance region (Figure 2.7–18). The precise shape of this ellipse depends on the scales of the axes and the specification limits. If the tolerances are equal and the x and y axes are on the same scale, this leads to the special case of a circular tolerance region. It is not permissible to express positional tolerances in the form of a rectangle or square, since values close to the corners could lead to problems in subsequent process steps. Hence, the values within the shaded area are considered out of tolerance. A capability study based on individual coordinates, performed in the manner that is customary for one-dimensional characteristics, would thus be incorrect. For a correct assessment, the model distribution used must either be a Rayleigh distribution (Section 2.7.4) or, as shown here, the two-dimensional normal distribution.

2.8 Numerical Test Procedures

Starting from the appropriate formulas (see Section 2.3), it is possible to calculate the corresponding statistics. From a purely *mathematical* point of view, these results will be correct, but it is by no means certain that they will accurately reflect the true state

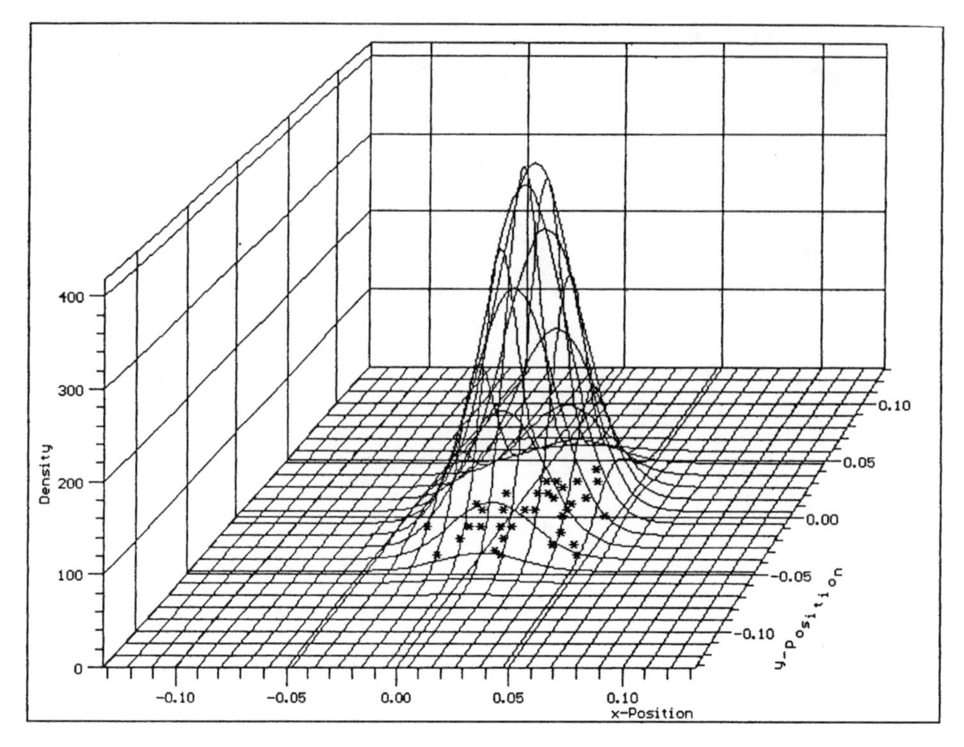

Figure 2.7–17. Value pairs with overlaid distribution model.

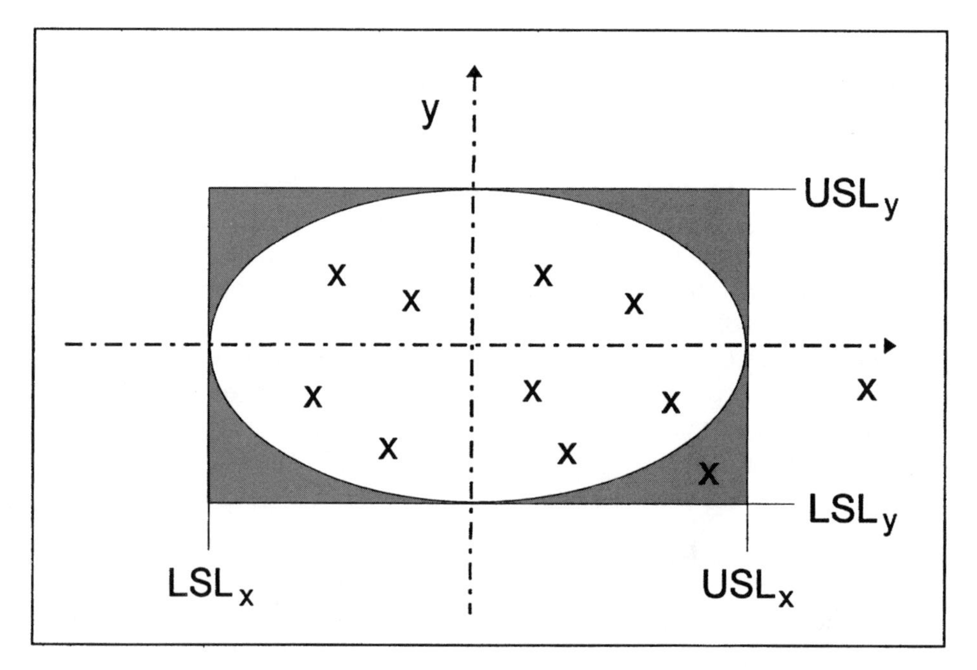

Figure 2.7–18. Elliptical and rectangular positional tolerance.

of affairs. For example, the capability index based on the methods of calculation shown in Section 2.3.1 will only give accurate results if the values

- Come from a normally distributed population
- Are not subject to any trends, but only subject to random variation
- Contain no outliers

In order to decide whether these conditions are satisfied, graphical representations may be supplemented by numerical test procedures. In their simplest form, the graphical representations used for this purpose are run charts of the individual values or plots on probability paper. The interpretation, in this case, depends on the observer and is not suitable for automatic assessment by a computer system. These disadvantages may be partially overcome by means of numerical test procedures. In computer-aided assessment, these will at least give the user some useful information as to whether a particular distribution model may be regarded as suitable or not. It is then the responsibility of the user to utilize this information as a starting point and further investigate the situation using additional tools and methods.

2.8.1 Assessment Criteria for Graphical Analyses

A widely used assessment procedure for testing whether a distribution is normal is to plot the measured values on probability paper (see Section 2.6.5). The following criterion is then applied:

If the measured values (classed or unclassed) lie on a straight line when plotted on normal probability paper, it may be inferred that the population from which the sample was taken has a normal distribution.

Figure 2.8–1 shows two differing sets of values to illustrate this point.

On the other hand, an extract from ISO 5479 (draft version) will serve to illustrate the limitations of this method:

It should always be borne in mind that such a graphical representation is in no way a rigorous test for normal distribution. For small sample sizes, even quite curved plots

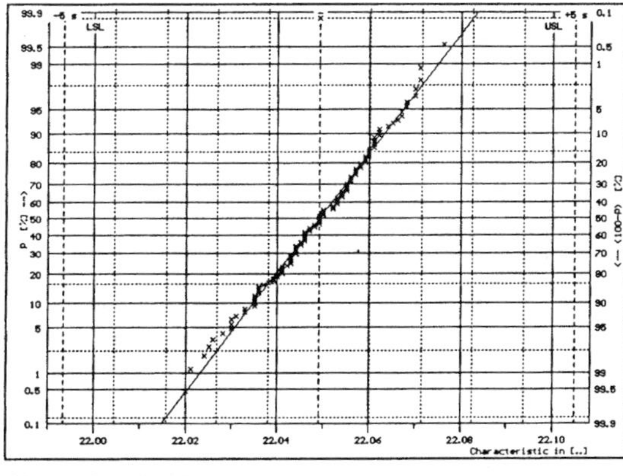

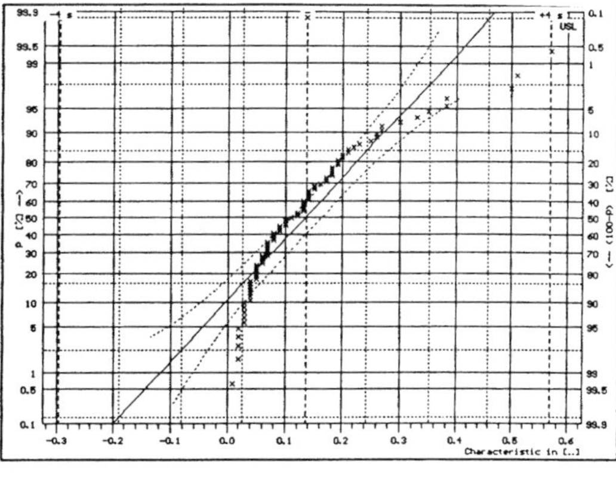

Normally distributed measurement series

Not normally distributed measurement series

Figure 2.8–1. Probability plots.

may still be compatible with the normal distribution model, whereas for large sample sizes even slight curvature may indicate a nonnormal distribution.

In addition to the assessment of a distribution model, the question frequently arises as to whether the set of measurements includes outliers, trends, or cyclical behavior, which could falsify the interpretation of the data. One simple method is to plot a run chart of the individual values over time (see Figure 2.8–2). The example shows upward trends that have been corrected periodically, perhaps by adjustment of a tool. In analyses of this type, it is of vital importance that technical explanations for the observed behavior are sought. This information then forms an integral part of the analysis.

2.8.2 Description of Numerical Test Procedures

This section covers the range of application of various test procedures as well as some general instructions for their use. The interpretation of the results is dependent on the type of test in question. For a detailed description of the procedures, the reader should consult the relevant literature [6, 7, 23, 37, 45, 46, 53].

General Concepts

A company controls the quality of its product by taking a sample every 30 minutes. The relevant characteristics (e.g., length, diameter, strength, etc.) are measured and sample statistics determined. These statistics are, however, subject to small random

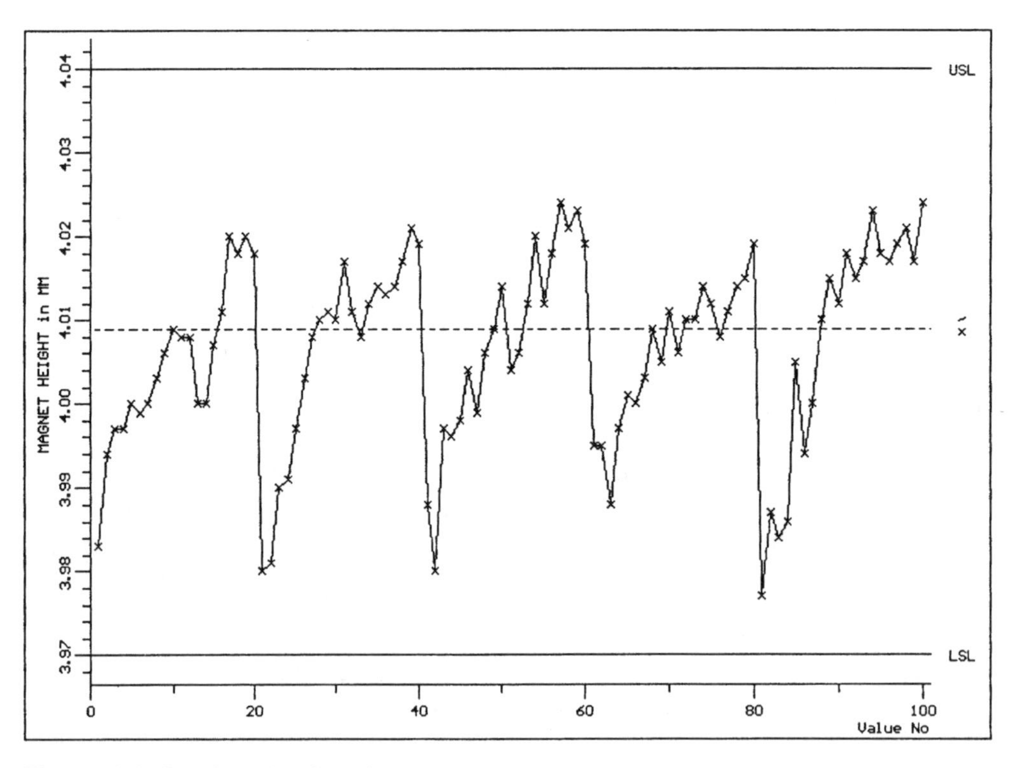

Figure 2.8–2. Actual value chart showing trend.

fluctuations arising from variation in the materials, variation in the machines used for production, and variation between different operators.

Let the required length of the finished part be 20 mm. This means that we shall have to test the hypothesis $\mu = 20$ mm. If now the mean $\bar{x}$ of the sample deviates *not too much* from $\mu = 20$ mm, then we accept the hypothesis and allow production to continue. If, however, the deviation is *too large,* we reject the hypothesis, stop production, and seek the cause of the deviation.

But where should we position the dividing line between small, purely random fluctuations and larger deviations that are no longer explicable by random variation? These latter are known as *significant* deviations. In order to obtain objective statements on this matter, numerical test procedures have been developed. All these procedures share the basic approach that follows.

Test procedures involve two hypotheses:

- H_0: Null hypothesis
- H_1: Alternative hypothesis

If in a test for normal distribution the null hypothesis is confirmed, this means that the values do indeed appear to originate from a normally distributed population. If H_1 is confirmed, this means that the population would not appear to be normally distributed. Hence, there are four possible combinations of test results and actual reality:

		REALITY	
		H_0 Confirmed	**H_1 Confirmed**
Test result	H_0	Correct decision!	Wrong decision! Probability for this error of type 2 is β.
	H_1	Wrong decision! Probability for this error of type 1 is α.	Correct decision!

If the assumptions on which the statistical procedure is based are not consistent with reality, we have a misapplication of the test.

Since the true state of affairs is not known when testing hypotheses, we can never be certain of our conclusions. Each decision is subject to a risk of error.

If the null hypothesis is accepted on the basis of the test results, it does not follow that the null hypothesis is necessarily true. If we choose a low-significance level, we can reduce the risk of a type 1 error (probability α), but at the same time, given identical test conditions, the risk of a type 2 error (probability β) will rise. We should attempt to minimize the probability of errors of both types. However, in many tests we are given only the significance level α, without any information about the magnitude of errors of the second type. In such cases, we should not say that the null hypothesis has been accepted, but only that it *could not be rejected* or *could not be disproved.*

If it appears that the probability of a type 2 error is too large, this may be remedied by repeating the test with a larger sample.

The formulation of hypotheses may be based on

- Requirements that must be satisfied
- Empirical knowledge of the relevant value
- A theory that requires verification
- Pure speculation, arising from desires or observation

The following procedure is usual in the technical sphere:

Three different significance levels are laid down. The threshold values of the higher significance levels serve as criteria for nonrejection *of the null hypothesis, and the threshold values of the lower ones as criteria for* rejection *of the null hypothesis. This procedure reduces errors both of the first type and of the second type. There is, however, an indeterminate zone, in which no decision may be derived from the results of the test. If the test statistic lies in this zone, a clear decision is possible only by repeating the test with a larger sample.*

In applied statistics, a common form of presenting test results is as follows:

xxxxx - Name of Test - xxxxx
Null Hypothesis : < statement >
Alternative Hypothesis : < statement >

Level of significance	Critical values Lower	Critical values Upper	Test statistic
$\alpha = 5$ % :	xxxxx	xxxxx	
$\alpha = 1$ % :	xxxxx	xxxxx	xxxxx <*,**,***>
$\alpha = 0.1$ % :	xxxxx	xxxxx	

Interpretation of the Table

If the calculated test statistic, which is compared with the critical values from the table, is not marked with any stars, then one may assume that the null hypothesis is true.

If the test statistic is marked with one or more stars, then the alternative hypothesis is correct:

- * with a probability of 95 percent
- ** with a probability of 99 percent
- *** with a probability of 99.9 percent

Summary: Procedure for Numerical Tests

The procedure may be summarized as follows:

1. Establish the null hypothesis H_0, e.g., set of measured values
   - Comes from a normally distributed population

- Contains no outliers
- Reflects only random variation
- etc.

2. Establish the alternative hypothesis H_1, e.g., set of measured values
   - Does not come from a normally distributed population
   - Does contain outliers
   - Reflects special causes of variation
   - etc.
3. Calculate the appropriate test statistic for the test concerned.
4. Compare the value of the test statistic with the appropriate critical values. These critical values are usually given for error probabilities (significance levels) $\alpha = 5$ percent, 1 percent, 0.1 percent.
5. Test result:
   If the test statistic falls short/exceeds one of the critical values and/or is outside the specified range, the null hypothesis is rejected in favor of the alternative hypothesis.

In the following chapters, a more detailed description of the individual test procedures as well as formulas for calculating test statistics and critical values will be given. We shall dispense with practical examples, since these problems are too time-consuming for manual assessment.

2.8.3 Test for Randomness

Sample-based statements about a given state of affairs in the population (e.g., proportion outside tolerance) are only possible if certain conditions are satisfied:

- The population conforms to the distribution model assumed (e.g., normal distribution).
- The sample does not include any values that originate from a different population (outliers).
- The sample values are representative of the population. It must be guaranteed that the results of the investigation are independent of sample size. This presupposes a random selection process for the sample items.

Swed-Eisenhart Test

Swed and Eisenhart developed a test for checking the last one of these assumptions, based on an analysis of the sequence of values in comparison with the median value. The median value is particularly suitable for this purpose, since, whatever the distribution, 50 percent of the values lie below and 50 percent lie above this value. The method involves subdividing the sequence of values into runs above and below the median. Figure 2.8–3 shows an example.

If the total number of runs falls short of a lower limit, the set of values has a trend. If, however, it exceeds an upper limit, this means that the behavior is oscillatory. One important advantage of this test is its independence of the population's distribution type.

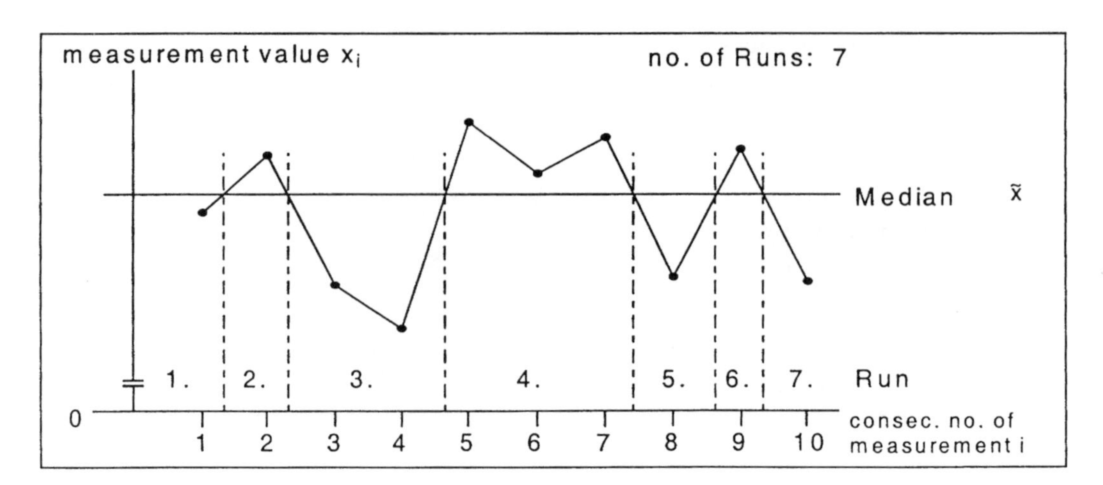

Figure 2.8–3. Determination of run.

Null Hypothesis H_0	Alternative Hypothesis H_1
Random arrangement of the values in the sample.	Nonrandom arrangement of the values in the sample.

After determining the number of runs r, the number of values above and below the median is determined:

$n_1 \Rightarrow$ number of values where $x_i > x > \tilde{x}$
$n_2 \Rightarrow$ number of values where $x_i \geq x < \tilde{x}$
$r \Rightarrow$ number of runs (sequences)

Alternative hypothesis	The null hypothesis H_0 is rejected in favor of the alternative hypothesis H_1 if		
	Test statistic		Critical value
Test version 1 $n_1, n_2 \leq 20$	r	$\leq$ $\geq$	$r_{\alpha/2;\, n_1;\, n_2}$ $r_{1-(\alpha/2);\, n_1;\, n_2}$
Test version 2 $n_1, n_2 > 20$	$\dfrac{\lvert r-\mu_r \rvert}{\sigma_r}$	$>$	$u_{1-(\alpha/2)}$
where	$\mu_r = \dfrac{2n_1n_2}{n_1+n_2}+1$ $\sigma_r^2 = \dfrac{2n_1n_2(2n_1n_2-n_1-n_2)}{(n_1+n_2)^2(n_1+n_2-1)}$		

Table 2.8–1. Critical values of the test for trend.

n	99.9%	99%	95%	n	99.9%	99%	95%
4	0.5898	0.6256	0.7805	33	1.0055	1.2283	1.4434
5	0.4161	0.5379	0.8204	34	1.0180	1.2386	1.4511
6	0.3634	0.5615	0.8902	35	1.0300	1.2485	1.4585
7	0.3695	0.6140	0.9359	36	1.0416	1.2581	1.4656
8	0.4036	0.6628	0.9825	37	1.0529	1.2673	1.4726
9	0.4420	0.7088	1.0244	38	1.0639	1.2763	1.4793
10	0.4816	0.7518	1.0623	39	1.0746	1.2850	1.4858
11	0.5197	0.7915	1.0965	40	1.0850	1.2934	1.4921
12	0.5557	0.8280	1.1276	41	1.0950	1.3017	1.4982
13	0.5898	0.8618	1.1558	42	1.1048	1.3096	1.5041
14	0.6223	0.8931	1.1816	43	1.1142	1.3172	1.5098
15	0.6532	0.9221	1.2053	44	1.1233	1.3246	1.5154
16	0.6826	0.9491	1.2272	45	1.1320	1.3317	1.5206
17	0.7104	0.9743	1.2473	46	1.1404	1.3387	1.5257
18	0.7368	0.9979	1.2660	47	1.1484	1.3453	1.5305
19	0.7617	1.0199	1.2834	48	1.1561	1.3515	1.5351
20	0.7852	1.0406	1.2996	49	1.1635	1.3573	1.5395
21	0.8073	1.0601	1.3148	50	1.1705	1.3629	1.5437
22	0.8283	1.0785	1.3290	51	1.1774	1.3683	1.5477
23	0.8481	1.0958	1.3425	52	1.1843	1.3738	1.5518
24	0.8668	1.1122	1.3552	53	1.1910	1.3792	1.5557
25	0.8846	1.1278	1.3671	54	1.1976	1.3846	1.5596
26	0.9017	1.1426	1.3785	55	1.2041	1.3899	1.5634
27	0.9182	1.1567	1.3892	56	1.2104	1.3949	1.5670
28	0.9341	1.1702	1.3994	57	1.2166	1.3999	1.5707
29	0.9496	1.1830	1.4091	58	1.2227	1.4048	1.5743
30	0.9645	1.1951	1.4183	59	1.2288	1.4096	1.5779
31	0.9789	1.2067	1.4270	60	1.2349	1.4144	1.5814
32	0.9925	1.2177	1.4354	∞	2.0000	2.0000	2.0000

2.8.4 Test for Trend

A simple test for trend (see Table 2.8–1), using the dispersion of sample values x_1, x_2, . . . , x_1, . . . , x_n, taken successively from a normally distributed population, is based on the ratio of the variance s^2 to the successive differential dispersion Δ^2:

$$\Delta^2 = \frac{1}{n-1}\left[(x_2 - x_2)^2 + (x_2 - x_3)^2 + (x_3 - x_4)^2 + \ldots + (x_i - x_{i+1})^2 + (x_{n-1} - x_n)^2\right.$$

$$\text{i.e., } \Delta^2 = \sum_{i=1}^{n} \frac{(x_i - x_{i+1})^2}{n-1}$$

If successive values are independent, then $\Delta^2 \cong 2s^2$ or $\Delta^2/s^2 \cong 2$. As soon as a trend appears, $\Delta^2 < 2s^2$, since adjacent values are more similar to each other than to those further removed, i.e., $\Delta^2/s^2 < 2$.

Null hypothesis H_0	Alternative hypothesis H_1
Sample is taken from the supposed distribution (e.g., normal distribution).	Sample is *not* taken from the supposed distribution.

Alternative hypothesis	The null hypothesis H_0 is rejected in favor of the alternative hypothesis H_1 if		
	Test statistic		Critical value
$n \le 60$	Δ^2/s^2	$\le$	see Table 2.8–1
$n > 60$	Δ^2/s^2	$\le$	$2 - u_{1-\alpha}\sqrt{\frac{n-2}{(n-1)(n+1)}}$

2.8.5 Tests for Normal Distribution

Many statistical procedures assume that the data have been taken from a normally distributed population. If *robust* statistical procedures (such as most mean-value comparisons) are used, deviations from the normal distribution have in most cases only a small effect on the test results.

In other statistical procedures, however, including above all sampling of quantitative characteristics (variables data testing), the inferences will only be correct if the values have actually been drawn from a normal distribution.

Here it is therefore *absolutely vital* to test the normal distribution hypothesis, using all the information available!

Advanced test procedures, which provide a high degree of discriminative ability, usually involve considerable calculation work. These methods are therefore suitable only for computer-aided analysis if meaningful results are to be obtained. The following test procedures will be discussed in more detail:

- The classical χ^2 goodness-of-fit test ($n \ge 50$)
- The d'Agostino test ($50 \le n \le 1000$)
- The Epps-Pulley test ($8 \le n \le 200$)
- The Shapiro-Wilk test ($3 \le n \le 50$)
- The extended Shapiro-Wilk test (small sample sizes)
- The test for asymmetry
- The test for kurtosis

Some tests overlap in their ranges of application. Comparison between test procedures shows that they may exhibit different levels of sensitivity for a given set of data. Depending on the *shape* of the distribution, one test may operate better than another and vice versa. Despite this, it is advisable to run tests in parallel to increase the reliability of the decision. Here it must, however, be noted that this approach increases the risk of type 1 errors (rejection of H_0 even though H_0 is correct).

These tests (with the exception of the χ^2 test and the d'Agostino test) are described in ISO 5479 (draft). This standard contains the coefficients and critical values of the tests.

The χ^2 Goodness-of-Fit Test

The basic principle of this test is very simple: We class the sample values, calculate from the hypothetical distribution function the theoretically expected frequency for each class (expected values), and then compare these with the actual class frequencies observed in the sample (observed values). Requirements are that

1. All expected values must be greater than 1.
2. No more than 20 percent of the expected values must be smaller than 5.

These conditions may in most cases be satisfied by combining the *marginal classes.*

Null hypothesis H_0	Alternative hypothesis H_1
Sample is taken from the supposed distribution (e.g., normal distribution).	Sample is *not* taken from the supposed distribution.

Alternative hypothesis	The null hypothesis H_0 is rejected in favor of the alternative hypothesis H_1 if		
	Test statistic		Critical value
s.a.	$\sum_{j=1}^{k} \frac{(\text{observed val.} - \text{expected val.})^2}{\text{expected value}}$	>	$\chi^2_{f;\,1-\alpha}$
where	$f = k - a - 1$ k = class number a = number of unknown parameters estimated with the help of the sample.		

In a test for normal distribution, a = 2 if the sample statistics for mean and standard deviation are also used as estimators for μ and σ of the population.

Note

It should be noted that the results of this test can be strongly affected by the way in which the values are classed.

The d'Agostino Test

The d'Agostino test was used for a long time as a test for normal distribution. More recent research has shown that this test is in fact not a rigorous test for normal distribution. The d'Agostino test is no longer contained in the new ISO 5479 draft (1994).

Null hypothesis H_0	Alternative hypothesis H_1
Sample is taken from a normally distributed population.	Sample is *not* taken from a normal distribution.

On the basis of the sorted values, we first calculate the auxiliary statistics S and D:

$$S = \sum \alpha_k \left[x_{(n+1-k)} - x_{(k)} \right] \quad \text{where } \alpha_k = \frac{n+1}{2} - k$$

$$D = S / \left(n^2 \sqrt{m_2} \right)$$

Here the index k runs from 1 to n/2 or to (n – 1)/2, depending on whether n is even or odd.

Alternative hypothesis	The null hypothesis H_0 is rejected in favor of the alternative hypothesis H_1 if		
	Test statistic		Critical value
s.a.	$\frac{\sqrt{n}(D - 0.28209479)}{0.02998598}$	< or >	$DA_{n;\alpha/2}$ $DA_{n;1-\alpha/2}$ see [46]

The Epps-Pulley Test

This test is included in the new ISO 5479 draft (tests for normal distribution) in place of the d'Agostino test.

Null hypothesis H_0	Alternative hypothesis H_1
Sample is taken from a normally distributed population.	Sample is *not* taken from a normal distribution.

The statistic is calculated as follows:

$$T_{EP} = 1 + \frac{n}{\sqrt{3}} + \frac{2}{n} \sum_{k=2}^{n} \sum_{j=1}^{k-1} \exp\left\{ \frac{-\left(x_j - x_k\right)^2}{2m_2} \right\} - \sqrt{2} \sum_{j=1}^{n} \exp\left\{ \frac{-\left(x_j - \bar{x}\right)^2}{4m_2} \right\}$$

Alternative hypothesis	The null hypothesis H_0 is rejected in favor of the alternative hypothesis H_1 if		
	Test statistic		Critical value
s.a.	T_{EP}	>	$EP_{n;1-\alpha}$ see Table 2.8–2
where			m_2 = moment of 2. order $m_2 = \frac{1}{n} \sum_{i=1}^{n} (x_i - \bar{x})^2$

Table 2.8–2. Critical values of the Epps-Pulley test.

n	0.90	0.95	0.97	0.99
		$1-\alpha$		
8	0.271	0.347	0.426	0.526
9	0.275	0.350	0.428	0.537
10	0.279	0.357	0.437	0.545
15	0.284	0.366	0.447	0.560
20	0.287	0.368	0.450	0.564
30	0.288	0.371	0.459	0.569
50	0.290	0.374	0.461	0.574
100	0.291	0.376	0.464	0.583
200	0.290	0.379	0.467	0.590

The Shapiro-Wilk Test

Null Hypothesis H_0	Alternative hypothesis H_1
Sample is taken from a normally distributed population.	Sample is *not* taken from a normal distribution.

On the basis of the sorted values, we first calculate the auxiliary statistic S:

$$S = \sum \alpha_k \left[x_{(n+1-k)} - x_{(k)} \right.$$

Here the index k runs from 1 to n/2 or to (n – 1)/2, depending on whether n is even or odd. The coefficients α_k are dependent on the sample size n.

Alternative hypothesis	The null hypothesis H_0 is rejected in favor of the alternative hypothesis H_1 if		
	Test statistic		**Critical value**
s.a.	$\frac{s^2}{n \times m_2}$	<	$SW_{n;1-\alpha}$
where			m_2 = moment of 2. order $m_2 = \frac{1}{n}\sum_{i=1}^{n}(x_i - \bar{x})^2$

The Extended Shapiro-Wilk Test

There is always great uncertainty in assessing normality on the basis of small samples (type 2 error). If we now take more samples from a population, the question arises as to the assessment of the combined set of samples. The frequently used method of combining individual samples into one overall sample is valid only if the mean of the population is constant. Since, however, the question of normality of the distribution of individual values frequently arises in connection with nonconstant means, in the context of process capability studies, the foregoing procedure is not suitable. The extended Shapiro-Wilk test is designed for combined assessment of a number of samples.

First, we establish the test statistic W for each sample. (For large deviations from normality, the rejection of the normal distribution hypothesis is already likely to occur at this stage.) From this statistic, we calculate the G_τ values from the following formula:

$$G_\tau = \tau + \delta \ln\left(\frac{W - \varepsilon}{1 - W}\right)$$

The values of τ, δ, and ε are functions of n and are given in Biometrika Tables II for $3 \leq n \leq 50$.

If the individual samples are taken from a normally distributed population, the G_τ values form a normal distribution with mean 0 and standard deviation 1. The result is a simple means of assessment by plotting the values on probability paper and carrying out a u test.

Note: This test procedure is used only for assessment of a number of separate samples. The aim of the test is to determine whether the samples taken come from a normally distributed population. In the extended Shapiro-Wilk test, test statistics are calculated and plotted on probability paper (see Figure 2.8–4). If the mean of the statistics is 0 and the standard deviation is 1, then the population may be regarded as normally distributed.

Tests for Skewness and Kurtosis

The test for skewness analyzes the degree of asymmetry of the distribution (whether it is skewed to the right, i.e., positively skewed, or skewed to the left, i.e., negatively skewed). The test for kurtosis analyzes the peakedness of the distribution (whether

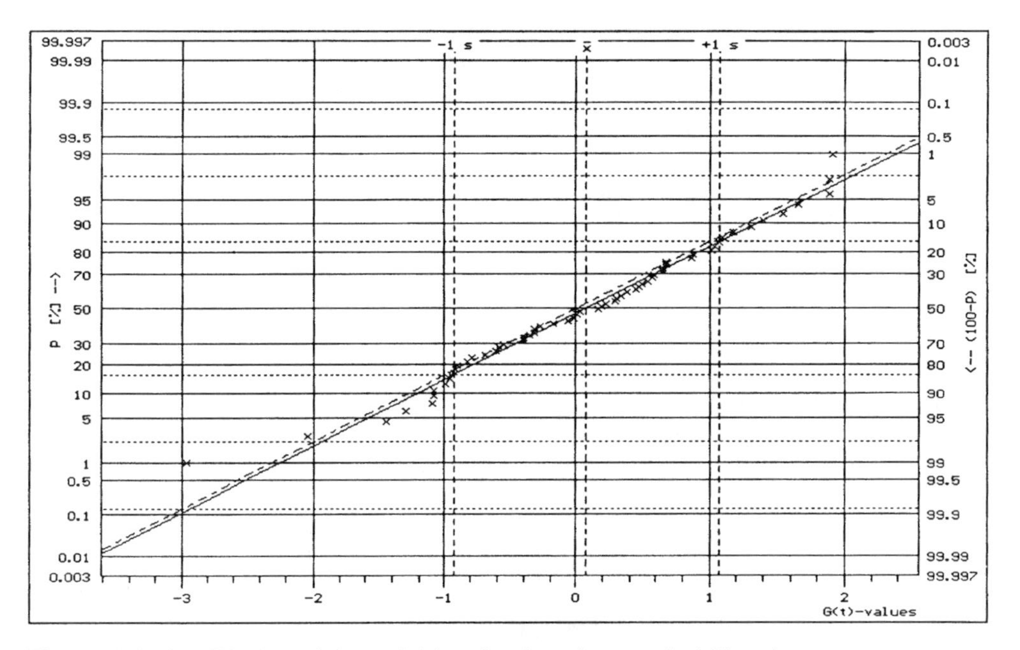

Figure 2.8–4. Display of the calculated values in a probability plot.

the distribution is flat-topped, i.e., platykurtic, or sharply peaked, i.e., leptokurtic). The test statistics are identical with the standard statistical indices for skewness and kurtosis (see Section 2.5). The decision whether the null or alternative hypothesis is true follows from comparison with the critical values. These can be found in ISO 5479 (1994).

2.8.6 Tests for Outliers

Outliers are extreme values in a sample that are at an unnatural distance from the other sample values and hence *must be assumed to originate from a different population than the other values.*

Outliers must be exceptions. If, in the assessment of a characteristic, *outliers* are repeatedly found, the reason for this is probably that the presumed distribution model is not valid.

The two following test procedures for outliers assume that the measurement values have been taken from a normally distributed population. This will be known from experience or can be tested by appropriate means.

Such test procedures should form an essential element of any analysis, even—or particularly—where fully computerized methods are used for collection and analysis of measurement values. Outliers may result from many causes:

- Premature operation of sensors in data collection
- Measurement sensor stuck in position
- Electrical interference
- Mechanical vibration
- Etc.

If the technical reasons for an outlier have been identified, it may be excluded from the analysis. The actual values should be retained and the technical explanation noted.

The David, Hartley, and Pearson Test for Outliers

Null hypothesis H_0	Alternative hypothesis H_1
Neither the maximum value x_{max} nor the minimum value x_{min} is an outlier.	Either the maximum value x_{max} or the minimum value x_{min} is an outlier.

Alternative hypothesis	The null hypothesis H_0 is rejected in favor of the alternative hypothesis H_1 if		
	Test statistic		Critical value
s.a.	$\frac{R}{s}$	>	$A_{1-\alpha; n}$ see [46]

If the null hypothesis is rejected, the extreme value with the greatest distance from the arithmetic mean is an outlier. Only when both extreme values are found to be the same distance from the mean can both values be recognized as outliers in a single test. If a value is eliminated as an outlier, the test is repeated with the remaining values.

The Grubbs Test for Outliers

Null hypothesis H_0	Alternative hypothesis H_1
The minimum or the maximum value is not an outlier.	The minimum or the maximum value is an outlier.

Alternative hypothesis	The null hypothesis H_0 is rejected in favor of the alternative hypothesis H_1 if		
	Test statistic		**Critical value**
s.a. Single-sided for minimum x_1	$\frac{\bar{x} - x_1}{s}$	>	$B_{1-\alpha;\,n}$
s.a. Single-sided for maximum x_n	$\frac{x_n - \bar{x}}{s}$	<	$B_{1-\alpha;\,n}$ see [46]

If an outlier is detected, the complete analysis should be repeated with the remaining values.

2.8.7 Comparing Variances and Means

The sets of measured values may represent different batches of material or batches of parts manufactured on different machines and/or at different times.

The question then may arise whether the population variance has or has not changed due to these different conditions. There are test procedures that investigate the question whether the population variance, as represented by two or more samples, may be assumed to have been constant throughout the period of observation.

In the same manner, one may wish to know whether the means of two populations are or are not equal. Here we distinguish scenarios where the variances of the population are

- Unknown and equal
- Unknown and unequal

Comparison of the Variances of Two Populations (F-Test)

The problem is to test whether the variances s^2 of two samples A and B differ significantly from one another. This test procedure is also known as the F-test. It is used in process analysis in order to determine whether the population variance may be assumed to have been constant over a period of time.

	Null hypothesis H_0		Alternative hypothesis H_1
	$\sigma_A^2 = \sigma_B^2$		$\sigma_A^2 \neq \sigma_B^2$
or	$\sigma_A^2 \leq \sigma_B^2$	or	$\sigma_A^2 > \sigma_B^2$

Alternative hypothesis	The null hypothesis H_0 is rejected in favor of the alternative hypothesis H_1 if			
	Test statistic		Critical value	
$\sigma_A^2 \neq \sigma_B^2$ Double - sided	$\frac{S_A^2}{S_B^2}$	$>$	$F_{1-\frac{\alpha}{2};f_1;f_2}$	$f_1 = n_A - 1$
$\sigma_A^2 > \sigma_B^2$ Single - sided	$\frac{S_A^2}{S_B^2}$	$<$	$F_{1-\alpha;f_1;f_2}$	$f_2 = n_B - 1$

Note: The indexing is always to be chosen such that s_A^2 is *the greater variance.*

Comparison of Variances of Several Populations (Bartlett Test)

Null hypothesis H_0	Alternative hypothesis H_1
$\sigma_1^2 = \sigma_2^2 = \ldots = \sigma_k^2 = \sigma^2$	$\sigma_i^2 \neq \sigma_j^2$ for at least one pair (i, j)

Precondition: $f_i \gtrsim 5$

Alternative hypothesis	The null hypothesis H_0 is rejected in favor of the alternative hypothesis H_1 if		
	Test statistic		Critical value
s.a.	$-\frac{1}{c}\sum_{i=1}^{k} f_i \ln \frac{s_i^2}{s^2}$	$>$	$\chi^2_{k-1;1-\alpha}$
where		$f_i = k_i$ $c = 1 + \frac{1}{3(k-1)}\left(\sum_{i=1}^{k}\frac{1}{f_i} - \frac{1}{f_g}\right)$ $f_g = \sum_{i=1}^{k} f_i$ $s^2 = \left(\sum_{i=1}^{k} f_i \times s_i^2\right) / f_g$	

Comparison of the Means of Two Populations Having Unknown but Equal Variances

Here three cases are distinguished:

Null hypothesis H_0	Alternative hypothesis H_1
$\mu_A = \mu_B$	$\mu_A \neq \mu_B$
$\mu_A \leq \mu_B$	$\mu_A > \mu_B$
$\mu_A \geq \mu_B$	$\mu_A < \mu_B$

Calculation of auxiliary statistic:

$$s_d = \sqrt{\frac{(n_A - 1)s_A^2 + (n_B - 1)s_B^2}{n_A + n_B - 2} \frac{n_A + n_B}{n_A n_B}}$$

if

$$n_A = n_B = n: \qquad s_d = \sqrt{\frac{s_A^2 + s_B^2}{n}}$$

Alternative hypothesis	The null hypothesis H_0 is rejected in favor of the alternative hypothesis H_1 if			
	Test statistic		**Critical value**	
$\mu_A \neq \mu_B$ Double-sided	$\frac{\lvert \bar{x}_A - \bar{x}_B \rvert}{s_d}$	$>$	$t_{1-\frac{\alpha}{2};f}$	$f = n_A + n_B - 2$
$\mu_A > \mu_B$ Single-sided	$\frac{\lvert \bar{x}_A - \bar{x}_B \rvert}{s_d}$	$>$	$t_{1-\alpha;f}$	$f = n_A + n_B - 2$
$\mu_A < \mu_B$ Single-sided	$\frac{\lvert \bar{x}_A - \bar{x}_B \rvert}{s_d}$	$<$	$-t_{1-\alpha;f}$	$f = n_A + n_B - 2$

Comparison of Means of Several Populations with Variances That Are Unknown but Assumed to Be Equal (See Bartlett Test)

Null hypothesis H_0	Alternative hypothesis H_1
$\mu_1 = \mu_2 = \ldots = \mu_k = \mu$	$\mu_i \neq \mu_j$ for at least one pair (i, j)

Alternative hypothesis	The null hypothesis H_0 is rejected in favor of the alternative hypothesis H_1 if			
	Test statistic		**Critical value**	
s.a.	$\frac{(n-k)\sum_{i=1}^{k}\left(\bar{x}_i - \bar{\bar{x}}\right)^2 n_i}{(k-1)\sum_{i=1}^{k} s_i^2 (n_i - 1)}$	$>$	$F_{f1;f2;1-\alpha}$	$f_1 = k - 1$, $f_2 = n - k$
where	$\bar{\bar{x}} = \frac{\sum_{i=1}^{k} n_i \bar{x}_i}{\sum_{i=1}^{k} n_i} = \frac{1}{n}\sum_{i=1}^{k} n_i \bar{x}_i$			

Comparison of Means of Two Populations Where the Variances Are Unknown and Unequal

Null hypothesis H_0	Alternative hypothesis H_1
$\mu_A = \mu_B$	$\mu_A \neq \mu_B$
$\mu_A \leq \mu_B$	$\mu_A > \mu_B$
$\mu_A \geq \mu_B$	$\mu_A < \mu_B$

Calculation of auxiliary statistic:

$$s_d = \sqrt{\frac{s_A^2}{n_1} + \frac{s_B^2}{n_2}}$$

Alternative hypothesis	The Null hypothesis H_0 is rejected in favor of the alternative hypothesis H_1 if			
	Test statistic			**Critical value**
$\mu_A \neq \mu_B$ Double-sided	$\frac{\lvert\bar{x}_A - \bar{x}_B\rvert}{s_d}$	$>$	$t_{1-\frac{\alpha}{2};f}$	$\frac{1}{f} = \frac{c^2}{n_A - 1} + \frac{(1-c)^2}{n_B - 1}$
$\mu_A > \mu_B$ Single-sided	$\frac{\lvert\bar{x}_A - \bar{x}_B\rvert}{s_d}$	$>$	$t_{1-\alpha;f}$	where
$\mu_A < \mu_B$ Single-sided	$\frac{\lvert\bar{x}_A - \bar{x}_B\rvert}{s_d}$	$<$	$-t_{1-\alpha;f}$	$c = \frac{\frac{s_A^2}{n_A}}{\left(\frac{s_A^2}{n_A}\right) + \left(\frac{s_B^2}{n_B}\right)}$

2.8.8 Representation of Test Results

The user wishes to obtain information about the significant properties of a set of measured values by means of numerical test procedures. As already mentioned, these procedures are very time-consuming and nowadays only considered suitable for computer-aided analysis. In computer-aided analysis, a number of test procedures are often applied before the analysis proper in order to gain a quick impression of the data. Clear representation of test results then becomes a matter of great importance. A distinction should be made between the representation of a single test result (Table 2.8–3) and a summary of several test results (Table 2.8–4).

The F-test produces the test statistic 0.931. This is compared with the critical value (see Table 2.8–3). The result of the test is: *The null hypothesis is not rejected.*

Table 2.8–3. Results: F-test.

Null hypothesis	**H_0**	$\sigma_1^2 = \sigma_2^2$ The variance between the samples is null.
Alternative hypothesis	**H_1**	$\sigma_1^2 \neq \sigma_2^2$ The variance is NOT equal to null.

Significance level	Critical value lower	Critical value upper	Test statistic
$\alpha = 5\%$	0.53	1.89	
$\alpha = 1\%$	0.43	2.32	0.931
$\alpha = 0.1\%$	0.34	2.96	

Table 2.8–4. Test results summary.

Test		Critical values $\alpha = 0.1\%$	$\alpha = 1\%$	$\alpha = 5\%$	Test statistic
Randomness	above	3.291	2.576	1.960	1.228
	below	—	—	—	
Normal distribution	above	—	1.85	1.50	–13.170 * *
d'Agostino	below	—	–3.30	–2.39	H0 rejected
CHI² test	above	24.32	18.48	14.07	68.382 * * *
equal classes	below	—	—	—	H0 rejected
Outliers	above	—	6.840	6.390	5.267
David/Hartley/Pearson	below	—	—	—	
Asymmetry test	above	—	0.400	0.280	1.625 * *
	below	—	—	—	H0 rejected
Kurtosis test	above	—	3.980	3.570	5.802 * *
	below	—	2.370	2.510	H0 rejected
Test for Trend	above	—	—	—	1.872
	below	1.5652	1.6727	1.7685	

2.9 Summary of Basic Procedures

The previous chapters have dealt with basic statistical procedures. These may be divided into three groups:

- Calculation of statistics
- Graphical representation
- Numerical test procedures

Each one of these procedures has its own specific sphere of application determined by its function. The results of these procedures present an overall picture by which the process may be described and assessed. Some procedures may lead to redundant and/or contradictory results. In the first case, the knowledge gained is reinforced; in the second, further analysis will be required.

Table 2.9–1 summarizes the basic procedures as well as their functions and applications.

Table 2.9–1. Basic procedures and their functions.

	Procedure	Function
Statistics	Average, Median	Estimators for Process location
	Standard deviation, Range	Estimators for Process spread
	Skewness, Kurtosis, Excess	Evaluation of distribution model
	Regression coefficient	Appraisal of the goodness of a model fit
	Capability indices, excess proportion	Performance and capability of a process
Graphics	Raw values chart	Recognition of peculiarities like: outliers, trends, periodicities, fluctuation of the averages, values outside specification
	Actual value plot	Estimation of distribution form, gauge resolution, number of values outside specification
	Histogram	Estimation of distribution form, estimation of capability indices
	Probability plot (individual values/classed)	Model distribution fit, estimation of excess proportions
	Cumulative frequency line	Model distribution fit
	Quality control chart (see Chapter 3)	Evaluation of process stability regarding location and spread: recognition of Run, Trend, Middle Third, and Violation of control limits
	Operation characteristic (see Chapter 3)	Sensibility of Quality control charts
	Probability plot averages	Recognition of Process disturbances regarding location
	CHI^2 plot	Recognition of Process disturbances regarding spread
	xy plot	Display of Position tolerances; Recognition of dependencies between two characteristics
	Box plot	Quick summary with the help of statistical values
Test Procedures	Randomness • Swed-Eisenhart • Trend	Recognition of nonrandom series of individual and average values; Recognition of peculiarities like Trends
	Adaptation test • Shapiro-Wilk • CHI^2 • d´Agostino • Kolmogoroff-Smirnoff • extended Shapiro-Wilk • Kurtosis • Asymmetry	Evaluation criteria for establishing deviations from the model distribution
	Outliers • David, Hartley, Pearson • Grubbs	Establishing occurrence of outliers
	Analysis of variance • F-test	Establishing existence of spread between samples; estimation of size
	Equality of variances • Bartlett	Establishing fluctuations of the variances

3
Control Charting

3.1 What Is a Control Chart?

A control chart enables a visual assessment of a process with regard to its location and dispersion. For this purpose, certain statistics (e.g., number of nonconforming items, number of nonconformities per unit, raw values, means, medians, standard deviations, and ranges) are plotted over time and compared with limit lines (known as control limits) to assess location and dispersion. This comparison enables a statement about the quality (stability) of the processes.

The horizontal axis (abscissa, x axis) of a control chart (Figure 3.1–1) may show

- The number of the sample
- The point in time (date, time) when the sample was taken
- The batch number or some similar identification

On manually prepared charts, 25 to 30 samples are shown. In computer-generated charts, substantially more samples may be shown, depending on screen resolution. The vertical axis (ordinate, y axis) shows the measurement scale of the characteristic.

In the case of *continuous* characteristics, the scaling depends on the statistic plotted:

- A scale for the original values
- A scale for the sample statistic (mean, median, standard deviation or range).

In the case of discrete characteristics, the scaling of the ordinate reflects either the *number of nonconforming items,* the *proportion of nonconforming items,* the *number of nonconformities,* or the *number of nonconformities per unit.*

Depending on the nature of the control chart,

- A center line (C)
- Warning limits (WL)
- Control limits (CL)

may be calculated and shown. From this, we have the picture as shown in Figure 3.1–1. The *values* are then plotted on this chart.

The calculation of the limits is explained in the following chapters. In current practice, warning limits are rarely used, if at all.

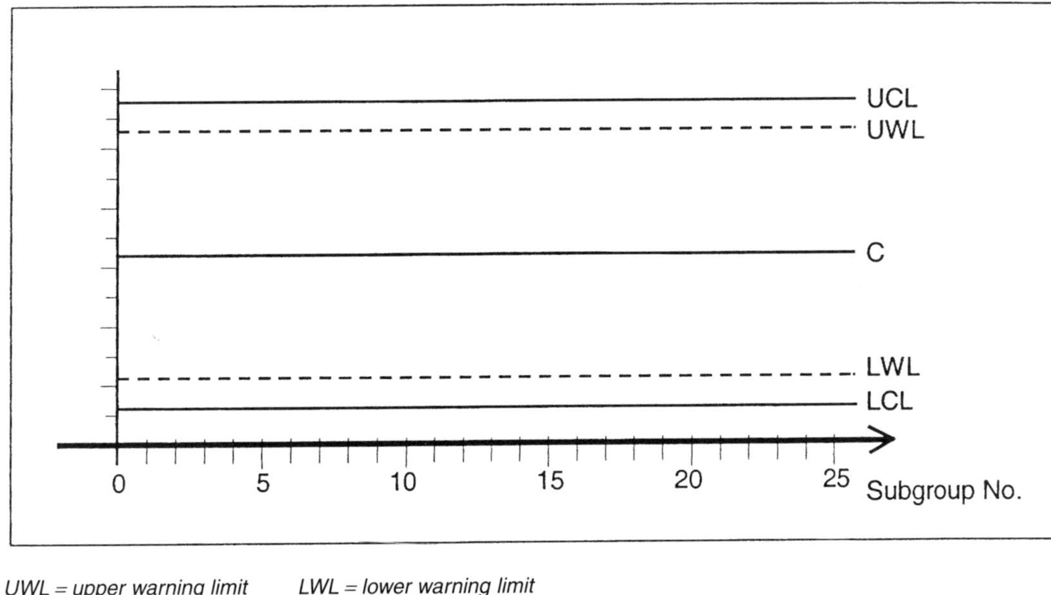

UWL = upper warning limit
UCL = upper control limit
C = center line
LWL = lower warning limit
LCL = lower control limit

Figure 3.1–1. Basic construction of a control chart.

Interpretation of a Control Chart

During ongoing production samples of size n are taken at (ideally) regular intervals. The characteristic to be monitored by the control chart is checked. When monitoring discrete (countable) characteristics, the proportion of nonconforming items or the number of nonconformities per unit is determined and plotted on the chart. Here, the sample size may vary from sample to sample. In contrast, in the case of continuous characteristics, the sample size must always remain constant. Incomplete samples must not be taken into account. It is customary to use a sample size of $n = 5$. This size has been found to give good results in practice. All parts within the sample are checked for one (or more) measurable characteristic(s). From the n original values, statistics such as $\bar{x}$, $\tilde{x}$, R, or s can be calculated. Depending on the type of control chart, the raw data themselves or the sample statistics are plotted on the chart. Figure 3.1–2 shows the so-called average chart ($\bar{x}$ chart), on which sample means are plotted to monitor a continuous characteristic.

The interpretation of a control chart is based on the following criteria:

Possibilities	Consequences
Raw or statistic values within the warning limits	Production continues as usual.
Raw or statistic values outside the warning limits, but within the control limits	Production continues as usual, more frequent testing; if need be, immediate taking of a new sample.
Raw or statistic values outside the control limits	Adjustment of production; if need be, sorting out of parts produced since the last test.

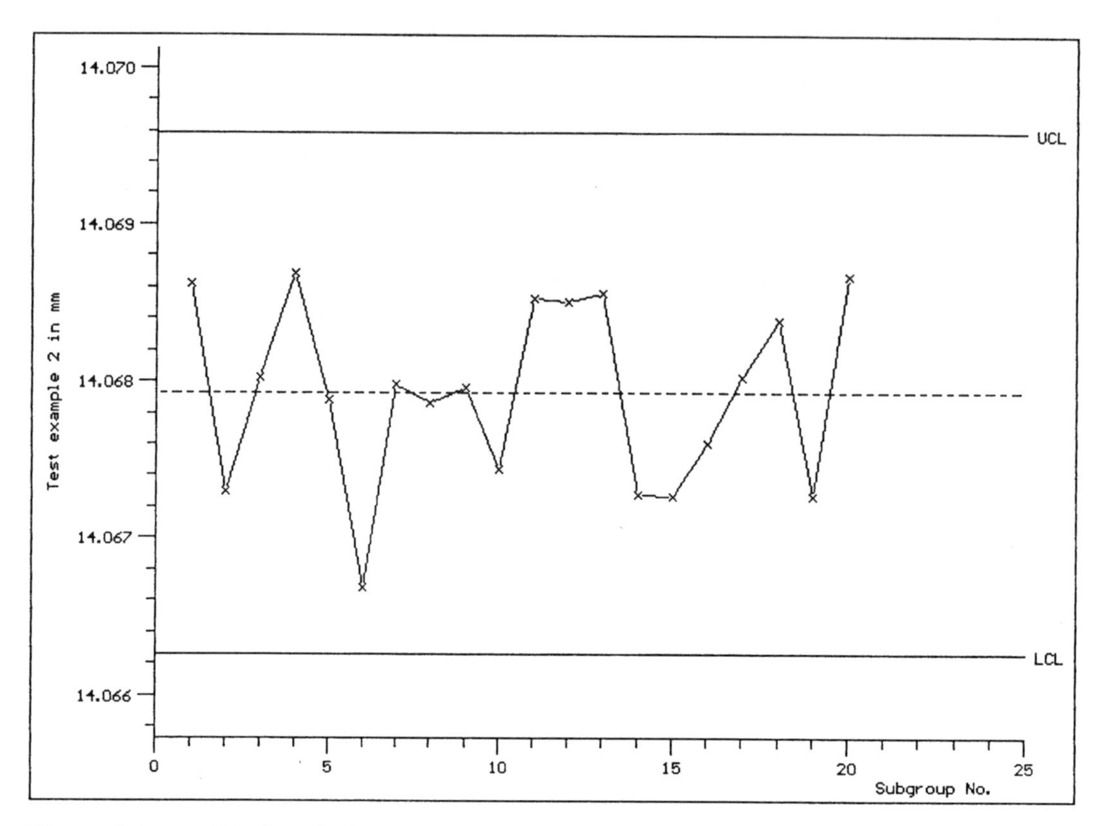

Figure 3.1–2. Display of 20 averages in an $\bar{x}$ chart.

Depending on the type of chart, additional criteria such as runs, trends, or *middle third* (see Section 3.5.3) may be used as well.

Basic Rules for Construction of a Control Chart

The following rules should be observed in control charting:

- *Use regular sampling.*
  The time interval may range from a few minutes to a few hours (days), depending on the type of manufacturing process and empirical knowledge as to its susceptibility to change. The time intervals should be laid down in the specification of the control chart.
- *Use a constant sample size.*
  For continuous characteristics, the sample size must be kept constant, since the statistically calculated limits are dependent on n for all control charts.
- *Record process events on control chart.*
  To have available, in subsequent analysis, a comprehensive record of process actions (tool adjustment, tool change, etc.) as well as process disruptions, these events should be recorded and linked to the relevant samples. For a given process, there is usually only a small number of possible process events. These may be given codes and cataloged. This will simplify the recording. Over longer periods of time, Pareto analyses may then be carried out to identify process weaknesses and take corrective action.

- *Record sample time and name of person taking the sample.*
  To satisfy the need for documentation, the name of the person taking the sample and the measuring device used should be recorded, in addition to the date and the time of day. Furthermore, it must be ensured that, once recorded, measurement results are not subsequently altered without written justification. An unambiguous attribution to the relevant manufacturing process—e.g., part number, characteristic number, and machine number—is also required. The basis for the calculation of the control limits must be shown.

General Comments on Monitoring Dispersion

The calculation of warning and control limits for the individuals, average, and median charts is based on an estimate of the dispersion of the population. In other words, the limits are based on a constant dispersion model. If it cannot be assumed that process dispersion is constant, it is necessary to monitor it. In addition, many, if not most, processes are susceptible to changes in dispersion. This is another reason why it is necessary to monitor process dispersion using a control chart. Any non-random increases of dispersion (points beyond the upper control limit) should be investigated and the causes corrected. As for the *normal* or expected level of dispersion (within the control limits), this can usually only be reduced by process improvement actions, such as overhaul of the machine or more careful selection of the materials used. If a significant reduction in process variation is noted on the control chart, the cause should be identified and incorporated in the standard process procedure. Following this, it will be necessary to recalculate all control limits, both for the location and the dispersion charts.

3.2 Common Types of Control Chart

Depending on the type of characteristic, control charts may be divided into control charts for discrete and continuous characteristics. For qualitative characteristics, multiple characteristics control charts are used. For continuous characteristics, an additional distinction is made between Shewhart control charts and acceptance control charts. The control limits on Shewhart charts are completely independent of the characteristic's tolerance and are determined solely on the basis of existing process data. The calculation is typically based on $k = 20$ samples, with a sample size of $n = 3$, 4, or 5. In contrast to this, the calculation of control limits on acceptance control charts is based on the specification limits. However, within the context of *never-ending improvement* and the philosophy of process performance aimed at the target value, the Shewhart chart is the only suitable type of chart. Hence, Shewhart charts constitute the main type of control chart in use today.

Shewhart control charts are used to monitor, on an ongoing basis, both process location and dispersion. For this purpose the control chart comprises both a *location chart* (showing individuals, means, or medians) and a *dispersion chart* (for plotting the sample standard deviation or the sample range).

Typical combinations are the x/s chart, the $\bar{x}/s$ chart, and the $\tilde{x}/R$ chart (Figure 3.2–1). Which combination is appropriate in a given case will be discussed in the following chapter.

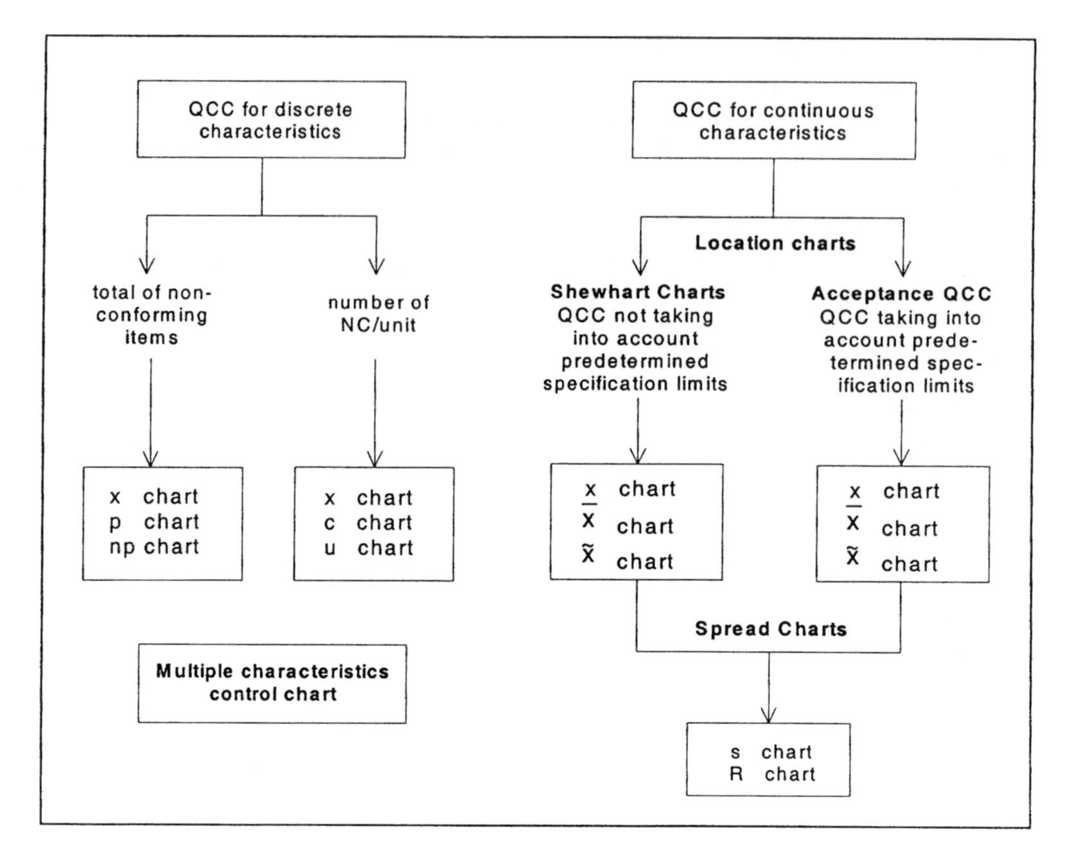

Figure 3.2–1. Summary of Quality Control charts.

3.3 Control Charts for Discrete Characteristics

As a matter of principle, quantitative assessment (collecting measurement data) is the preferred approach for controlling quality characteristics. But if this is not possible, or not economically viable, then an assessment may be made on the basis of qualitative observations. This leads to statistics such as the *number of nonconforming items,* the *proportion of nonconforming items,* the *number of nonconformities,* or the *number of nonconformities per unit.* These constitute discrete characteristics (count data).

Advantages and Limitations

Control charts for discrete characteristics are based on the existence and discovery of nonconforming items or nonconformities. Unlike control charts for continuous characteristics, they do not give warnings of undesirable process changes until a number of nonconformities have already occurred. Hence, one operates on the basic assumption that there will always be an unavoidable number, large or small, of nonconforming items. The control charts for continuous characteristics show the direction of changes in location (up or down) and dispersion (e.g., larger or smaller) as well as the current process average and the current level of process variation. Attribute charts show only the proportion of product that is *unsatisfactory* (*not okay*).

Conformance Criteria

Qualitative (attribute) data are frequently used where measurement is not practicable. However, it should be noted that subjective assessments of conformity or nonconformity will be subject to variation between different persons and even time-to-time variation for a single person. To produce an effective control charting system, it is necessary to minimize the scope of this variation. For this purpose, conformance criteria should be defined:

- Provide reference standards.
- Provide visual aids and define methods of assessment.
- Ensure that personnel are qualified.
- Train personnel and develop defect detection and decision-making skills.
- Provide suitable working conditions (e.g., lighting).

Attribute Control Charts

Depending on the situation at hand, the following control charts may be applied:

- x chart for the number of nonconforming items or number of nonconformities per unit
- np chart for the number of nonconforming items
- p chart for the proportion of nonconforming items
- c chart for the proportion of nonconformities
- u chart for the number of nonconformities per unit.

Table 3.3–1 shows examples of cases where these charts could be used.

Table 3.3–1. Application examples of quality control charts with discrete characteristics.

Discrete characteristics	Control chart used
• Existence/nonexistence of necessary screws • Electric current open/closed • Shaft diameter correct/wrong	x chart for the number of nonconforming items, p chart for the proportion of nonconforming items or np chart for the number of nonconforming items (equal sample size)
• Blisters in a windshield • Paint error on a door • Longitudinal material crack	x chart for the number of nonconforming items, c chart for the proportion of nonconformities (equal sample size) or u chart for the number of nonconformities per unit

Action Following Violation of the Control Limits

- If the *upper control limit* is exceeded, this is evidence of a deterioration in the manufacturing process. The cause must be found and corrected.
- If the results (frequently) fall short of the *lower control limit,* this is evidence of an improvement in the manufacturing process. The reasons should be established, made permanent, and, as far as possible, transferred to other processes. The center line and control limits should then be recalculated.

If the expected number of nonconformities is small, the chances of obtaining a zero sample (without any nonconformities) under normal process conditions may be so large that the lower control limit ceases to exist. A quality improvement is then indicated when the distance from the upper limit is becoming increasingly large and/or there are long runs below the average line; i.e., samples with low numbers of nonconformities are found more frequently than expected.

In order to obtain unambiguous decisions (*in control* or *not in control*) from control charts for discrete characteristics (count data), the limits for x, np, and c charts should always be set at values lying between two whole numbers, e.g., upper control limit = 13.5. This means that 13 nonconformities would not exceed the upper control limit. Process action would only be indicated if there were 14 or more nonconformities.

3.3.1 Calculation of Control Limits

The control limits for a control chart are based on a double-sided random dispersion interval. Determination of this interval must be based on a suitable distribution model. For discrete characteristics, this is either the binomial distribution (number of nonconforming items) or the Poisson distribution (number of nonconformities). Under certain conditions (see Figure 3.3–1), both distributions may be approximated by the normal distribution. As the figure shows, p and n must be quite large to permit the transition to another distribution model, and, in this day and age, where we are frequently dealing with parts per million (ppm) or pursuing a *zero defects* goal, this requirement is in most cases not given. Hence, the use of the normal distribution as an approximation may lead to considerable errors, especially in the case of small percentages of nonconformities and small sample sizes. Table 3.3–2 shows possible methods for determining the control limits on the basis of the distribution model concerned, using the x chart as an example. In noncomputerized applications, the use of graphical tools or tables (if available) is required for the binomial and Poisson distribution, whereas the limits for the normal distribution may easily be calculated manually. Hence, in the past it was customary to fall back on the approximation.

3.3.2 x Chart for the Number of Nonconforming Items

If the true proportion of nonconforming items encountered in production is unknown, the first step is to calculate the average number of nonconforming items from a trial run. Here $\hat{p}$ is the estimated proportion of nonconforming items in the population.

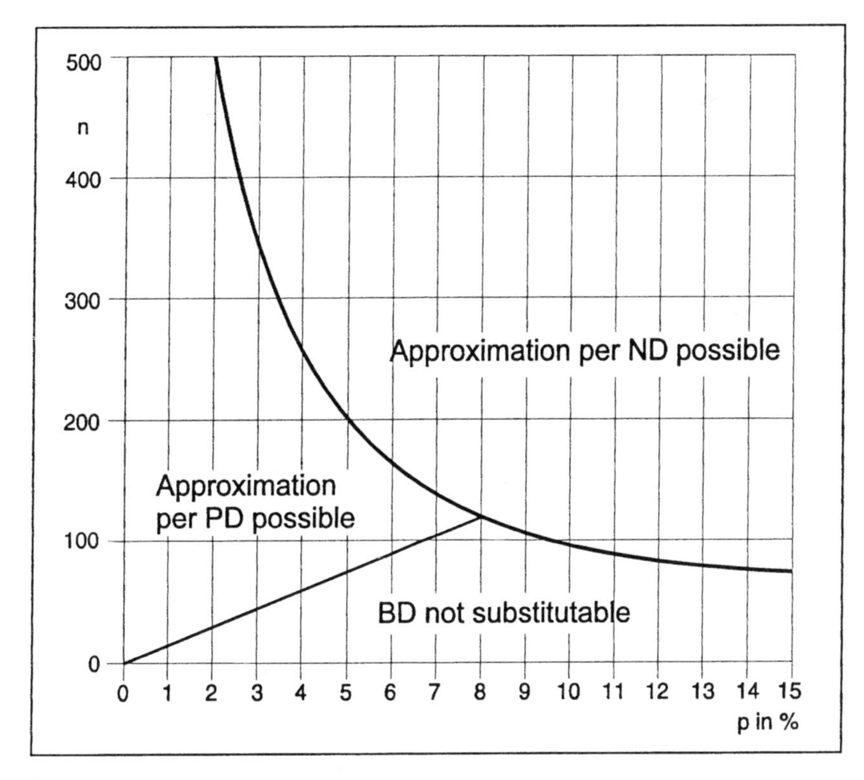

Figure 3.3–1. Approximation by means of the distribution model.

Table 3.3–2. Limited determination of an x chart through the number of defective units.

Possibilities	Calculation of limits x Chart
Random spread region of the binomial distribution where P = 95% or 99% $G(x_{lo};n,p) \leq \alpha/2$ $G(x_{up};n,p) \leq 1-\alpha/2$	1. Tables or computer programs 2. Larson nomogram 3. Nomogram for the spread limits of the Binomial distribition
Approximation per Poisson distribution if $\hat{p} \leq 10\%$ and $n \geq 80$	1. Tables or computer programs 2. Thorndike nomogram 3. Nomogram for the spread limits of the Poisson distribution
Approximation per normal distribution if $n \ \hat{p} \ (1-\hat{p}) \geq 9$ $\hat{p}$ = proportion of defective units n = sample size	UWL, LWL $= n\hat{p} \pm u_{1-\alpha/2}\sqrt{n\hat{p}(1-\hat{p})}$ where $\alpha = 5\%$ LCL, UCL $= n\hat{p} \pm u_{1-\alpha/2}\sqrt{n\hat{p}(1-\hat{p})}$ where $\alpha = 1\%$

$$\hat{p} = \frac{\sum_{i=1}^{k} x_1}{\sum_{i=1}^{k} n_1} \quad \text{where} \quad \begin{array}{l} x_i = \text{number of nonconforming items per sample} \\ n_i = \text{sample size} \\ k = \text{number of samples} \end{array}$$

On the basis of this estimate, we may determine the upper and lower warning limits (UWL/LWL) and the upper and lower control limits (UCL/LCL) as indicated in Table 3.3–2.

The determination of the limits based on the two-sided random dispersion interval of the binomial or Poisson distribution was covered in Chapter 2, Section 2.2.1.

If the normal distribution model (Table 3.3–2) is used to determine the control limits, the chart is known as the np chart (see Section 3.3.3). In this type of chart, the control limits are usually based on a nonintervention probability of 99.73 percent (meaning that, for an undisturbed process, 99.73 percent of the sample results will be expected to lie within the limits), corresponding to $\pm 3\sigma$. This implies, of course, that the sample size should remain constant. For the x chart, this is not necessary, but if the sample size does vary, then the control limits must be recalculated for each sample, which is inconvenient for manual charting. If the sample size fluctuates by less than ±25 percent, the deviations are deemed too small to justify a recalculation from the practical point of view.

Example: x Chart with Constant Sample Size

On k = 50 successive days, samples of size n = 200 of an assembly were subjected to a final inspection. A quantity of x components was rejected each day. Find the control limits for P = 99 percent and plot the values in Table 3.3–3 on the control chart.

Solution

Mean number of nonconforming items $\quad \bar{x} = \frac{1}{k}\sum_{i=1}^{k} x_i = 8.6$

Percentage of nonconforming items $\quad \hat{p} = \frac{8.66}{200} = 4.33\%$

Table 3.3–3. Defective units with constant subgroup size.

Day	x_i	Day	x_i	Day	x_i	Day	x_i	Day	x_i
1	10	11	8	21	6	31	12	41	7
2	11	12	7	22	13	32	9	42	9
3	7	13	9	23	13	33	13	43	5
4	6	14	9	24	4	34	10	44	9
5	6	15	12	25	10	35	10	45	10
6	9	16	4	26	10	36	7	46	9
7	9	17	9	27	10	37	8	47	9
8	9	18	7	28	10	38	6	48	9
9	4	19	10	29	9	39	7	49	12
10	8	20	10	30	10	40	4	50	9

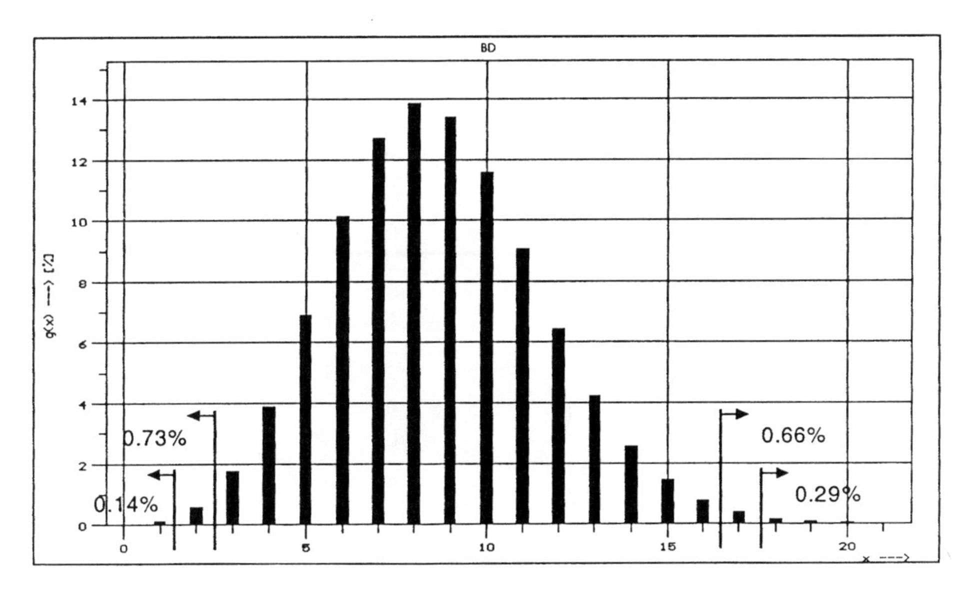

Figure 3.3–2. Binomial distribution for p = 4.33% and n = 200.

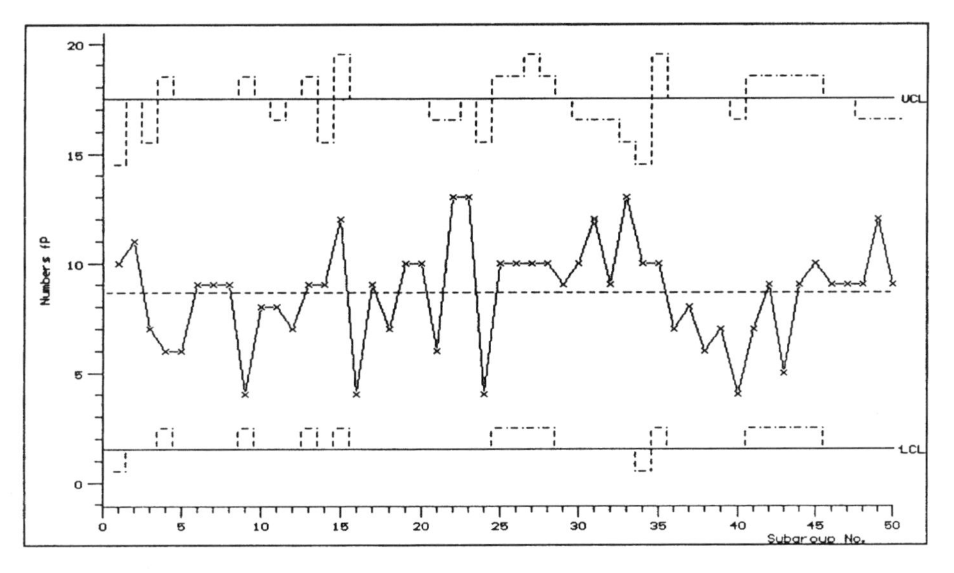

Figure 3.3–3. x chart with n ≠ constant.

The double-sided random dispersion interval for P = 99 percent of the corresponding binomial distribution is shown in Figure 3.3–2. This gives upper and lower control limits of 17.5 and 1.5 respectively. The control chart is illustrated in Figure 3.3–3. The control limit adjustments (dotted lines) would, of course, not apply in this case, as we are dealing with a constant sample size.

Example: x Chart with Varying Sample Sizes

Solution

average sample size $\bar{n} = 198{,}1$

estimated percentage of nonconforming items $\hat{p} = 4.3715\%$

Table 3.3–4. Number of nonconforming units with differing subgroup size.

Sample Size	Defective Units	Sample Size	Defective Units	Sample Size	Defective Units
150	10	200	7	240	10
200	11	200	10	200	7
175	7	200	10	200	8
210	6	190	6	200	6
200	6	180	13	200	7
205	9	195	13	190	4
200	9	175	4	210	7
195	9	210	10	220	9
220	4	220	10	215	5
200	8	230	10	220	9
185	8	210	10	210	10
195	7	200	9	205	9
210	9	190	10	195	9
170	9	180	12	190	9
230	12	185	9	185	12
200	4	175	13	180	9
200	9	160	10		

From this we have the upper control limit UCL = 17.5 and the lower control limit LCL = 1.5. These are the same as for the data in Table 3.3–3. Figure 3.3–3 shows adjusted control limits (dotted lines) for the various sample sizes as per Table 3.3–4.

3.3.3 p Chart for the Proportion of Nonconforming Items

The p chart shows the proportion of nonconforming items in a sample. We might, for instance, take two samples of 75 parts per day, or inspect 100 percent of the production on an hourly or daily basis, or use some other sampling scheme. The proportion of nonconforming items may be based on the evaluation of a single characteristic (was a particular component missing?) or on simultaneous evaluation of several characteristics (did the electrical test show any faulty components?).

It is important that

- Each part or vehicle inspected should be recorded as conforming or nonconforming (if a part has several faults it is still only counted as one nonconforming part).
- The inspection results should be summarized in a meaningful format and the number of nonconforming items entered on the control chart as a proportion of the sample size.

Before data collection begins, the sample size, sampling frequency, and the number of samples should be determined:

- *Sample size*
  Charts for discrete characteristics (count data) in general require a large sample size (i.e., 50 or more) to be able to identify small changes in process performance. To observe patterns on p charts, the sample size must be defined such that several nonconforming items are likely to occur within each sample. Large

sample sizes may, however, be a disadvantage if each sample represents a long period of production. It is very much recommended that the sample size be kept constant, or at least not allowed to fluctuate more than ±25 percent.

- *Sampling frequency*
  The choice of sampling frequency should be based on practical considerations, consistent with natural production periods, in order to assist with the analysis and correction of any problems identified. Short sampling intervals allow for fast reaction but may be at odds with the requirement for large samples.
- *Number of samples*
  The sampling period should be long enough to include all likely sources of variation that may affect the process. Furthermore, it is desirable to take at least 20 samples to obtain a good test for stability, and, if stability has been achieved, to obtain a valid estimate of process performance.

On the p chart, the number of nonconforming items np is divided by the sample size n.

Hence, we determine the proportion

$$p_i = \frac{x_i}{n_i}$$

where x_i = number of nonconforming items in sample no. i
n_i = size of sample no. i

This value is entered as a percentage on the p chart.

The mean number of nonconforming items $\overline{p}$ in k samples is given by

$$\overline{p} = \frac{x_1 + x_2 + \ldots + x_k}{n_1 + n_2 + \ldots + n_k}$$

Hence, $\overline{p}$ is an estimator of the proportion of nonconforming items in the whole population. Based upon this, the control limits may be determined, as for the x chart, from the random dispersion intervals of the binomial distribution.

For the sake of simplicity, the control limits may also be based on the normal distribution model (±3σ limits). This requires the following formulas:

$$UCL = \hat{p} + 3\sqrt{\frac{\hat{p}(1-\hat{p})}{\overline{n}}}$$

$$LCL = \hat{p} - 3\sqrt{\frac{\hat{p}(1-\hat{p})}{\overline{n}}}$$

where $\hat{p} = \frac{\sum_{i=1}^{k} x_i}{\sum_{i=1}^{k} n_i}$

x_i = number of nonconforming items
k = number of samples
n_i = sample size
$\overline{n}$ = average sample size

Notes

When p and/or n is small, the lower control limit may be negative. In this case, no lower control limit is plotted. A disadvantage of the p chart as compared with the x chart is that an additional calculation step is necessary to arrive at the proportions; furthermore, proportions are more difficult to plot than integers.

Example

The sample results shown in Table 3.3–3 and Table 3.3–4 give, based on the binomial distribution with P = 99 percent, a p chart as shown in Figure 3.3–4 and, based on the normal distribution with P = 99.73 percent, a p chart as shown in Figure 3.3–5.

3.3.4 np Chart for the Number of Nonconforming Items

The np chart shows the number of nonconforming items in a sample. It is very similar to the p chart, but with the difference that the number of nonconforming items is plotted, rather than the proportion of nonconforming items in the sample. The field of application of the p and np charts is the same; the np chart is preferred if

- The sample size remains constant.
- The number of nonconforming items is easier to report than the proportion.

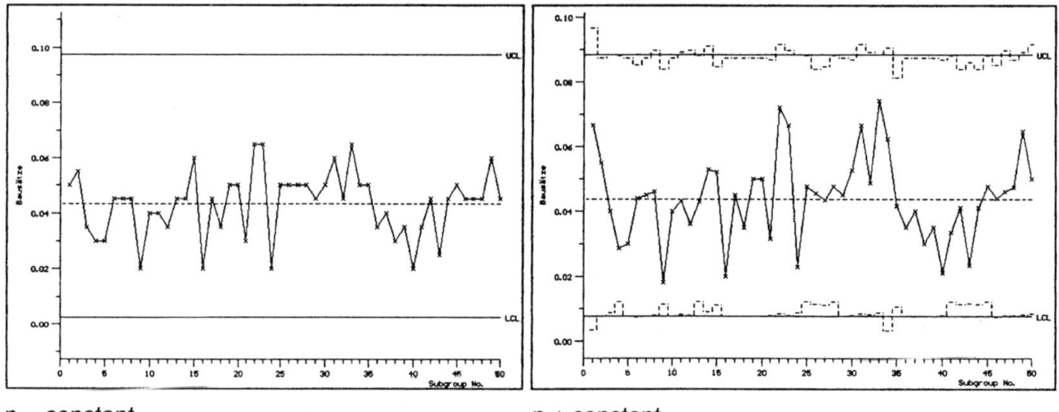

n = constant *n ≠ constant*

Figure 3.3–4. p chart from binomial distribution P = 99%.

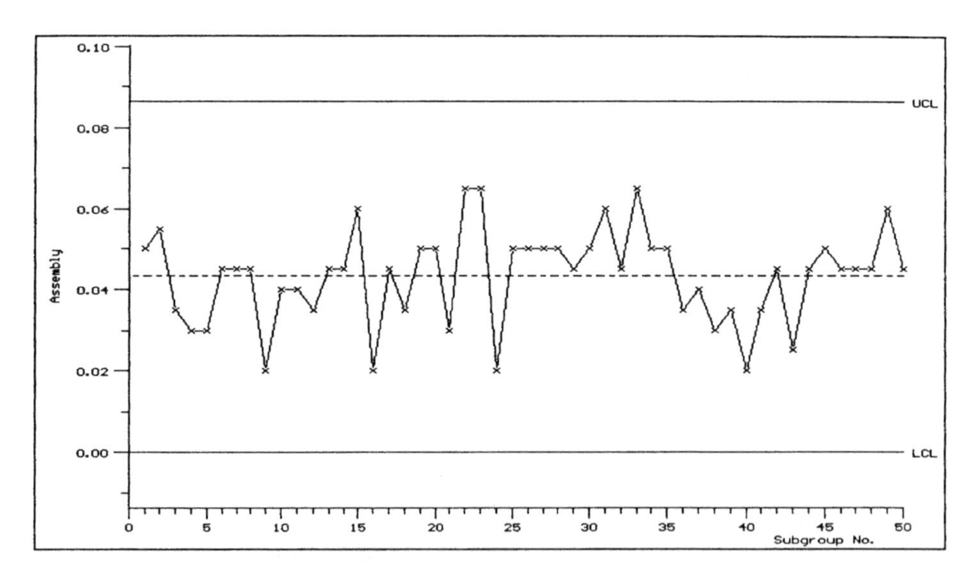

Figure 3.3–5. p chart from normal distribution with P = 99.73%.

The np chart is comparable to the x chart. The difference is that in the np chart, the control limits are based on the normal distribution model:

$$\begin{matrix}UCL \\ LCL\end{matrix} = n\hat{p} \pm u_{1-\alpha/2}\sqrt{n\hat{p}(1-\hat{p})}$$

Based on ±3σ limits, i.e., P = 99.73 percent, $U_{1-\alpha/2} = 3$. This presupposes a constant sample size.

Example

Based on the sample results in Table 3.3–3, the control limits are calculated as follows:

$$UCL = 200 \times 0.0433 + 3\sqrt{200 \times 0.0433(1-0.0433)} = 17.295$$

$$LCL = 200 \times 0.0433 - 3\sqrt{200 \times 0.0433(1-0.0433)} = 0.025$$

The lower control limit is close to zero and is therefore not shown on the control chart. Figure 3.3–6 shows the resulting control chart.

3.3.5 x Chart for the Number of Nonconformities per Unit

In contrast to the x chart described in Section 3.3.2, the number of nonconformities is here seen in relation to the inspection unit (e.g., a product) or in relation to the number of production units inspected. Typical examples of nonconformity counts are

- Number of casting defects per carburetor housing
- Number of gas pockets per meter of welded joint
- Number of enamel defects per coil of enameled wire
- Number of yarn defects per mile of yarn
- Number of nonconformities per m² of paper

Experience shows that these characteristics may be regarded as following a Poisson distribution. Hence, the control limits on the control charts are given by the

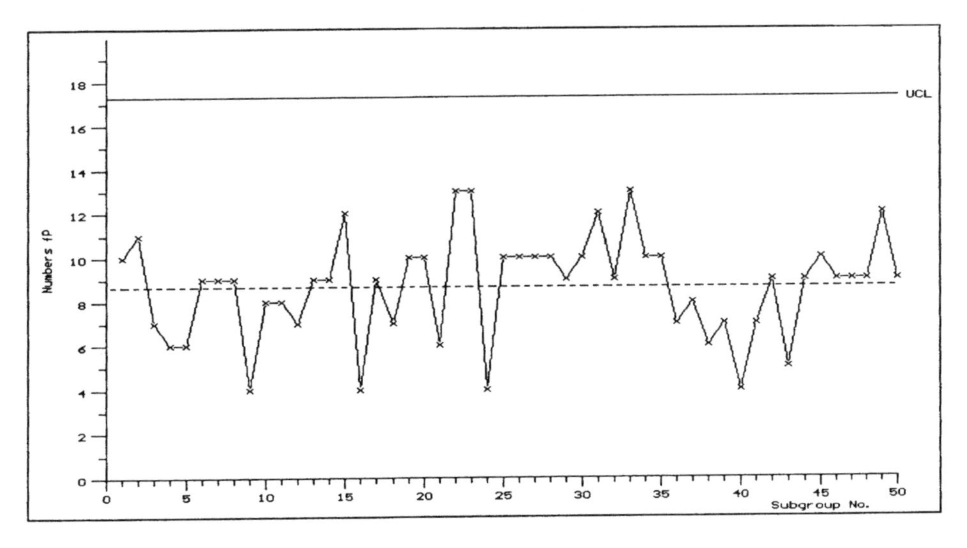

Figure 3.3–6. np chart.

Table 3.3–5. x chart control limits.

Possibilities	Calculation of Limits
Calculation of the limits using the Poisson distribution	1. Tables 2. Thorndike diagram 3. Nomogram for the spread limits of PV 4. Computer program
Formal treatment of Poisson distribution as a normal distribution if: $\mu \geq 9$	UWL, LWL $= \mu \pm u_{1-\alpha/2}\sqrt{\mu}$ where $\alpha = 5\%$ LCL, UCL $= \mu \pm u_{1-\alpha/2}\sqrt{\mu}$ where $\alpha = 1\%$

Table 3.3–6. Subgroup results: Defects per unit.

1	2	3	4	5	6	7	8	9	10	11	12	13	14	15	16	17	18	19	20
4	6	4	3	6	4	6	1	2	2	1	7	6	4	5	3	1	8	3	4

double-sided random dispersion interval of the Poisson distribution. Alternatively, for $\mu \geq 9$, the calculation may, as an approximation, be based on the normal distribution model (see Table 3.3–5).

Typical probabilities of nonintervention are 95 percent for warning limits and 99 percent for control limits. If $\pm 3\sigma$ (i.e., 99.73 percent) limits are used, these charts are known as u charts (see Section 3.3.7).

If the average number of nonconformities per unit is not known, it may be estimated using data from a trial run. For example: Let k be the number of wire coils inspected, and let x_i (i = 1, 2, . . ., k) represent the number of nonconformities observed in the number i wire coil:

$$\mu \Leftarrow \bar{x} = \frac{\sum^{k} x_i}{k} \qquad \begin{array}{l} k = \text{number of inspection unit} \\ x_i = \text{number of defects in unit no. i} \end{array}$$

$\bar{x}$ will be the center line of the corresponding chart. It is assumed that the same reporting unit is used as the basis for all samples.

Example: x Chart

The number of yarn breaks is recorded for successive similar runs of a yarn-twisting machine (see Table 3.3–6). The reporting unit is 20 runs in each case. Calculate control limits based on P = 99 percent.

Solution

The estimate for the mean μ is $\bar{x} = 4.0$. This gives the Poisson distribution shown in Figure 3.3–7. Hence, we have an upper control limit of 10.5 and no lower control limit. Figure 3.3–8 shows the corresponding control chart.

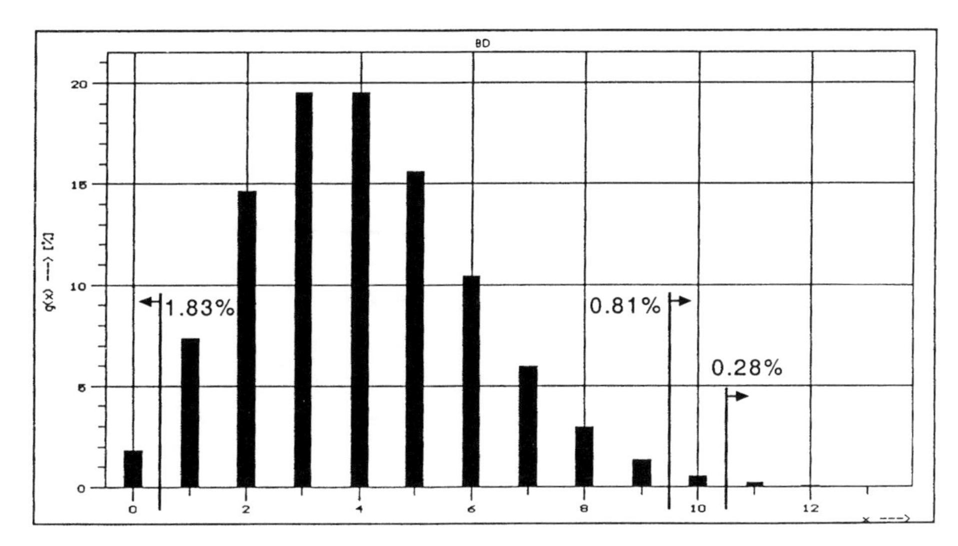

Figure 3.3–7. Poisson distribution with $\mu = 4.0$.

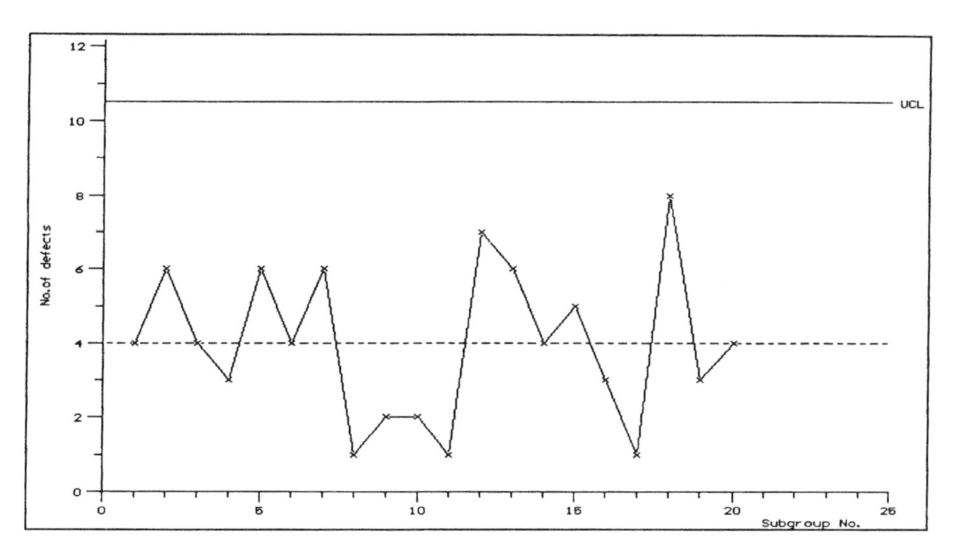

Figure 3.3–8. x chart for number of defects per unit.

3.3.6 c Chart for the Number of Nonconformities

The c chart shows the number of nonconformities in a sample (as opposed to the number of nonconforming items on an x or np chart). The c chart requires a constant sample size or a constant quantity of test material. It is used in two types of situations.

- If nonconformities are spread out over a more or less continuous flow of production (e.g., tears along a length of cloth, bubbles in glass, places of thin insulation along a wire, etc.)
- If nonconformities in the sample under investigation may arise from many different causes (e.g., records of a repair station, where each individual vehicle or part may have one or more nonconformities of very different types)

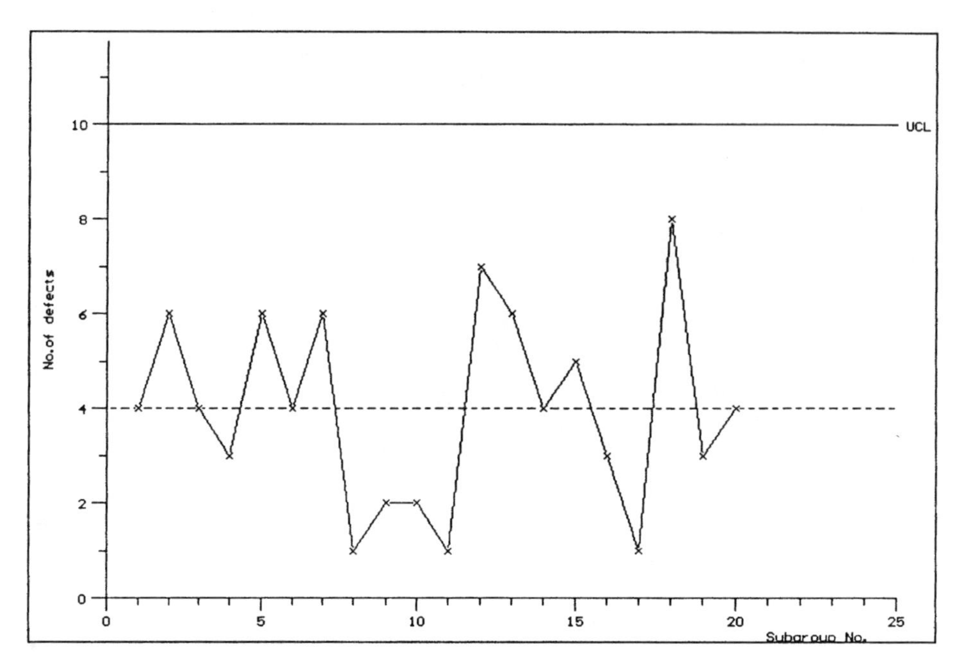

Figure 3.3–9. c chart.

Based on the normal distribution model, ±3σ control limits are calculated as shown in Table 3.3–5, i.e.,

$$UCL = \mu + 3\sqrt{\mu} \qquad \text{where} \quad \mu = \frac{\sum_{i=1}^{k} X_i}{k}$$
$$LCL = \mu - 3\sqrt{\mu}$$

Example

Based on the sample results shown in Table 3.3–6 with $\mu = 4.0$, we find the following control limits:

$$UCL = 4.0 + 3\sqrt{4} = 10$$

$$LCL = 4.0 - 3\sqrt{4} = -2$$

If the lower limit is negative, it is not shown on the chart. Figure 3.3–9 shows the resulting control chart.

3.3.7 u Chart for the Number of Nonconformities Per Unit

The u chart shows the number of nonconformities per unit in a sample, where the sample size (or the quantity of material investigated) should ideally remain constant. As for the p chart, the sample size should at least not be allowed to fluctuate by more than ±25 percent. The u chart is similar to the c chart, with the difference that the number of nonconformities is expressed on a per-unit basis. Both the u chart and the c chart are suitable for the same basic situation. Samples should, however, contain more than one unit. The typical application is the multiple characteristics control chart (see Section 3.4, which gives the required calculations and an example). The u chart is the only available choice if nonconformities are counted and the sample size may vary. Application of the u chart is otherwise similar to the procedure for p charts.

The control limits for the u chart with P = 99.73 percent (±3σ limits) are calculated on the basis of the normal distribution model, as follows:

$$UCL_u = \bar{x}(NC) + 3\sqrt{\frac{\bar{x}(NC)}{\bar{n}}}$$

$$LCL_u = \bar{x}(NC) - 3\sqrt{\frac{\bar{x}(NC)}{\bar{n}}}$$

$$\text{where } \bar{x}(NC) = \frac{x_1 + x_2 + \ldots + x_k}{n_1 + n_2 + \ldots + n_k}$$

$$\text{and } \bar{n} = \frac{1}{k}\sum_{i=1}^{k} n_i$$

3.4 Multiple Characteristics Control Charts

In many processes, there are several qualitative characteristics that must be monitored. Sometimes it is advantageous to collect the data and enter them on a single chart. This chart is then known as a multiple characteristics control chart.

Multiple characteristics charts are units (assemblies, parts, equipment, or processes) that may have a number of different nonconformities. Subgroups of size u = 1, 2, 3, … are taken in order to control the running production. The number of nonconformities of a given type must be observed and recorded in the multiple characteristics control charts. Subgroup size may vary with every subgroup. Subgroup frequency and size depend on the current product or production. This may be: fixed intervals, complete batches, per shift, or 100 percent test. Qualitative assessment (OK/n. OK) is used if quantitative assessment (measuring) is impossible or would be prohibitively expensive.

The result is a compressed but still clearly detailed overview of

- The quality history of a unit
- The mean number of nonconformities in products or materials supplied
- Process performance
- Supplier quality

3.4.1 Construction of a Multiple Characteristics Control Chart

The multiple characteristics chart (Figure 3.4–1) may be subdivided into 3 sections: A, B, and C.

Top Section A: Divided into three subsections:

Left: List of possible nonconformities, based on existing quality data (NC 1, NC 2, NC 3, . . .)

Center: Number of nonconformities of a given type observed in the sample

Right: Nonconformity statistics in terms of total number and percentage contribution.

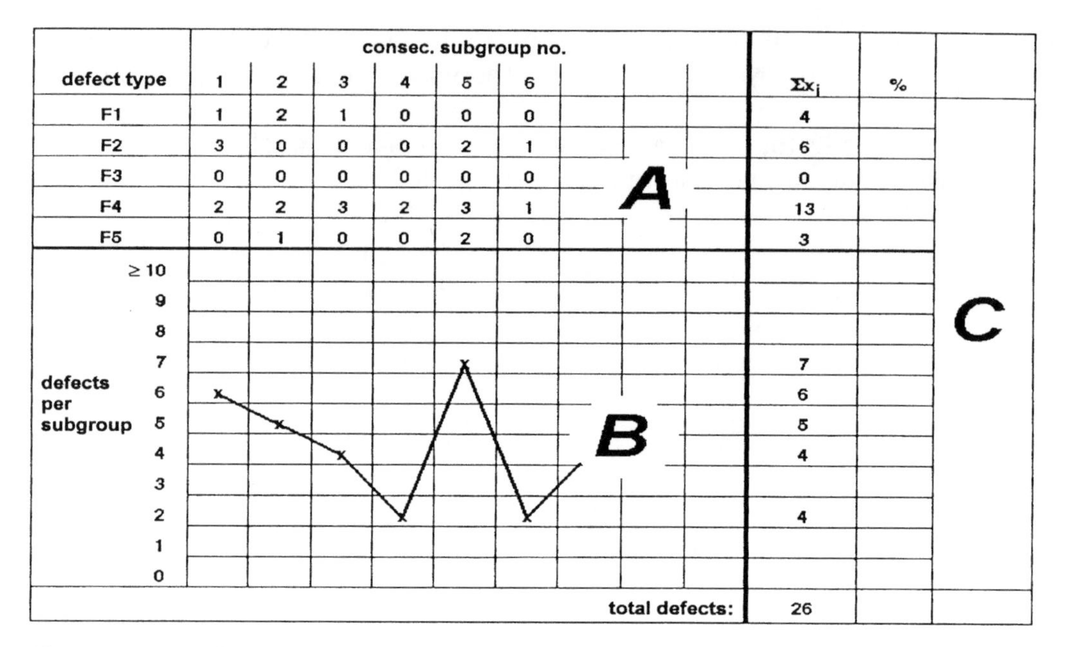

Figure 3.4–1. Standard layout of a multiple characteristics control chart.

Bottom Section B: Graphical representation of the number of nonconformities per sample

Right-Hand Section C: Bar chart for average number of nonconformities per one unit

Figure 3.4–2 shows a typical blank form for a multiple characteristics chart.

Using this method of collecting and representing data, it is possible to track an individual type of nonconformity on the control chart and also to monitor the sum total of all nonconformities encountered in the samples. By carrying out a Pareto analysis, the most frequently occurring types of nonconformity can then be identified and corrective action taken. u charts in particular (see Section 3.3.7) are frequently used for multiple characteristics charting.

3.4.2 Construction of a Multiple Characteristics u Chart

Collection and Graphical Representation of Data

Below the current sample number, we indicate

- In section A, the number of nonconformities of each type observed in the sample
- In section B, the sum total of all nonconformities in this sample by placing a cross (x) at the appropriate vertical position.

Evaluation of a Multiple Characteristics Chart

Following inspection of the samples, the sum total of nonconformities of each individual type is calculated and entered in the Σx_i column.

From this we calculate, for each type of nonconformity, $\bar{x}$(NC), the average number of nonconformities *per single unit*.

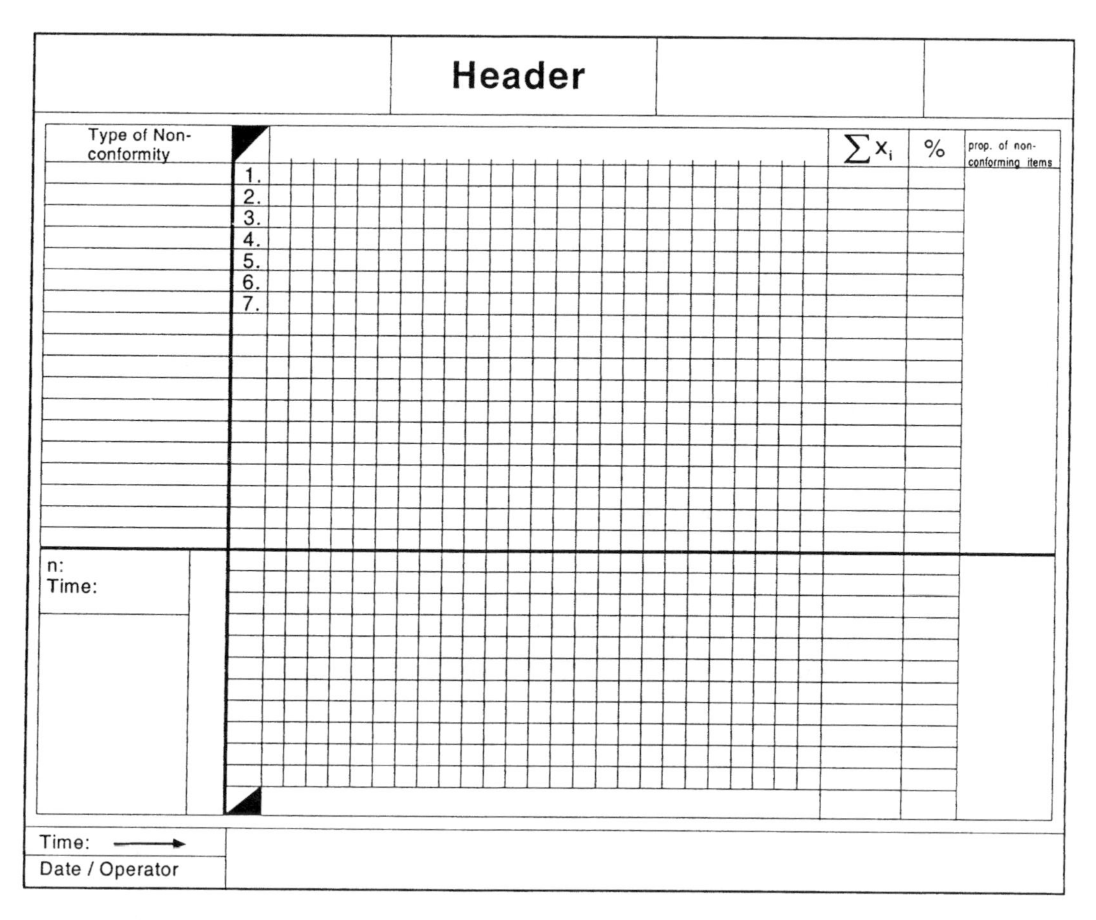

Figure 3.4–2. Blank form for a multiple characteristics chart.

$$\bar{x}(NC) = \frac{1}{kn}(x_1 + x_2 + \ldots + x_k) = \frac{1}{kn}\sum_{i=1}^{k} x_i \quad \text{where}$$

$\bar{x}(NC)$ = average number of nonconformities of type NC per unit

$\sum x_i$ = sum total of all nonconformities of type NC

k = number of samples

n = sample size

If the sample size is not constant, $\bar{x}(NC)$ is given by

$$\bar{x}(NC) = \frac{x_1 + x_2 + \ldots + x_k}{n_1 + n_2 + \ldots + n_k} = \frac{\sum_{i=1}^{k} x_i}{\sum_{i=1}^{k} n_i}$$

If the sample size is not constant, the control limits will vary from sample to sample. Figure 3.4–3 gives an example of the statistics used in evaluating a multiple characteristics chart with respect to a single type of nonconformity. The figures are based on the *incorrect angle* type of nonconformity in Table 3.4–1.

Evaluated	k(act)	:	40		
from subgroup no.	k(sta)	:	1 to subgroup no.	k(end) :	40
Model distribution		:	Poisson distribution		
Estimator of μ		:	0.20000		
Estimator of sigma		:	0.44721		
Total reference units		:	608.00		
Average val. of reference unit		:	15.20		
Minimum val. of reference unit		:	10.00		
Maximum val. of reference unit		:	24.00		
Total number of defects		:	8		
Average number fail. of 100 u.		:	1.3158		
Confidence interval μlo..μup		:	0.5681 . . 2.5926		
Min. number failures of 100 u.		:	0.0000		
Max. number failures of 100 u.		:	13.3333		

Figure 3.4–3. Evaluation of a nonconformity.

The control limits for the u chart may be calculated from the Poisson distribution, using the calculated estimate of μ and the double-sided random dispersion intervals. Alternatively, the control limits may also be based on the normal distribution model (see Section 3.3.7).

Example: Multiple Characteristics u Chart

Every 40 minutes, punched components were taken with differing sample sizes from continuous production and examined for seven types of nonconformity as given on a multiple characteristics chart. Table 3.4–1 shows the results of $k = 40$ samples. The values are to be entered into the multiple characteristics chart and analyzed.

Based on the data from Table 3.4–1, the frequency of the individual types of nonconformities may be represented in a Pareto diagram (see Figure 3.4–4). In addition, the u chart monitors the maximum acceptable number of nonconformities per sample. In all, 100 nonconformities were detected in 608 inspections. The control limits for a nonintervention probability of 99.73 percent (corresponding to $\pm 3\sigma$ limits) are given by (see Section 3.3.7).

$$UCL = \bar{x}(NC) + 3\sqrt{\frac{\bar{x}(NC)}{\bar{n}}} = 0.165 + 3\sqrt{\frac{0.165}{15.2}} = 0.474$$

$$LCL = \bar{x}(NC) + 3\sqrt{\frac{\bar{x}(NC)}{\bar{n}}} = 0.165 - 3\sqrt{\frac{0.165}{15.2}} = -0.147$$

$$\text{where } \bar{x}(NC) = \frac{\sum_{i=1}^{k} x_i}{\sum_{i=1}^{k} n_i} = \frac{100}{608} = 0.165$$

$$\text{where } \bar{n} = \frac{1}{k}\sum_{i=1}^{n} n_i = \frac{608}{40} = 15.2$$

Table 3.4–1. Sample results.

Nonconformity Type	1	2	3	4	5	6	7	8
pressure mark	1	0	0	3	0	0	1	0
twisted	0	0	0	0	0	1	0	0
torn	0	0	0	0	2	0	0	0
wrong angle	0	0	1	1	0	0	0	0
edge	0	0	0	0	0	3	0	0
corrosion	0	0	0	0	0	0	0	0
flaws	0	1	0	0	0	0	0	0
No. of parts	15	10	20	15	10	12	15	20
Tot. nonconformities	1	1	1	4	2	4	1	0
NC's/100 parts in %	6.7	10.0	5.0	26.7	20.0	33.3	6.7	0.0

Nonconformity Type	9	10	11	12	13	14	15	16	17	18	19
pressure mark	0	2	0	0	0	0	3	0	4	0	0
twisted	0	0	0	0	1	0	0	0	0	0	0
torn	0	1	0	0	0	2	0	0	0	0	1
wrong angle	0	0	0	0	0	0	0	0	0	0	0
edge	0	0	0	2	1	0	0	0	0	1	0
corrosion	0	0	0	0	0	0	0	0	1	0	0
flaws	1	0	0	0	0	0	0	3	4	0	2
No. of parts	15	15	10	15	12	20	15	12	15	10	15
Tot. nonconformities	1	3	0	2	2	2	3	3	9	1	3
NC's/100 parts in %	6.7	20.0	0.0	13.3	16.7	10.0	20.0	25.0	60.0	10.0	20.0

Nonconformity Type	20	21	22	23	24	25	26	27	28	29	30
pressure mark	0	2	0	0	1	3	4	0	0	0	0
twisted	0	0	0	0	0	0	0	0	0	0	1
torn	1	0	0	0	0	0	0	0	0	0	0
wrong angle	0	0	0	0	1	0	0	0	0	0	0
edge	0	0	2	0	0	0	3	0	0	0	1
corrosion	1	0	1	2	0	2	0	0	0	0	0
flaws	0	0	0	0	0	0	0	0	1	0	0
No. of parts	17	18	20	18	20	15	12	10	15	12	15
Tot. nonconformities	2	2	3	2	2	5	7	0	1	0	2
NC's/100 parts in %	11.8	11.1	15.0	11.1	10.0	33.3	58.3	0.0	6.7	0.0	13.3

Nonconformity Type	31	32	33	34	35	36	37	38	39	40	Σ	NC/100E
pressure mark	0	4	0	0	2	0	0	0	3	0	33	5.4276
twisted	2	0	0	0	0	1	0	0	0	0	6	0.9868
torn	0	2	0	1	1	0	0	0	1	0	12	1.9737
wrong angle	0	0	0	0	0	0	2	1	2	0	8	1.3158
edge	0	0	4	0	0	0	1	0	0	3	21	3.4539
corrosion	0	0	0	0	0	0	0	0	0	0	7	1.1513
flaws	0	0	0	0	0	1	0	0	0	0	13	2.1382
No. of parts	15	18	13	15	16	18	15	16	15	24	608	
Tot. nonconformities	2	6	4	1	3	2	3	1	6	3	100	
NC's/100 parts in %	13.3	33.3	30.8	6.7	18.8	11.1	20.0	6.3	40.0	12.5	16.4	

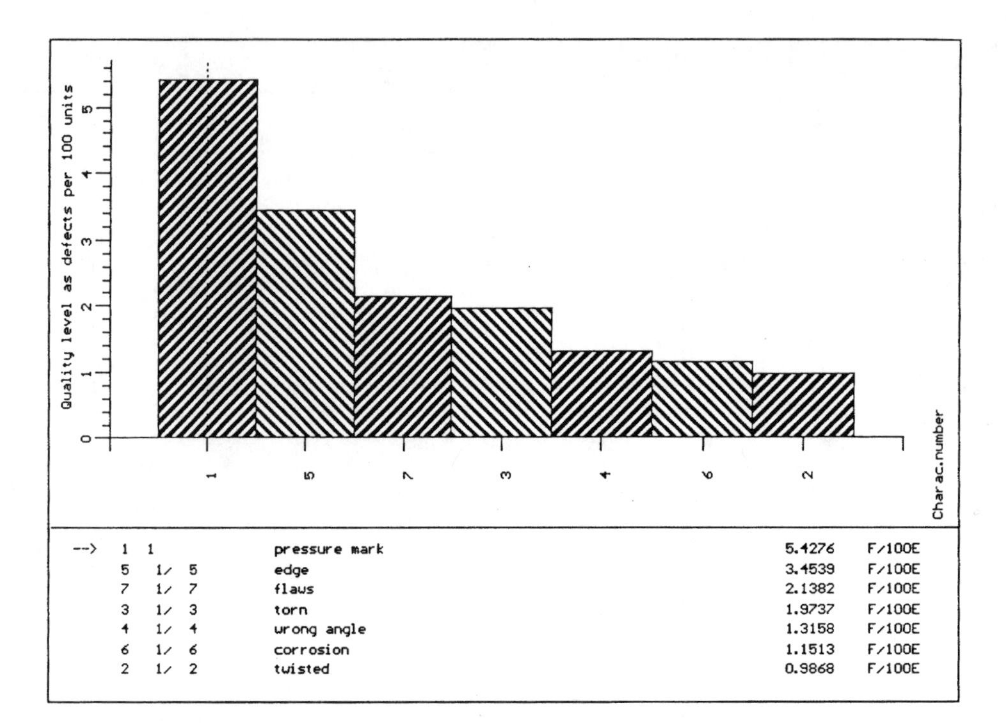

Figure 3.4–4. Pareto analysis of nonconformity types.

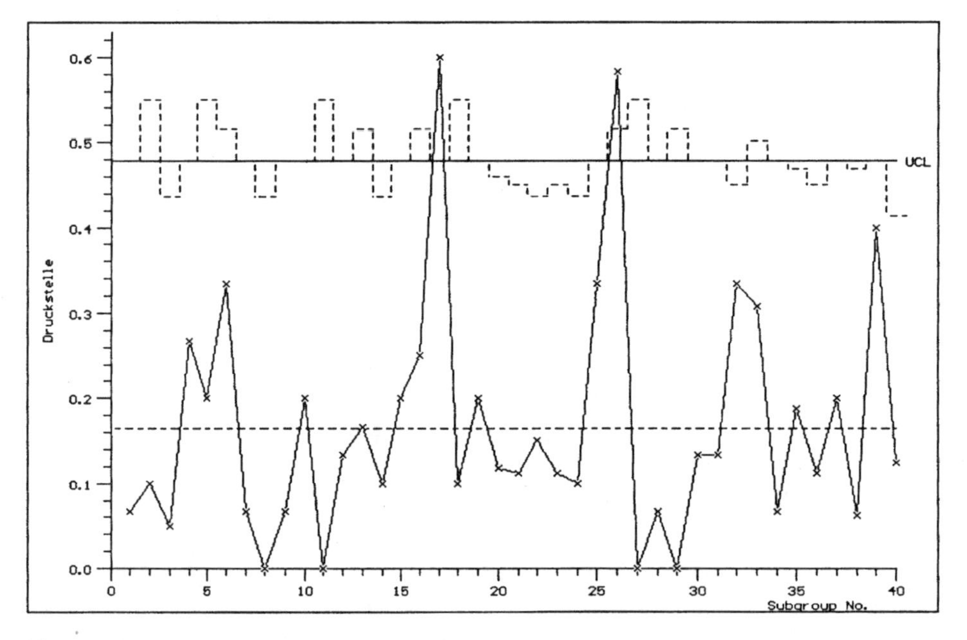

Figure 3.4–5. u chart.

Hence, there is no lower control limit. The upper control limit is adjusted (dotted lines in Figure 3.4–5) to account for differing sample sizes. The control chart shows that samples 17 and 26 violate the control limit.

Each type of nonconformity could, of course, also be monitored separately by means of a control chart as described in Section 3.3.

3.5 Control Charts for Continuous Characteristics

3.5.1 Construction of Control Charts

The control chart is at the heart of all SPC applications. The graphical representation of the measured values in itself enables a very good assessment of process behavior and, hence, of process performance. Additional analyses, based on the measurement values recorded on the control chart, enable a meaningful evaluation of the quality that the process delivers.

A fact of absolutely critical importance, however, is that in all control charting applications the control limits must be based on the actual performance of the process concerned (*not* the specification). A simple assessment as to whether the process is *in statistical control* (i.e., stable) is only possible using control charts with performance-based control limits. Being able to demonstrate process stability is an essential prerequisite for determining capability indices. The capability formulas can of course also be applied when stability has not yet been demonstrated, but the numerical results of such a calculation can in no way be regarded as capability indices. The purpose of capability indices is to enable a prediction of future process performance. If the data used in assessment of the process originate from a start-up or assessment period in which the process was unstable, then any conclusions drawn from these data may perhaps give an accurate assessment of what happened in the past, but, given the lack of process stability, they will not provide a reliable prediction of what is going to happen in the future.

A simple example may serve to illustrate the kinds of errors in judgment that may result from an improper procedure. A survey to establish the population's interest in classical music is carried out at the entrances to opera houses and concert rooms. In this case, too, it is quite possible to give an answer, *based on actual data,* to the question at hand. But without questioning a representative sample of the whole population, the answer will be unreliable and, in fact, misleading.

So, let us establish this point once and for all: The calculation of process capability indices may only be carried out if the data clearly demonstrate that the process is stable.

In monitoring continuous production, the type of control chart used should be that which is most suitable for the specific process conditions. The decision as to the best-suited control chart depends on the specific application; in all cases, long-term experience should be taken into account. For the start of an investigation, it is suggested that the control chart used should show process location and process dispersion separately. Such a location/dispersion chart offers two key advantages: The distribution of sample means will conform well to the normal distribution model; in accordance with the central limit theorem, this usually holds true for all sample sizes $n \geq 5$, even if the individual values themselves are not normally distributed. In addition, the chart provides information about both the location and dispersion of the process to be monitored.

For these reasons, it is always advisable at the start not to get lost in theoretical considerations, but quite simply to launch SPC by introducing the easily understandable and easy-to-use tool of control charting.

As a first step, an $\bar{x}/R$, $\tilde{x}/R$, or $\bar{x}/s$ control chart may be used. If there are only few values available, an individuals chart is advised. The graphical representation on the control chart gives a clear, visual indication of process behavior. In many cases, it is possible, without any further effort, to make use of the information on the chart to stabilize and improve process performance.

Since in most cases the data used are now measured by automatic means, it certainly makes sense to establish a direct linkage to computer-aided control charting. In these cases, the system used should make it possible to store all the data and make them available for any further analysis that might become necessary later on.

For the foregoing reasons, the following discussions center on Shewhart's $\bar{x}/s$ control chart. If, however, other location/dispersion charts are used, then the same procedure will still apply, except that the control chart formulas will need to be adapted accordingly.

From a purely economic viewpoint, it makes little sense to carry out such calculations manually based on mathematical tables and textbooks. Today, there are many computer programs available that are able to perform such calculations with great speed, precision, and efficiency. However, before introduction of such computer systems, proof should be obtained that all calculations are performed correctly.

As already stated, it is important that only performance-based (rather than specification-based) control charts are used for SPC.

3.5.2 Procedure for Using an $\bar{x}/s$ Chart

The procedure described here is actually an example explaining the principle. The subject is generalized in the next chapter. Data from a trial run are plotted in the sections of the control chart provided for them. Figure 3.5–1 shows the $\bar{x}$ chart and Figure 3.5–2 the s chart.

Based on the individual measurement values, the sample statistics (mean and standard deviation) to be plotted in the appropriate sections of the control chart are calculated for the individual samples in accordance with the formulas that follow.

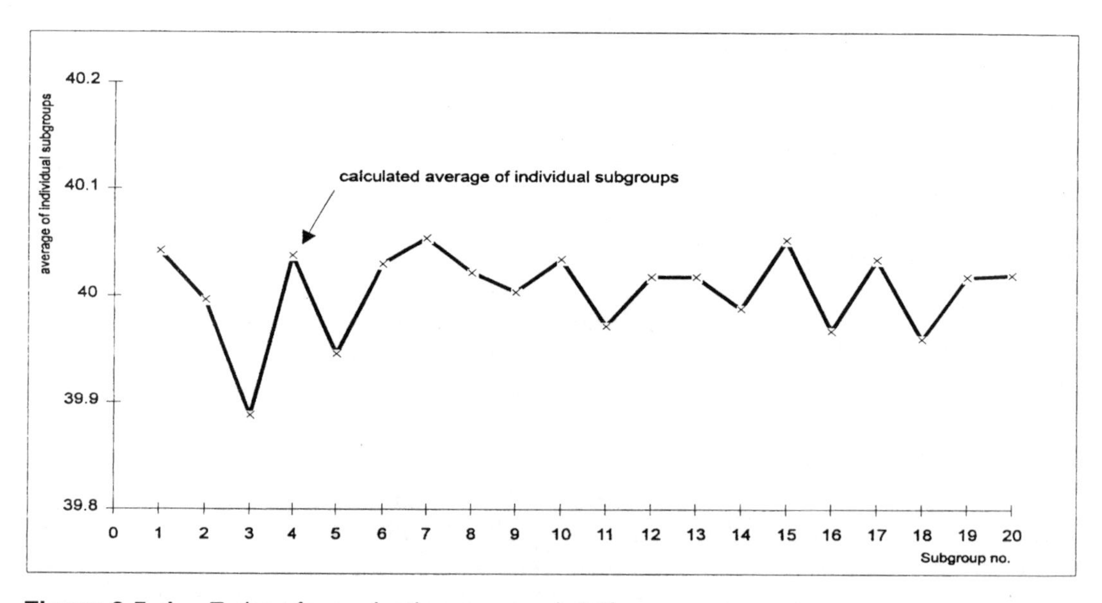

Figure 3.5–1. $\bar{x}$ chart for evaluating process location.

The mean $\bar{x}_k$ of sample no. k is calculated from all the measurement values in the sample of size n:

$$\bar{x}_k = \frac{x_1 + x_2 + x_3 + \cdots x_n}{n} = \frac{1}{n}\sum_{i=1}^{n} x_i \quad \text{where } x_i = \text{no. i measurement value}$$

Similarly, the standard deviation s_k is given by

$$s_k = +\sqrt{\frac{\sum_{i=1}^{n}(x_i - \bar{x}_k)^2}{n-1}}$$

From all k samples it is possible to obtain the mean $\bar{\bar{x}}$ of all sample means. This is given by

$$\bar{\bar{x}} = \frac{1}{k}\sum_{j=1}^{k}\bar{x}_j \quad \text{where } \bar{x}_j = \text{mean of sample no. j}$$

In a similar manner, the average $\bar{s}$ of the k sample standard deviations can serve as an estimator of dispersion.

$$\bar{s} = \frac{1}{k}\sum_{j=1}^{k} s_j \quad s_j = \text{standard deviation of sample no. j}$$

When about 20 samples of size n, with $3 \leq n \leq 5$, have been entered in each plot, the performance-based control limits for the chart are calculated from the formulas that follow. If the calculation is based on a smaller amount of data, the results will be less reliable and should be treated with great caution (see Chapter 2, Section 2.4, on confidence intervals).

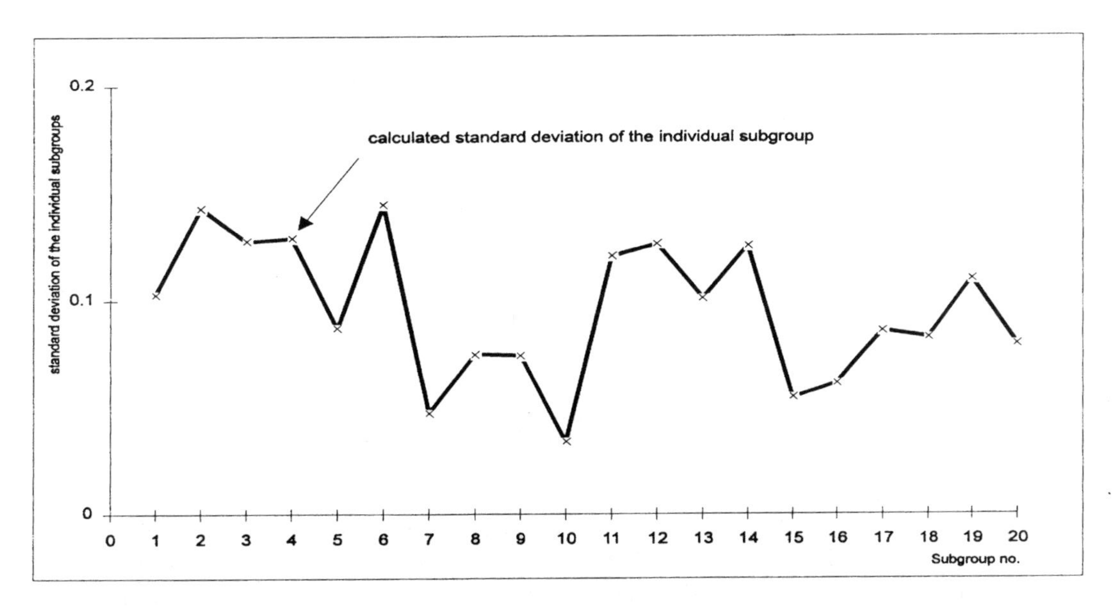

Figure 3.5–2. s chart for evaluating process spread.

Control Limits for the Average Chart

Upper control limit $UCL_{\bar{x}} = \bar{\bar{x}} + A_3 \times \bar{s}$

Lower control limit $LCL_{\bar{x}} = \bar{\bar{x}} - A_3 \times \bar{s}$

Control Limits for the Standard Deviation Chart

Upper control limit $UCL_s = B_4 \times \bar{s}$
Lower control limit $LCL_s = B_3 \times \bar{s}$ (only applies for sample sizes $n \geq 6$)*

Here $\bar{\bar{x}}$ represents the average of all the sample means and $\bar{s}$ the average of all the sample standard deviations plotted on the chart. The constants A_3, B_3, and B_4 are numerical factors [4] dependent on sample size, assuming a nonintervention probability of 99.73 percent (corresponding to $\pm 3\sigma$ limits). These factors are also given in Table 10.7.

Figure 3.5–3 and Figure 3.5–4 show an $\bar{x}/s$ control chart with control limits based on the aforementioned formulas. This procedure is based solely on estimates of process location and dispersion and the concept of 99.73 percent control limits. A general outline follows, using the example of the $\bar{x}$ location chart.

The general formula for determining the control limits on a Shewhart control chart for sample means (Figure 3.5–3) is as follows:

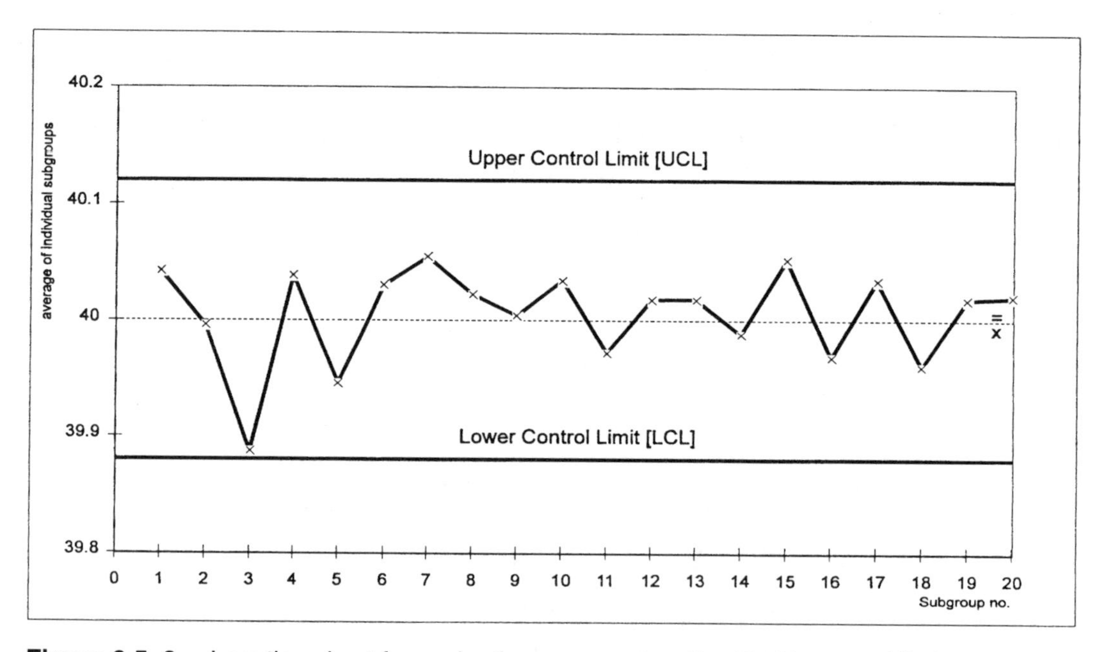

Figure 3.5–3. Location chart for evaluating process location $\bar{x}$ with control limits.

**The control limits on the s chart would theoretically have to be calculated on the basis of the χ^2 distribution. For large sample sizes, it is possible to use an approximation based on the normal distribution. The factors B_3 and B_4 are determined on this basis. The approximation is not valid for $n \leq 5$.*

$$UCL = \mu + u_{1-\alpha/2} \frac{\hat{\sigma}}{\sqrt{n}}$$

α = error probability; typical values for control limits are 1% or 0.27%.

$$LCL = \mu - u_{1-\alpha/2} \frac{\hat{\sigma}}{\sqrt{n}}$$

Estimator of process location μ:

$$\mu \Leftarrow \bar{\bar{x}} = \frac{1}{k}\sum_{i=1}^{k} \bar{x}_i \qquad \text{(total average)}$$

$$\mu \Leftarrow \bar{\tilde{x}} = \frac{1}{k}\sum_{i=1}^{k} \tilde{x}_i \qquad \text{(average of the median values)}$$

Estimator of process dispersion $\hat{\sigma}$:

$$\hat{\sigma} \Leftarrow \sqrt{\overline{s^2}} = \sqrt{\frac{1}{k}\sum_{i=1}^{k} s_i^2}$$

$$\hat{\sigma} \Leftarrow \frac{\bar{s}}{a_n} = \frac{1}{a_n} \frac{\sum_{i=1}^{k} s_i}{k}$$

$$\hat{\sigma} \Leftarrow \frac{\bar{R}}{d_n} = \frac{1}{d_n} \frac{\sum_{i=1}^{k} R_i}{k}$$

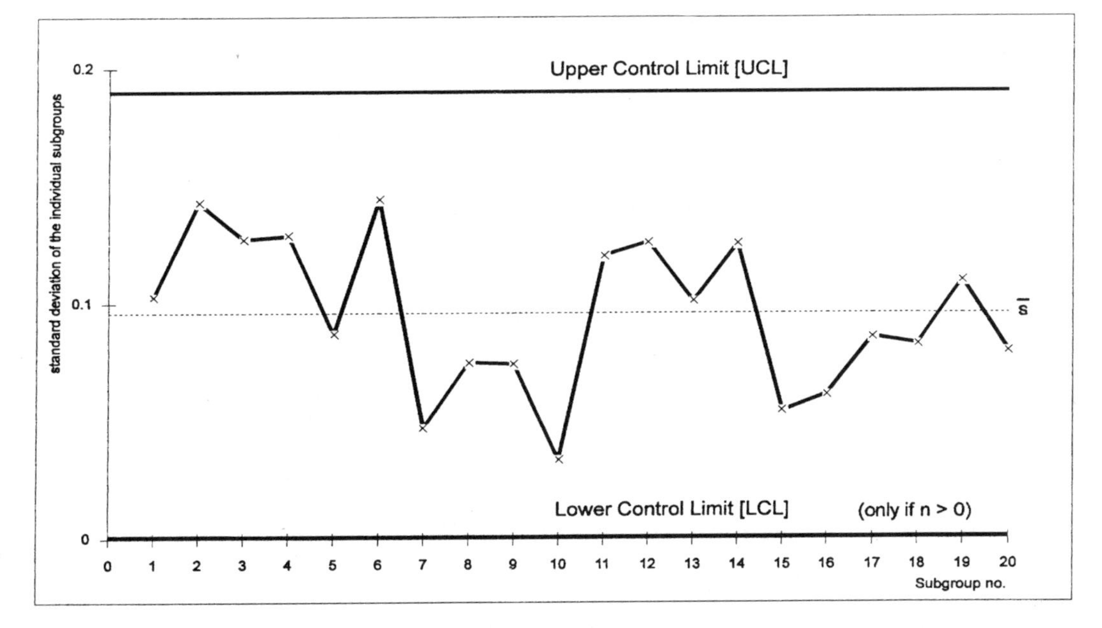

Figure 3.5–4. Spread chart for evaluating process spread (s) with control limits.

$$\hat{\sigma} \Leftarrow s_{ges} = +\sqrt{\frac{1}{nk-1}\sum_{i=1}^{nk}(x_i - \bar{x})^2}$$

n	=	sample size
k	=	number of samples
a_n	=	0.940 for $n = 5$ (s. table in Chapter 10)
d_n	=	2.326 for $n = 5$ (s. table in Chapter 10)
$u_{1-\alpha/2}$	=	3.0 for $\alpha = 0.27\% \mathrel{\hat{=}} 99.73\%$ nonintervention probability
$u_{1-\alpha/2}$	=	2.578 for $\alpha = 1\% \mathrel{\hat{=}} 99.00\%$ nonintervention probability
$u_{1-\alpha/2}$	=	1.96 for $\alpha = 5\% \mathrel{\hat{=}} 95.00\%$ nonintervention probability

The following conditions must be satisfied for correct determination of the parameters:

- The process must have been stable (in statistical control) during the period of observation.
- The number of individual measurements collected should be $n\,k \geq 100$ (n = sample size, k = number of samples).

Figure 3.5–5 gives another illustration of the construction of the control chart and the various estimators that can be used to establish the control limits.

If $\hat{\sigma}$ is estimated from $\sqrt{\overline{s^2}}$, $\bar{s}/a_n$, or $\bar{R}/d_n$, and the control limits are calculated on this basis, then the position of the control limits only reflects short-term variation (variation within the samples). Hence, the limits are very close together, and there is little play for any fluctuations and trends inherent in the process design. Processes that are subject to such unavoidable fluctuations and trends are discussed in Section 3.9. Section 3.5.4 deals with the remaining Shewhart control charts for x, $\tilde{x}$, R, and s.

Example: $\bar{x}$ Chart

In a process for coiling springs, it is desired to monitor the strength of the wires after coiling. A preliminary run gave the breaking loads shown in Table 3.5–1 (in daN).

If we take $\mu = x$ as an estimate of process location and $\hat{\sigma} = \bar{s}/a_n$, we arrive, based on $P = 99.73$ percent, at the following control limits for the $\bar{x}$ chart (cf. Table 3.5–6)*:

$$\begin{matrix} UCL_{\bar{x}} \\ LCL_{\bar{x}} \end{matrix} = \bar{\bar{x}} \pm 3\frac{\hat{\sigma}}{\sqrt{n}} = 133.1 \pm 3\frac{3.530}{\sqrt{5}} = \begin{matrix} 137.84 \\ 128.36 \end{matrix}$$

In comparison, an $\bar{x}$ chart with $P = 99$ percent has the following control limits*:

$$\begin{matrix} UCL_{\bar{x}} \\ LCL_{\bar{x}} \end{matrix} = \bar{\bar{x}} \pm 2.578\frac{\hat{\sigma}}{\sqrt{n}} = 133.1 \pm 2.578\frac{3.530}{\sqrt{5}} = \begin{matrix} 137.17 \\ 129.03 \end{matrix}$$

Figure 3.5–6 shows the two charts with $P = 99.73$ percent and $P = 99$ percent.

For reasons of simplicity, our example calculations are based on only ten samples.

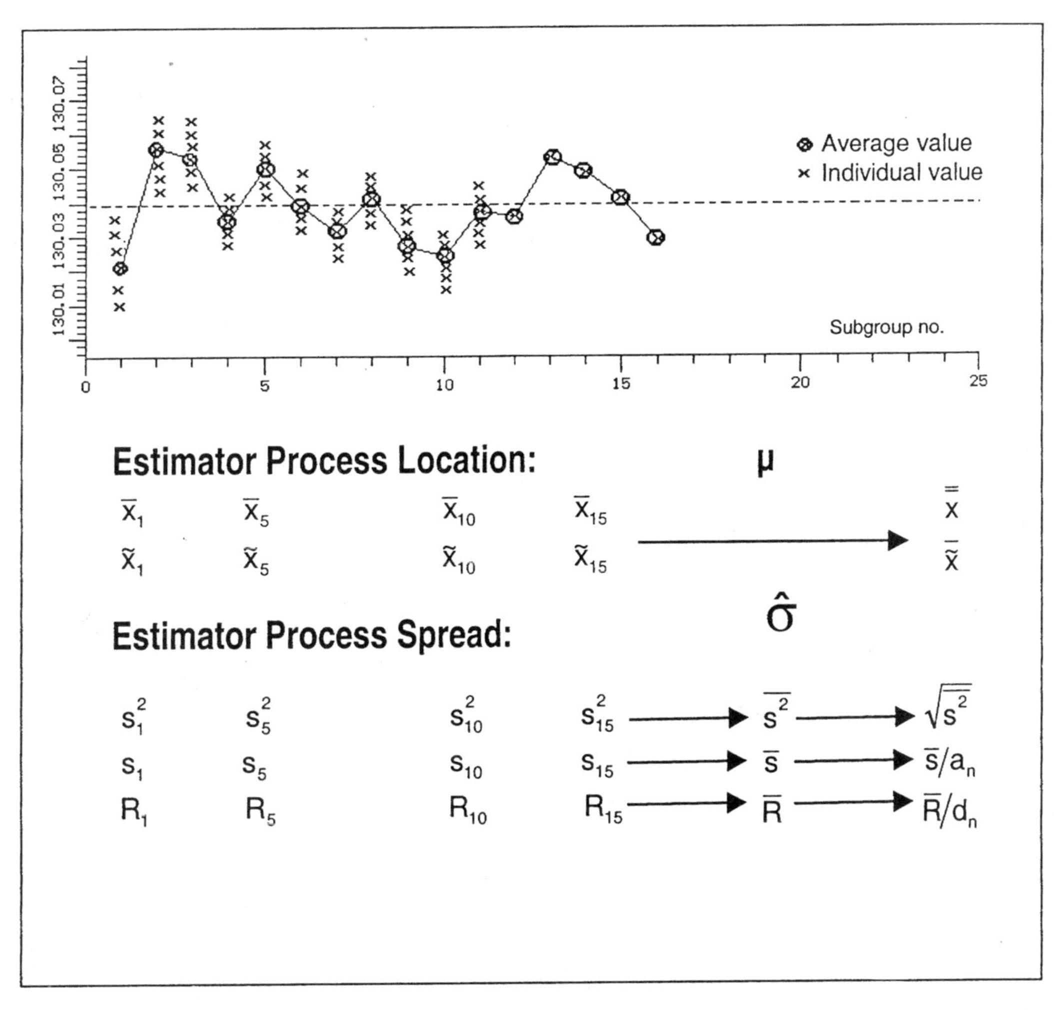

Figure 3.5–5. Different calculation methods for determining the estimator parameters of location and spread.

Table 3.5–1. Tensile loads in daN with subgroup statistics.

Day:	1	2	3	4	5	6	7	8	9	10	Estimator Process Location and Spread
1	135	138	136	134	131	134	137	131	131	133	
2	135	126	135	133	128	131	139	140	130	135	
3	130	129	135	135	130	131	131	132	132	138	
4	132	131	133	127	131	137	133	131	137	133	
5	126	138	138	126	131	135	140	128	138	135	
$\bar{x}_i$	131.6	132.4	135.4	131.0	130.2	133.6	136.0	132.4	133.6	134.8	$\Rightarrow \bar{\bar{x}} = 133.10 \hat{=} \mu$
$\tilde{x}_i$	132	131	135	133	131	134	137	131	132	135	$\Rightarrow \bar{\tilde{x}} = 133.10 \hat{=} \mu$
s_i	3.78	5.41	1.82	4.18	1.30	2.61	3.87	4.51	3.65	2.05	$\Rightarrow \bar{s}/a_n = 3.530 \hat{=} \hat{\sigma}$
s_i^2	14.30	29.27	3.31	17.50	1.70	6.77	15.00	20.30	13.30	4.20	$\Rightarrow \sqrt{\overline{s_2}} = 3.545 \hat{=} \hat{\sigma}$
R_i	9	12	5	9	3	6	9	12	8	5	$\Rightarrow \bar{R}/d_n = 3.535 \hat{=} \hat{\sigma}$
											$\Rightarrow s_{tot} = 3.694 \hat{=} \hat{\sigma}$

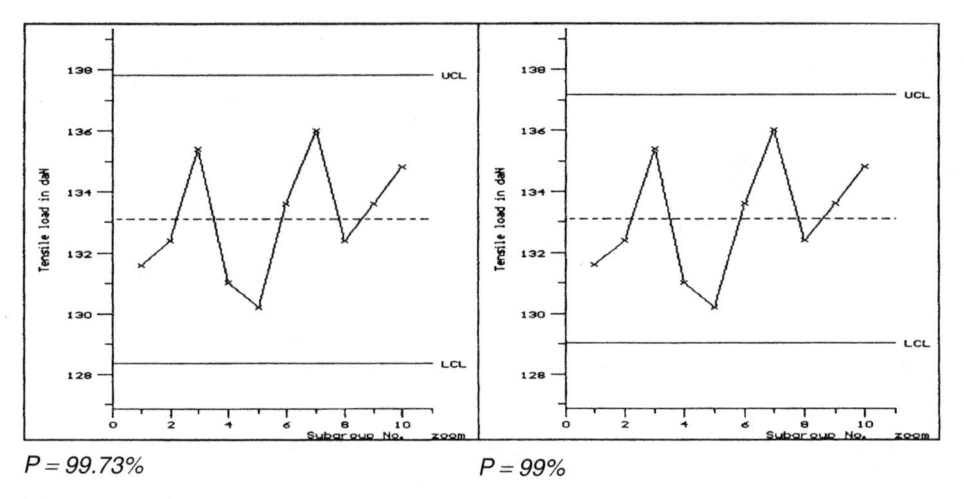

P = 99.73% *P* = 99%

Figure 3.5–6. $\bar{x}$ chart with different noninterference limits.

3.5.3 Stability Criteria

Based on the control limits calculated, it is now possible to apply statistical criteria to determine whether the process is stable. On a Shewhart control chart with $\bar{x}$ control limits based on P = 99.73 percent (i.e., $\pm 3\sigma$ of the variation of the sample means), it is possible to distinguish two types of process variation:

- Common cause variation
- Special cause variation

If in process assessment based on a control chart it is found that all stability criteria have been met, this may be regarded as evidence that the process is affected only by common causes of variation. This analysis then also indicates the magnitude of the common cause variation. The facts provided by the chart, i.e., the evidence of stability and the magnitude of the process variation, then allow a simple and reliable estimate of process performance.

If there are special (assignable) causes affecting the process, these lead to special cause variation. Special cause variation simply means that one or more of the stability criteria have been violated.

The following are evidence of instability (process out of control):

- Point(s) above the upper control limit
- Point(s) below the lower control limit
- A run of seven points above the center line
- A run of seven points below the center line
- Seven successive connecting lines between points showing an uninterrupted upward trend
- Seven successive connecting lines between points showing an uninterrupted downward trend
- More than 90 percent of the points within the middle third of the area between the control limits*
- Forty percent or less of the points within the middle third of the area between the control limits*

**These percentages apply to 25 successive plot points on the control chart.*

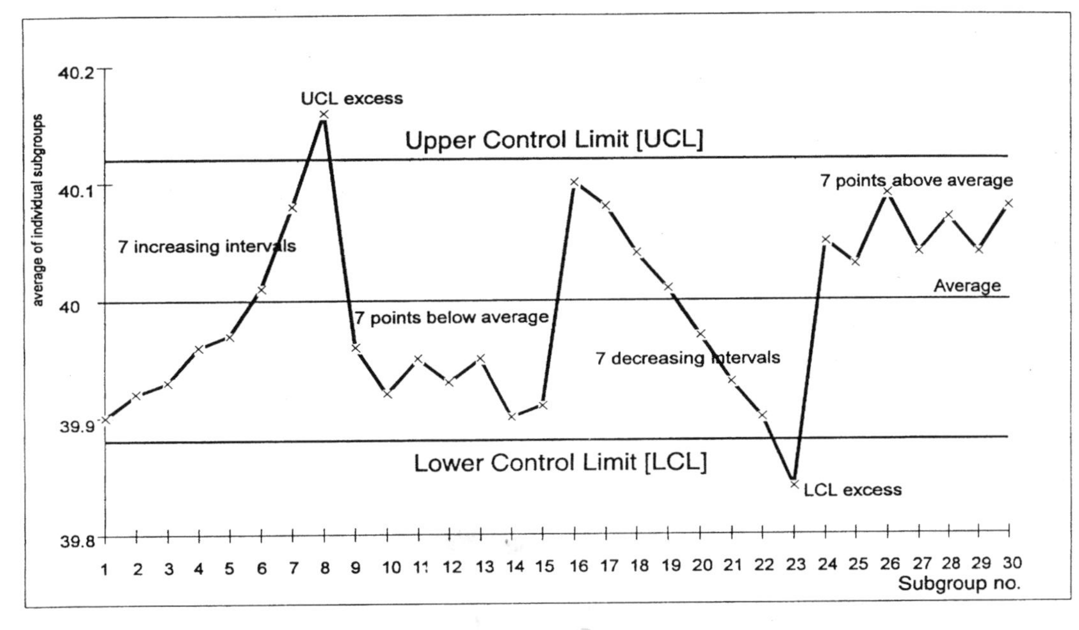

Figure 3.5–7. Shewhart chart with stability violation.

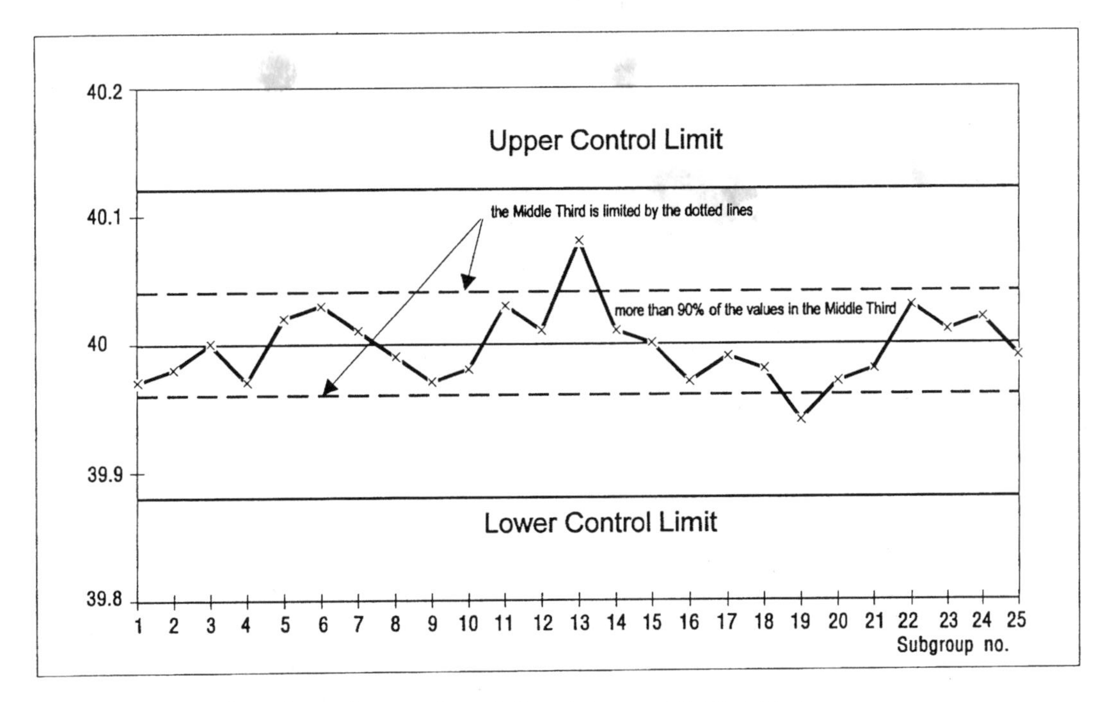

Figure 3.5–8. Shewhart chart with stability violation *Middle Third.*

These stability criteria are valid without restriction in Shewhart control chart applications for monitoring process location ($\overline{x}$). In monitoring dispersion s, these rules are valid only for sample sizes $n \geq 25$. (It is only for samples of 25 or more items that the χ^2 distribution may be regarded as symmetrical.)

Figure 3.5–7 and Figure 3.5–8 illustrate the various out-of-control indicators, using a Shewhart average chart as an example.

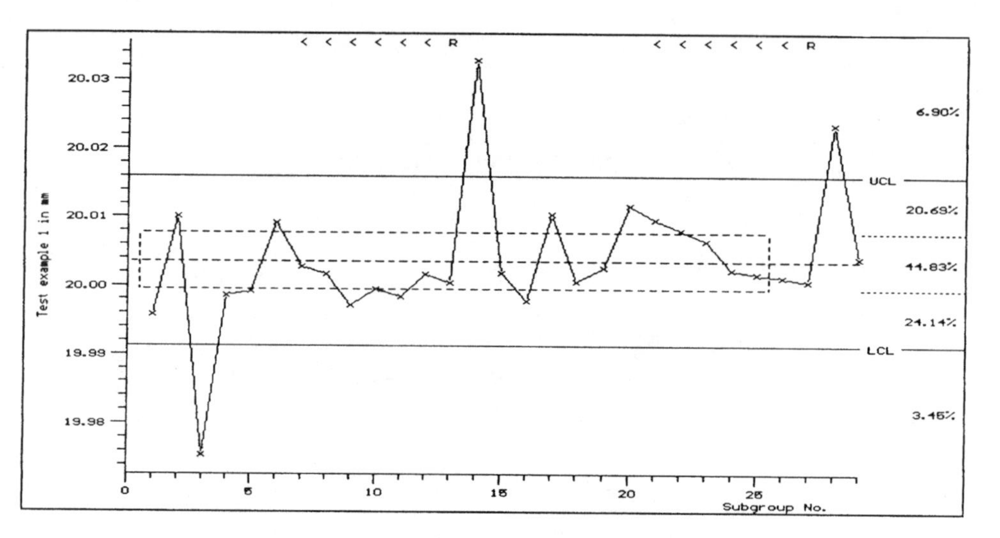

Figure 3.5–9. $\overline{x}$ chart with run between subgroup 7 and 13.

Table 3.5–2. Probabilities for run display in location charts.

No. of Values above (or below) the average k	Type 1 Errors α in % 0.5^k	Probability for no Run display 1 – α in %
1	50 %	50 %
2	25 %	75 %
3	12.5 %	87.5 %
4	6.25 %	93.75 %
5	3.125 %	96.875 %
6	1.5625 %	98.4375 %
7	0.78125 %	99.21875 %
8	0.390625 %	99.609375 %

The case where 40 percent or less of the points lie in the middle third is not illustrated in the percent figure.

What Is a Run?

The term *run* on a control chart refers to a series of seven successive plot points (Figure 3.5–9) lying either above or below the center line. This is valid both for an $\overline{x}$ chart and an s chart. What is the reason for choosing the number *seven?*

Runs on Location Charts

If we have a stable, normally distributed process, then we can say for average charts and median charts that we should expect 50 percent of the plot points to lie above the center line, and 50 percent below. For a stable process, the probability of a *false alarm* should be less than or equal to 1 percent (type 1 error). From this requirement, we can determine the number of values that will signal a run (Table 3.5–2).

Table 3.5–3. Probabilities for run display.

No. of Values above resp. below the center line	Region below the center line		Region above the center line	
	Probability for no Run display	Type 1 Errors	Probability for no Run display	Type 1 Errors
k	1 – α in %	α in %	1 – α in %	α in %
1	47.2665	52.7335	47.2665	52.2665
⋮	⋮	⋮	⋮	⋮
5	95.9221	4.0779	97.6408	2.3592
6	97.8496	2.1504	98.8849	1.1151
7	98.8660	1.1340	99.4729	0.5271
8	99.4020	0.5980	99.7509	0.2401

As is seen from the values in the table, the smallest run length that fulfills the requirement $\alpha \leq 1\%$ is seven values. Hence, we say that a run has occurred whenever seven successive plot points lie below or above the center line.

Runs on Dispersion Charts

In contrast to sample statistics measuring location, sample statistics measuring dispersion are not symmetrically distributed even if the population of individual measurement values is normally distributed. They have a skewed distribution, with the degree of skewness dependent on sample size: the greater the sample size, the smaller the departures from symmetry.

If we have an s chart with a sample size of n = 5, then 52.7335 percent of the points will lie below the center line $\bar{s}$, and only 47.2665 percent will lie above.

$$\chi^2_{f;G} = f(s a_n)^2$$

where $f = 4 (n = 5)$ and $\sigma = 1$

is $\chi^2_{f;G} = 4 \times 0.94^2 = 3.5344$

and $G = 52.7335\%$

Hence, the probabilities for runs are as shown in Table 3.5–3.

From these values, it is apparent that for this s chart, with n = 5, a run below the center line should be indicated only after eight values, and a run above the center line after seven values. But, for reasons of convenience, this distribution is regarded in practice as *sufficiently symmetrical*, and a run is defined in both cases as seven or more values. Differentiation between run rules above and below the center line is usually only deemed necessary for dispersion charts with n < 5.

Trend

A trend occurs when a control chart shows a development in which the plotted values are consistently increasing or decreasing (see Figure 3.5–9, samples 20 to 27).

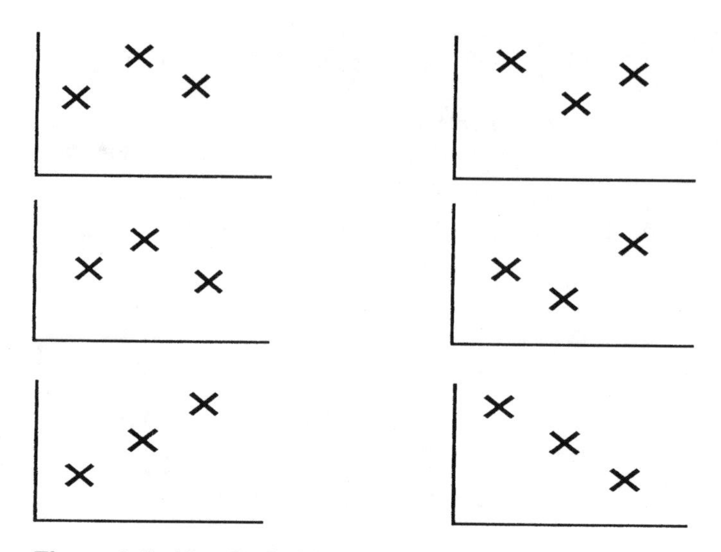

Figure 3.5–10. Probabilities for trend.

Table 3.5–4. Probability for trend.

k	Probability for single-sided trend			
2	1/2	=	50	%
3	1/6	=	16.667	%
4	1/24	=	4.167	%
5	1/120	=	0.833	%
6	1/720	=	0.1388	%
7	1/5040	=	0.01984	%
8	1/40320	=	0.00248	%

If we start from a random time sequence of values, then for k values there are k! (k factorial) possible sequences. One of these sequences will form a consistently increasing series, and one will form a consistently decreasing series. (The possibility of two values being identical is not taken into consideration.)

For k = 3 samples we have, for example, 3! = 6 possibilities, which are set out in Figure 3.5–10.

For three values, the probability of an increasing or decreasing series is therefore $1/6 = 0.1\overline{6}$. Table 3.5–4 shows the probabilities from k = 2 to k = 8.

If this probability is required to be < 1 percent, then five successive values behaving in this manner would constitute a *significant* trend. But, in general, this rule is only used to signal an *extreme* trend. Hence, the criterion is frequently applied that a trend spanning eight values, i.e., seven connecting lines all going up or all going down, must be observed.

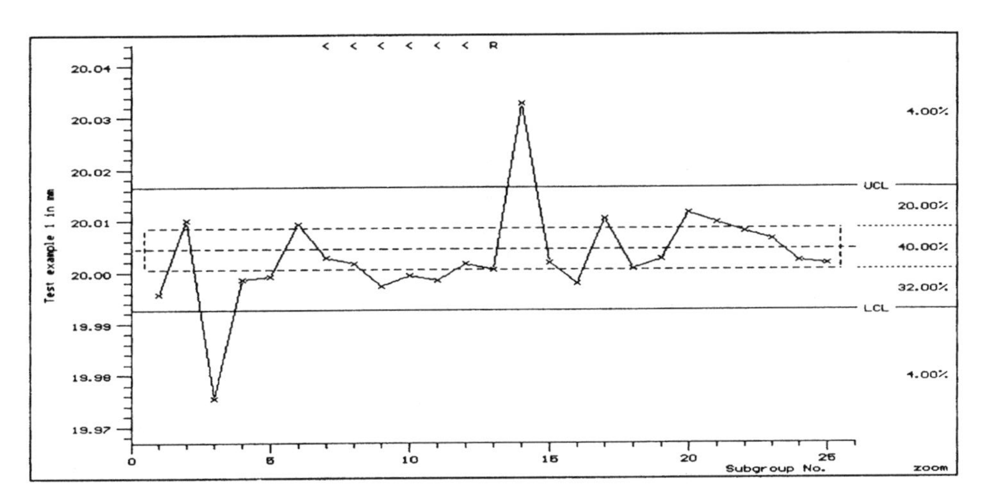

Figure 3.5–11. $\bar{x}$ with the violation *Middle Third* ≤ 40%.

Middle Third

Two further criteria for assessment of process behavior are based on the middle third of the area between the control limits (Figure 3.5–11). For 25 samples, more than 40 percent of the values, but no more than 90 percent of the values, should lie within this interval.

The application of these criteria in general assumes that

- The values are drawn from a normally distributed population whose mean is not subject to fluctuations.
- The control limits correspond to the $\pm 3\sigma$ limits of the sample distribution of the plotted statistic.
- Each time the same number of samples (in general, $k = 25$) are taken for assessment.

This is illustrated for an average chart in Figure 3.5–12. For an undisturbed process, 68.27 percent of the sample averages are expected to lie within the middle third of the area between the control limits ($\mu \pm 1\sigma_{\bar{x}}$).

For a control chart with 25 samples, we would expect to find 17 sample averages within the central interval. The binomial distribution enables us to assess what number of samples may be considered a *significant occurrence.*

For $k = 25$ averages and $p = 68.27$ percent, the single-sided upper limit x_{up} of the random dispersion interval with $1 - \alpha = 99$ percent is found to be 22. If not more than 22 sample averages lie within this interval, the results may be assumed to be random; from 23 average values, $23/25 \times 100$ percent = 92 percent, there is a middle third violation.

The lower limit x_{lo} of the random dispersion interval is given for $1 - \alpha = 99$ percent as 11, i.e., for 10 and fewer average values ($10/25 \times 100$ percent = 40 percent), there is a middle third violation.

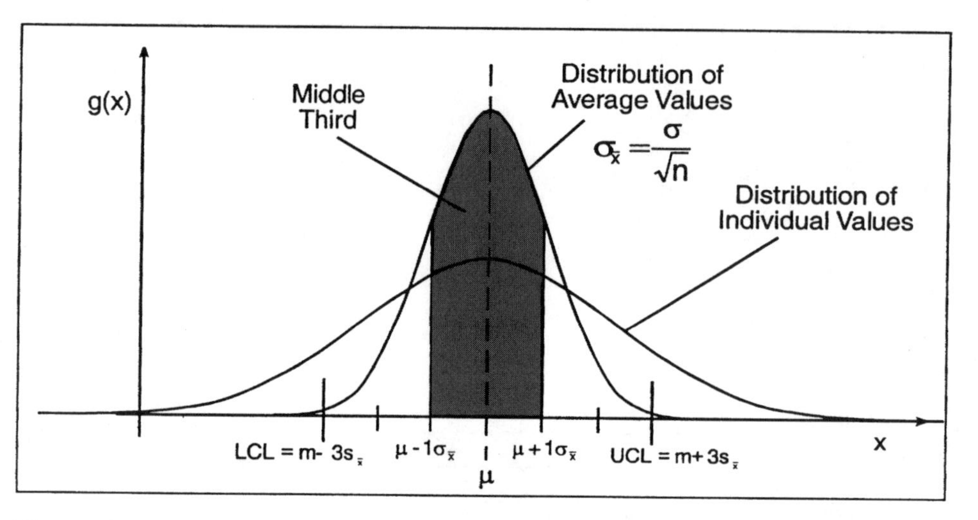

Figure 3.5–12. Distribution of individual and average values.

For asymmetrical distributions (e.g., dispersion charts) or charts with control limits determined on a different basis, x_{up} and x_{lo} would of course have to be adjusted. The same applies for numbers of samples other than $k = 25$.

Stability Criteria for Shewhart Charts with Extended Limits

In many processes, the process mean is subject to unavoidable fluctuations (process models B and C). In these cases, Shewhart charts with extended limits (see Section 3.9) are used. If the process monitored requires the use of these charts, then the application of the *run, trend,* and *middle third* rules is no longer appropriate.

Significance of Out-of-Control Indicators in a Large Number of Samples

If stability on the control chart is a prerequisite for assessment of capability and for calculation of capability indices, then the following problem arises whenever a large number of samples have been taken:

> *If only one or only a few of the sample results are outside the control limits, then, by definition, the process is not stable and the capability indices* must not *be calculated. On the other hand, Table 3.5–5 shows that the probability of all points being within the control limits actually becomes fairly small for larger numbers of samples, even if the process is perfectly stable and normally distributed. If the other stability criteria—run, trend, and middle third—are also taken into account, the probability that a truly stable process will not generate* any *out-of-control signals falls to still lower levels.*
>
> *This begs the question of to what extent occasional violations of the control limits can be regarded as* normal *(due to common causes) and acceptable.*

This problem may be solved in the following manner:

On a Shewhart location chart with $\pm 3\sigma$ limits, 0.27 percent of values may be expected to lie outside the control limits if the process is stable. The question as to

Table 3.5–5. Probabilities for control limit violations.

Number of Subgroups	Probability of violation of Control Limits	Probability of violation of Control Limits	Probability of violation of Control Limits
	Location Chart	Spread Chart	Total
1	99.73	99.6105	99.3416
10	97.33	93.6072	91.1103
25	93.4642	90.7043	84.7761
50	87.3556	82.2727	71.8700
100	76.3100	67.6880	51.6528
200	58.2323	46.2326	26.9223
400	33.9100	20.9917	7.1182
600	19.7466	9.8820	1.9513
800	11.4999	4.4065	0.5067
1000	6.6961	2.0189	0.1352
2000	0.0407	0.0407	0.00018

how many sample results must be outside the control limits before we consider this as evidence of instability, rather than as the effect of common cause variation, can be answered on the basis of a binomial distribution model.

The sum total of cases where the control limits have been violated (number of values outside the control limits) must not be outside the random dispersion limits of the binomial distribution (99 percent, double-sided).

Example

One hundred samples are to be analyzed. What is the expected range for the number of control limit violations for a Shewhart location chart (99.73 percent control limits), given that the process is in fact stable?

For $p = 0.27$ and 100 samples the limits of the 99 percent dispersion interval are $x_{up} = 2$ and $x_{lo} = 0$.

This means that up to two violations of the control limits are acceptable.

3.5.4 Shewhart Charts for Process Model A

The construction and application of a control chart was illustrated in Section 3.5.2, using an $\overline{x}/s$ chart with a nonintervention probability of 99.73 percent as an example. There follows a summary of the most important Shewhart control chart formulas. Control charting procedures and stability criteria are as described in Sections 3.5.2 and 3.5.3.

Tables 3.5–5 to 3.5–7 show, for various Shewhart control charts, the appropriate formulas for process model A (see Chapter 1, Section 1.3) and various nonintervention probabilities.

Table 3.5–6. Calculation formulas for process location (warning and control limits).

Calculation Formulas for Shewhart Charts Process Location with a Noninterference Probability of 95% or 99%	
Average chart	$UCL = \mu + u_{1-\alpha/2} \frac{\sigma}{\sqrt{n}} = \mu + A_C \hat{\sigma}$
	$UWL = \mu + u_{1-\alpha/2} \frac{\sigma}{\sqrt{n}} = \mu + A_W \hat{\sigma}$
	$LWL = \mu - u_{1-\alpha/2} \frac{\sigma}{\sqrt{n}} = \mu - A_W \hat{\sigma}$
	$LCL = \mu - u_{1-\alpha/2} \frac{\sigma}{\sqrt{n}} = \mu - A_C \hat{\sigma}$
Median chart	$UCL = \mu + u_{1-\alpha/2} \frac{c_n \sigma}{\sqrt{n}} = \mu + C_C \hat{\sigma}$
	$UWL = \mu + u_{1-\alpha/2} \frac{c_n \sigma}{\sqrt{n}} = \mu + C_W \hat{\sigma}$
	$LWL = \mu - u_{1-/2} \frac{c_n \sigma}{\sqrt{n}} = \mu - C_W \hat{\sigma}$
	$LCL = \mu - u_{1-\alpha/2} \frac{c_n \sigma}{\sqrt{n}} = \mu - C_C \hat{\sigma}$
Raw values chart	$UCL = \mu + u_{\left(\sqrt[n]{1-\alpha}\right)_{zw}} \cdot \sigma = \mu + E_C \cdot \hat{\sigma}$
	$UWL = \mu + u_{\left(\sqrt[n]{1-\alpha}\right)_{zw}} \sigma = \mu + E_W \hat{\sigma}$
	$LWL = \mu - u_{\left(\sqrt[n]{1-\alpha}\right)_{zw}} \cdot \sigma = \mu - E_W \cdot \hat{\sigma}$
	$LCL = \mu - u_{\left(\sqrt[n]{1-\alpha}\right)_{zw}} \sigma = \mu - E_C \hat{\sigma}$

Here, process location is estimated using

$$\mu = \bar{\bar{x}} \text{ or } \bar{\tilde{x}}$$

and process dispersion is estimated using

$$\sigma = \sqrt{\bar{s^2}} \text{ or } \frac{\bar{s}}{a_n} \text{ or } \frac{\bar{R}}{d_n}$$

The factors A, B, C, and D are found in the tables in Chapter 10.

The formulas in Table 3.5–7 allow for the asymmetry of the distribution of s and R. Hence, the upper and lower limits are not symmetrical with respect to the center line. The formulas and values in Table 3.5–8 do not take this asymmetry into account. They assume, for reasons of convenience, a symmetrical distribution (normal distribution model). For a sample size of $n > 6$, this assumption is more or less justified. In the s chart, the lower control limit is set to zero for sample sizes $n < 6$; in the R chart, the lower control limit is set to zero for sample sizes $n < 7$.

The factors A_2, $\tilde{A}_2$, B_3, B_4, D_3, D_4, and E_2 are given in Table 10–7.

Table 3.5–7. Calculation formulas for process spread (warning and control limits).

Calculation Formulas for Shewhart Charts Process Spread Noninterference Probability of 95% or 99%	
Standard deviation	$UCL = \sqrt{\frac{\chi^2 f; 1-\alpha/2}{f}}\,\hat{\sigma} = B_{E_{up}}\,\hat{\sigma}$ $UWL = \sqrt{\frac{\chi^2 f; 1-\alpha/2}{f}}\,\hat{\sigma} = B_{W_{up}}\,\hat{\sigma}$ $LWL = \sqrt{\frac{\chi^2 f; \alpha/2}{f}}\,\hat{\sigma} = B_{W_{lo}}\,\hat{\sigma}$ $LCL = \sqrt{\frac{\chi^2 f; \alpha/2}{f}}\,\hat{\sigma} = B_{E_{lo}}\,\hat{\sigma}$
Range	$UCL = w_{n;1-\alpha/2}\hat{\sigma} = D_{E_{up}}\,\hat{\sigma}$ $UWL = w_{n;1-\alpha/2}\hat{\sigma} = D_{W_{up}}\,\hat{\sigma}$ $LWL = w_{n;\alpha/2}\hat{\sigma} = D_{W_{lo}}\,\hat{\sigma}$ $LCL = w_{n;\alpha/2}\hat{\sigma} = D_{E_{lo}}\,\hat{\sigma}$

Table 3.5–8. Calculation formulas for location and spread charts (control limits 99.73%).

Calculation Formulas for Shewhart Charts Process Location with a Noninterference Probability of 99.73% ($\triangleq \pm 3\,\sigma$)	
Average/Range chart $\bar{x}$/R chart	$UCL_{\bar{x}} = \bar{\bar{x}} + A_2\bar{R}$ $LCL_{\bar{x}} = \bar{\bar{x}} - A_2\bar{R}$ $UCL_R = D_4\bar{R}$ $LCL_R = D_3\bar{R}$
Median/Range chart $\tilde{x}$/R chart	$UCL_{\tilde{x}} = \bar{\tilde{x}} + \tilde{A}_2\bar{R}$ $LCL_{\tilde{x}} = \bar{\tilde{x}} - \tilde{A}_2\bar{R}$ $UCL_R = D_4\bar{R}$ $LCL_R = D_3\bar{R}$
Average/Standard deviation $\bar{x}$/s Chart	$UCL_{\bar{x}} = \bar{\bar{x}} + A_3\bar{s}$ $LCL_{\bar{x}} = \bar{\bar{x}} - A_3\bar{s}$ $UCL_s = B_4\bar{s}$ $LCL_s = B_3\bar{s}$
Individuals chart	$UCL_x = \bar{\bar{x}} + E_2\bar{R}$ $LCL_x = \bar{\bar{x}} - E_3\bar{R}$

Table 3.5–9. Characteristics values of 20 samples.

1	2	3	4	5
130.04	130.07	130.06	130.07	130.11
129.98	130.04	130.07	129.99	130.05
130.02	130.07	130.05	130.05	130.06
130.03	130.01	130.07	130.04	130.07
130.02	130.09	130.05	130.03	129.99
129.98	130.07	130.05	130.03	130.03
130.08	130.04	130.02	130.03	130.04
6	**7**	**8**	**9**	**10**
130.04	130.03	130.08	130.01	130.05
130.03	129.97	130.04	130.05	129.98
130.04	129.99	130.05	130.00	130.03
130.04	130.04	130.03	130.07	130.03
130.03	130.02	130.02	130.01	129.99
130.06	130.09	130.04	130.03	130.04
130.03	130.08	130.03	130.02	130.05
11	**12**	**13**	**14**	**15**
129.99	130.04	130.05	130.08	130.03
130.02	129.99	130.03	130.06	130.01
130.08	130.07	130.08	130.11	130.02
130.05	130.06	130.00	130.05	130.04
130.05	130.01	130.06	130.00	130.07
130.04	130.04	130.08	130.05	130.05
130.03	130.04	130.07	129.99	130.07
16	**17**	**18**	**19**	**20**
129.97	129.99	130.02	130.03	130.07
130.04	130.07	130.04	130.09	130.07
130.04	130.01	130.03	130.06	130.03
130.04	130.06	130.02	130.08	130.05
129.98	130.09	130.05	130.03	130.06
130.06	130.07	130.03	130.06	130.07
130.07	130.03	130.04	130.03	130.03

3.5.5 Shewhart Charts: Examples

Based on the data in Table 3.5–9 and the formulas given in Section 3.5.4, the control limits are calculated for various charts and plotted accordingly (see Figure 3.5–13 to Figure 3.5–17). Various estimators of process location and process dispersion are used. The nonintervention probability varies between 99 percent and 99.73 percent. The givens and the results reached from them are shown underneath each chart.

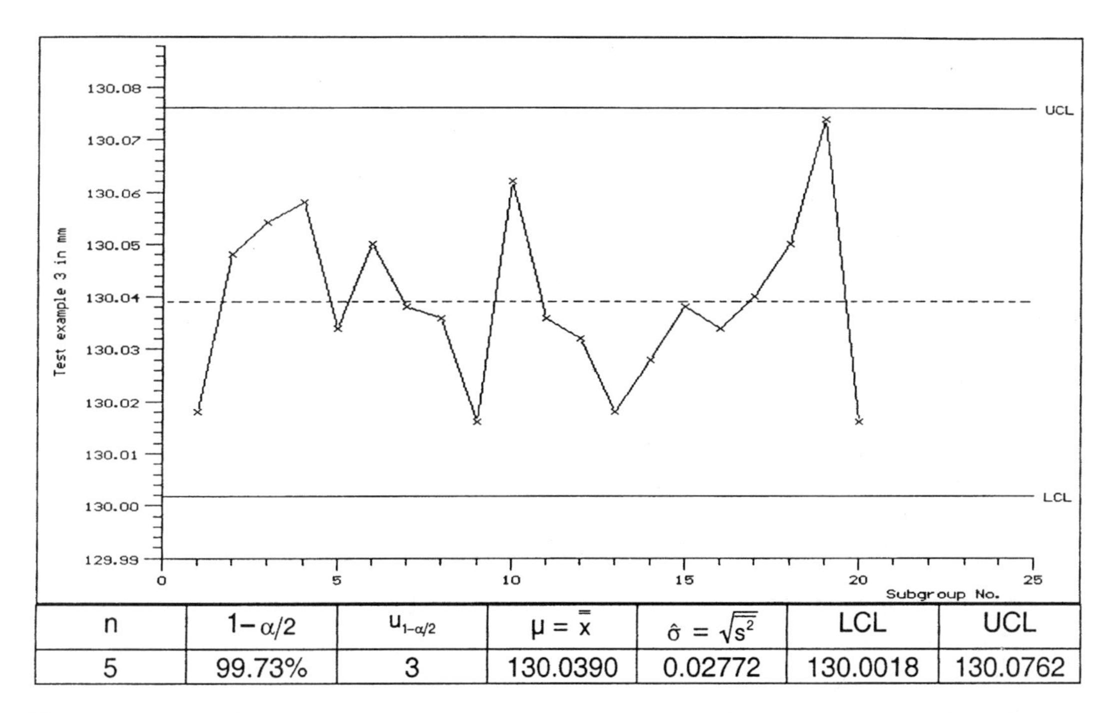

n	$1-\alpha/2$	$u_{1-\alpha/2}$	$\mu = \bar{\bar{x}}$	$\hat{\sigma} = \sqrt{\overline{s^2}}$	LCL	UCL
5	99.73%	3	130.0390	0.02772	130.0018	130.0762

Figure 3.5–13. Average chart.

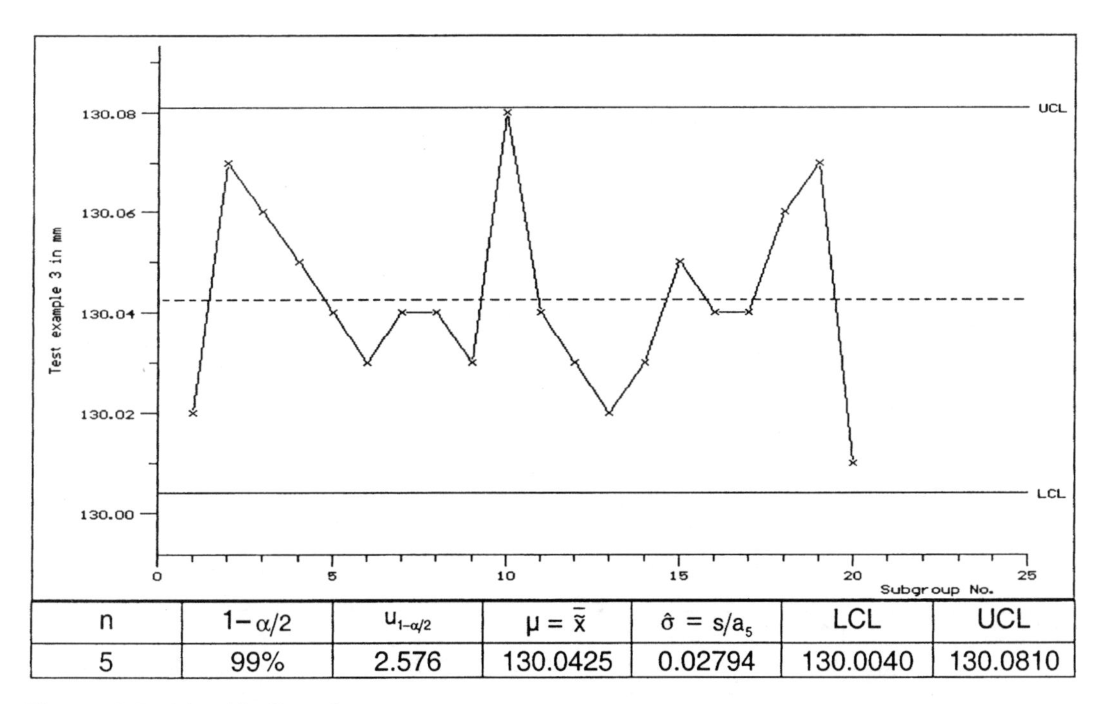

n	$1-\alpha/2$	$u_{1-\alpha/2}$	$\mu = \bar{\tilde{x}}$	$\hat{\sigma} = s/a_5$	LCL	UCL
5	99%	2.576	130.0425	0.02794	130.0040	130.0810

Figure 3.5–14. Median chart.

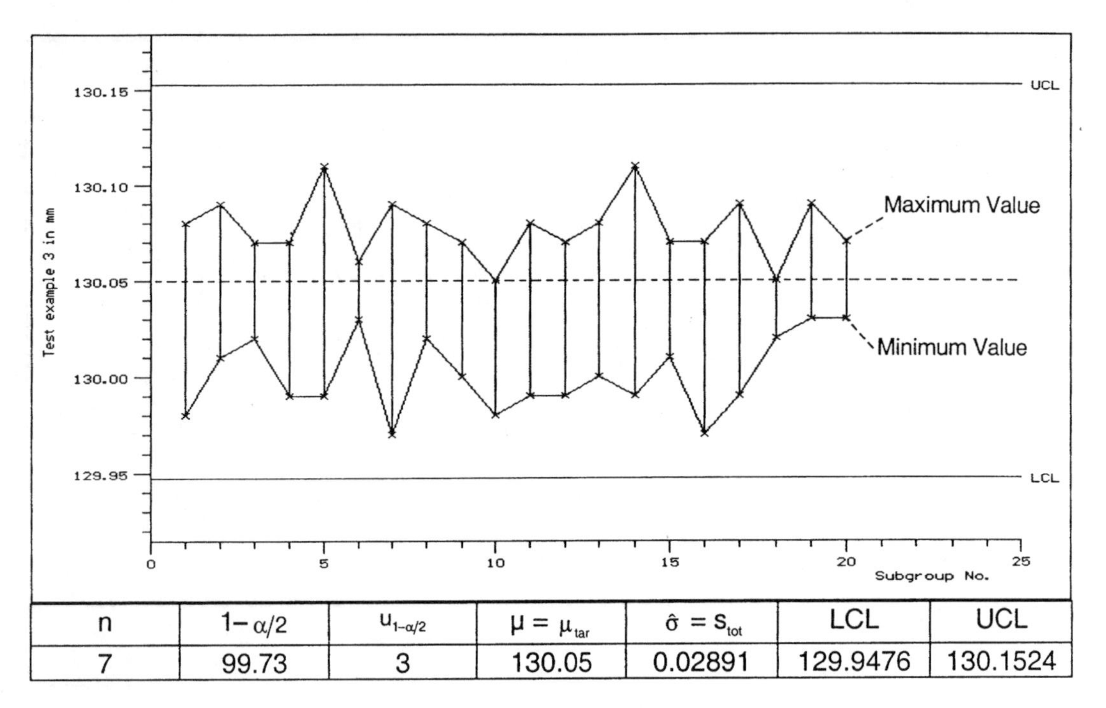

n	$1-\alpha/2$	$u_{1-\alpha/2}$	$\mu = \mu_{tar}$	$\hat{\sigma} = s_{tot}$	LCL	UCL
7	99.73	3	130.05	0.02891	129.9476	130.1524

Figure 3.5–15. Raw values chart.

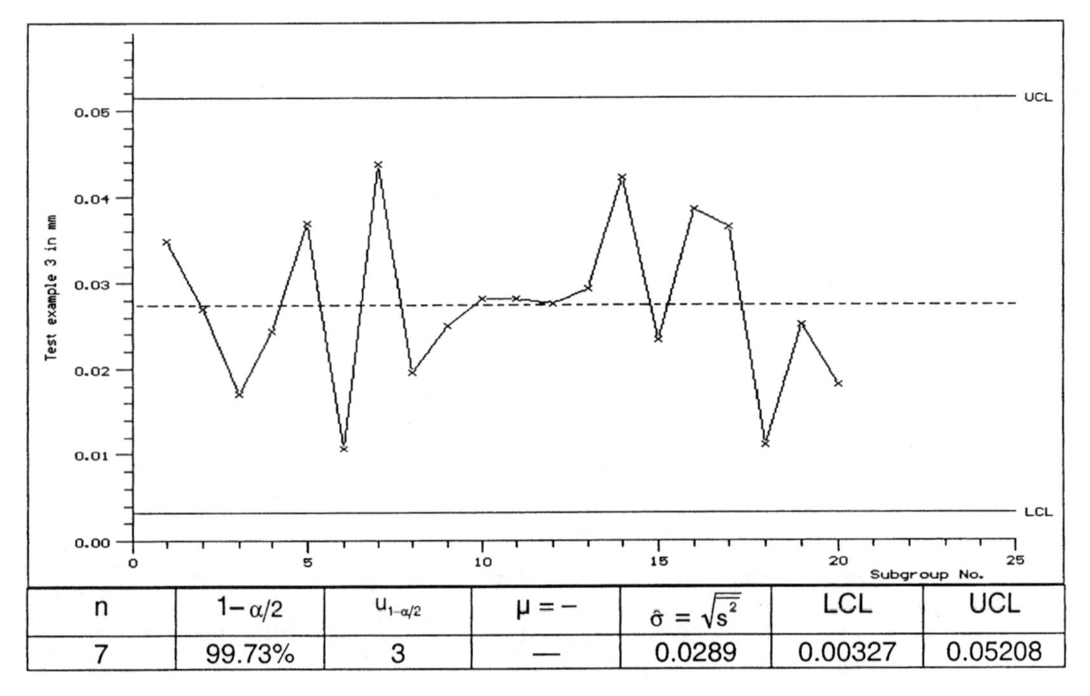

n	$1-\alpha/2$	$u_{1-\alpha/2}$	$\mu = -$	$\hat{\sigma} = \sqrt{\overline{s^2}}$	LCL	UCL
7	99.73%	3	—	0.0289	0.00327	0.05208

Figure 3.5–16. Standard deviations chart.

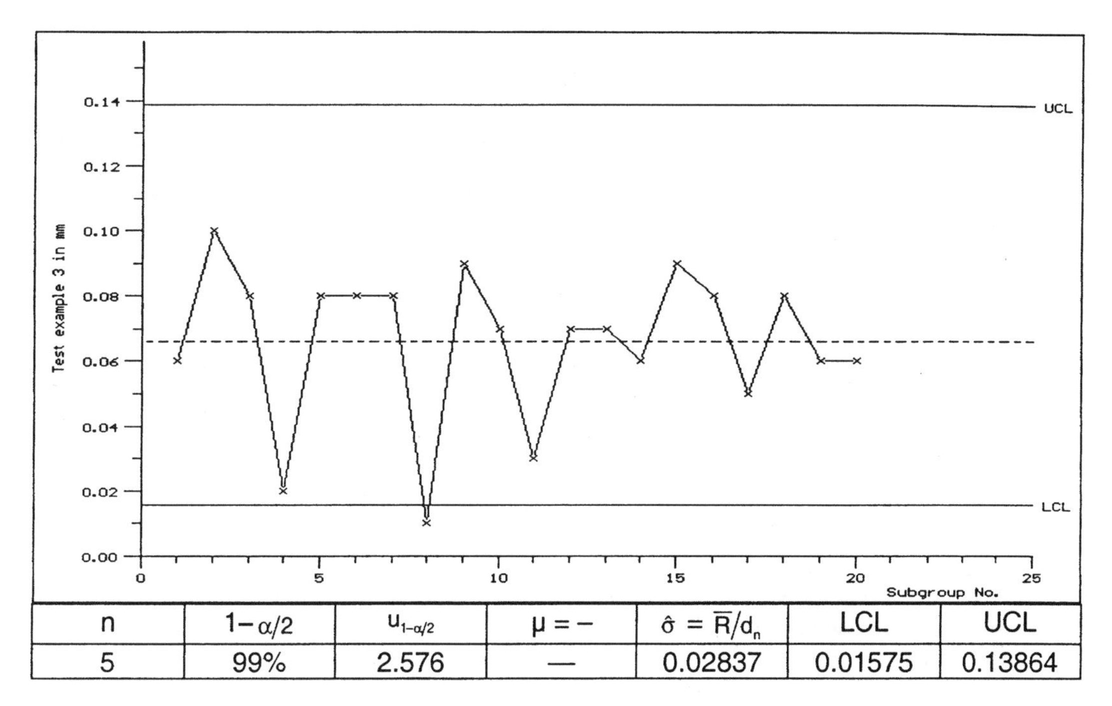

n	$1-\alpha/2$	$u_{1-\alpha/2}$	$\mu = -$	$\hat{\sigma} = \overline{R}/d_n$	LCL	UCL
5	99%	2.576	—	0.02837	0.01575	0.13864

Figure 3.5–17. Range chart.

3.5.6 Pros and Cons of the Various Location and Dispersion Charts

Table 3.5–10 shows advantages and disadvantages of the various location charts.

Comparison of the dispersion charts (s and R charts) shows that the s chart is more sensitive than the R chart. On the other hand, the s chart is not really suited to manual charting. In contrast, calculation and plotting of the range can usually be performed manually without a computer.

Due to this, $\bar{x}$/s charts have become the preferred charting method for computerized systems. For manual charting, $\tilde{x}$/R charts are the preferred choice.

3.6 Acceptance Control Charts

3.6.1 Construction of an Acceptance Chart

The limits on acceptance charts are calculated on the basis of the tolerance limits. This gives a control chart in which the tolerance range may be *fully utilized.* This is contrary to the principle of process control aimed at the target value.

The underlying rationale for the control limits on an acceptance chart for averages is illustrated in Figure 3.6–1 on page 165.

In the calculation, we start from a proportion p of items outside specification at a particular time and a probability of intervention P_a for the plotted statistics (sample means). These values may be freely chosen. Once they have been fixed, they define the control limits as shown in Figure 3.6–1.

Table 3.5–10. Location charts in comparison.

	Advantages	Disadvantages
Average Chart	Higher sensibility than raw value charts. Deviations from the normal distribution are much less critical than in the case of the raw values chart. Ideal for computer use.	The individual values are not documented in this QCC. In a computer-supported system, this is uncritical if the raw values are saved to disk. Calculation of the mean is difficult without a calculator.
Raw Values Chart	The raw values may be recorded without the need for calculation. This makes it understandable even for untrained personnel after a short introduction. All measurement values are recorded and thus documented. The raw values chart shows—conditional—an increase in process spread. However, it may not be regarded as a full replacement for a range or standard deviation chart. Used with small sample sizes (especially around n = 1).	A normal distributed process is required. Minimal sensibility. Chart will quickly become unclear if all values are recorded.
Median Chart	Easy chart keeping with uneven sample sizes. The raw values are sorted automatically during recording. The statistic value *median* may easily be determined by counting and marked. Is frequently applied with manual chart keeping and recording through optical voucher readers.	Less sensibility compared to the average chart. May become unclear when individual values are recorded.

However, care should be taken that the distance between the control limits does not become less than that given for a corresponding Shewhart chart. The within-subgroup variation $\hat{\sigma}$, as determined from a trial run, is used to estimate the dispersion of the individual values. The variation of the means is given by

$$\hat{\sigma}_{\bar{x}} = \frac{\hat{\sigma}}{\sqrt{n}}$$

Acceptance charts are not in line with the principle of never-ending improvement, since with reduced variation the distance between the control limits will become greater rather than smaller. An example will serve to illustrate this point.

For the process shown in Figure 3.6–2, control limits based on Shewhart's formulas are to be determined and compared with those of an acceptance chart (see Section 3.6.2). If the limits for the first 120 values (24 samples) are calculated, the control charts shown in Figure 3.6–3 are obtained. The control limits for the two charts are approximately equal (see Table 3.6–1). After this point, the process variation becomes smaller. Hence, the control limits for values 121 to 240 were recalculated

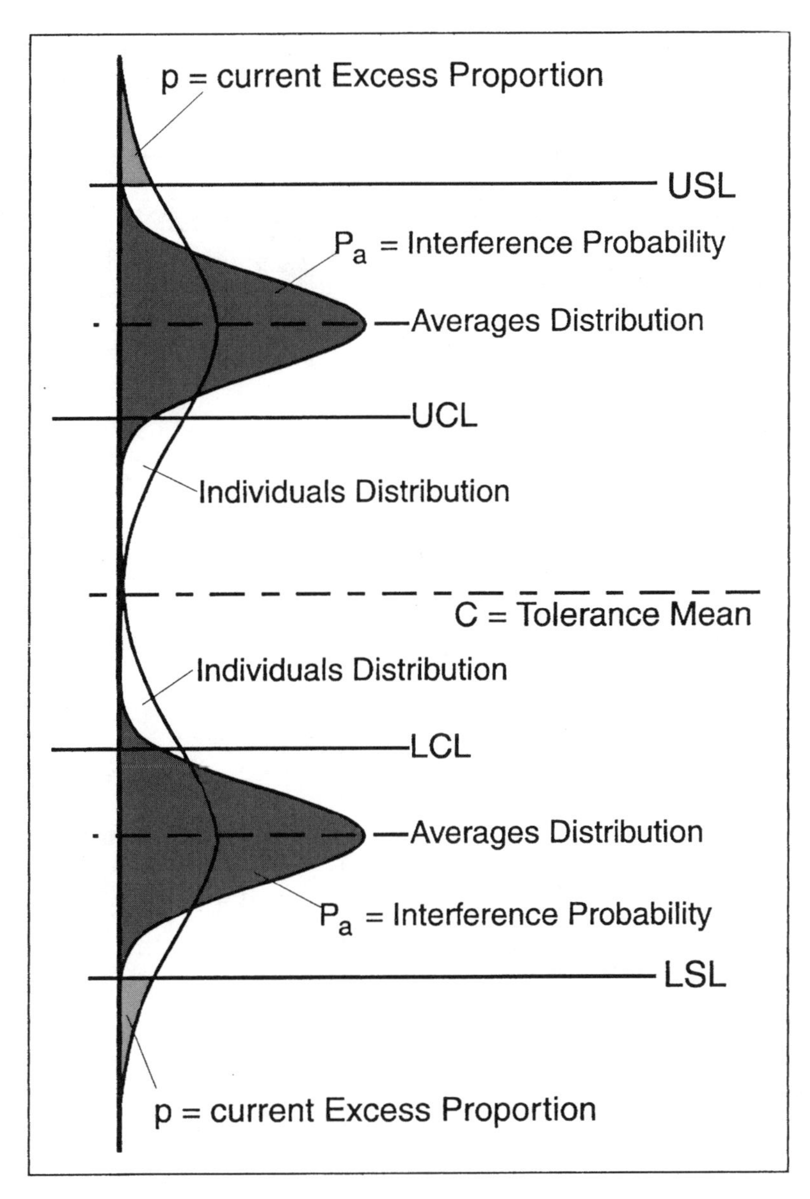

Figure 3.6–1. Control limits determination of an $\bar{x}$ acceptance chart.

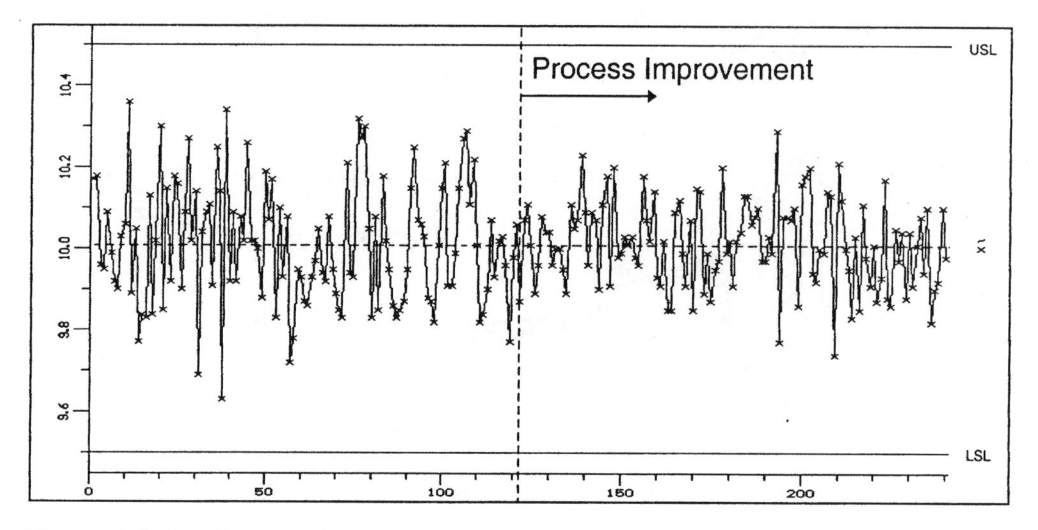

Figure 3.6–2. Process development with decreasing process spread.

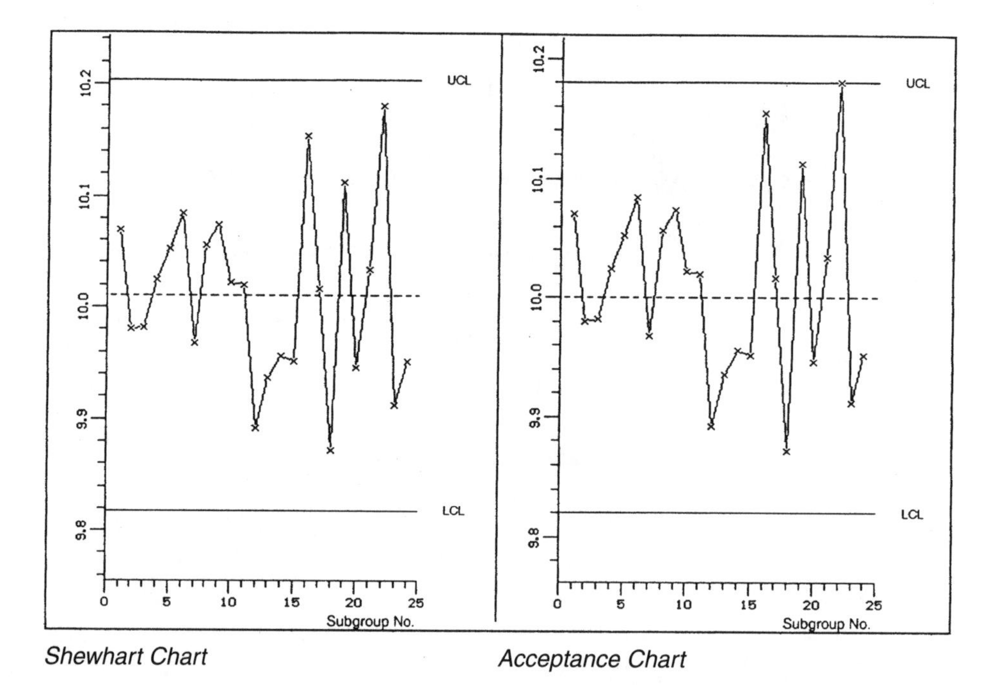

Shewhart Chart *Acceptance Chart*

Figure 3.6–3. Comparison of two control charts with process spread $\hat{\sigma} = 0.144$.

Table 3.6–1. Control limits in comparison.

Value Ranges	Estimator Location and Spread	Excess Proportion Interference Probability	Shewhart Chart	Acceptance Chart
1 – 120	$\mu = 10.0102$	$p = 5\%$	UCL = 10.2	UCL = 10.18
	$\bar{s}/a_n = 0.144$	$1 - P_a = 90\%$	LCL = 9.82	LCL = 9.82
121 – 240	$\mu = 10.0123$	$p = 5\%$	UCL = 10.16	UCL = 10.26
	$\bar{s}/a_n = 0.109$	$1 - P_a = 90\%$	LCL = 9.86	LCL = 9.74

Table 3.6–2. Calculation formulas for the control limits of acceptance charts.

Calculation Formulas for Acceptance-QCC Process Location	
Average chart	$UCL = USL - \left(u_{1-p} + \frac{u_{1-P_a}}{\sqrt{n}}\right)\hat{\sigma} = USL + k_A\hat{\sigma}$
	$LCL = LSL + \left(u_{1-p} + \frac{u_{1-P_a}}{\sqrt{n}}\right)\hat{\sigma} = LSL + k_A\hat{\sigma}$
Median chart	$UCL = USL - \left(u_{1-p} + \frac{c_n u_{1-P_a}}{\sqrt{n}}\right)\hat{\sigma} = USL - k_C\hat{\sigma}$
	$LCL = LSL + \left(u_{1-p} + \frac{c_n u_{1-P_a}}{\sqrt{n}}\right)\hat{\sigma} = LSL + k_C\hat{\sigma}$
Raw values chart	$UCL = USL - \left(u_{1-p} - u_{\sqrt[n]{P_a}}\right)\hat{\sigma} = USL - k_E\hat{\sigma}$
	$LCL = LSL + \left(u_{1-p} - u_{\sqrt[n]{P_a}}\right)\hat{\sigma} = LSL + k_E\hat{\sigma}$

(see Figure 3.6–4 and Table 3.6–1). The new limits on the Shewhart chart are closer together, those on the acceptance chart further apart.

In other words, the control limits for a Shewhart chart become narrower as the variation is reduced. The distance between the control limits on the acceptance chart, on the other hand, increases!

This means that the only meaningful application of the acceptance chart is in processes that are subject to trends (process model C). However, this scenario can also be addressed using a Shewhart chart with extended limits (see Section 3.9).

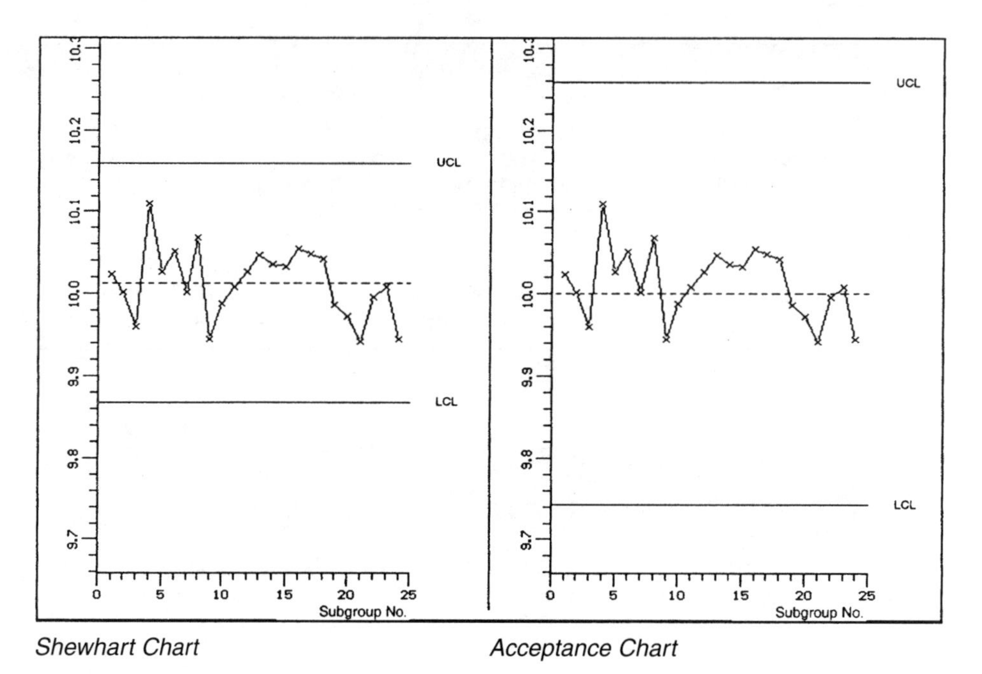

Shewhart Chart *Acceptance Chart*

Figure 3.6–4. Comparison of two control charts with process spread $\hat{\sigma} = 0.109$.

3.6.2 Control Limits for Acceptance Charts

In contrast to the Shewhart chart, the control limits for an acceptance chart may be calculated without a trial run, simply by specifying

- A short-term proportion outside specification (based on the agreed specification limits)
- The probability of intervention $1 - P_a$

It is also possible to draw up control charts for monitoring process location using individual values, means, or medians. The formulas for calculation of the control limits are summarized in Table 3.6–2.

One calculates the center line as

$$C = \frac{USL + LSL}{2}$$

The u values are calculated as described in Chapter 2, Section 2.2.2.1.

The estimate for the process dispersion is determined from $\hat{\sigma}$;

$$\sqrt{\overline{s^2}} \text{ or } \frac{\bar{s}}{a_n} \text{ or } \frac{\bar{R}}{d_n} \text{ or } \sigma \text{ may be known.}$$

The factors k_A, k_C, and K_E are taken from the Wilrich nomogram.

Table 3.6–3. Characteristics with trend.

1	2	3	4	5	6	7	8	9	10
15.08	15.10	15.08	15.12	15.14	15.10	15.12	15.13	15.13	15.17
15.09	15.05	15.09	15.08	15.12	15.12	15.11	15.12	15.15	15.17
15.06	15.11	15.08	15.10	15.13	15.15	15.15	15.13	15.15	15.12
15.07	15.05	15.07	15.09	15.15	15.11	15.12	15.14	15.13	15.14
15.04	15.09	15.08	15.13	15.13	15.14	15.13	15.12	15.12	15.14

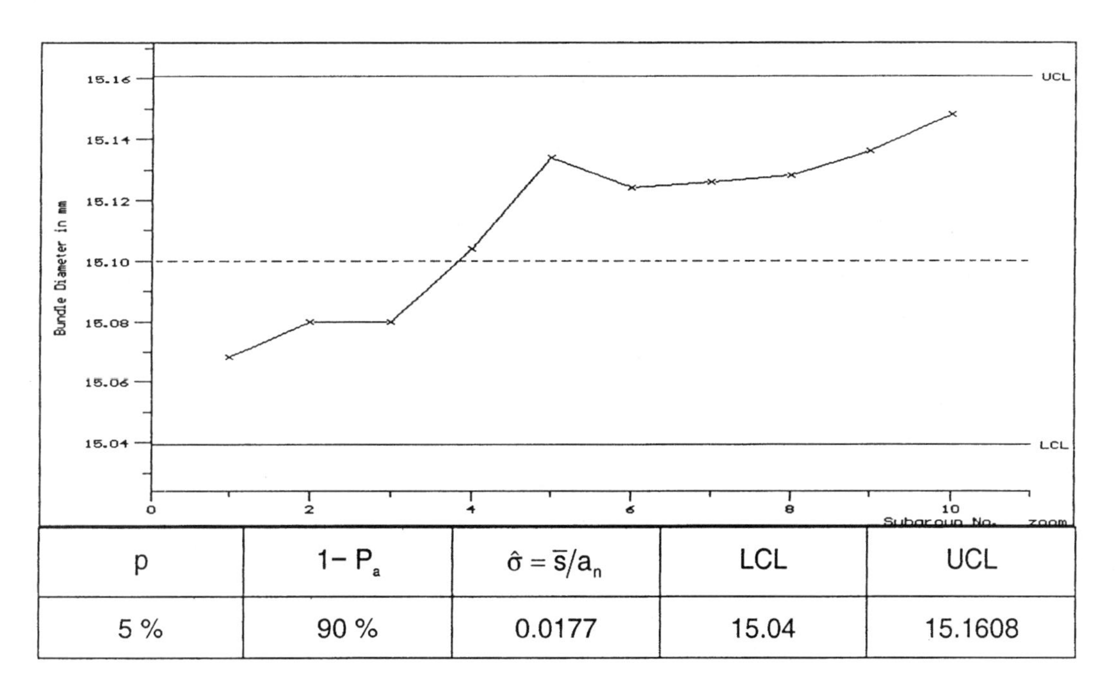

p	$1-P_a$	$\hat{\sigma} = \bar{s}/a_n$	LCL	UCL
5 %	90 %	0.0177	15.04	15.1608

Figure 3.6–5. Acceptance chart: Averages.

Examples Acceptance Charts

The desired diameter of a turned component is $d = 15.10 \pm 0.1$ mm. At regular half-hour intervals, five components are removed and measured. This led to the values given in Table 3.6–3.

Based on various specifications for the proportion outside tolerance p and the probability of intervention, $1 - P_a$, control limits were calculated in accordance with the formulas in Table 3.6–2 for mean, median, and individuals charts. The values or statistics of the ten samples are plotted on the relevant acceptance charts (Figure 3.6–5 to Figure 3.6–7).

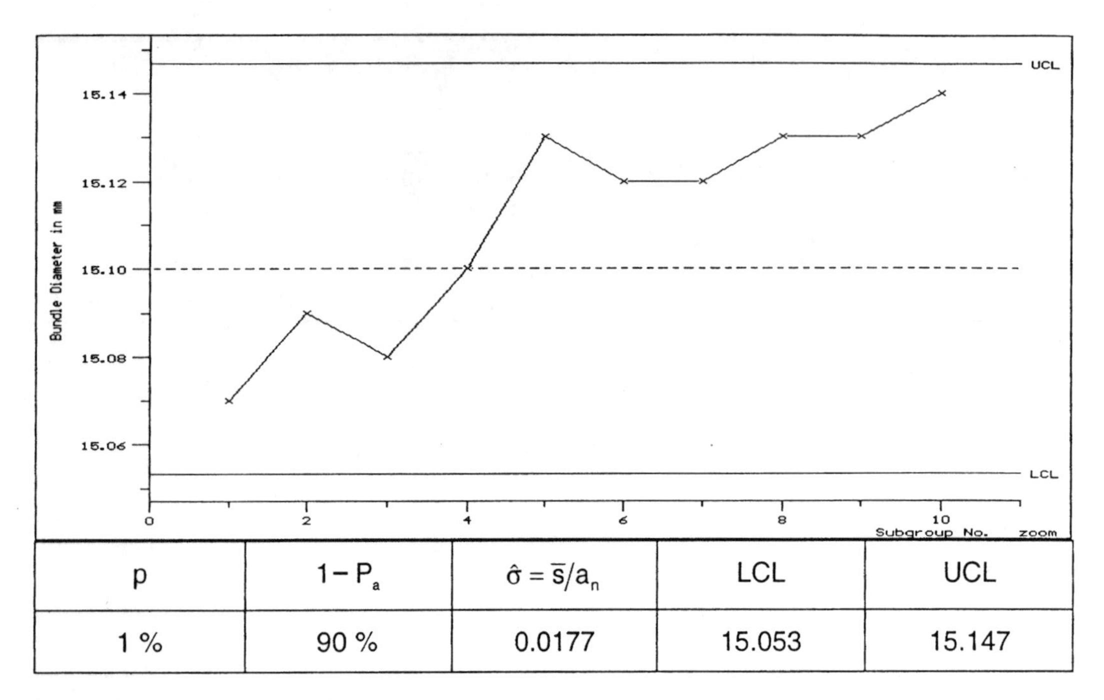

p	1− P_a	$\hat{\sigma} = \bar{s}/a_n$	LCL	UCL
1 %	90 %	0.0177	15.053	15.147

Figure 3.6–6. Acceptance chart: Median values.

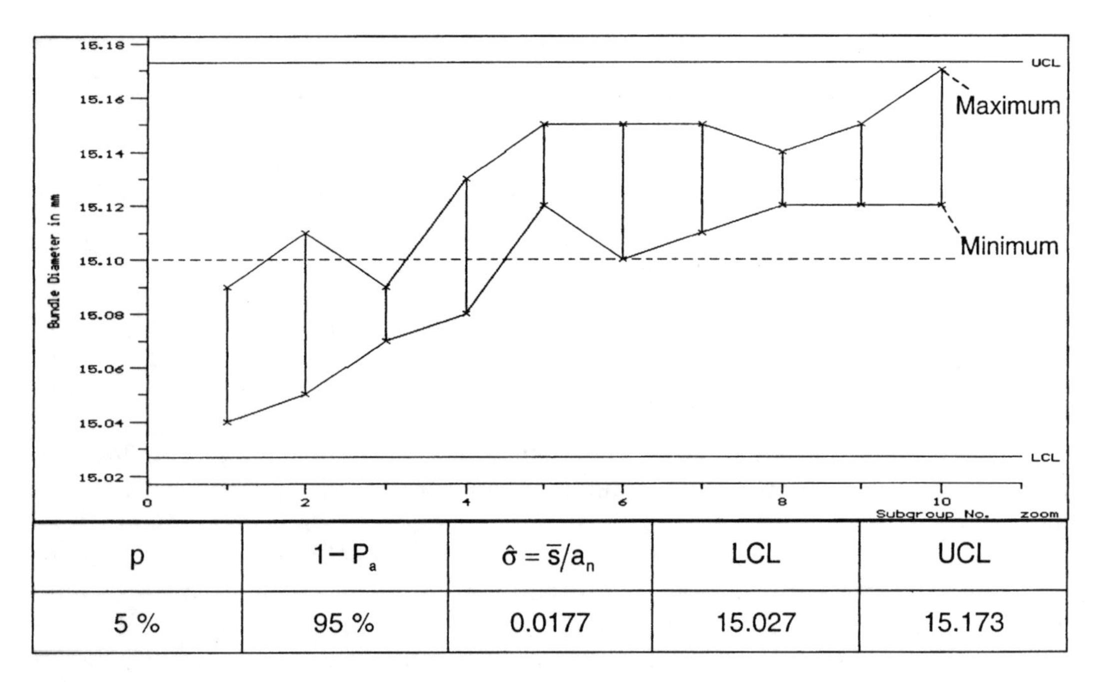

p	1− P_a	$\hat{\sigma} = \bar{s}/a_n$	LCL	UCL
5 %	95 %	0.0177	15.027	15.173

Figure 3.6–7. Acceptance chart: Raw values.

3.7 Shewhart Charts with Moving Characteristics

There are processes where for various reasons the sample size is n = 1. Typical examples are

- Destructive testing
- Monitoring of parameters
- High cost of inspection/testing
- Short production runs

These processes may be monitored either by individual value charts or control charts with moving statistics. For example, in the moving average chart, two, three, or more successive values are combined to yield an average. These values thus constitute a sample of size 2, or 3, or 4, etc. Based on these *pseudo samples,* it is also possible to arrive at a measure of dispersion, which is used to calculate control limits in accordance with the Shewhart formula given in Table 3.5–6. This dispersion chart—an R or s chart—is also based on the subgroups formed in this manner. Figure 3.7–1 illustrates the construction of such a chart.

The control limits for a moving average chart are calculated from the following formulas:

$$UCL = \mu + u_{1-\alpha/2} \frac{\hat{\sigma}}{\sqrt{n}} \quad \text{with} \quad n = \textit{Pseudo} \text{ subgroup size}$$

$$\mu = \overline{\overline{x}} \text{ or } \overline{\tilde{x}}$$

$$LCL = \mu - u_{1-\alpha/2} \frac{\hat{\sigma}}{\sqrt{n}} \qquad \hat{\sigma} = \sqrt{\overline{s^2}},\ \overline{s}/a_n,\ \text{or } \overline{R}/d_n$$

In similar manner, the limits for x, $\overline{x}$, $\tilde{x}$, R, and s charts may be calculated, as shown in Tables 3.5–6 to 3.5–8.

Example

One hundred three successive values (see Table 3.7–1) are taken from a manufacturing process. The resulting control charts are shown in Figures 3.7–2 and 3.7–3.

3.8 Pearson Control Charts for Averages

Pearson charts are recommended for processes that may be described by a unimodal, skewed distribution model, such as

- Lognormal distribution
- Folded normal distribution
- Rayleigh distribution
- Weibull distribution
- Johnson transformation
- Pearson function
- Mixed distribution

This applies in particular where the sample means do not follow a normal distribution. The skewness of the means is calculated from the skewness of the individual values divided by $\sqrt{n}$. The kurtosis of the means is calculated from the kurtosis of the individual values divided by n. Another possibility is to calculate the skewness and kurtosis of the means directly.

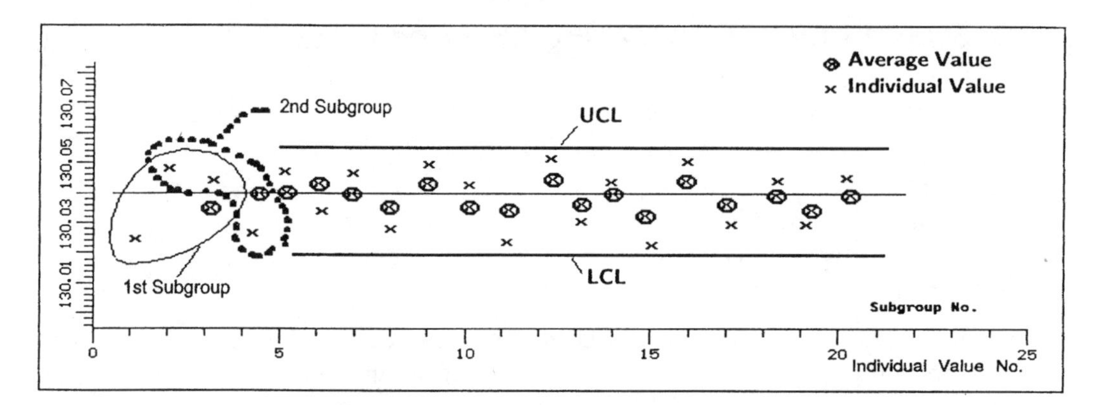

Figure 3.7–1. Shewhart chart with moving averages (*pseudo* subgroup n = 3).

Table 3.7–1. Ford test example 10.

64.5	63.3	65.1	64.7	65.4	65.1	65.9	66.1	62.8	63.3
64.1	65.8	65.9	66.9	64.2	63.7	62.8	65.2	64.4	63.3
65.6	66.1	64.0	63.3	65.1	63.6	64.5	64.5	65.5	65.8
66.0	65.4	64.6	64.9	66.5	66.2	65.5	64.6	65.7	
63.0	66.5	65.2	64.9	63.4	65.1	62.8	65.1	63.4	
65.5	66.4	63.2	65.7	64.3	65.8	66.3	64.6	65.2	
65.0	63.6	65.0	64.1	65.3	64.5	65.5	66.8	63.8	
65.9	63.3	65.3	65.4	64.6	62.0	64.6	64.7	66.1	
65.7	66.8	65.8	65.2	64.0	65.3	64.6	64.9	65.4	
65.8	65.5	63.6	67.9	66.4	64.3	65.2	65.2	65.1	
64.4	63.8	64.4	65.5	64.9	65.7	62.6	64.5	67.1	

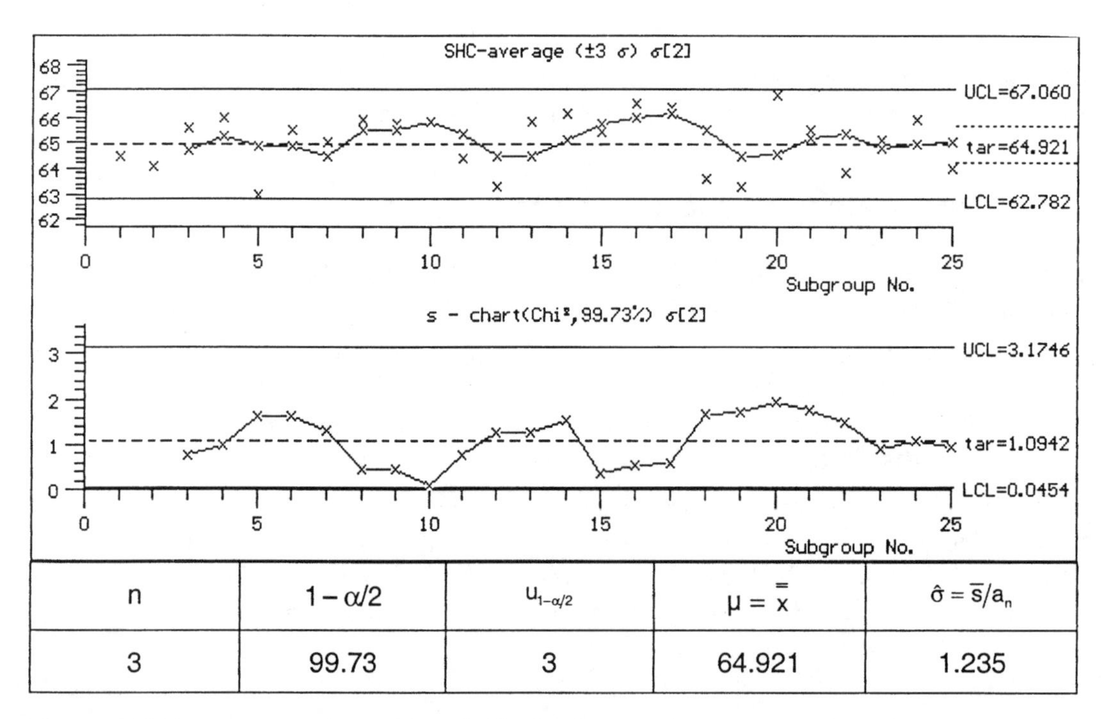

n	$1-\alpha/2$	$u_{1-\alpha/2}$	$\mu = \bar{\bar{x}}$	$\hat{\sigma} = \bar{s}/a_n$
3	99.73	3	64.921	1.235

Figure 3.7–2. Average chart with individual values and moving standard deviations.

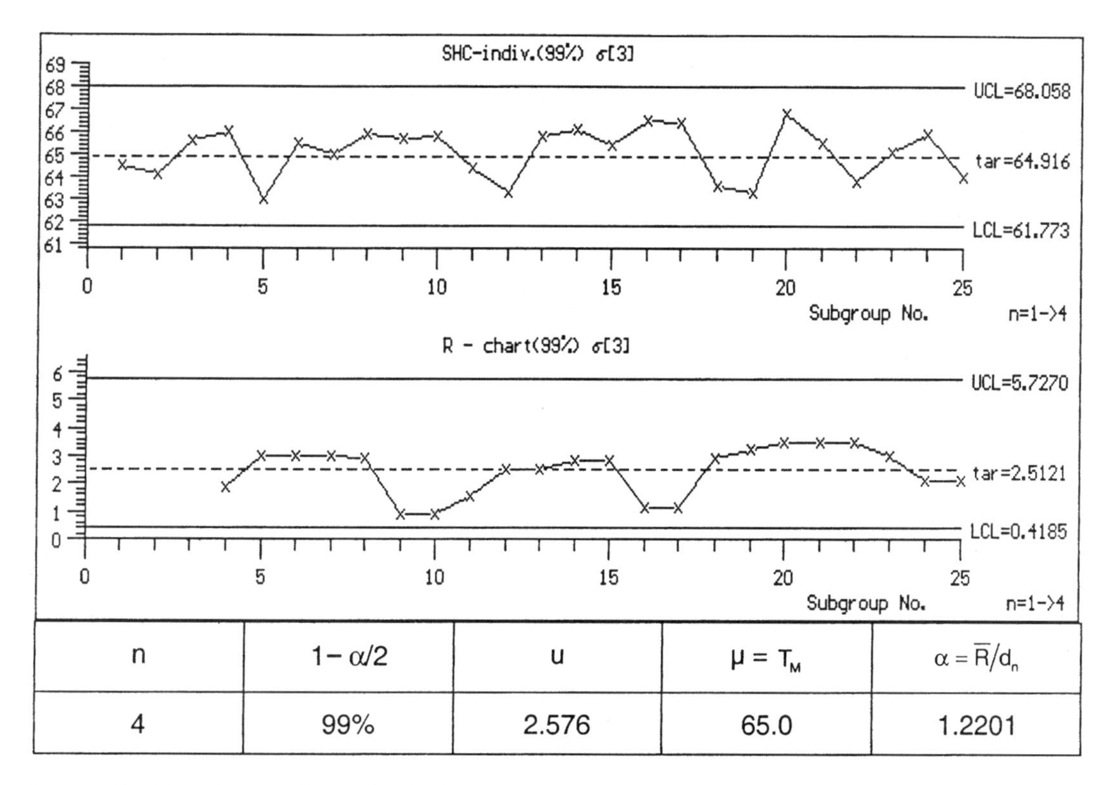

n	$1-\alpha/2$	u	$\mu = T_M$	$\alpha = \bar{R}/d_n$
4	99%	2.576	65.0	1.2201

Figure 3.7–3. Raw values chart with moving ranges.

In the context of control charting, the Pearson distribution system is a suitable tool for calculating dispersion intervals for the distribution of sample means. By means of this system, the appropriate control limits (random dispersion limits) may be calculated for any skewness and kurtosis.

Figure 3.8–1 shows a series of measurements (lognormal distribution). The data are as per the Ford Motor Company test example 4 (see Chapter 11).

When using Pearson average charts to monitor process location, it is in general possible to dispense with the dispersion chart, since a change in process variation also gives rise to a change in the distribution of sample means. Hence, changes in process variation will show up in the location chart.

Figure 3.8–2 clarifies this fact. The histogram shows a data set of n = 1000 values coming from two different populations with regard to location. Only a slight change of the averages from $\mu_1 = 0.5$ to $\mu_2 = 0.8$ with equal variation $\sigma_1 = \sigma_2 = 0.3$ of the population leads to a large change in variation. This is valid vice versa as well.

The Pearson quality control charting technique is not limited to calculation of the control limits for the averages chart, but may be used universally for the calculated statistical values like median, standard deviation, and range. Calculation of the control limits is based on the current values.

Figure 3.8–3 displays in part the course of the standard deviation in a Shewhart or Pearson chart derived from the set of data shown in Figure 3.8–2 (lognormal distribution). Whereas the Shewhart chart based on symmetrical control limits shows several violations, the Pearson chart adapts to the *skewness.* This results in unsymmetrical control limits. The total number of control limit violations is reduced!

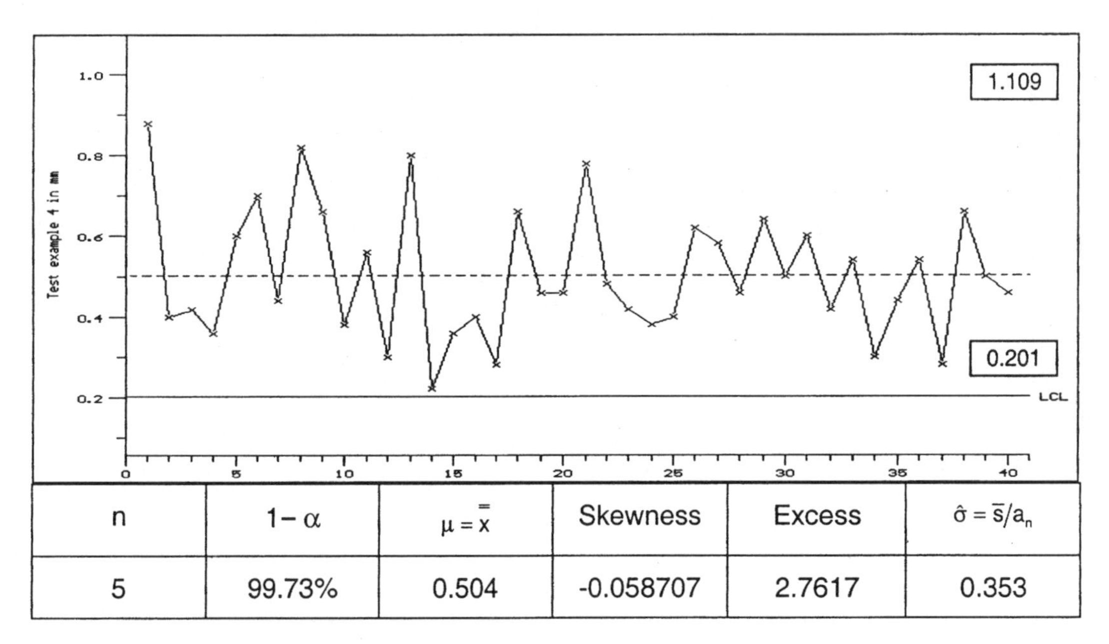

n	1− α	$\mu = \overline{\overline{x}}$	Skewness	Excess	$\hat{\sigma} = \overline{s}/a_n$
5	99.73%	0.504	-0.058707	2.7617	0.353

Figure 3.8–1. Pearson quality control chart.

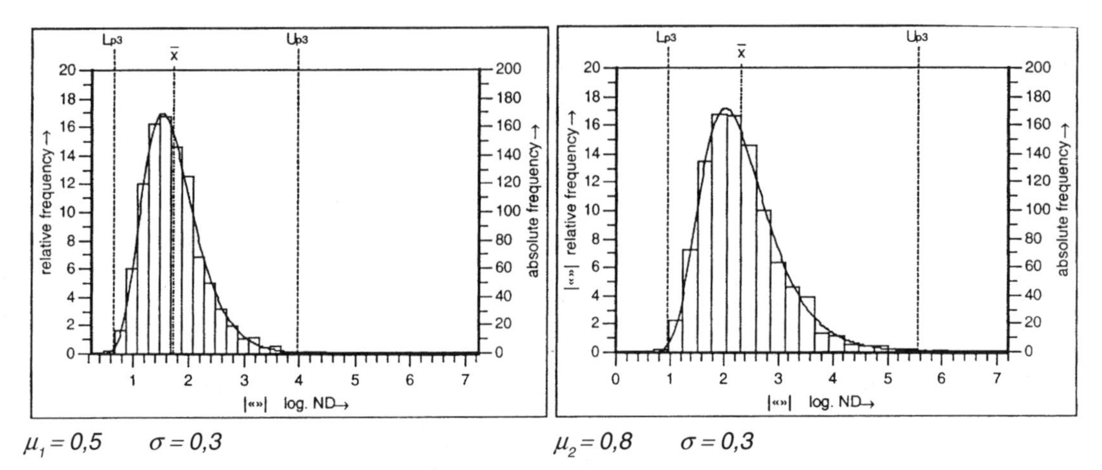

Figure 3.8–2. Different lognormal distributions (n = 1000, σ = 0).

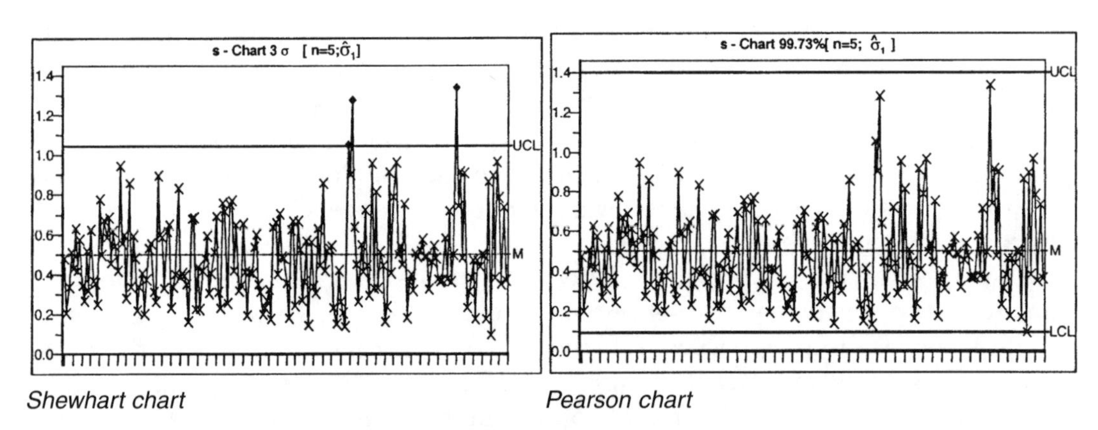

Figure 3.8–3. Shewhart and Pearson charts in comparison.

3.9 Shewhart Charts for Process Models B and C

3.9.1 Process Model B

There are processes that, even in stable operation, are subject to unavoidable fluctuations of the process average. A characteristic of these processes is large between-subgroups variation in comparison with the variation within the subgroups.

The reasons for this may be

- Batch-to-batch variation
- Equipment variation
- Time-to-time variation
- Variation in raw materials
- Tool variation
- Variation in environmental conditions
- Variation in process parameters
- Etc.

Figure 3.9–1 shows an example of such a situation. The measurement values are given in Table 3.9–3 (on page 179). The chart shows very little variation within the subgroups, but the variation of the subgroup means is disproportionately large. Situations encountered in practice may generate different patterns of variation, but the method described here may be used for all of them.

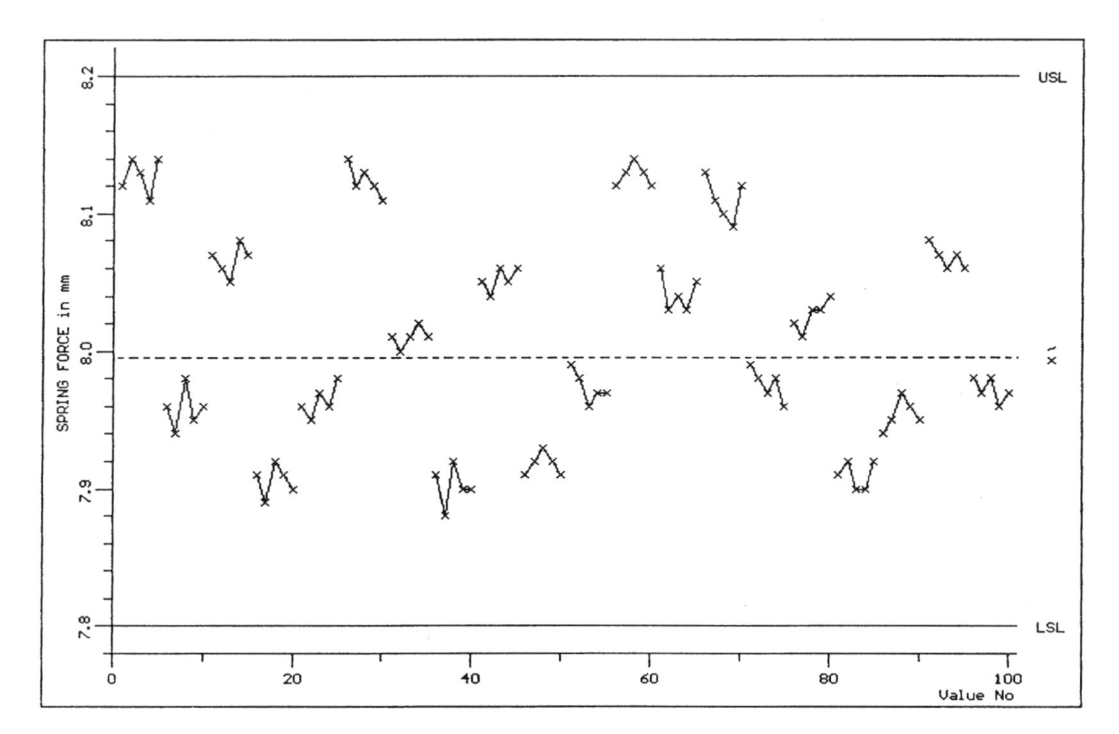

Figure 3.9–1. Individuals chart from a process with fluctuating averages.

If the control limits for a Shewhart average chart are calculated without taking the additional variation of the averages into account, these limits are found from

$$\begin{matrix}UCL\\LCL\end{matrix} = \mu \pm u_{1-\alpha/2}\frac{\hat{\sigma}}{\sqrt{n}} \qquad \text{where} \qquad \begin{matrix} n = \text{subgroup size} \\ \mu = \text{average} \\ \hat{\sigma} = \text{estimate of } \sigma \end{matrix}$$

Here $\hat{\sigma}$ represents the estimated process variation and $1 - \alpha$ the probability of non-intervention, e.g., 99 percent or 99.73 percent.

If we choose $\bar{\bar{x}}$ as an estimate of process location and $\bar{s}/a_n$ (see Section 3.5–3), we obtain a control chart whose control limits are too narrow (Figure 3.9–2).

$$UCL = 8.0094 + 3\frac{0.0011791}{\sqrt{5}} = 8.025$$

$$LCL = 8.0094 - 3\frac{0.0011791}{\sqrt{5}} = 7.994$$

We have in this instance also shown the specification limits (USL and LSL) on the chart (this would not normally be done on an average chart!), and we have given a rough indication of the individual measurement values within the samples, as well as their associated normal distribution curves. It will be seen that, in spite of the control limit violations, all values are within the set tolerances. If the shifts in average are an inherent and accepted feature of the process (e.g., reflecting tool variation), then the chart shown cannot be used for controlling this process, since it would lead to constant intervention due to incessant violations of the stability criteria.

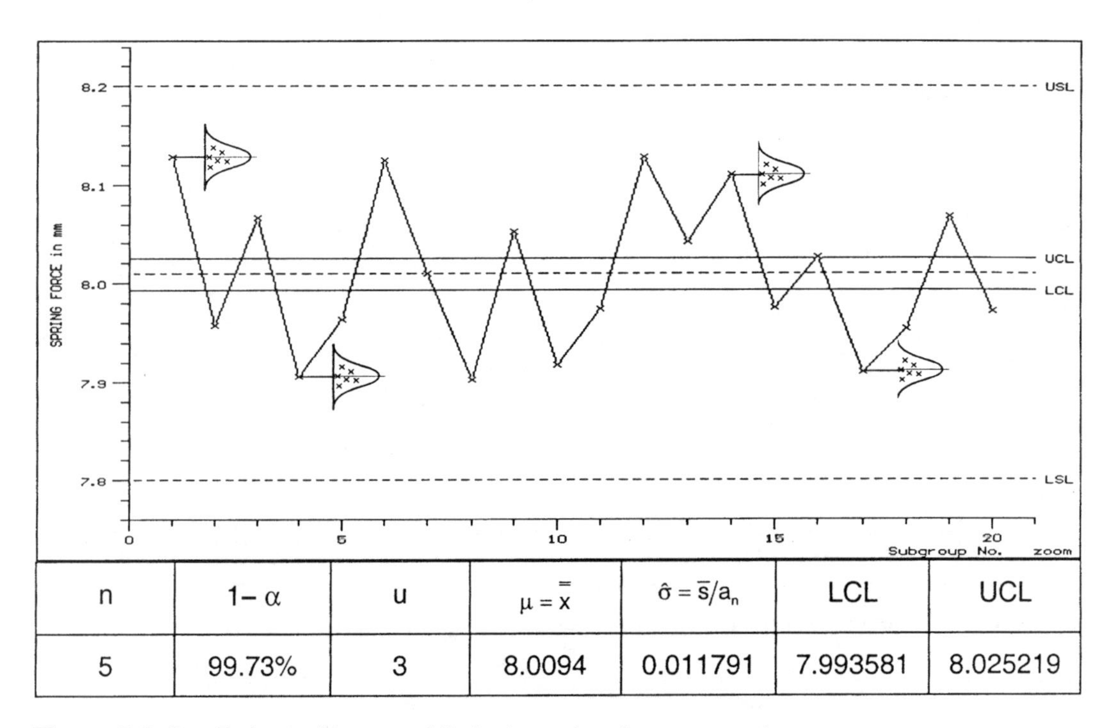

n	1 − α	u	$\mu = \bar{\bar{x}}$	$\hat{\sigma} = \bar{s}/a_n$	LCL	UCL
5	99.73%	3	8.0094	0.011791	7.993581	8.025219

Figure 3.9–2. $\bar{x}$ chart with control limits based on inner spread.

The question as to whether the process is subject to additional variation of the mean may be answered either graphically, e.g., from the control chart, or numerically, e.g., by means of an analysis of variance (ANOVA)/F-test. The object of a simple ANOVA is to identify the various factors contributing to the overall variation and to quantify them. In inferential statistics, there are two ANOVA models, *Model I* with nonrandom (fixed) factor levels and *Model II* with random factor levels. Chapter 9, Section 9.2 gives further descriptions of both models.

To establish numerically if *the between-subgroups variance is zero,* i.e., the process average does not vary, an F-test is performed. The test result is shown in Table 3.9–2. The F-test procedure, including the interpretation of results, is described in Chapter 2, Section 2.8. In our case, the result indicates that *the between-subgroups variation is not zero.*

Hence, it is confirmed graphically from the control chart and also numerically from the F-test that the mean of the population is subject to fluctuations. If these are due to known technological causes and are deemed acceptable, then the control limits on the control chart may be widened. There are a number of different methods for determining the amount by which the control limits should be widened:

- *Analysis of variance*
  The 86.64 percent interval (corresponding to $u = 1.5$) of the normally distributed population of subgroup averages is a recommended interval for adapting the chart to the fluctuations in process location. This is an empirical value that has proved useful in practice. The parameters for the location and dispersion of the *distribution of the averages* are determined by ANOVA (see Chapter 9, Section 9.2). This provides a numerical method for determining the variation of the averages. The variation interval thus calculated is added to the calculated control limit spacing (Shewhart chart).
- *Visual estimation*
  The extent of variation of the averages may be estimated from the control chart (location chart), from a run chart of the subgroup averages, or on the basis of empirical process knowledge. The variation interval thus calculated is added to the calculated control limit spacing (Shewhart chart).
- *Variation of the averages*
  Calculation of the control limits is based on the variation of the subgroup averages. However, a 99 percent or 99.73 percent interval based on this, rather than the within-subgroup variation, usually leads to control limits that are too far apart. A value of $u = 2.0$ (corresponding to 95.45 percent) has proved useful in practice.
  Requirement: The averages must be normally distributed. This can be checked by plotting the averages on normal probability paper.
- *Giving separate estimates of* Upper Process Location *and* Lower Process Location
  Technical considerations may indicate that calculating process location from the measurement values would not lead to the best results. Also, the variation pattern of the means may be asymmetrically distributed about the overall process average. In this case, it is recommended that separate values be specified for the parameters *process location* μ_{upper} and *process location* μ_{lower}. Apart from that, the usual Shewhart criteria apply.

Graphical methods are always subjective and thus not suitable for automatic processing by computer software. Hence, numerical procedures are to be preferred. The set of measurement values in Table 3.9–1 (on page 178) is used as follows to illustrate

Table 3.9–1. Characteristics with random average fluctuations.

1	2	3	4	5	6	7	8	9	10
8.12	7.96	8.07	7.91	7.96	8.14	8.01	7.91	8.05	7.91
8.14	7.94	8.06	7.89	7.95	8.12	8.00	7.88	8.04	7.92
8.13	7.98	8.05	7.92	7.97	8.13	8.01	7.92	8.06	7.93
8.11	7.95	8.08	7.91	7.96	8.12	8.02	7.90	8.05	7.92
8.14	7.96	8.07	7.90	7.98	8.11	8.01	7.90	8.06	7.91
11	**12**	**13**	**14**	**15**	**16**	**17**	**18**	**19**	**20**
7.99	8.12	8.06	8.13	7.99	8.02	7.91	7.94	8.08	7.98
7.98	8.13	8.03	8.11	7.98	8.01	7.92	7.95	8.07	7.97
7.96	8.14	8.04	8.10	7.97	8.03	7.90	7.97	8.06	7.98
7.97	8.13	8.03	8.09	7.98	8.03	7.90	7.96	8.07	7.96
7.97	8.12	8.05	8.12	7.96	8.04	7.92	7.95	8.06	7.97

Table 3.9–2. F-test results.

	Critical Values		
	lower	**upper**	**Test Statistic**
$\alpha = 5\ \%$	---	1.72	
$\alpha = 1\ \%$	---	2.14	**231.777*** **
$\alpha = 0.1\%$	---	2.73	

Result: Null hypothesis is rejected in favor of the alternative hypothesis.

Table 3.9–3. Subgroup results.

j	Average	Std. Deviation
1	8.12800	0.013038
2	7.95800	0.014832
3	8.06600	0.011402
4	7.90600	0.011402
5	7.96400	0.011402
6	8.12400	0.011402
7	8.01000	0.007071
8	7.90200	0.014832
9	8.05200	0.008367
10	7.91800	0.008367
11	7.97400	0.011402
12	8.12800	0.008367
13	8.04200	0.013038
14	8.11000	0.015811
15	7.97600	0.011402
16	8.02600	0.011402
17	7.91000	0.010000
18	7.95400	0.011402
19	8.06800	0.008367
20	7.97200	0.008367
$\bar{\bar{x}}, \bar{s}$	8.00940	0.011084

the ANOVA method. First, the *mean* and *standard deviation* are calculated for each sample (Table 3.9–3). For Model II (see Chapter 9, Section 9.2), the overall standard deviation $\hat{\sigma}_{tot}$ of the values in Table 3.9–3 is given by

Between-subgroups standard deviation	$\hat{\sigma}_A$	: 0.0770128
Within-subgroup standard deviation	$\hat{\sigma}_\varepsilon$	: 0.0113358
Overall standard deviation as per ANOVA model	$\hat{\sigma}_{tot} = \sqrt{\hat{\sigma}_A^2 + \hat{\sigma}_\varepsilon^2}$	: 0.0778426

To arrive at meaningful control limits, both components of variation must be taken into account. Hence, the limits on the Shewhart chart must be widened by the amount of between-subgroups variation observed. Figure 3.9–3 illustrates the establishment of the control limits from the two components of variation.

The control limits are thus calculated from the following formula:

$$\begin{matrix} UCL \\ LCL \end{matrix} = \mu \pm \left(u_{1-\alpha/2} \frac{\hat{\sigma}}{\sqrt{n}} + \hat{u}_{1-\alpha/2} \hat{\sigma}_A \right)$$

where $\mu = \bar{\bar{x}}$

$\hat{\sigma} = \sqrt{s^2}$, $\bar{s}/a_n$, or $\bar{R}/d_n$

$\hat{\sigma}_A$ = between-subgroups standard deviation as per Model II ANOVA

$u_{1-\alpha/2} = 3$ for $P = 99.73\%$

$\hat{u}_{1-\alpha/2} = 1.5$ for $P = 86.6\%$ (empirical value)

Hence, the extension of the limits will be dependent on $\hat{u}_{1-\alpha/2}$. 86.64 percent has been found to be a suitable value for $1 - \alpha = 86.64$ percent ($\hat{u} = 1.5$).

Hence, we arrive at the following results for the values in Table 3.9–1:

where $\mu = \bar{\bar{x}} = 8.0094$; $u_{1-\alpha/2} = 3$ $\hat{\sigma} = \sqrt{s^2} = 0.0113358$;

$\hat{u}_{1-\alpha/2} = 1.5$ $\hat{\sigma}_A = 0.0770128$ (ANOVA)

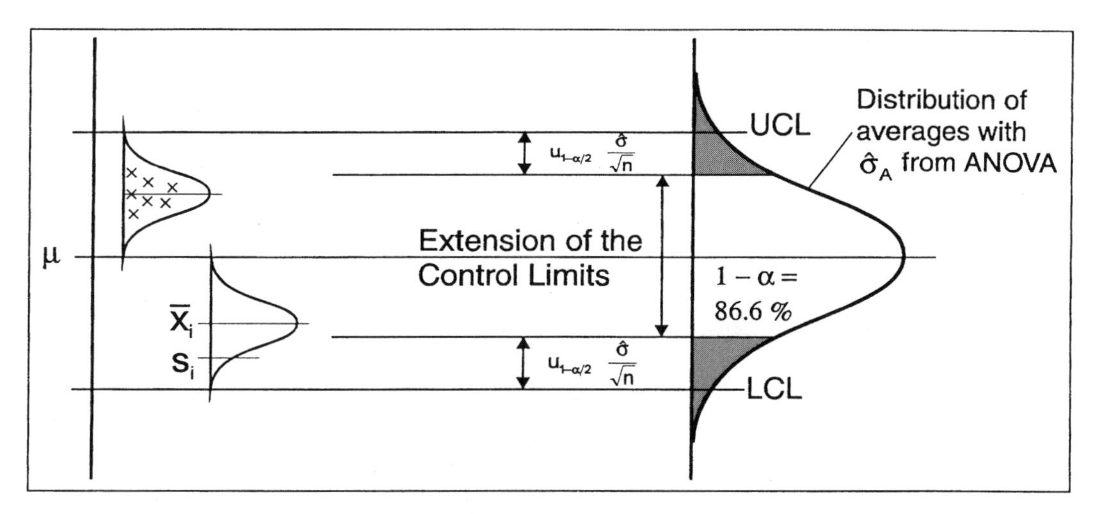

Figure 3.9–3. Extension of the control limits.

$$\text{UCL} = 8.0094 + 3\frac{0.0113358}{\sqrt{5}} + 1.5 \times 0.0770128 = 8.1401$$

$$\text{LCL} = 8.0094 - 3\frac{0.0113358}{\sqrt{5}} - 1.5 \times 0.0770128 = 7.8787$$

The new control chart is shown in Figure 3.9–4. The distance between the control limits has been increased by 0.231.

The widening of the control limits also changes the distribution model (see Figure 3.9–5). The left-hand side and the right-hand side correspond to the normal distribution based on the $\hat{\sigma}$ estimate obtained from the within-subgroup variation. The constant interval in the center corresponds to $\bar{x}_{zus}$ (= amount by which the control limits on the chart have been widened). Thus, there is a firm link between the control chart and the distribution model for calculation of the capability indices (see Chapter 6).

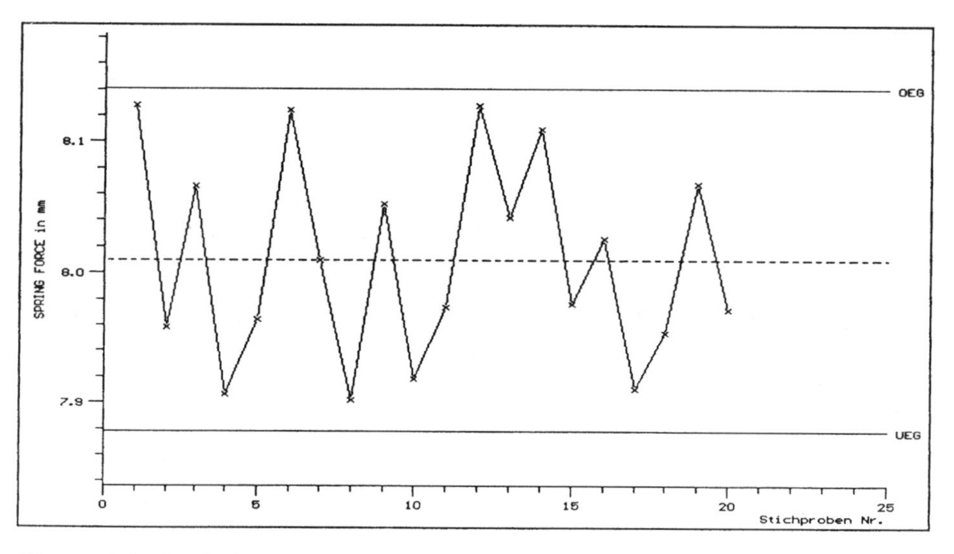

Figure 3.9–4. $\bar{x}$ chart with extended limits.

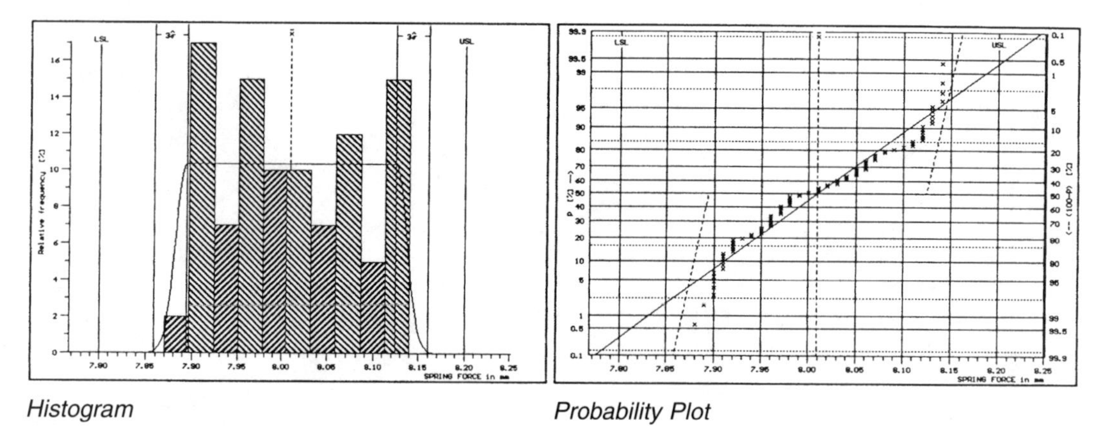

Histogram *Probability Plot*

Figure 3.9–5. Processes with $\mu_{(t)} \neq$ constant.

*"Stichproben Nr." (referred to in Figure 3.9–4) means *subgroup number.*

3.9.2 Process Model C

Wear and tear, process fluctuations, etc., may result in a process that is subject to trends. As a rule, we are dealing here with special causes of variation whose elimination is not feasible from an economic point of view. The object of any production is for every tool to have the longest possible operating life. This implies, contrary to the target-oriented philosophy, the greatest possible use of the tolerances. This fact must be taken into account in determining the control limits. In processes subject to trends, the mean of the population also changes over time. The procedures described under *Process Model B* (see Section 3.9.1) are therefore also valid in this case. This is true both for the recognition of the *trend process* situation and for the calculation of the control limits. The existence of a trend may also be discovered by means of the *test for trend* (see Chapter 2, Section 2.8.4). The *test for randomness* (see Chapter 2, Section 2.8.3) will usually also given an indication of the situation. In situations corresponding to *Process Model C,* either acceptance charts or Shewhart charts may be used.

Figure 3.9–6 illustrates two different types of trend behavior.

Calculation of the control limits based on the within-subgroup variation, as with process model B, leads to control limits that are too narrow. Hence, Shewhart charts with extended limits or acceptance charts should be used. The following example will be used to illustrate various methods of calculation.

Example: Trend Process

Twenty samples of size n = 5 are taken from a process that is subject to trends. Table 3.9–4 shows the results. The left-hand side of Figure 3.9–6 shows a run chart of the values. Determine the control limits for P = 99.73 percent.

The methods outlined in Section 3.9.1 may be used for a Shewhart chart with extended control limits.

Analysis of Variance (ANOVA)

With $\mu = \bar{\bar{x}} = 4.00657$

$\hat{\sigma} = \bar{s}/a_n = 0.005284$

$\hat{\sigma}_A = 0.010159$ Standard deviation between the subgroups from ANOVA - Model II

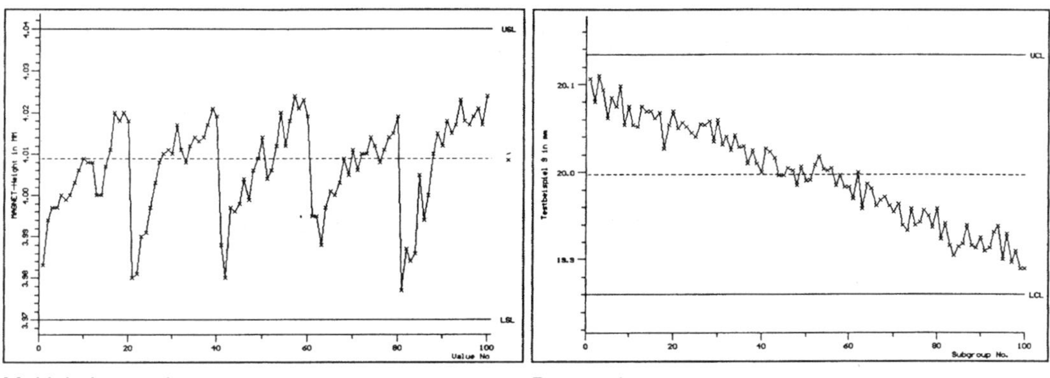

Multiple increasing *Decreasing*

Figure 3.9–6. Trend charts.

Table 3.9–4. Subgroup results: Trend process.

1	2	3	4	5	6	7	8	9	10
3.983	3.999	4.008	4.011	3.980	4.003	4.017	4.013	3.988	4.004
3.994	4.000	4.008	4.020	3.981	4.008	4.011	4.014	3.980	3.999
3.997	4.003	4.000	4.018	3.990	4.010	4.008	4.017	3.997	4.006
3.997	4.006	4.000	4.020	3.991	4.011	4.012	4.021	3.996	4.009
4.000	4.009	4.007	4.018	3.997	4.010	4.014	4.019	3.998	4.014
11	**12**	**13**	**14**	**15**	**16**	**17**	**18**	**19**	**20**
4.004	4.018	3.995	4.000	4.006	4.008	3.977	3.994	4.018	4.017
4.006	4.024	3.995	4.003	4.010	4.011	3.987	4.000	4.015	4.019
4.012	4.021	3.988	4.009	4.010	4.014	3.984	4.010	4.017	4.021
4.020	4.023	3.997	4.005	4.014	4.015	3.986	4.015	4.023	4.017
4.012	4.019	4.001	4.011	4.012	4.019	4.005	4.012	4.018	4.024

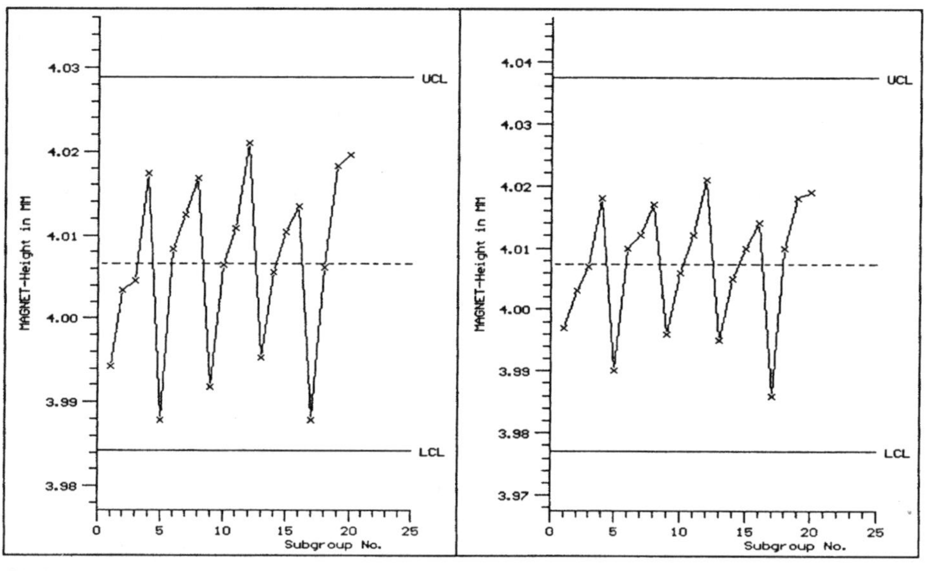

CL from Analysis of Variance *CL from Spread of the Averages*

Figure 3.9–7. $\bar{x}$ chart with extended limits.

we have

$$\text{UCL} = \mu + 3\frac{\hat{\sigma}}{\sqrt{n}} + 1.5\hat{\sigma}_A = 4.00657 + 3\frac{0.00528}{\sqrt{5}} + 1.5 \times 0.010159 = 4.0289$$

$$\text{LCL} = \mu - 3\frac{\hat{\sigma}}{\sqrt{n}} - 1.5\hat{\sigma}_A = 4.00657 - 3\frac{0.00528}{\sqrt{5}} - 1.5 \times 0.010159 = 3.984$$

Figure 3.9–7 (left-hand side) shows the resulting control chart. The amount by which the control limits have been widened is 0.030476.

Variation of the Averages

$$\text{With} \quad \mu = \bar{\bar{x}} = 4.00657$$
$$\hat{\sigma} = s_{\bar{x}} = 0.01044$$

we have

$$UCL^* = \mu + u_{1-\alpha/2}\hat{\sigma} = 4.00657 + 2 \times 0.0104 = 4.028$$
$$LCL^* = \mu - u_{1-\alpha/2}\hat{\sigma} = 4.00657 - 2 \times 0.0104 = 3.99$$

Figure 3.9–7 (right-hand side) shows the resulting control chart. This corresponds to a widening of the control limits by 0.027592.

Giving Separate Estimates of Upper Process Location *and* Lower Process Location. In processes subject to trends, it is possible to obtain an estimate for an upper and lower nominal value (μ_{upper} and μ_{lower}, respectively), based on a trend peak value or the mean of a number of such peak values. For P = 99.73 percent, we then obtain the following control limits:

$$\text{With} \quad \mu_{upper} = 4.018$$
$$\mu_{lower} = 3.992$$
$$\hat{\sigma} = \bar{s}/a_n = 0.005284$$
$$UCL = \mu_{upper} + 3\frac{\hat{\sigma}}{\sqrt{n}} = 4.018 + 3 \times \frac{0.005284}{\sqrt{5}} = 4.025$$
$$LCL = \mu_{lower} - 3\frac{\hat{\sigma}}{\sqrt{n}} = 3.992 - 3 \times \frac{0.005284}{\sqrt{5}} = 3.985$$

The resulting control chart is shown in Figure 3.9–8 (left-hand side). The widening of the control limits is given by $\mu_{upper} - \mu_{lower} = 0.026$.

Overall Standard Deviation. In processes where the population average fluctuates, it is sometimes possible to determine the overall standard deviation (i.e., the standard deviation calculated from all the values) and to determine the control limits on this basis. It is, of course, necessary to check, e.g., using probability paper or by inspection of the frequency histogram, that the process model assumed is correct. Based on these assumptions, we then find the following control limits:

$$\text{With} \quad \mu = \bar{\bar{x}} = 4.00657$$
$$\hat{\sigma} = s_{tot} = 0.01133$$

we have

$$UCL = \mu + 3\frac{\hat{\sigma}}{\sqrt{n}} = 4.00657 + 3 \times \frac{0.01133}{\sqrt{5}} = 4.022$$
$$LCL = \mu - 3\frac{\hat{\sigma}}{\sqrt{n}} = 4.00657 - 3 \times \frac{0.01133}{\sqrt{5}} = 3.991$$

**A u value of $u_{1-\alpha/2} = 2$ has been found to give good results in practice.*

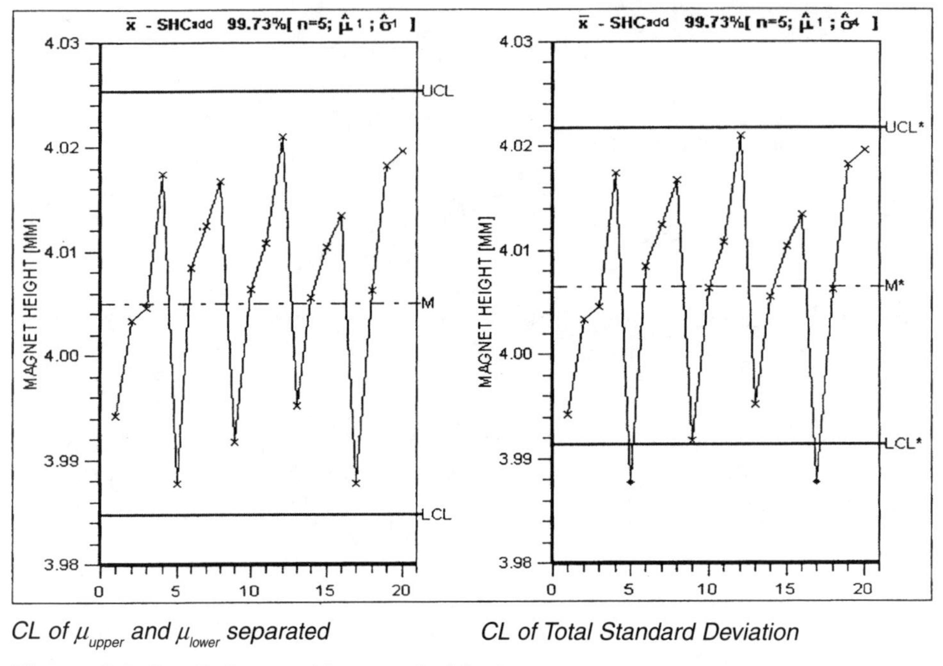

CL of μ_{upper} and μ_{lower} separated *CL of Total Standard Deviation*

Figure 3.9–8. $\bar{x}$ charts with extended limits.

The resulting control chart is shown in Figure 3.9–8 (right-hand side). The control limits are narrower than with the other calculation methods. Two samples fall below the lower control limit, which is of minor importance here, since the trend in our example is an increasing one.

Acceptance Control Charts. While acceptance control charts are not in line with the philosophy of performance aimed at the target value, their use does make sense for processes subject to trends, where the acceptable trend component is predefined by economic considerations. The control limits for different nonconformity proportions are found as follows (see Table 3.6–2):

$$USL = 4.04 \quad \text{and} \quad LSL = 3.97$$

$$\text{with} \quad 1 - P_a = 90\% \quad \text{and} \quad \hat{\sigma} = \sqrt{s^2} = 0.00541$$

for p = 1 percent

$$UCL = USL - \left(u_{1-p} + \frac{u_{1-P_a}}{\sqrt{n}} \right) \hat{\sigma} = 4.04 - \left(2.3265 + \frac{1.28155}{\sqrt{5}} \right) 0.00541 = 4.024$$

$$LCL = LSL + \left(u_{1-p} + \frac{u_{1-P_a}}{\sqrt{n}} \right) \hat{\sigma} = 3.97 + \left(2.3265 + \frac{1.28155}{\sqrt{5}} \right) 0.00541 = 3.986$$

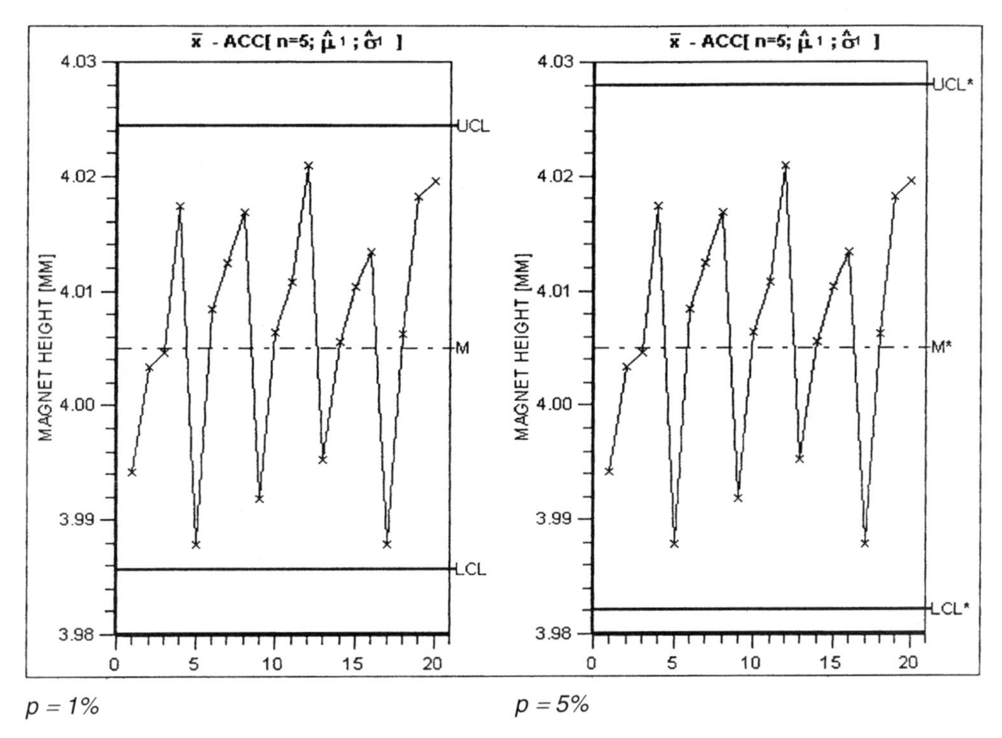

p = 1% p = 5%

Figure 3.9–9. $\bar{x}$ acceptance chart with different error proportions.

for p = 5 percent

$$UCL = USL - \left(u_{1-p} + \frac{u_{1-P_a}}{\sqrt{n}} \right)\hat{\sigma} = 4.04 - \left(1.6448 + \frac{1.28155}{\sqrt{5}} \right)0.00541 = 4.028$$

$$LCL = LSL + \left(u_{1-p} + \frac{u_{1-P_a}}{\sqrt{n}} \right)\hat{\sigma} = 3.97 + \left(1.6448 + \frac{1.28155}{\sqrt{5}} \right)0.00541 = 3.982$$

The resulting control charts are shown in Figure 3.9–9.

3.10 Control Charts: Summary

If the process type is known, a suitable control chart can be selected as shown in Table 3.10–1.

Figure 3.10–1 illustrates typical patterns of variation and shows the appropriate control charts and control limit formulas in each case. The examples for the Shewhart chart with extended limits and the associated formulas should be regarded as suggestions only. Other combinations are also possible.

Table 3.10–1. Process type with appropriate quality control chart.

Process type	Quality control chart
No fluctuations of averages ($\Delta\mu = 0$)/normally distributed (Process model A, B, or E)	Shewhart chart
No fluctuations of averages ($\Delta\mu = 0$)/not normally distributed (Process model A1)	Shewhart chart alternately Pearson chart[1]
Processes with random and accepted fluctuations of averages ($\Delta\mu \neq 0$) (Process model B, F, or G)	Shewhart chart with extended limits
Processes with Trend ($\Delta\mu \neq 0$) (Process model C)	Shewhart chart with extended limits alternately Acceptance charts[2]
Sample size n = 1 • Destroying test • Parameter control • Small batch sizes	Shewhart chart with moving averages

[1] *Whether or not a Pearson chart is appropriate depends on the distribution of averages. The Pearson chart is recommended for extremely skew distributions and small sample sizes, where the averages cannot be considered to follow the normal distribution. This should be tested by plotting the averages on normal probability paper.*

[2] *Acceptance charts are not consistent with the philosophy of never-ending improvement and are thus subject to certain limitations.*

Variation charts usually are not dependent on process type, since a constant variation over time is assumed. As an example, this assumption may be controlled in the CHI^2 plot for the standard deviation. If this is not applicable, then the Pearson method may be used for calculation of the control limits.

If computer systems are used, then the s chart should be preferred to the R chart. The user must differentiate whether he wants to use the exact calculation method for the control limits (see Table 3.5–7) or, for simplicity's sake, the approximation calculation.

Control charts with extended control limits are used in special process situations. The amount by which the control limits are widened should be agreed between supplier and purchaser. The modification of the limits is only permissible if stability is not confirmed by a standard Shewhart chart and there are known technical reasons for the additional variation, whose reduction or elimination is not feasible on technical and/or economic grounds.

Section 3.5.3 dealt with the stability criteria for process model A only. The stability criteria for process models B and C are summarized in Table 3.10–2, along with those for Shewhart charts with moving statistics and Pearson charts. Additional criteria, e.g., relating to the pattern of variation of the individual values (such as whether there are individual measurement values that are outside specification), should be taken into consideration as well. Level 2 is permissible only if there is a large quantity of data and the control chart gives a *visual appearance* of stability. The rationale for this can be found in Section 3.5.3.

Only the violation of the control limits should be taken into account as stability criterion for variation charts (s or R charts).

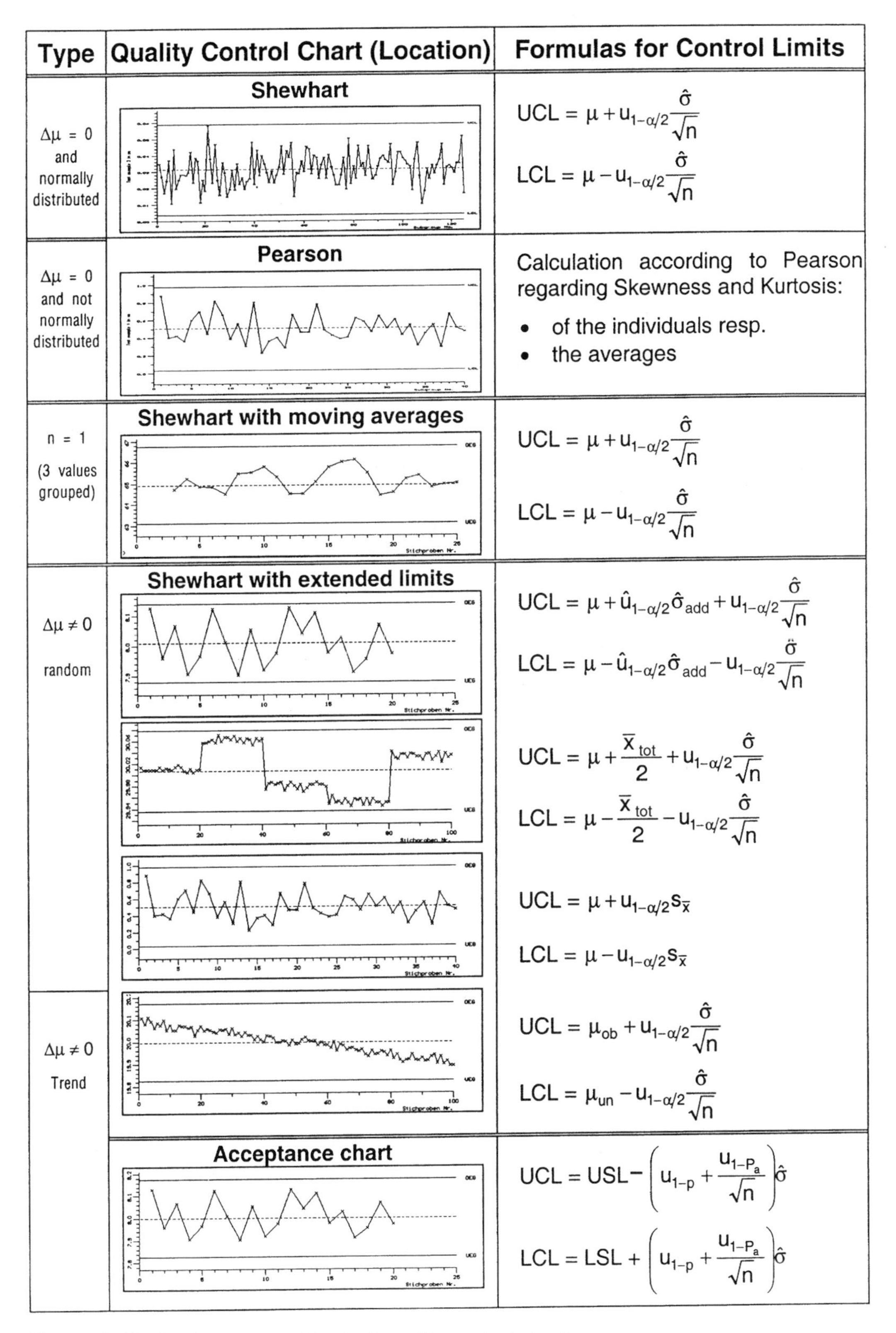

Type	Quality Control Chart (Location)	Formulas for Control Limits
$\Delta\mu = 0$ and normally distributed	Shewhart	$UCL = \mu + u_{1-\alpha/2}\frac{\hat{\sigma}}{\sqrt{n}}$ $LCL = \mu - u_{1-\alpha/2}\frac{\hat{\sigma}}{\sqrt{n}}$
$\Delta\mu = 0$ and not normally distributed	Pearson	Calculation according to Pearson regarding Skewness and Kurtosis: • of the individuals resp. • the averages
n = 1 (3 values grouped)	Shewhart with moving averages	$UCL = \mu + u_{1-\alpha/2}\frac{\hat{\sigma}}{\sqrt{n}}$ $LCL = \mu - u_{1-\alpha/2}\frac{\hat{\sigma}}{\sqrt{n}}$
$\Delta\mu \neq 0$ random	Shewhart with extended limits	$UCL = \mu + \hat{u}_{1-\alpha/2}\hat{\sigma}_{add} + u_{1-\alpha/2}\frac{\hat{\sigma}}{\sqrt{n}}$ $LCL = \mu - \hat{u}_{1-\alpha/2}\hat{\sigma}_{add} - u_{1-\alpha/2}\frac{\ddot{\sigma}}{\sqrt{n}}$
		$UCL = \mu + \frac{\overline{x}_{tot}}{2} + u_{1-\alpha/2}\frac{\hat{\sigma}}{\sqrt{n}}$ $LCL = \mu - \frac{\overline{x}_{tot}}{2} - u_{1-\alpha/2}\frac{\hat{\sigma}}{\sqrt{n}}$
		$UCL = \mu + u_{1-\alpha/2}s_{\overline{x}}$ $LCL = \mu - u_{1-\alpha/2}s_{\overline{x}}$
$\Delta\mu \neq 0$ Trend		$UCL = \mu_{ob} + u_{1-\alpha/2}\frac{\hat{\sigma}}{\sqrt{n}}$ $LCL = \mu_{un} - u_{1-\alpha/2}\frac{\hat{\sigma}}{\sqrt{n}}$
	Acceptance chart	$UCL = USL - \left(u_{1-p} + \frac{u_{1-P_a}}{\sqrt{n}}\right)\hat{\sigma}$ $LCL = LSL + \left(u_{1-p} + \frac{u_{1-P_a}}{\sqrt{n}}\right)\hat{\sigma}$

Figure 3.10–1. Summary of typical quality control charts.

*"Stichproben Nr." (referred to in Figure 3.10–1) means *subgroup number.*

Table 3.10–2. Stability criteria depending on process type.

Process type	Stability	
	Level 1	Level 2
$\Delta\mu = 0$ *and* normally distributed *or* not normally distributed (Shewhart Chart)	**no infringement of tolerance limits** **no infringement of control limits** **no Run** **no Trend** **no Middle Third**	**no infringement of control limits** **The sum total of the control limit infringements (number of values outside the control limits, i.e., $\bar{x}$) does not exceed the random spread region of the Binomial distribution (99% double-sided)** (see example on p. 158)
$\Delta\mu = 0$ *or* not normally distributed (Pearson Chart) *or* QCC with moving Averages	**no infringement of tolerance limits** **no infringement of control limits**	

3.11 Sensitivity of a Control Chart

The sensitivity of a control chart is a performance measure for a given control chart application and can be assessed by means of the operating characteristic (OC) curve (see Figure 3.11–1). Here, the percentage of nonconforming items, p, is plotted on the x axis and the intervention probability, $1 - P_a$, on the y axis. If the process is not centered on the tolerance interval, there will be separate operating characteristics for the upper and lower specification limits. For a given nonconformity proportion p, the probability of intervention may be read off on the y axis. It is thus possible to answer the question: *What is the probability that this nonconformity proportion will be detected by the present control chart?* The aim must be to find control limits that lead to the steepest possible operating characteristic, i.e., ensure a high degree of sensitivity.

In addition, the sensitivity of a control chart is dependent on the sample size. Figure 3.11–2 shows operating characteristic curves for upper and lower specification limits and various sample sizes n.

This tool can help assess whether sample sizes could be reduced and, if so, by how much.

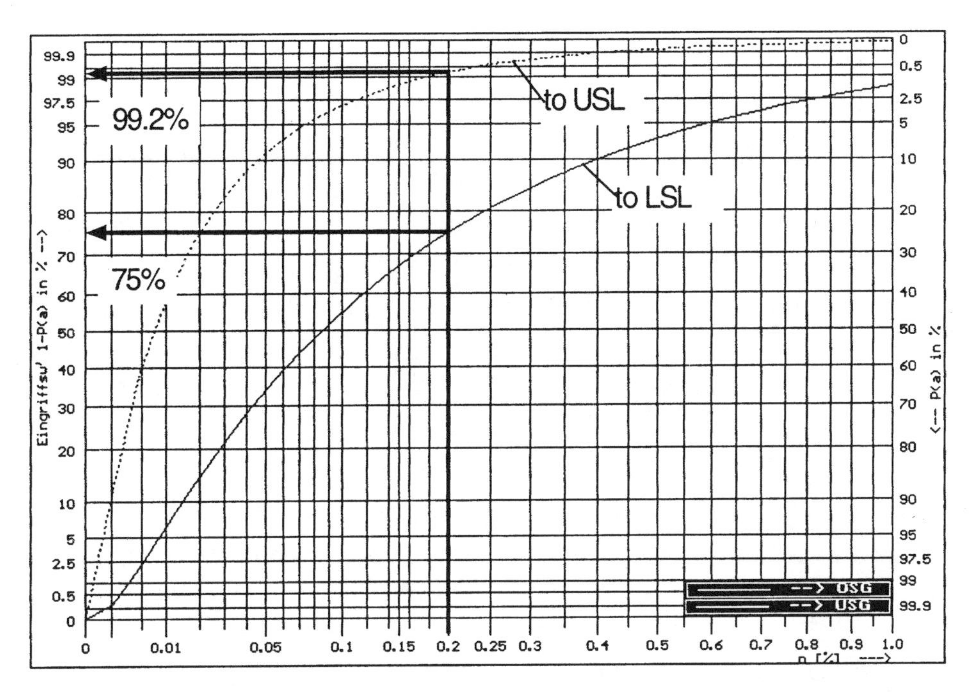

Figure 3.11–1. Operation characteristic.

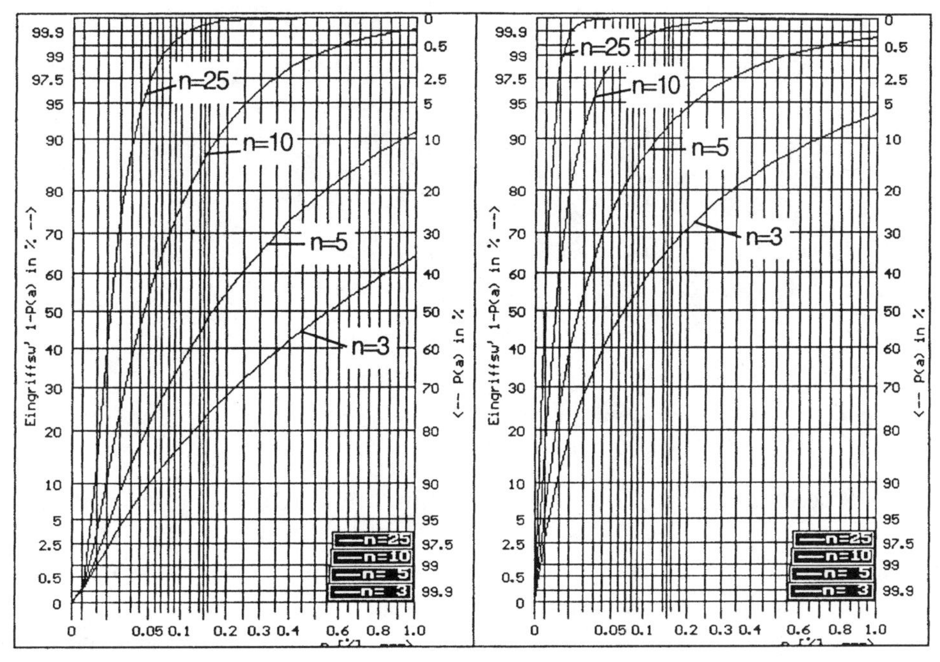

To the lower specification *To the upper specification*

Figure 3.11–2. OC for different subgroup sizes.

3.12 Other Types of Control Charts

3.12.1 PRE-Control Charts

The PRE-control method makes no a priori assumptions about the capability of the process with regard to the specified tolerances. Though single-sided tolerances and attribute characteristics would require a modified procedure, the procedure for a characteristic with a double-sided tolerance follows:

- The tolerance range of the characteristic is divided into four equal parts.
- The two quarters in the center form the green zone.
- The two outer quarters form the yellow zone.
- The regions beyond the tolerance limits form the red zone.

Notes

1. Different definitions for the zones are sometimes used.
2. This type of control chart should only be used when a preliminary trial run of the process demonstrates process stability and sufficient capability (C_p, $C_{pk} \geq 2.0$).

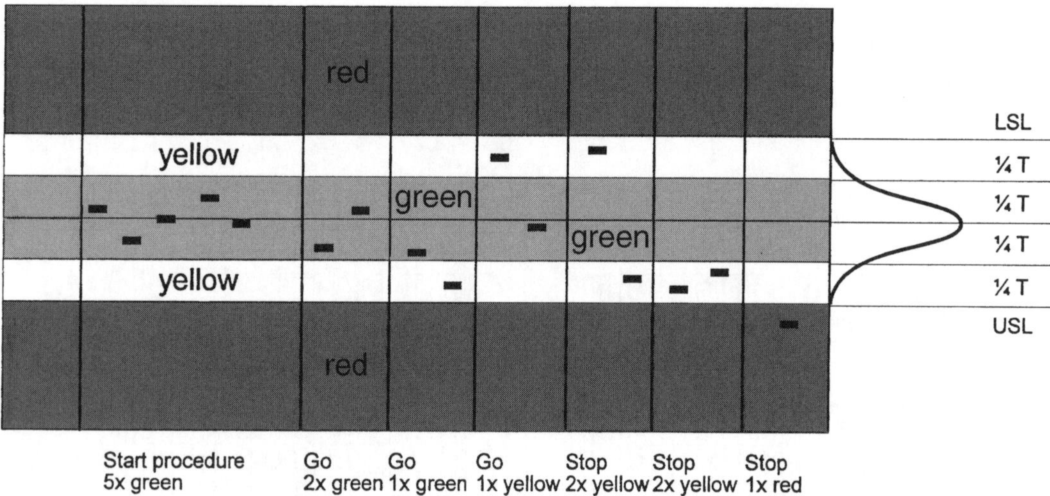

Thus we have preset control limits. Practical application is also very straight forward:

- Setup procedure: After setting up a machine, five successive parts are checked.
- If one or more parts lie in the yellow or even the red zone, the machine should be checked and adjusted and the setup procedure repeated.
- If all 5 parts are *green,* i.e., lie in the green zone, then the process is capable, the distribution is narrow and centered, and it is possible to change over to the normal two-piece sampling procedure.

- The two-piece samples are taken at intervals corresponding to one-sixth of the interval between two machine setups, e.g., every 40 minutes for a four-hour setup interval.
- Decision rules for the two-piece sample:

two green	Go
one green, one yellow	Go
two yellow	Stop, adjust machine
one red	Stop, adjust machine

- Following a machine adjustment, the setup procedure must be repeated.

This is a very simple procedure for process control, which is highly suited for application by operators.

The disadvantages of the method are:

- The method is based on the specified tolerances and thus does not support the never-ending improvement concept.
- Due to the small sample size, the sensitivity of the PRE-Control chart is low.
- Since one value in the red means that the process must be stopped, different sample sizes (in the above example, n = 2 and n = 1) may result. Hence, these data are not very suitable for a subsequent reevaluation of process capability.

The last disadvantage may be avoided by taking two-piece samples during the manufacturing process.

3.12.2 CUSUM Control Charts

The CUSUM (*cu*mulative *sum*) control chart enables us, by summing the deviations of inspection data or sample statistics from a target value, to detect small changes in the mean level of a parameter (in the region of σ to 1.5 σ) at a relatively early stage, even if such changes are observed over only a small number of sampling intervals (see Figure 3.12–1). The commonly used Shewhart chart does not respond well to such small-scale changes in mean level. On the other hand, the Shewhart is more sensitive to larger-scale changes in mean level (≥ 2 σ) than the CUSUM chart. Whether or not the CUSUM chart should be preferred to the Shewhart chart in any

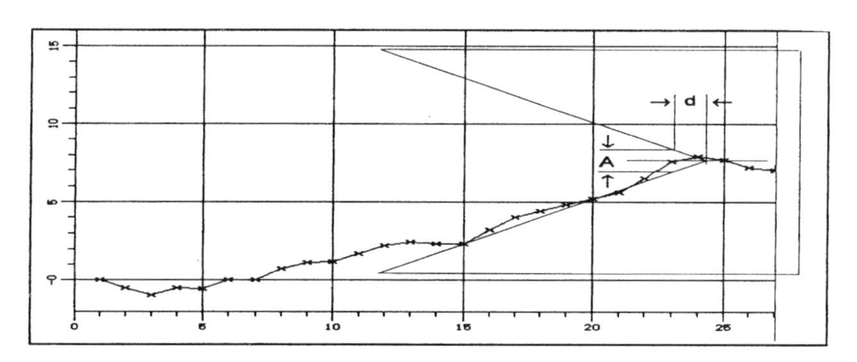

Figure 3.12–1. CUSUM chart.

given application therefore depends very much on the type of process to be monitored and on the monitoring objectives.

The CUSUM chart is suitable for both measurement and count data, including individual measurement values. It has a completely different structure from the charts described up to now, although it is used for the same kind of data. Each newly plotted point holds information about *all the values* that have occurred in the past, including the most recent sample result. The CUSUM chart could therefore be described as a *control chart with unlimited memory.*

The *control limits* for the CUSUM chart usually take the form of a superimposed V mask. The datum point of the V mask is located at a lead distance d from the point of the V. To evaluate the chart, the mask is positioned such that the datum point coincides with the latest plot point. If the entire chart trace is found to lie within the limbs of the mask, then no significant change in mean level has occurred. If, however, the chart trace crosses either the upper or the lower limb of the mask, then a significant departure from the target value is indicated.

Disadvantages of the CUSUM chart:

- The interpretation of the slope of the curve, rather than the ordinate, as indicating mean level often causes conceptual difficulties to the uninitiated.
- The interpretation of the chart is more complicated; computer-based operation is therefore recommended.
- The calculations are simple, but the relationship between the plotted results and the actual process data is much less obvious.

Calculation method for CUSUM control chart:

1. Subtract the target value of the process, T (or denoted by μ_0), from each sample value ($\bar{x}$ or x).
2. Add the result to the cumulative sum of all the previous differences, and plot the total sum on the CUSUM chart. (The result for the first sample is added to the neutral starting point 0.) The addition of negative results leads to a downward slope in the curve.

Using a CUSUM chart with a V mask is not the only way the CUSUM method can be applied. There is also a mathematically equivalent, purely computational CUSUM method using tabulation. This involves subtracting reference values, rather than the target value, from each sample value, and the application of a *decision interval* that the CUSUM results must not exceed.

The sensitivity of a given CUSUM application is usually measured in terms of the ARL (average run length), which represents the average number of samples taken between occurrences of an out-of-control signal (this is analogous to the OC curve described in Section 3.11).

3.12.3 EWMA Control Charts

EWMA (Exponentially Weighted Moving Average) charts in some respects take up an intermediate position between CUSUM charts and Shewhart charts with moving statistics. They have an unlimited memory, but each time a new sample value

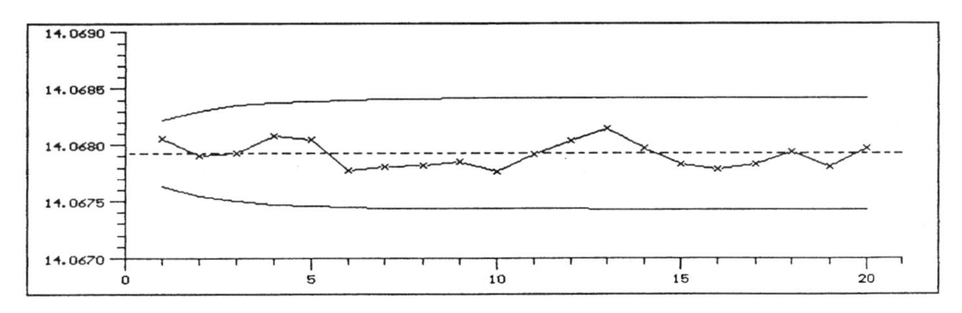

Figure 3.12–2. EWMA chart with λ = 0.2 and k = 2.7.

becomes available, the historical data are retrospectively weighted using an exponential factor. In other words, the old values gradually *fade away,* but they still influence the result sufficiently to give the chart some predictive capabilities as to the likely position of the next plot point. This control chart method is considered to provide particularly sensitive control for complex processes (see Figure 3.12–2).

The memory function may be influenced by the weighting factor λ. The Shewhart chart (without memory) and the CUSUM chart are available as borderline cases. The average chart is recalculated for every sample using

$$\begin{matrix} UCL = \mu_o + k\,\sigma_{(t)} \\ LCL = \mu_o - k\,\sigma_{(t)} \end{matrix} ;\quad \sigma_{(t)} = \sqrt{\frac{\lambda}{n(2-\lambda)}\left[1-(1-\lambda)^{2t}\sigma_0^2\right]}$$

with λ and k from the Crowder nomogram [51].

Consequently, two parameters (λ and k) are required for calculation of the quality control charts. These two parameters determine the average run length (ARL) of the quality control chart, i.e., the average number of samples without process intervention.

Disadvantage: As the EWMA chart is suited to show minor deviations, this may lead to considerable problems in a process with moving averages.

4

Assessment Criteria for Selection of a Distribution Model

It was already stated in the introduction that the validity of any statistical evaluation and its results is totally dependent on the use of the correct model for describing the data. Ever since statistical methods for describing data were first developed, there have been continual discussions as to how an appropriate distribution model may be identified and whether the description afforded by the selected model is sufficiently accurate. The purpose of this chapter is to set out the assessment criteria to be applied in searching for a suitable distribution model. To illustrate the point, various sets of data are assessed with the aid of these criteria and the results discussed.

Figure 4.1–1 shows a number of different frequency distributions with probability functions superimposed. In *a* and *c*, the data are described by the normal distribution model, but there is an obvious lack of fit, and the normal distribution model is clearly not appropriate. The correct models are given in figures *b* and *d*. The numerical values for the capability index C_{pk} show marked differences. As the examples show, the application of the correct model may, depending on the distribution type at hand, lead to an increase or a decrease in the capability indices. This illustrates again the importance of selecting the correct model for describing the distribution of the data.

As a rule, description methods for manual evaluation are very time-consuming and limited to plotting the values on probability paper. Hence, it is not usually possible to compare several alternative methods. Numerical procedures are also used very rarely on account of the large amount of time required. Computer-assisted procedures, however, can make new numerical assessment criteria available and provide the possibility of quickly comparing several alternative models to optimize the evaluation. These possibilities are discussed in the following sections.

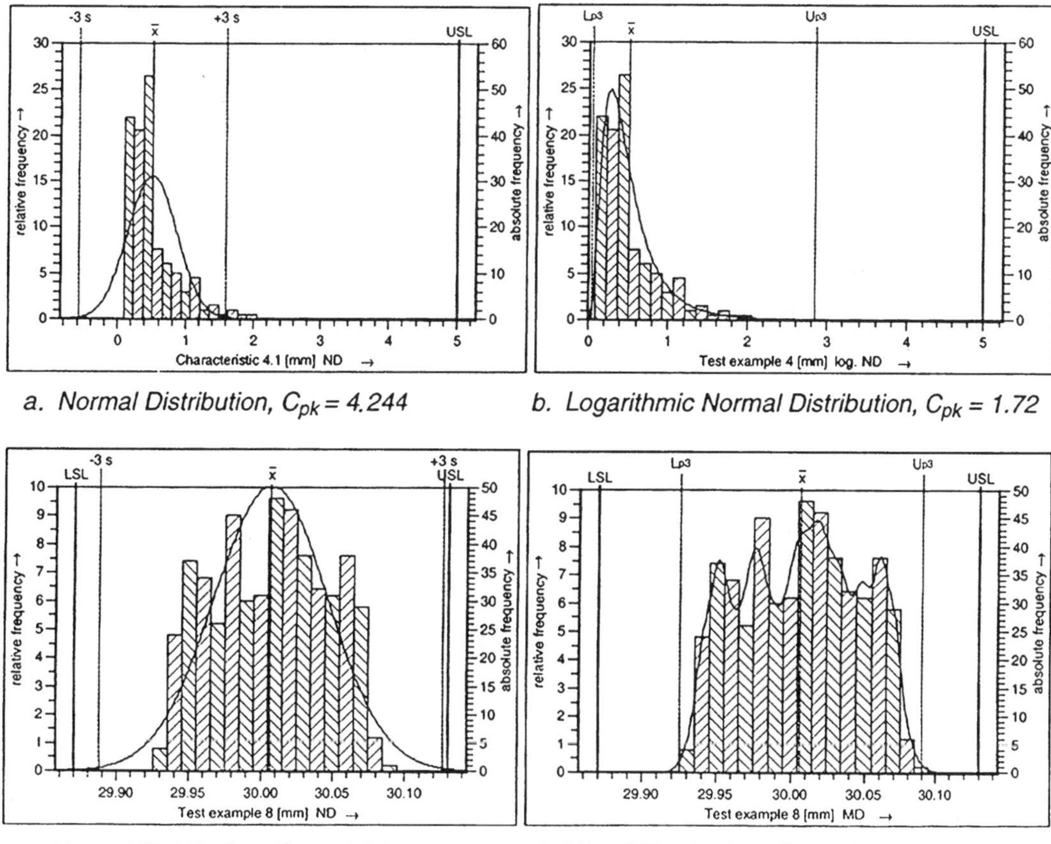

a. *Normal Distribution, C_{pk} = 4.244*

b. *Logarithmic Normal Distribution, C_{pk} = 1.72*

c. *Normal Distribution, C_{pk} = 1.04*

d. *Mixed Distribution, C_{pk} = 1.48*

Figure 4.1–1. Model adaptation.

4.1 Selection of a Distribution Model

The first rule in selecting a distribution model should be *no assessment without technical knowledge.* It is inadvisable to subject a set of data to mathematical procedures without considering the process that generated the data. It is better to identify the distribution model that would be expected for this type of process and to test for confirmation. For example, characteristics that are limited on one side, such as shape and positional tolerances, can be expected to follow a skew distribution. Several types of technical processes and their relevant distribution models are described in bibliography reference [5].

If a model is not confirmed, this usually means that there has been a disturbance in the process or that the process is affected by other external or internal factors, which in many cases cannot be eliminated for technical and/or economic reasons. At any rate, the causes should be identified and explained. In this case the set of measurements deviates from the *ideal model.* To find a statistical description of the measurements, it is then necessary to select the model that gives the best approximation. The distribution models discussed in Chapter 2, Section 2.7, are available for computer-assisted evaluation:

Table 4.1–1. Possible distribution models.

Process type	Distribution model
Process type A (B, E) normally distributed	Normal distribution
Process type A1* not normally distributed	Rayleigh distribution Folded normal distribution Weibull distribution
Process type B, C (D, F, G)	Mixed distribution Johnson transformation Pearson function

**This process type may also be described using the mixed distribution, Johnson transformation, or Pearson function.*

Model	Distribution	Bibliography*
1.	Normal distribution	[13]
2.	Logarithmic normal distribution	[13]
3.	Folded normal distribution/folded at 0	[44]
4.	Rayleigh distribution/folded at 0	[44]
5.	Folded normal distribution/foldover point $\neq 0$	[5/6]
6.	Rayleigh distribution/foldover point $\neq 0$	[5/6]
7.	Weibull distribution	[13]
8.	Johnson transformations	[33]
9.	Pearson functions	[33]
10.	Mixed distribution	

Using suitable computer software, a *numerical assessment* of the possibilities described here may be achieved and a graphical comparison made of the different distribution models. In addition, in models 2, 3, 4, and 7, the values can be subjected to a linear offset in either direction. This offset value may be determined by means of an iterative process. In other words, one begins with an estimated value, which is then adjusted in small increments until an optimum fit has been achieved. In models 5 and 6, the task consists of identifying the appropriate foldover point. In contrast, models 8 and 9 will automatically adapt to the correct shape.

From a theoretical point of view, a distribution model may be allocated to every process type (see Table 4.1–1).

Conformity of the distribution model with the process type must be established with the help of graphic or numeric methods for each case. Suitable evaluation criteria are discussed in Section 4.2.

**These distributions and their practical application are described in more detail in the literature references indicated.*

4.2 Assessment Criteria

4.2.1 Graphical Representations

The best-known method for assessing a distribution model is to plot the data, or their cumulative class frequencies, on probability paper. Preprinted paper is available for different distribution models. If the points plotted conform to the expected straight-line (or curved) probability plot (Figure 2.6–11), then the data do not contradict the assumed distribution model. To support the decision process, confidence limits can be shown as well. In addition to probability plots, cumulative frequency curves (Figure 4.2–1) and histograms may be generated to reach a decision. These graphical methods are extremely subjective and have always been the subject of much debate, especially where results show similarities. Only experienced specialists are able to select the correct distribution model on this basis.

To reach a more objective decision, numerical procedures should be used as an additional tool in the assessment (Sections 4.2.2 and 4.2.3).

4.2.2 Regression Methods

Calculation of a regression coefficient has been found to be a very good criterion for assessment. The regression coefficient compares the theoretical model with the true values and gives a measure of the agreement between the measurement values and the selected distribution model. Figure 4.2–1 illustrates how the regression coefficient is determined.

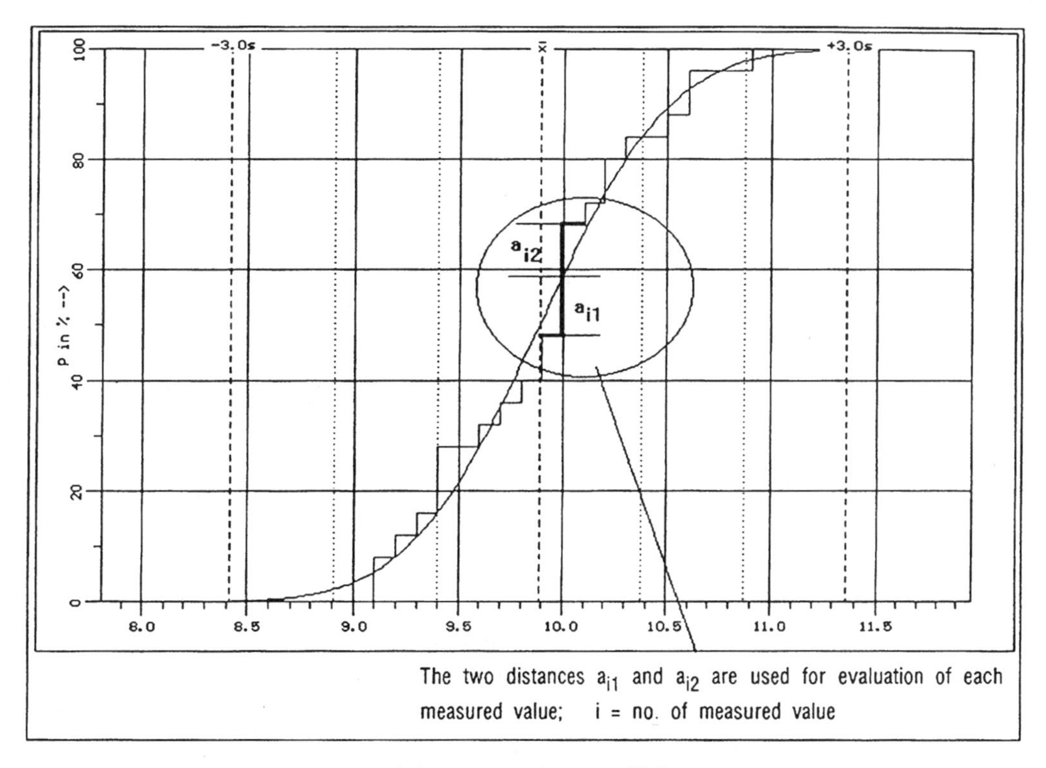

Figure 4.2–1. Determination of the regression coefficient.

To obtain the best possible fit, the regression coefficient is calculated not only for all values (r_{total}), but also for the 25 percent of values lying closest to the critical specification limit ($r_{25\%}$). The figure of 25 percent is an empirically determined value that has been found to give good results in practice. The reason for the additional consideration of these 25 percent of values is that the expected proportion of values outside specification, and hence the associated capability index, is very much dependent on the distribution of the measurement values that are closest to the critical specification limit. Hence, it makes sense to examine this region more closely to ensure that it is accurately described by the model.

The sum of the two regression coefficients (r_{total} and $r_{25\%}$) should be used as a criterion for deciding on the distribution model that provides the best fit. This procedure assigns to each distribution model a numerical value that indicates the goodness of fit. From the numerical viewpoint, the model with the highest numerical value is the one providing the best fit. If several distribution models give the same or very similar results, one should select the model that is most in line with the expectations formed on the basis of the technical features of the process.

These procedures are based on many years of practical experience, and in the authors' opinion, they have proven their worth.

4.2.3 Numerical Test Procedures

Numerical test procedures provide a further means of assessing the goodness of fit of a distribution model. Most of these tests can only be used to test for normal distribution. These include

- Kolmogoroff-Smirnoff-Lillefors test
- Shapiro-Wilk test
- d'Agostino test (limited application; see Chapter 2, Section 2.8)
- Epps-Pulley test
- Test for normal distribution with skewness and kurtosis

This means that these tests can only be used to determine whether the normal distribution model is appropriate or not. The only case where these tests may be used for nonnormally distributed data is where the values may be transformed by means of a suitable function in such a manner that a normal distribution may be assumed for the transformed data. The above-named test procedures will then be used to establish whether the original values themselves or the transformed values are normally distributed. One typical example is the logarithmic transformation of the values. In this case, calculation of the statistics must be carried out on the transformed values and the results transformed back by a suitable method.

One of the few tests that can be applied to any kind of distribution is the χ^2 goodness-of-fit test. The basic concept of the χ is very simple:

First, the sample values are subdivided into classes. Then the probabilities for these classes are calculated based on the assumed distribution function (expected frequencies), and these are compared with the actual class frequencies observed in the sample (observed frequencies). It is assumed that all expected frequencies are greater than one and that no more than 20 percent of all expected frequencies are less than five. This assumption can usually be satisfied by combining the marginal classes. *By using the classified set of sample data as the basis for the test, the goodness of fit of each distribution model may be tested.*

Due to different expected class frequencies, especially in the tails of the distribution, the classes may need to be combined in different ways, depending on the distribution model, to satisfy the requirements for application of the χ^2 test. This, along with differing numbers of required estimated parameters in the various distribution models, leads to differences in the degrees of freedom. Consequently, the test results for different distribution models are not, strictly speaking, comparable. This problem can be solved, however, by determining the significance level α (probability of type one error) based on the value of the test statistic and the appropriate number of degrees of freedom. This significance level can then be used to assess a distribution model and to compare models against each other.

A good fit is indicated by a high significance level. A significance level below five percent would already be enough to cast doubt on the appropriateness of the distribution model.

4.3 Selecting an Appropriate Distribution Model

As we have seen in the previous chapters, there are various distribution models available to describe a set of data, and there are also a number of different methods for assessing the goodness of fit of these models. To arrive quickly at a suitable choice, the following procedure is recommended (Figure 4.3–1):

Identify the expected distribution model, based on the technical properties of the process. Calculate the sum of the coefficients r_{total} and $r_{25\%}$ for various distribution models, using the regression method described earlier, and determine the significance level of the χ^2 test.

Decision A

If this approach clearly identifies only one suitable model, verify the result on probability paper and, if possible, by application of numerical test procedures. In addition, check if the distribution model matches the expected distribution type for this kind of process.

Decision B

If the numerical results suggest that several different distribution models might be suitable, determine the magnitude of the deviations (i.e., the difference between the regression coefficients, and the significance level). If these show only minor differences for the various models, select the model that would be expected for this kind of process, and check for fit by plotting the values on probability paper and/or plotting a cumulative frequency curve. If the result is not confirmed, further analysis of the process is required.

Decision C

If there is a large discrepancy between the expected model and the model identified in the analysis, it is advisable to check the assumptions made. If there is still a discrepancy after this, then the causes must be found and the reasons for the difference established.

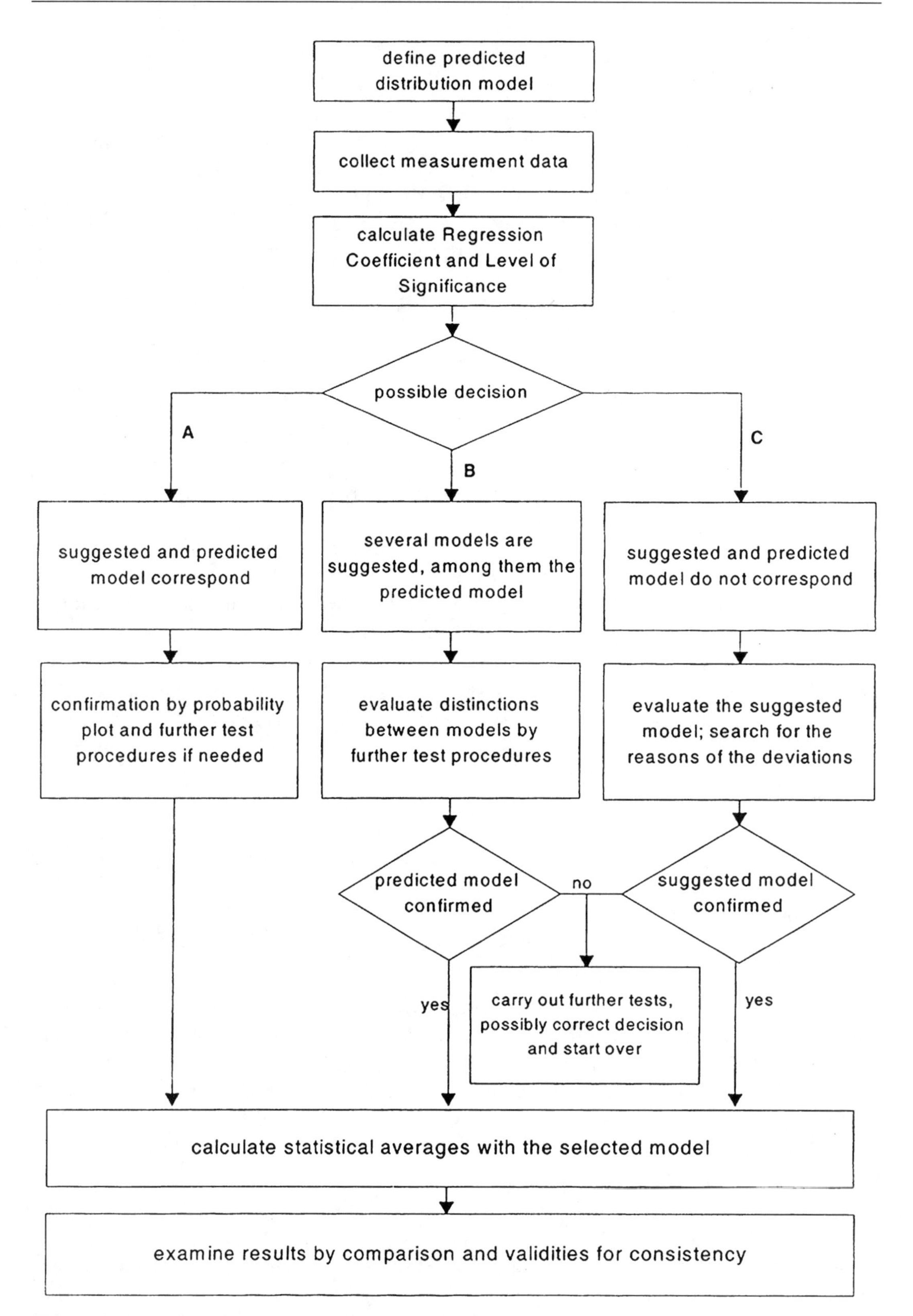

Figure 4.3–1. Procedure for selecting an appropriate distribution model.

Finally, the statistical parameters of the distribution model in question should be calculated. These results should never be regarded as *the absolute truth,* but should always be cross-checked for logical consistency. For example: *If ongoing production has a nonconformity level of 0.3 percent, then it's not plausible for the capability index to be at 2.5, as indicated by the model.* If such inconsistencies appear, the analysis will need to be looked at again.

Examples

To illustrate the foregoing, the Ford test examples 2 to 9 (see Appendix and [40]), as well as the examples from bibliography reference [4], were subjected to the aforementioned assessment criteria and the capability index determined on the basis of the model identified. The results are summarized in Table 4.3–1. The individual columns mean:

Column 1: Number of test.
Column 2: Model distributions used, including any transformations applied to the data. (The model with the gray background is suggested by the authors as the most suitable model.)
Column 3: Regression coefficient as in Figure 4.2–1 for all values.
Column 4: Regression coefficient as in Figure 4.2–1 for the 25 percent of values closest to the critical specification limit.
Column 5: Result of χ^2 test.
In addition, the second line gives an indication as to whether the alternative hypothesis was preferred to the null hypothesis. Here,
H_0 = null hypothesis not rejected
* alternative hypothesis preferred with $\alpha \leq 5\%$
** alternative hypothesis preferred with $\alpha \leq 1\%$
*** alternative hypothesis preferred with $\alpha \leq 0.1\%$
Column 6: An indication as to which distribution models have received the best assessment results.
The plus signs have the following meaning:
In the first line, a + indicates that the model achieved the highest value in column 2, 3, or 4. In the second line, ++ indicates that this model provides the best overall agreement with the actual data.
Column 7: Critical capability indices as determined on the basis of the various distribution models.

Notes

For some examples, certain distribution models are not represented. In these cases, the models concerned were either not deemed compatible with the technical properties of the process, or the application of numerical methods was not possible with the procedures described here.

The results based on the *mixed distribution* are included as supplement, however, were not subjected to the evaluation criteria described for the *most suitable distribution model.*

Table 4.3–1. Assortment of suitable distribution models.

Test No.	Model Distribution Transformation	r (total)	r (25%)	α χ^2 Test	Notes	Critical C value
2	Normal Distribution	0.99880426	0.97532639	0.90769184	+ + +	**1.90**
	No Transformation			H0	++	
	lognormal Distribution	0.99822735	0.97458783	0.89032684		
	g(x) = ln(105363 – x)			H0		1.99
	Mixed Distribution	*0.99904*	*0.99040*			*2.13*
	No Transformation			H0		
3	Normal Distribution*	0.99517910	0.95785667	0.67488242	+	**1.42**
	No Transformation					
	lognormal Distribution	0.99523271	0.95807775	0.58474274	+ +	1.39
	g(x) = ln(131.8142 – x)			H0	++	
	Mixed Distribution	*0.99525*	*0.96069*			*1.49*
	No Transformation					
4	Normal Distribution	0.94583964	0.46215323	0.00000000		4.15
	No Transformation			***		
	lognormal Distribution	0.99022790	0.96491493	0.36408993	+ + +	**1.92**
	g(x) = ln(0.030 + x)			H0	++	
	Folded Normal Distribution	0.97240934	0.81746940	0.00000389		3.04
	g(x) = –0.015 + x			***		
	Rayleigh Distribution	0.96189159	0.67026064	0.00000000		3.23
	g(x) = 0.227 + x			***		
	Folded Norm.Distrib. ≠ 0	0.96922827	0.75950550	0.00000000		3.38
	No Transformation			***		
	Weibull Distribution	0.98636123	0.92286168	0.01833671		2.79
	g(x) = –0.065 + x			*		
	Mixed Distribution	*0.99085*	*0.98389*			*2.94*
	No Transformation					
5	Normal Distribution	0.99000941	0.90252047	0.00745065		1.10
	No Transformation			**		
	Weibull Distribution	0.99769900	0.99024385	0.86775062	+ + +	**0.82**
	g(x) = 782.89 + x			H0	++	
	Mixed Distribution	*0.99825*	*0.96359*			*0.97*
	No Transformation					
6	Normal Distribution	0.99578939	0.99062517	0.00045416		1.84
	No Transformation			***		
	lognormal Distribution	0.99815468	0.98998057	0.53965441		1.34
	g(x) = ln(0.03095 + x)			H0		
	Folded Normal Distribution	0.97192798	0.70182023	0.00000000		1.02
	g(x) = –0.00100 + x			***		
	Rayleigh Distribution	0.99929597	0.99688434	0.95476993	+ + +	**1.51**
	g(x) = 0.00080 + x			H0	++	
	Folded Norm.Distrib. ≠ 0	0.99845098	0.99568060	0.06502316		1.75
	No Transformation			H0		
	Weibull Distribution	0.99905306	0.99636819	0.92049135		1.50
	g(x) = 0.00002 + x			H0		
	Mixed Distribution	*0.99948*	*0.99461*			*1.56*
	No Transformation					
7	Normal Distribution	0.97701938	0.91237632	0.00000000		2.26
	No Transformation			***		
	lognormal Distribution	0.99442409	0.97127704	0.00065632		1.09
	g(x) = ln(0.00045 + x)			***		

**Although the lognormal distribution would appear to provide the best fit, the normal distribution was confirmed for this example, as the lognormal distribution approaches the normal distribution for $\mu/\sigma \Rightarrow \infty$.*

Test No.	Model Distribution Transformation	r (total)	r (25%)	α χ^2 Test	Notes	Critical C value
	Folded Normal Distribution g(x) = –0.00200 + x	0.99600935	0.98432659	0.13610794 H0	+ + ++	1.67
	Rayleigh Distribution g(x) = 0.00081 + x	0.99094538	0.96777528	0.00010895 ***		1.84
	Folded Normal Distrib. ≠ 0 No Transformation	0.98381880	0.93378716	0.00000000 ***		2.11
	Weibull Distribution g(x) = –0.00168 + x	0.99601817	0.98270093	0.08717876 H0	+	1.55
	Mixed Distribution *No Transformation*	*0.99642*	*0.98971*			*1.69*
8	Normal Distribution No Transformation	0.99695556	0.89211887	0.00000000 ***		1.04
	lognormal Distribution g(x) = –ln(30.92880 – x)	0.99153190	0.85729883	0.00000000 ***		1.08
	Johnson	0.99695556	0.96353154	0.07202874 H0	+ + + ++	**1.64**
	Pearson	0.99695556	0.96344909	0.01719898 *		1.69
	Mixed Distribution *No Transformation*	*0.99722*	*0.96556*			*1.48*
9	Normal Distribution No Transformation	0.99852080	0.95793254	0.00098354 ***		1.55
	lognormal Distribution g(x) = ln(21.2254 – x)	0.99606678	0.95811031	0.00077521 ***		1.45
	Johnson	0.99852080	0.98661201	0.13998326 H0	+ + + ++	**1.88**
	Pearson	0.99852080	0.98575944	0.12768573 H0		1.89
	Mixed Distribution *No Transformation*	*0.99872*	*0.98195*			*1.69*
BV1	Normal Distribution No Transformation	0.98388154	0.85640969	0.61347993 H0	+	1.17
	Folded Normal Distribution No Transformation	0.97242800	0.80782695	0.08144864 H0		0.71
	Rayleigh Distribution No Transformation	0.97879983	0.82493459	0.06622697 H0		1.13
	Folded Normal Distr. ≠ 0 No Transformation	0.98679442	0.88967245	0.36309868 H0	+ + ++	**1.06**
	Mixed Distribution *No Transformation*	*0.9944*	*0.94022*			*0.57*
BV2	Normal Distribution No Transformation	0.99221152	0.91177247	0.00000000 ***		0.45
	lognormal Distribution g(x) = ln(x)	0.98907859	0.92599946	0.00000000 ***	+	0.34
	Folded Normal Distribution No Transformation	0.64063877	0.00000000	0.00000000 ***		0.08
	Rayleigh Distribution No Transformation	0.79510338	0.00000000	0.00000000 ***		0.13
	Folded Norm.Distrib. ≠ 0 No Transformation	0.99221152	0.91177247	0.00000000 ***		0.45
	Folded Norm.Distrib. ≠ 0*** No Transformation	0.99221270	0.91212313	0.00000000 ***	+ ++	**0.45**
	Weibull Distribution No Transformation	0.98906110	0.86849528	0.00000000 ***		0.55
	Mixed Distribution *No Transformation*	*0.99418*	*0.94461*			*0.51*

Figures 4.3–2 to 4.3–11 illustrate the agreement between the selected distribution models and the actual measurement values.

The results in Table 4.3–1 show clearly that despite wide-ranging numerical investigations, it is not always possible to identify a distribution model without ambiguity. Further tests, e.g., plotting on probability paper, are necessary, taking into account process and environmental conditions. The results should always be tested for logical consistency. Where appropriate, the model suggestion should be revised.

It should be noted that the choice of distribution model has a significant impact on the capability index. The substantial discrepancies observed in some examples underscore once more the vital importance of choosing the correct distribution model.

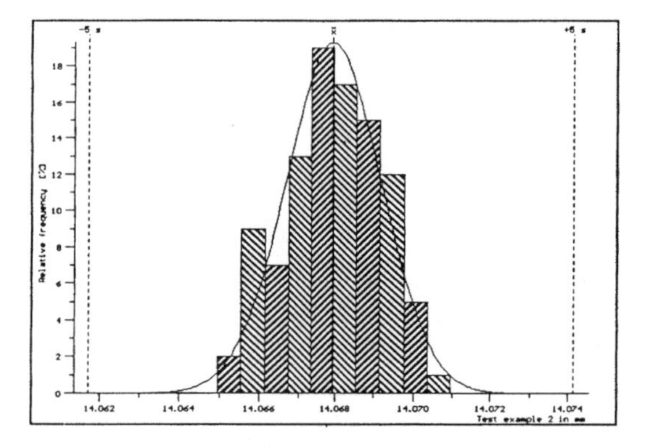

Figure 4.3–2. Test 2: Histogram with normal distribution.

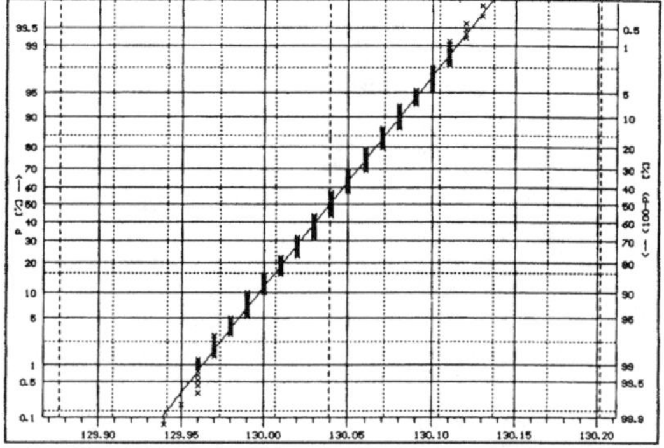

Figure 4.3–3. Test 3: Probability plot for normal distribution.

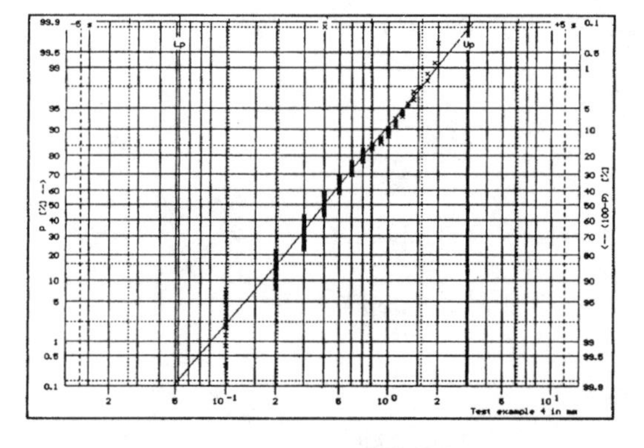

Figure 4.3–4. Test 4: Probability plot for logarithmic normal distribution.

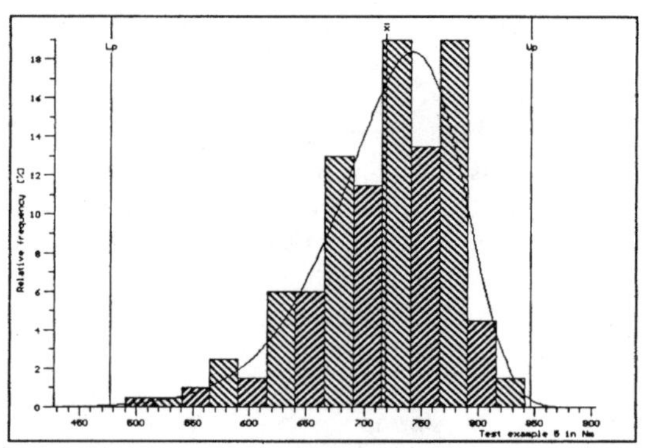

Figure 4.3–5. Test 5: Histogram with overlaid density function of the Weibull distribution.

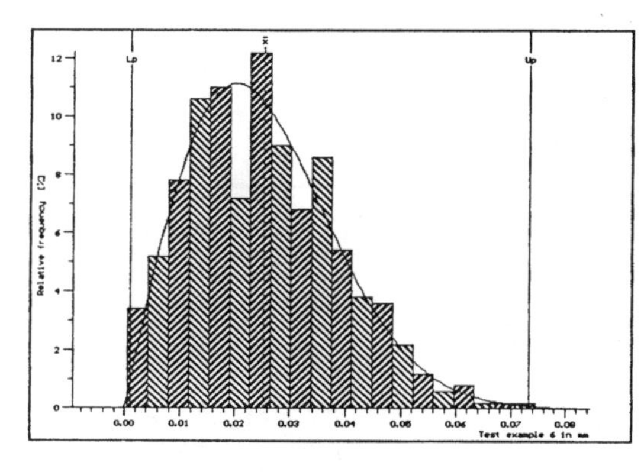

Figure 4.3–6. Test 6: Histogram with overlaid Rayleigh distribution.

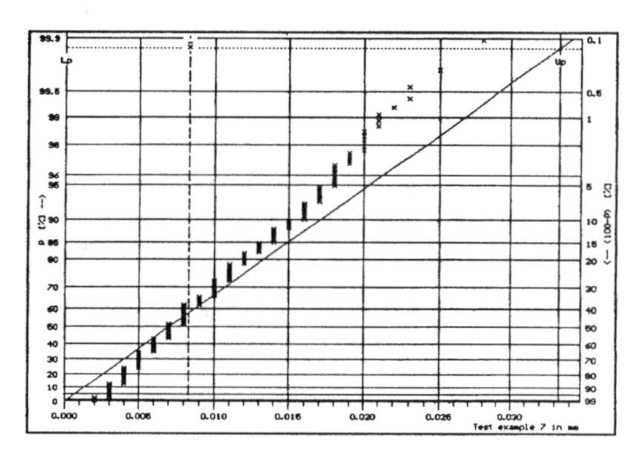

Figure 4.3–7. Test 7: Probability plot for folded normal distribution.

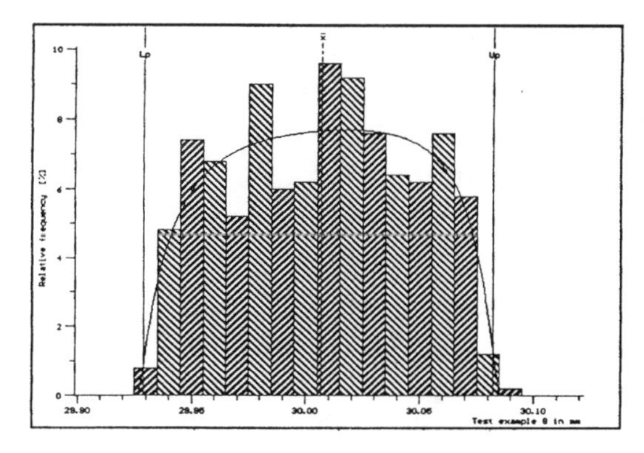

Figure 4.3–8. Test 8: Histogram with overlaid distribution function based on the Johnson transformation.

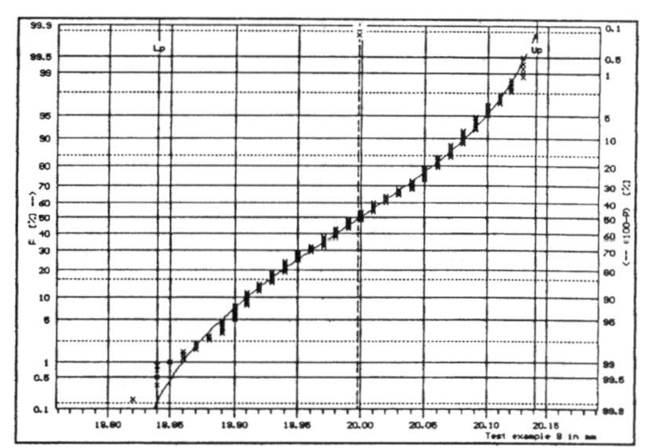

Figure 4.3–9. Test 9: Probability plot for Johnson transformation.

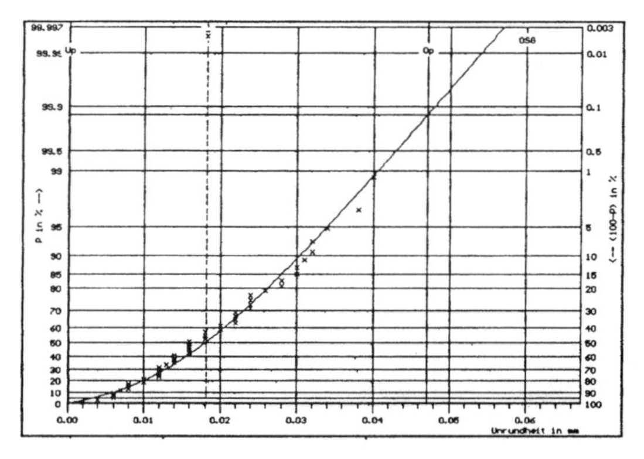

Figure 4.3–10. Test BV-1: Probability plot for folded normal distribution folded at ≠ 0.

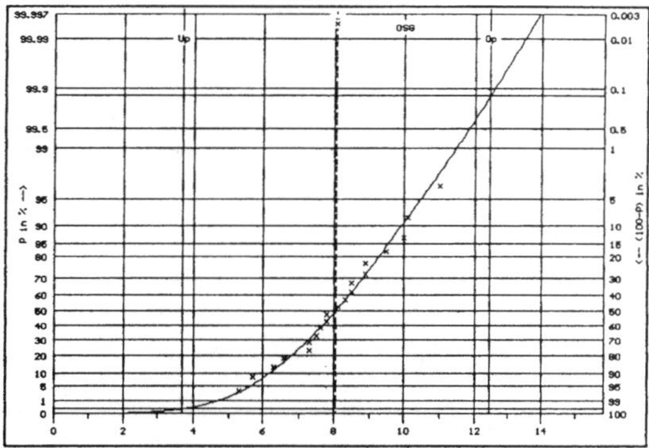

Figure 4.3–11. Test BV-2: Probability plot for Rayleigh distribution folded at ≠ 0.

5 Procedures for Process Assessment

5.1 Choice of Process Model A, B, or C

Proceeding from the description of basic statistical methods in Chapters 2, 3, and 4, this chapter aims to demonstrate the role of these methods in process analysis. Due consideration will also be given to processes with additional variation of the mean (such processes are frequently encountered in practice) and to processes generating nonnormally distributed data.

To simplify the procedure, we have provided flowcharts for process modeling using the appropriate statistical procedures and for establishing the correct process capability indices based on the selected model. This provides the reader with an easy-to-follow set of instructions that will guide him or her automatically through the appropriate steps.

Data Collection and Assessment of Process Stability

All assessments follow the same procedure (see Figure 5.1–1):

- Collect measurement values from the process, using a capable measurement system.
- Group the measurement values to form samples (also called subgroups) of size n = 5, i.e., the first five values form subgroup 1, the next five values form subgroup 2, etc.
- Calculate the mean and standard deviation for each subgroup.
- Plot the mean and standard deviation of each subgroup in the appropriate section of the control chart, above the corresponding subgroup number.
- When at least 20 subgroups have been plotted in both sections of the chart, calculate the control limits, based on the appropriate Shewhart formulas, and plot the limits on the chart.
- Assess process stability in accordance with the standard guidelines for Shewhart control charts.

If the Shewhart chart shows that the process is *not stable,* the special causes of variation must be identified and appropriate corrective action taken.

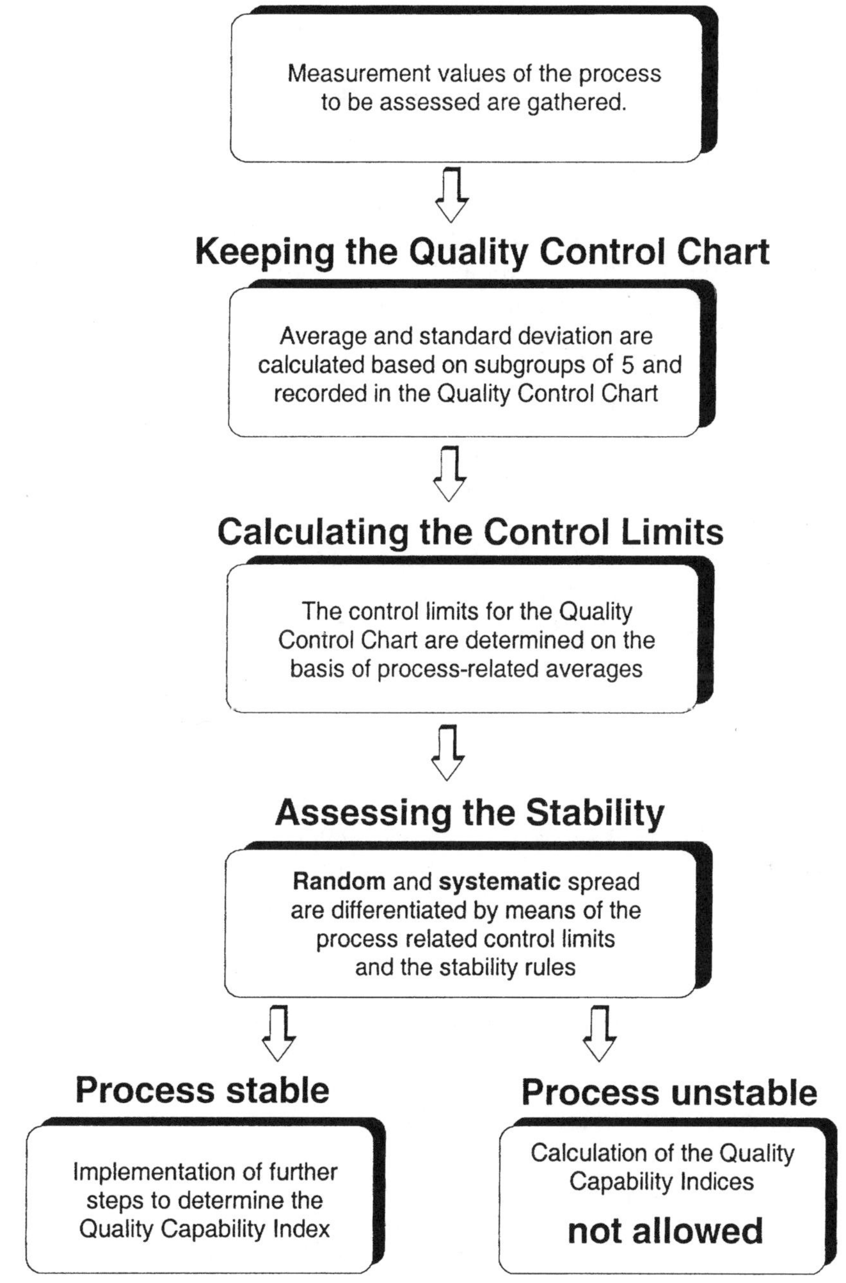

Figure 5.1–1. Data collection and stability assessment.

Initial assessment of stability should always be carried out in strict accordance with this procedure. If stability has to be demonstrated first, we can in many cases stop right there and then, because all processes whose process average varies over time will not be considered stable according to these criteria. Three different methods, corresponding to process models A, B, and C, are available for further assessment.

Note

To get an overview of the process, even in the *process unstable* situation, capability indices should be calculated using the standard deviation or the percentile method (see Chapter 6, Section 6.4.1) based on the mixed distribution. For better identification, the abbreviation of this capability index must be T.

Once stability has been confirmed, further steps may be taken to determine process capability, based on the most appropriate process model.

To proceed further, it is important to know whether process stability has been shown to imply a constant process mean, or whether the process mean is subject to regular and known changes over time. In other words, we need to find out whether process model A, B, or C is appropriate.

Example: Process Model A

If the process mean does not vary and may therefore be regarded as constant over time, the next step consists in finding the appropriate distribution model (see Figure 5.1–2) for description of the process result.

If process analysis confirms that process behavior is consistent with model A, capability indices may be determined in accordance with Chapter 6, Section 6.2. If the normal distribution is confirmed as a valid approximation, the indices will be based on the parameters of this normal distribution. If $\hat{\sigma} \approx \sigma_{total}$, which should be the case here, the calculation of the indices will lead to very similar results regardless of whether $\hat{\sigma}$ or σ_{total} is used as the basis of the calculation.

If a different process model is found that provides a better description of process output (see Chapter 4), determination of capability indices proceeds on the basis of the distribution pattern observed, using either the percentile method (see Chapter 6, Section 6.4.1) or the percentage-outside-specification method (see Chapter 6, Section 6.4.2).

Example: Process Models B and C

There are many industrial processes where the mean is subject to known changes over time. Efforts must always be made to identify the causes of these changes and to eliminate or at least minimize them. But technological limitations or economic considerations frequently render it impossible to eliminate such changes in the mean completely.

If additional variation of the mean due to inherent process conditions is found to be the reason for apparent special-cause variation, it is possible to use a suitable type of control chart, which takes this variation of the mean into account, as a basis for confirming stability.

If the technical properties of the process do not give an adequate indication of the extent of variation of the process means, the ANOVA (analysis of variance) method can be used to establish whether or not the mean does vary over time (see F-test in Chapter 2, Section 2.8.7) and to determine the magnitude of the variation.

To confirm the stability of such a process, the control limits on the control chart must be adjusted and recalculated to take the additional variation of the mean into

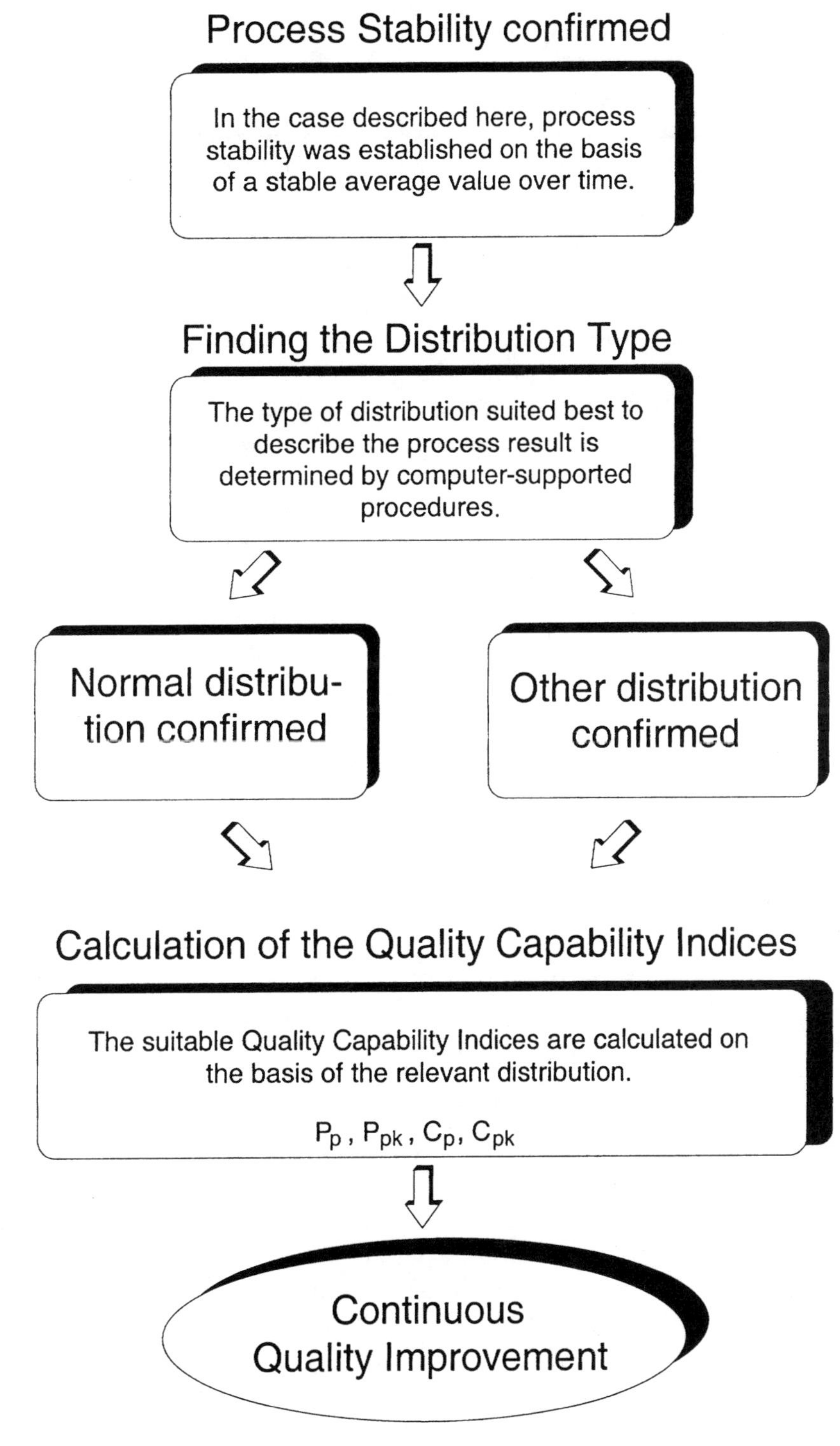

Figure 5.1–2. Procedure for process model A.

account. For the sake of simplicity, it will first be assumed that process models B and C can be sufficiently well described by a normal distribution. This assumption can be tested using numerical test procedures or by plotting the individual values on normal probability paper. The parameters of the normal distribution describing the total population of all individual values then form the basis for calculating process capability indices. Figure 5.1–3 shows an outline of the procedure used in this situation. If the assumption is not validated, then a different distribution model that takes the additional variation of the mean into account will have to be found.

The additional variation of the mean usually arises from random fluctuations in the case of process model B and from linear process fluctuations in the case of process model C. The parameters of the normal distribution observed within the subgroups are used as the basis for calculating the capability indices, but the variation of the process mean that is observed over time is taken into account in the calculation as an additional component of variation. This procedure is shown in Figure 5.1–4.

Process models B and C should be regarded as approximate models for describing processes with additional variation of the mean, and it should be kept in mind that for most such processes a truly exact description of the pattern of variation is not possible by either model. However, for the practical purposes of determining capability indices, the process model that fits the observed data best provides an acceptable basis for assessing process behavior.

5.2 Summary of the Procedure

Before assessment of any process, the capability of the measurement system should be demonstrated (see Chapter 8). If the capability of the measurement system is not confirmed, the system must be improved. Process analysis should be based on measurement values obtained using a capable measurement system. The observed values are grouped to form subgroups of a predefined size and entered on a control chart. The control limits and capability indices are calculated only when a sufficiently large number of samples or values are available. In line with the quest for never-ending quality improvement, the calculated control limits should always reflect actual process performance (see Chapter 3, Section 3.5.4).

The first step is to assess the stability of the process. One needs to establish, using the most efficient method available, whether the process is subject to special causes of variation. For process location, this can be done by analyzing the $\bar{x}$ chart and/or plotting the means on normal probability paper; for process dispersion, this is done by analyzing the s chart and/or plotting the sample variances on χ^2 probability paper. Figure 5.2–1 shows a process subject to special cause variation. If a first look at the process already shows evidence of instability, any further analysis aimed at evaluating process capability is inappropriate. Instead, corrective action is required to stabilize the process.

The next step in the evaluation of a process that is not subject to special cause variation centers on the assessment of any variation of the mean. This variation must be diagnosed and its magnitude estimated. Hence, we must distinguish between model A, *mean constant over time,* and models B and C, *mean subject to changes over time.*

Process Stability confirmed

In the case described here, process stability was confirmed by taking into account an additional variation of the averages over time.

⇩

Choosing the Process Model

Suitable computer-supported procedures determine whether Process Model B or C describe the process result correctly.

⇩

Normal Distribution confirmed

The Normal distribution may be accepted as a sufficient description of process performance using process model B and C.

⇩

Calculation of the Quality Capability Indices

Capability Indices are calculated on the basis of the Normal Distribution.

P_p, P_{pk}, C_p, C_{pk}

⇩

Continuous Quality Improvement

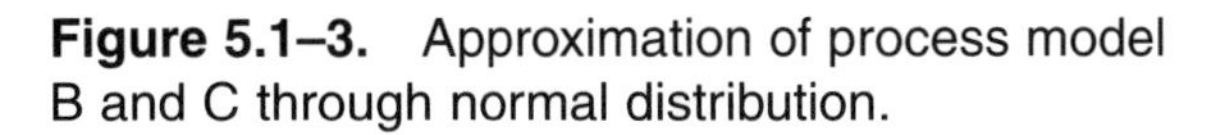

Figure 5.1–3. Approximation of process model B and C through normal distribution.

Process Stability confirmed

In the case presented here, process stability was confirmed taking into account the additional variation of the averages over time.

Choosing the Process Model

Suitable computer-supported procedures are used to determine whether process model B or C describes the process result correctly.

Normal distribution **not** confirmed

With process model B and C, the normal distribution may be accepted as a sufficiently correct description of process performance

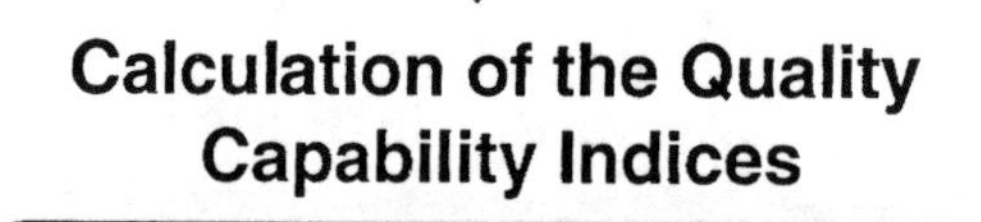

Calculation of the Quality Capability Indices

Capability indices are calculated considering the additional variation of the averages.

P_p, P_{pk}, C_p, C_{pk}

Continuous Quality Improvement

Figure 5.1–4. Approximation of process model B and C through normal distribution with additional variation of averages.

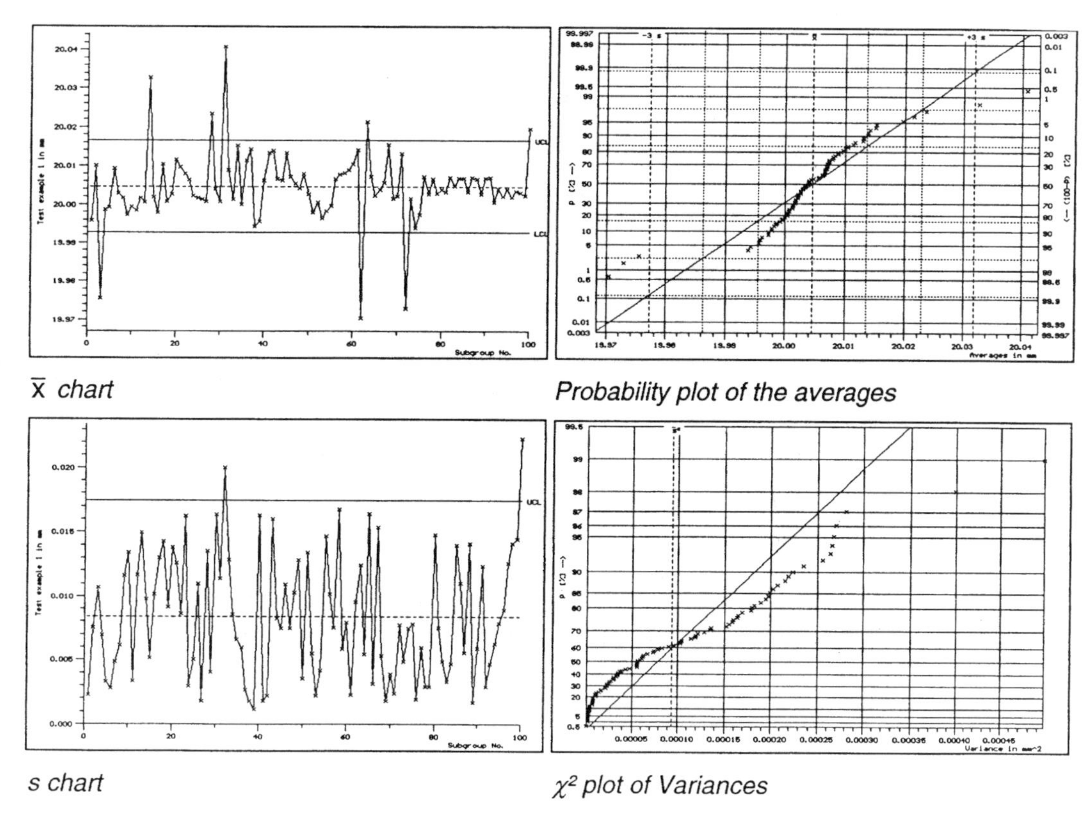

$\bar{x}$ chart

Probability plot of the averages

s chart

χ^2 plot of Variances

Figure 5.2–1. Disturbed process.

The overall process flow of the assessment is summarized in Figure 5.2–2.

Note

To get an overview of the process in this situation (instability), capability indices should be calculated using the standard deviation or the percentile method (see Chapter 6, Section 6.4.1) based on the mixed distribution. For better identification, the abbreviation of this capability index must be T.

Process Model A

First, a suitable distribution model must be found. Statistics, numerical tests, and graphical representations will indicate to what extent a given model is appropriate. If no suitable distribution model is found, capability indices should not be calculated, or the results should be treated with extreme caution.

Process Model B and C

If a process is subject to variation of the mean, this is referred to as a *special* process situation. In this case, we may make a further distinction between systematic influences such as trends (process model C) and unavoidable random variation of the mean (process model B).

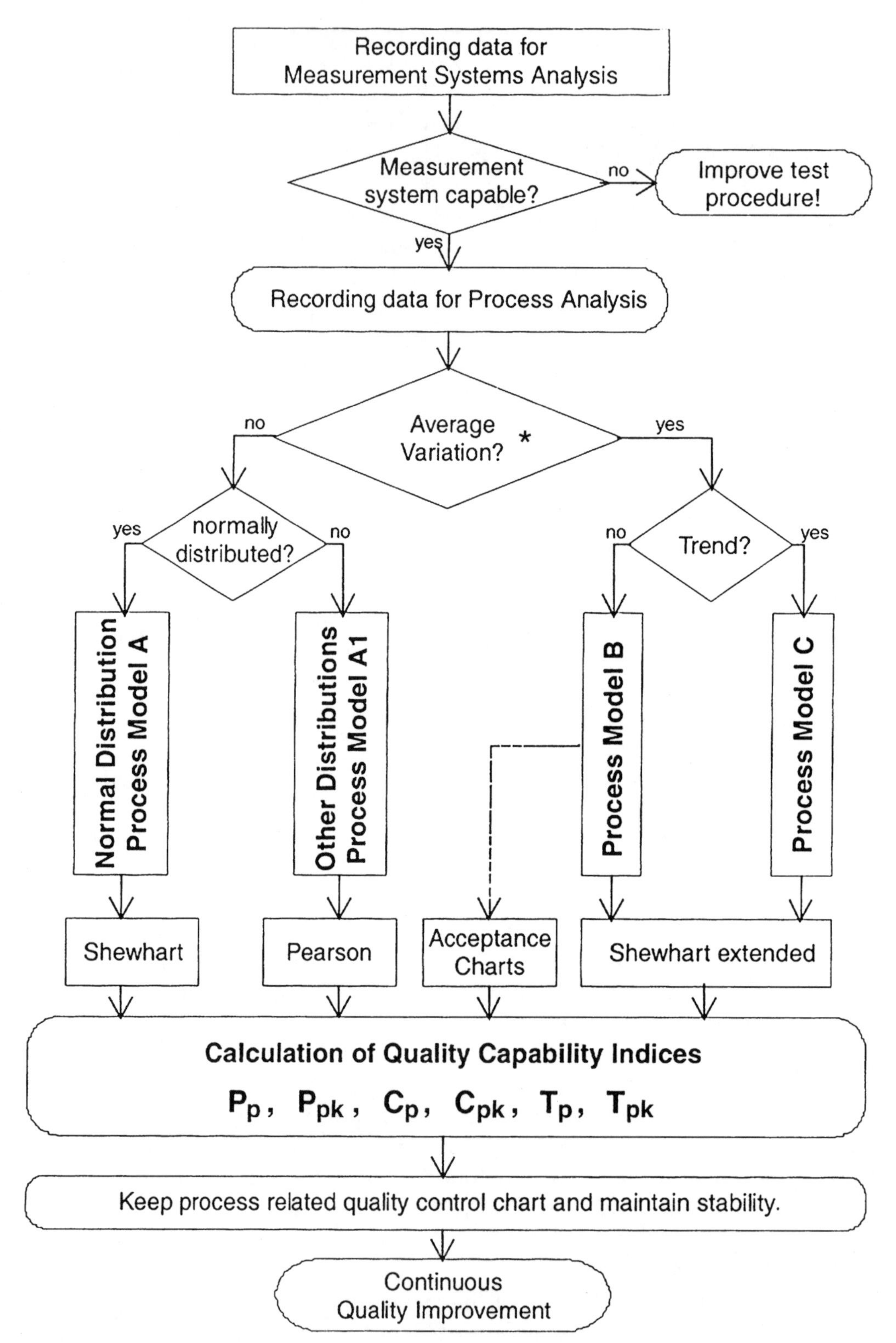

**The F-test may be used for establishing whether fluctuations of the averages occurred.*

Figure 5.2–2. Process analysis.

The type of process will be known at this stage in the analysis. The suitability of the control chart that was applied earlier may be confirmed, or a different chart devised according to process model B or C. Then the control limits are calculated, using the appropriate formulas (see Chapter 3, Section 3.9).

Before calculation of capability indices based on the corresponding formulas, stability needs to be demonstrated again for process models B and C, using the appropriate type of control chart. Comparison of the capability index with a predetermined minimum requirement will then indicate whether a process may or may not be declared capable. After this, there follows the stage of continuous on-site monitoring with a view to *never-ending improvement.*

Examples

At this stage, we should like to refer the reader to the Ford Motor Company test examples for SPC systems (see Appendix). These examples illustrate the following:

- Test_1 Process subject to special cause variation
- Test_3 Process model A, with normal distribution
- Test_4–7 Process model A, with nonnormal distribution
- Test_8 Process model B
- Test_9 Process model C

These examples thus cover the full range of possibilities for process assessment.

5.3 Time Sequence of Capability Studies

Figure 1.4–1 illustrated various phases within a process study. We shall now cover these phases in more detail.

Short-Term Capability/Machine Capability

The aim of a short-term capability study is, as far as possible, to assess only the behavior of the machine or production device itself. This study plays an important role both when acquiring the machine from the supplier and when putting the machine into service for the first time. For reasons of cost, only a few parts ($n \approx 50$) are produced and measured. An attempt is made to find a distribution model based on the measurement values, and capability indices are calculated on this basis. These are denoted by C_m and C_{mk}. The conclusions that can be drawn from these figures will, however, be somewhat vague due to the small sample size (see Chapter 2, Section 2.1.4).

Preliminary Process Capability

As soon as normal production conditions have been established, a preliminary process capability study should be carried out. Samples of size $n = 3, 4, 5, \ldots$ are taken from the process and plotted on a control chart. The number of samples and/or the study period must be chosen such that a sufficient number of data will be available to provide a good overview of process behavior. In general, at least 20 samples, i.e., more than a hundred individual values, are required. This leads to the chart shown in Figure 5.3–1.

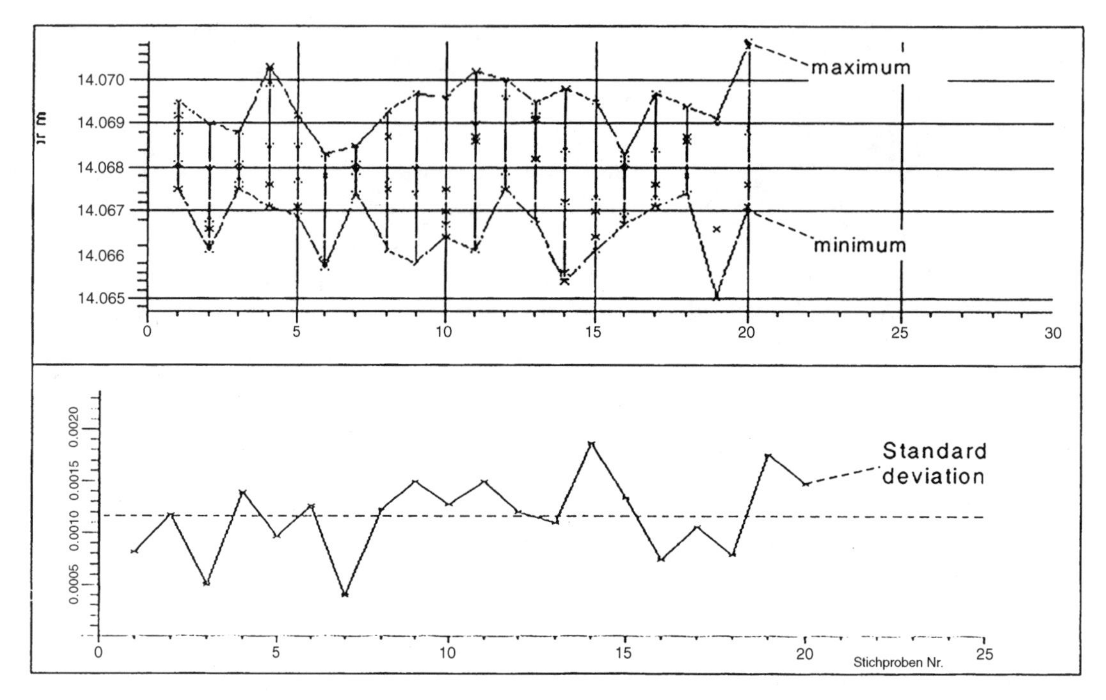

Figure 5.3–1. Recorded samples with individuals and standard deviation.

Finally, the upper and lower control limits for the process are determined, based on the collected data, and entered on the control chart. If the stability criteria are satisfied, the indices of *preliminary process capability*, P_p and P_{pk}, may be calculated. If the minimum requirements (e.g., 1.67) are met, this provides the basis for continuing process operation. Further data must be collected for assessment of *ongoing process capability* (Figure 5.3–2).

Ongoing Process Capability without Recalculation of Control Limits

Process operation continues based on the preliminary capability study and the control limits determined in the course of this study. For an assessment of *ongoing process capability,* the study period should be chosen such that all common causes of variation that are likely to affect the process during normal production are included in the observation period. Only then do the calculated control limits and capability indices provide a realistic forecast and description of future process behavior. The duration of the period of observation will depend on the manufacturing or production process; the usual requirement is 20 days of normal production. If the process remains stable throughout this period, it is not necessary to recalculate the capability indices, and the control limits determined in the preliminary study remain valid. The previously determined values for the capability indices P_p and P_{pk} are retained to become C_p and C_{pk}, and the calculated control limits form the basis of continuous on-site monitoring of the process. Figure 5.3–3 illustrates this situation.

*"Stichproben Nr." (referred to in Figure 5.3–1) means *subgroup number.*

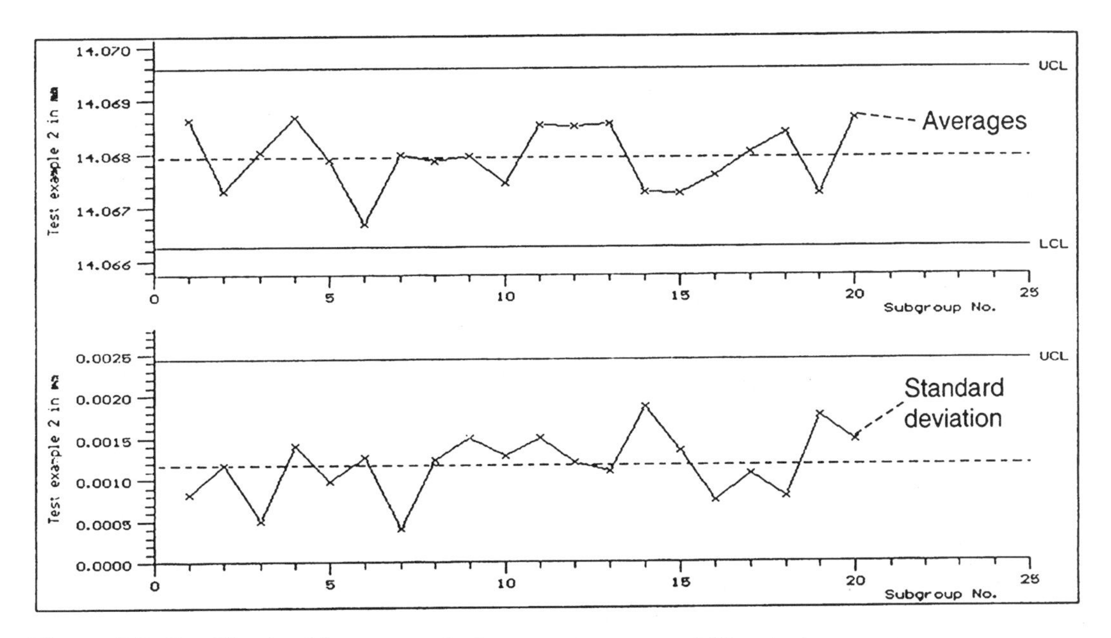

Figure 5.3–2. $\bar{x}$/s chart from a preliminary process capability study.

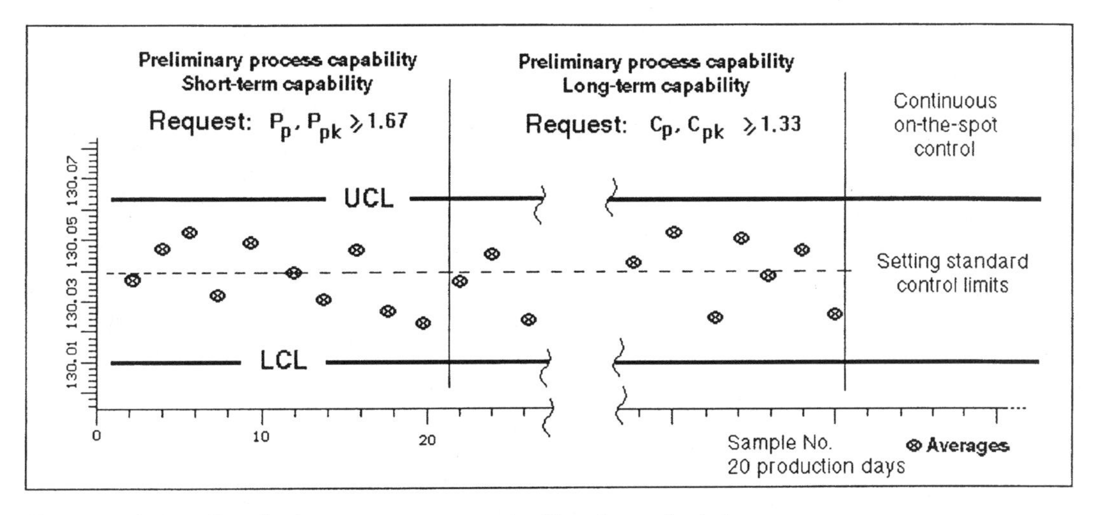

Figure 5.3–3. Continuing process control without recalculation.

Ongoing Process Capability with Recalculation of Control Limits

If the stability attained in the preliminary study could not be maintained (e.g., violation of control limits), improvements must be planned and corrective action taken. Where the greater amount of variation observed in the ongoing capability study can be fully explained from the limited amount of observations used in the preliminary study, a reanalysis of process behavior is permissible. This leads to new control limits and (if stability with respect to the new control limits is confirmed)

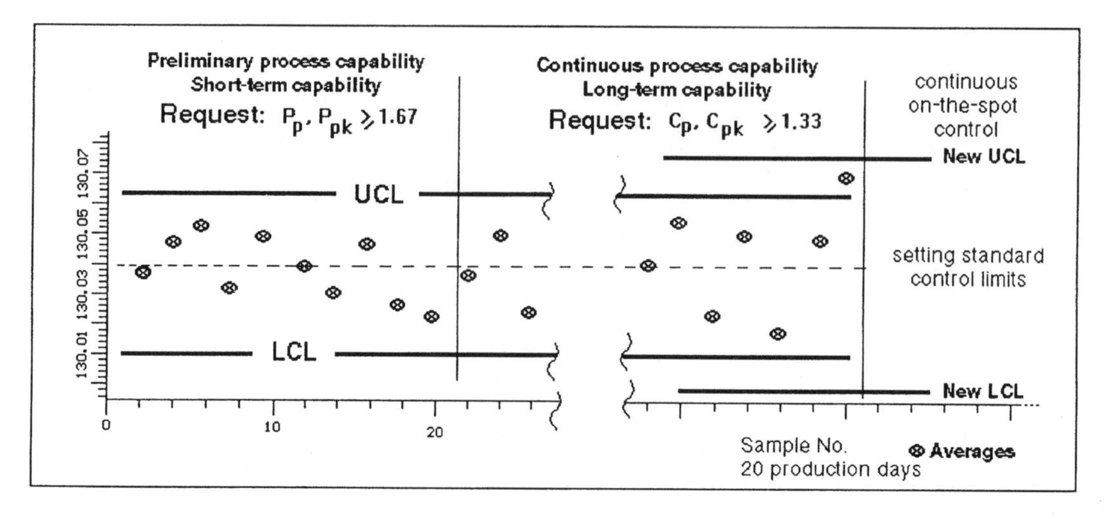

Figure 5.3–4. Continuing process control with recalculation.

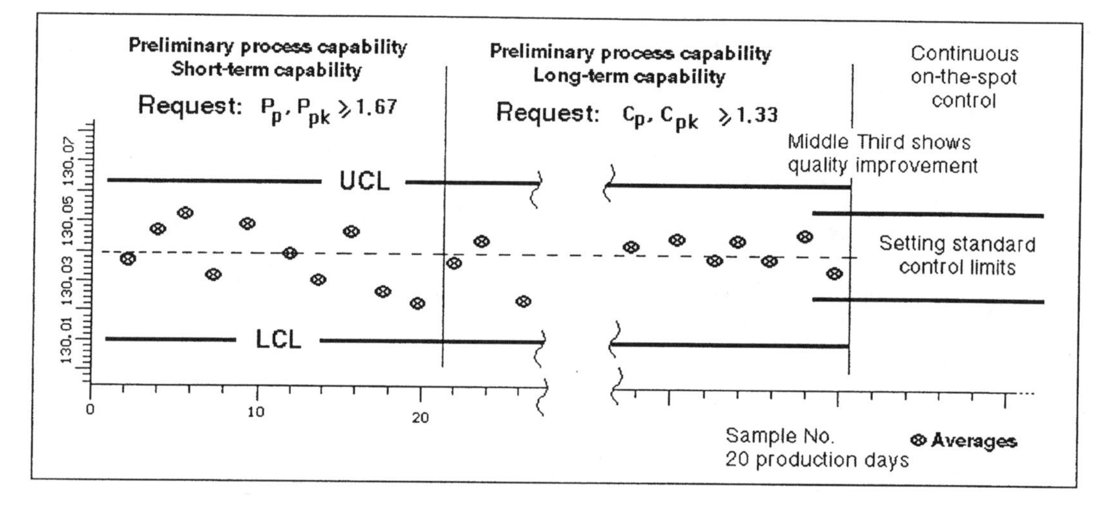

Figure 5.3–5. Shewhart average chart with closer control limits.

new capability indices. Finally, the process is put into operation again under these conditions. Figure 5.3–4 illustrates this situation.

If it is subsequently possible to eliminate sources of variation and a corresponding quality improvement is observed (e.g., indicated by more than 90 percent of plot points in the middle third), the control limits should again be recalculated. Figure 5.3–5 illustrates this.

Completion of the *ongoing capability study* signals the end of the analysis stage, which is then followed by continuous on-site monitoring.

However, in line with the policy of never-ending improvement, all sources of variation and their effects should continue to be analyzed. The information so obtained will permit the introduction of corrective action to reduce process variation further, and, if such a reduction has been achieved, the capability indices should be recalculated. This procedure forms the basis of continuous quality improvement.

6
Process Capability Indices

6.1 Introduction

The first priority in applying SPC should be given to monitoring and evaluating the process. To support the concept of never-ending improvement, appropriate process-specific control charts should be used that will allow the user to accumulate knowledge, assure quality levels, and improve quality wherever this is technically and economically feasible.

In addition, the correct application of SPC provides a basis for describing the current and future (expected) quality level of the process by means of a clear and simple capability index. But what is important, of course, is that this capability index should have a universally accepted definition that is easy to understand.

While it is today generally accepted in industry that process stability is a prerequisite for calculating process capability, there are still various opinions about the methods of calculation to be used. In the discussion as to whether this or that formula is appropriate, there are, unfortunately, no clear technical or mathematical arguments to prove that any particular method is *correct* or *wrong*. And since there is no standard method of calculation that is recognized and uniformly applied throughout the whole of industry, a great deal of time is wasted in the discussion of the pros and cons of the various methods and in comparisons between the capability indices arrived at using the various procedures.

Unfortunately, the study group concerned with the revision of the VDA Volume 4 mentioned in the introduction was also unable to establish a definite standard on the subject of capability calculation that would be binding for all sectors of industry.

There is, however, fairly good agreement on how to calculate capability indices for processes that conform to the normal distribution model and are not subject to any process-inherent additional variation of the mean.

But since the normal distribution is not a suitable model for certain processes, it would clearly be desirable for industry to arrive as soon as possible at a binding decision in favor of one or another method of calculation for these processes, too. The following discussions begin with an explanation of the basic principles underlying the calculation of capability indices. Then, the uncontested method for normal distributions is described, followed by a description and evaluation of the competing calculation methods in use for other types of processes.

Stated in quite general terms, a capability index is a numerical value that describes the relationship between process performance and the specified tolerance interval.

Starting from this general definition, the capability indices for *process potential* and *process capability* are defined as follows:

> *The index of process potential describes the* potential capability *of a process to generate a specific characteristic consistently within specification. Process potential compares the spread of the process with the width of the specified tolerance interval (see Figure 6.1–1).*

Process potential

- Is a measure of the performance the process would produce if it were exactly centered on the tolerance interval
- Depends on the process variation or *spread;* here the process spread is defined as an interval containing 99.73 percent ($\pm 3\sigma$) of all output
- Depends on the tolerance interval
- Compares the process spread with the specified tolerance interval

According to this definition, process potential is independent of process location. In general, process location is easier to influence or adjust than process variation. In this sense, process potential is more indicative of process behavior than process capability. Process potential is always greater than (or, for a perfectly centered process, equal to) process capability.

The following symbols are used for process potential, depending on the type of capability study carried out, and sometimes also depending on the definitions used by individual companies:

Process Potential = C_m, P_p, or C_p

C_m is used in machine capability studies.
P_p is used in preliminary process capability studies.
C_p is used in ongoing process capability studies.

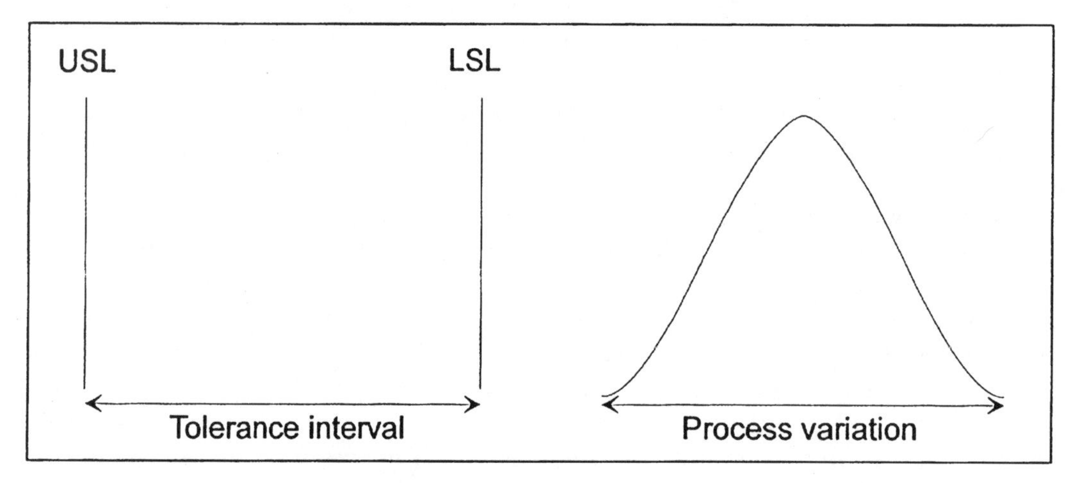

Figure 6.1–1. Process potential.

This means that there is, at present, no universally accepted symbol for process potential. For example, Ford, General Motors, and Chrysler use the symbol P_p on a worldwide basis for the result of a preliminary capability study, whereas other firms and institutions use C_p, both in preliminary and ongoing capability studies. However, there is widespread agreement on the use of the symbol C_p for the result of a long-term study carried out on a stable process running under normal production conditions. Since the mathematical relationships are completely independent of the duration of the observation period, the number of pieces inspected, and any deviations from normal production conditions, we shall henceforth use the symbol C_p in all formulas illustrating the basic method of calculation.

The general formula for calculating process potential is as follows:

$$C_p = \frac{\text{Total Specified Tolerance}}{\text{Process Spread}}$$

The index of process capability describes the actual capability *of a process to generate a specific characteristic consistently within specification. Process capability compares the spread of the process with the width of the specified tolerance interval but, at the same time, also takes process location into account (see Figure 6.1–2).*

Process capability

- Assesses process spread against the specified tolerance interval
- Assesses the location of the process average against the center of the tolerance interval
- Depends upon process location *and* spread

The following symbols are used for process capability, depending on the type of capability study carried out, and sometimes also depending on the definitions used by individual companies:

$$\text{Process capability} = C_{mk}, P_{pk}, \text{ or } C_{pk}$$

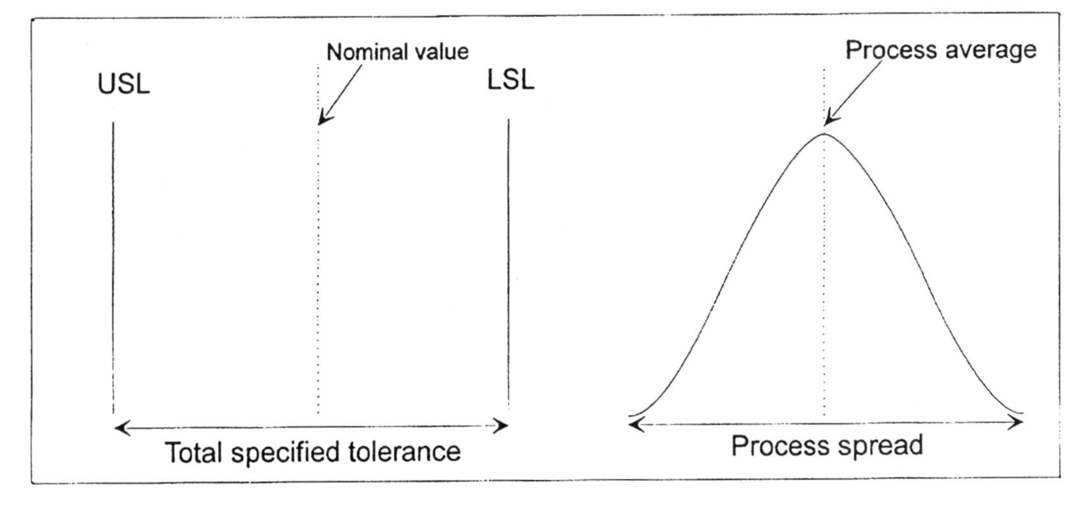

Figure 6.1–2. Process capability.

C_{mk} is used in machine capability studies.
P_{pk} is used in preliminary process capability studies.
C_{pk} is used in ongoing process capability studies.

Again, usage of these symbols varies, and the comments given under process potential apply equally to process capability. However, there is again widespread agreement on the use of the symbol C_{pk} for the result of a long-term study carried out on a stable process running under normal production conditions. Since the mathematical relationships are again completely independent on the duration of the observation period, the number of pieces inspected, and any deviations from normal production conditions, we shall henceforth use the symbol C_{pk} in all formulas illustrating the basic method of calculation.

The general formulas for calculating process capability are

$$C_{po} = \frac{\text{distance between upper specification limit and process average}}{\text{half the process spread}}$$

$$C_{pu} = \frac{\text{distance between lower specification limit and process average}}{\text{half the process spread}}$$

$$C_{pk} = \min\left\{c_{po}\,;\, c_{pu}\right\}$$

This means that the capability index takes both the process spread (i.e., the process variation or dispersion) and the process average (location or central tendency of the process) into account. The more precisely these parameters are determined, using appropriate statistical procedures, the more accurate will be the calculated indices.

In determining the spread of the process, we distinguish, as outlined in Chapter 5, between

- Process model A (normal distribution)
- Process model B (nonnormal distribution)
- Process model B or C (normal distribution)
- Process model B or C (additional $\overline{x}$ variation).

In *process model A* (Figure 1.3–1), we have $\sigma_{total} = \hat{\sigma}$; i.e., the overall standard deviation is identical to the short-term standard deviation of the process. The basic formulas for calculating capability indices are valid without restriction, and it makes no difference whether the calculation is based on $\hat{\sigma}$, as estimated from the individual subgroups, or on s_{total}, the estimate of the standard deviation of the total population (σ_{total}).

Where the normal distribution is not appropriate for process model A, the capability indices are calculated using the percentile method or a method based on the proportion outside specification (see Section 6.4). For *process model B*, we have $\sigma_{total} \gg \hat{\sigma}$; i.e., the overall standard deviation is very much *greater* than the short-term standard deviation $\hat{\sigma}$ of the process.

Thus, only the spread determined from the total production of individual values σ_{total} must be used in the basic capability formulas, and we must not use any estimate based only on the variation $\hat{\sigma}$ observed within the subgroups.

To ensure the comparability of the capability indices so determined, one should always first carry out appropriate tests to establish whether the normal distribution is a suitable model for describing the process. Where control charts with extended limits are used, the additional variation of the mean must be taken into account in determining the indices (see also process model C).

In *process model C,* the short-term variation of the process does conform to the normal distribution, and the standard deviation observed within the individual subgroups is used in calculating the control limits; but, as the process is subject to trends over time, the additional $\bar{x}$ variation must also be taken into account in calculating the control limits on the chart and in calculating capability (see Section 6.3).

6.2 Index Calculation for Normal Distribution

The general principles of capability calculation can very easily be related to the Gaussian normal distribution. This means that if the process can adequately be described by a normal distribution model, the calculation of capability indices is very straightforward, and, more importantly, it is possible to make direct and meaningful comparisons between the resulting indices.

Hence, as long as no other distribution model can be clearly shown to give a better description of the process, capability indices should always be calculated on the basis of the normal distribution to ensure this unambiguous comparability.

The process spread used in calculating capability indices is defined as the width of the central interval that brackets 99.73 percent of the total population. In the case of the normal distribution, the width of this interval is most conveniently measured in terms of standard deviations. Because 99.73 percent of a normally distributed population lies within a $\pm 3\sigma$ interval about the mean, the process spread is taken to be 6σ.

We thus arrive at the following formulas for calculating capability indices for normally distributed characteristics.

Process Potential

Process potential is defined as the *ratio of the specified tolerance interval to the 99.73 percent process spread.* Note that we have, by definition, a C_p value of 1.0 if the tolerance interval is equal to 6σ. Table 6.6–1 shows typical values of C_p.

$$c_p = \frac{USL - LSL}{6\sigma}$$

Process Capability

Since the index of process capability takes into account the location of the process average relative to the two specification limits, in the first instance two calculations are required, i.e., one for each specification limit.

$$C_{po} = \frac{USL - \mu}{3\sigma}$$

$$C_{pu} = \frac{\mu - LSL}{3\sigma}$$

The critical value, i.e., the smaller one of the two, is used to describe process performance and is recorded as C_{pk}, the index of process capability.

$$C_{pk} = \text{minimum of } C_{po} \text{ and } C_{pu}$$

In the ideal scenario where the process average is equal to the nominal value (assuming that this is located at the center of the tolerance interval), we have

$$C_p = C_{pk}$$

For many processes, the normal distribution provides a suitable basis for calculating capability indices. But it must always be kept in mind that the model used to describe the process is only a very good approximation, close enough to allow the use of SPC methods, but never an absolutely accurate description of process behavior. For these reasons, there is no point in calculating capability indices to more than two decimal places.

Confidence Intervals for Capability Indices

Without at this point entering into the details of calculating confidence intervals (see Chapter 2, Section 2.1.4), we should note that the confidence interval attached to a capability index is strongly dependent on the subgroup size and the number of subgroups taken. For example, the 99 percent confidence interval of a capability index is the interval that 99 percent of times will contain the true value of the index.

For example, in the case of the normal distribution, we have the following 99 percent confidence intervals for C_p:

for	10 subgroups of size	$n = 5$	$0.75 \cdot \hat{C}_p \leq C_p \leq 1.26 \cdot \hat{C}_p$
for	25 subgroups of size	$n = 5$	$0.82 \cdot \hat{C}_p \leq C_p \leq 1.18 \cdot \hat{C}_p$
for	50 subgroups of size	$n = 5$	$0.87 \cdot \hat{C}_p \leq C_p \leq 1.13 \cdot \hat{C}_p$

A capability index calculated to two decimal places may be regarded as a *precise* description of process performance only if it was based on at least 1000 measurement values.

In the discussion of the formulas used for calculating capability indices based on the normal distribution, process output was up to now tacitly considered to constitute a homogeneous population. But the shape of the distribution alone does not give a full description of process behavior. The manner in which the distribution of the characteristic varies over time is also an important factor that must always be taken into consideration in the assessment of process performance.

In assessing process performance, the size of the confidence interval attached to the capability index will depend on the selected distribution function, the confidence level, the process model, and the sample size. The accuracy of the calculated index is dependent on

- The process model
- The distribution function
- The subgroup size
- The number of subgroups

Process behavior over time is of special importance, because, without an assessment of this behavior, it is not possible to say whether or not a process is in statistical control (stable). It is not permissible to calculate capability indices without evidence of stability. Hence, it is vital to demonstrate stability, e.g., using a control chart, prior to calculating capability indices. In calculating capability, process variation is estimated in the same way as in calculating the control limits. There is, therefore, a direct link between the control chart and the capability indices.

6.3 Index Calculation for Processes with Additional $\bar{x}$ Variation

The calculation of capability indices should be based on the stable short-term standard deviation of the process. However, it must at the same time allow for the additional variation of the mean where this additional variation is assignable to an inherent feature of the process and its elimination is not feasible from an economic point of view.

Hence, the calculation formulas must be adapted in an appropriate way. However, at present there is still disagreement within industry as to what form this adaptation should take. The percentile method is currently used for calculating indices for this process model (see Section 6.4). As an alternative, there are currently two different approaches. These are explained and discussed in the following paragraphs. The difference between the two methods of calculation is that in one case the additional $\bar{x}$ variation is added to the short-term (within-subgroup) variation, and in the other case the additional $\bar{x}$ variation is subtracted from the tolerance of the characteristic.

Although it makes logical sense to regard the additional variation of the mean over time as a component of total process variation, there are still many companies today who calculate capability by subtracting the additional $\bar{x}$ variation from the tolerance of the characteristic.

6.3.1 Index Calculation by Adding the Components of Variation

To calculate the process spread used in the capability formula, the additional $\bar{x}$ variation is added to the within-subgroup variation (see Figure 6.3–1).

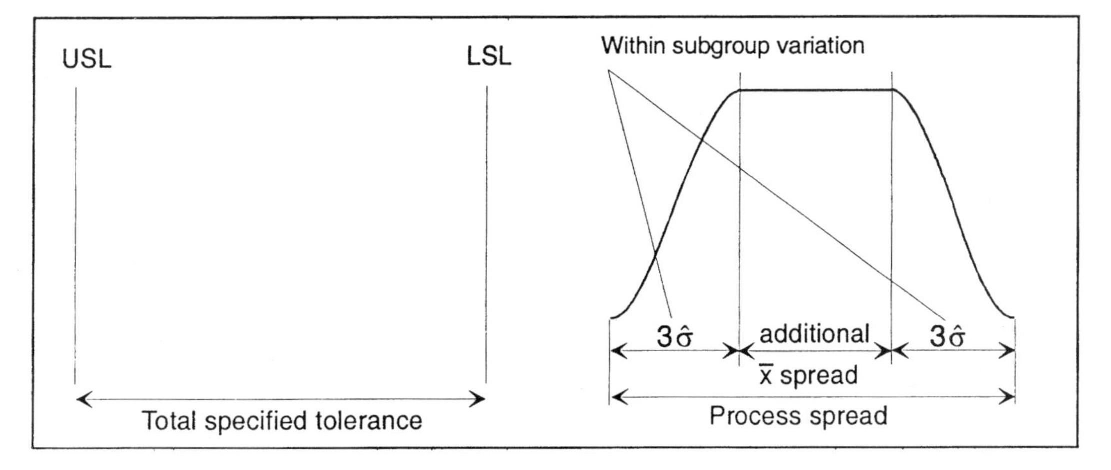

Figure 6.3–1. Process potential with additional $\bar{x}$ variation.

Process Potential

The process spread used in calculating process potential is made up of three intervals:

$$\text{Process spread} = 3\,\hat{\sigma} + \text{additional } \bar{x} \text{ variation} + 3\,\hat{\sigma}$$

Thus, we have the following formula for calculation of C_p:

$$C_p = \frac{\text{Total Specified Tolerance}}{6\hat{\sigma} + \text{additional } \bar{x} \text{ variation}}$$

Process Capability

The halved process spread that is used in calculating process capability is the sum of two intervals—half the additional $\bar{x}$ variation is added to half the within-subgroup variation (see Figure 6.3–2).

$$\text{Process spread} = 3\,\hat{\sigma} + \text{half the additional } \bar{x} \text{ variation}$$

Hence, we have the following formulas for calculation of C_{pk}:

$$C_{pu} = \frac{\text{USL} - \text{Process average}}{3\hat{\sigma} + \frac{1}{2}(\text{additional } \bar{x} \text{ variation})}$$

$$C_{pl} = \frac{\text{Process average} - \text{LSL}}{3\hat{\sigma} + \frac{1}{2}(\text{additional } \bar{x} \text{ variation})}$$

The critical value, i.e., the smaller one of the two, is again used to describe process performance and recorded as C_{pk}, the index of process capability.

$$C_{pk} = \text{minimum of } C_{pu} \text{ and } C_{pl}$$

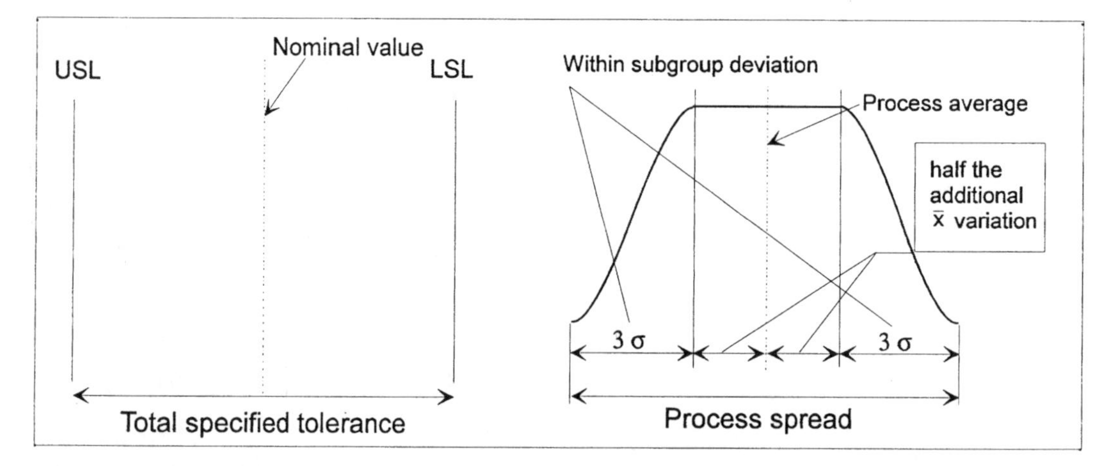

Figure 6.3–2. Process capability with additional $\bar{x}$ variation.

In the ideal scenario, where the process average is equal to the nominal value (assuming that this is located at the center of the tolerance level), we again have

$$C_p = C_{pk}$$

6.3.2 Index Calculation by Reducing the Tolerance Interval

In calculating capability, the additional $\bar{x}$ variation is subtracted from the tolerance of the characteristic (see Figure 6.3–3).

Process Potential

In this method of calculating process potential, only the within-subgroup variation is included in the process spread, but the additional $\bar{x}$ variation is subtracted from the tolerance of the characteristic.

Total specified tolerance used in calculation =

width of tolerance interval minus additional $\bar{x}$ variation

Thus, we have the following formula for calculation of C_p:

$$C_p = \frac{\text{Total specified tolerance} - (\text{additional } \bar{x} \text{ variation})}{6\hat{\sigma}}$$

Process Capability

In this calculation method for process capability, half the within-subgroup variation is taken to represent the process spread, but on the tolerance side, in the numerator of the fraction, the difference between the process average and the appropriate specification limit is reduced by half the additional $\bar{x}$ variation (see Figure 6.3–4).

Tolerence width used in calculation =

difference between process average and appropriate specification limit minus additional $\bar{x}$ variation

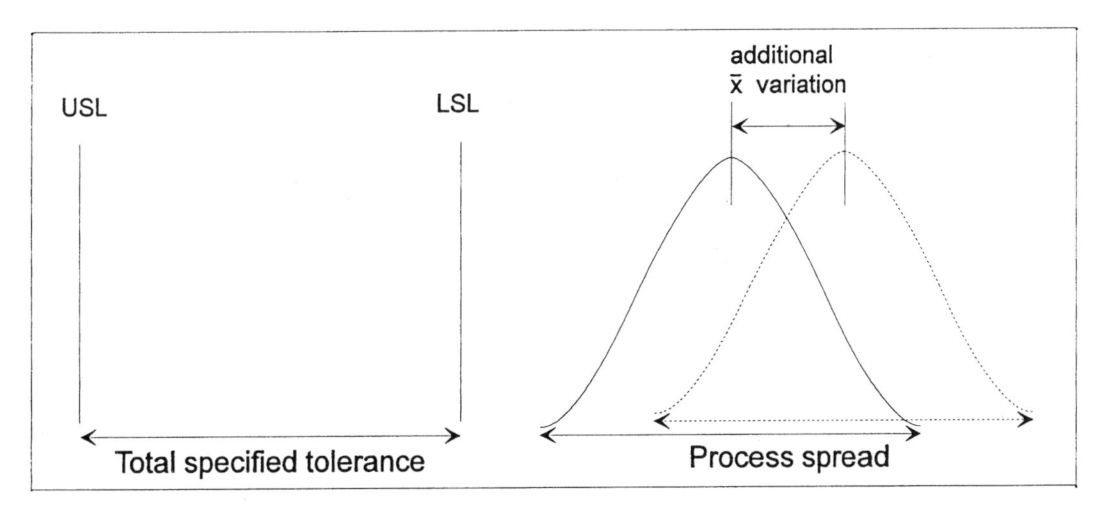

Figure 6.3–3. Process potential of the specified tolerance minus additional $\bar{x}$ variation.

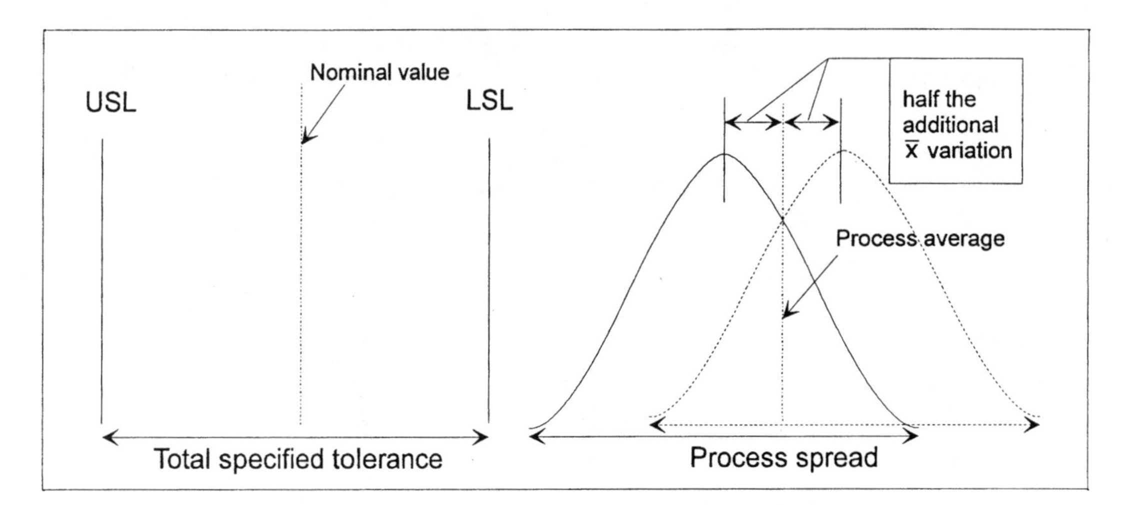

Figure 6.3–4. Process capability of tolerance width minus additional $\bar{x}$ variation.

Hence, we have the following formulas for calculation of C_{pk}:

$$C_{pu} = \frac{\text{USL} - \text{Process average} - \frac{1}{2}(\text{additional } \bar{x} \text{ variation})}{3s}$$

$$C_{pl} = \frac{\text{Process average} - \text{LSL} - \frac{1}{2}(\text{additional } \bar{x} \text{ variation})}{3s}$$

The critical value, i.e., the smaller one of the two, is again used to describe process performance and recorded as C_{pk}, the index of process capability.

$$C_{pk} = \text{minimum of } C_{po} \text{ and } C_{pu}$$

In the ideal scenario, where the process average is equal to the nominal value (assuming that this is located at the center of the tolerance interval), we again have

$$C_p = C_{pk}$$

6.3.3 Comparison of the Two Methods

A worked example serves to illustrate the different results of the two methods of calculation. The tolerance for the characteristic in our example is given by USL = 40.12 mm and LSL = 39.88 mm:

Within-subgroup standard deviation	$\hat{\sigma}$	= 0.01 mm
Additional $\bar{x}$ variation	$\bar{x}_{total}$	= 0.09 mm
Process average	$\bar{\bar{x}}$	= 39.983 mm

Figure 6.3–5 illustrates the relevant aspects of the example. The capability calculations and their results are given in Table 6.3–1.

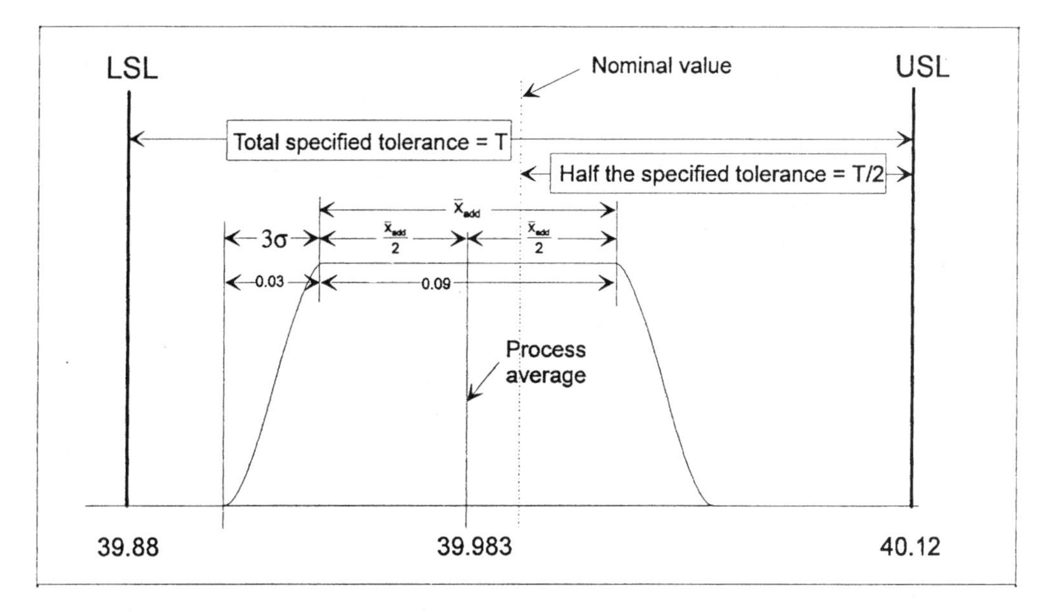

Figure 6.3–5. Example with additional $\bar{x}$ variation.

Table 6.3–1. C values depending on the calculation methods.

	Adding the variation	Reducing the tolerance
Process potential	$C_p = \frac{\text{Tolerance}}{6\hat{\sigma} + \bar{x}_{add}}$ $C_p = \frac{0.24}{0.06 + 0.09} = \frac{24}{15}$ $\mathbf{C_p = 1.6}$	$C_p = \frac{\text{Tolerance} - \bar{x}_{add}}{6\hat{\sigma}}$ $C_p = \frac{0.24 - 0.09}{0.06} = \frac{15}{6}$ $\mathbf{C_p = 2.5}$
Process capability	$C_{pk} = \frac{\bar{\bar{x}} - LSL}{3\hat{\sigma} + \frac{1}{2}\bar{x}_{add}}$ $C_{pk} = \frac{39.985 - 39.88}{0.03 + 0.045} = \frac{10.5}{7.5}$ $\mathbf{C_{pk} = 1.4}$	$C_{pk} = \frac{\bar{\bar{x}} - LSL - \frac{1}{2}\bar{x}_{add}}{3\hat{\sigma}}$ $C_{pk} = \frac{39.985 - 39.88 - 0.045}{0.03} = \frac{6}{3}$ $\mathbf{C_{pk} = 2.0}$

6.4 Index Calculation for Nonnormal Distributions

As for assessment of processes with additional $\bar{x}$ variation, there also are no universally accepted capability calculation methods for nonnormal distributions (see Chapter 2, Section 2.7). Here, too, there are two methods in common use. This chapter will provide a description and comparison of the two methods.

Both methods are based on the desire to maintain, as far as possible, an analogy to the normal distribution method. In one method, capability calculation is again based on comparing the 99.73 percent interval (corresponding to 6σ in the normal distribution model) with the tolerance interval of the characteristic. In the other method, the capability calculation is based on determining the proportion outside specification and relating this back to a normal distribution.

According to their basis of calculation we shall refer to the two methods as follows:

- Percentile method
- Method based on the proportion outside specification

It should be noted that the two methods, which may be applied to any distribution pattern, regardless of the distribution model, will, for identical data, lead to different capability indices. The two methods only lead to the same result in the case where the capability of the process is described by an index of $C_p = 1$.

Notes

1. Today, the percentile method is primarily used for calculation of the capability indices because it may be implemented for all process models! In this case, the 99.73 percent region is derived from the applicable distribution model.
2. The process model *normal distribution* is a special case of the percentile method. Here, the 99.73 percent region may be determined using six times the standard deviation.

6.4.1 Index Calculation Using the Percentile Method

After confirmation that the selected distribution model provides the best possible description of process output, an interval containing 99.73 percent of the population is defined as representing process spread, just as in the case of the normal distribution. The limits of this interval are known as the 0.135 percentile and the 99.865 percentile of the distribution. The interval defined by these two percentiles will contain 99.73 percent of the total population (see Figure 6.4–1).

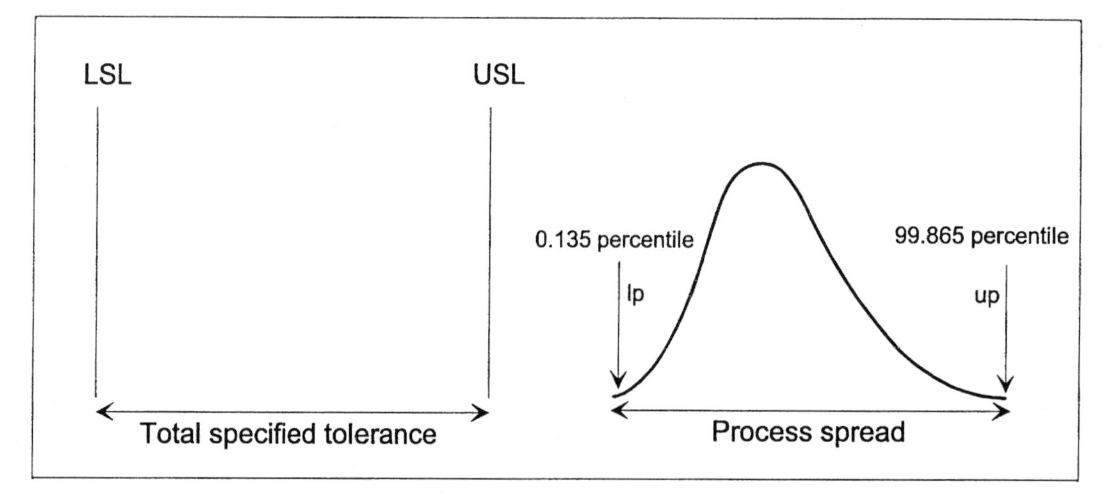

Figure 6.4–1. Process potential calculated with the percentile method, where up = upper percentile and lp = lower percentile.

This leads to a formula for calculating the C_p index that is directly comparable to the corresponding normal distribution formula.

The formula is as follows:

$$C_p = \frac{\text{Total specified tolerance}}{99.865\text{ percentile} - 0.135\text{ percentile}} = \frac{USL - LSL}{u_p - l_p}$$

If this approach is taken to its logical conclusion, we also arrive at formulas for C_{pk} that are directly comparable to the corresponding formulas for normal distributions (see Figure 6.4–2).

The formulas for calculating process capability are

$$C_{up} = \frac{LSL - \bar{\bar{x}}}{99.865\text{ percentile} - \bar{\bar{x}}} = \frac{USL - \bar{\bar{x}}}{u_p - \bar{\bar{x}}}$$

$$C_{lp} = \frac{\bar{\bar{x}} - LSL}{\bar{\bar{x}} - 0.135\text{ percentile}} = \frac{\bar{\bar{x}} - LSL}{\bar{\bar{x}} - I_p}$$

As for the normal distribution, the critical value, i.e., the smaller one of the two, is used to described process performance and recorded as C_{pk}, the index of process capability.

$$C_{pk} = \text{minimum of } C_{pu} \text{ and } C_{pl}$$

6.4.2 Index Calculation Based on the Proportion Outside Specification

Unlike the percentile method, the method based on the proportion outside specification does not take a 99.73 percent interval as its basis. Instead, the method is based on the percentage of output that is beyond the critical specification limit (see Figure 6.4–3). One then *pretends,* so to speak, that one has a normal distribution with the same proportion outside this specification limit. The capability index found by this method for a given distribution then corresponds to that of a normal distribution in the sense that,

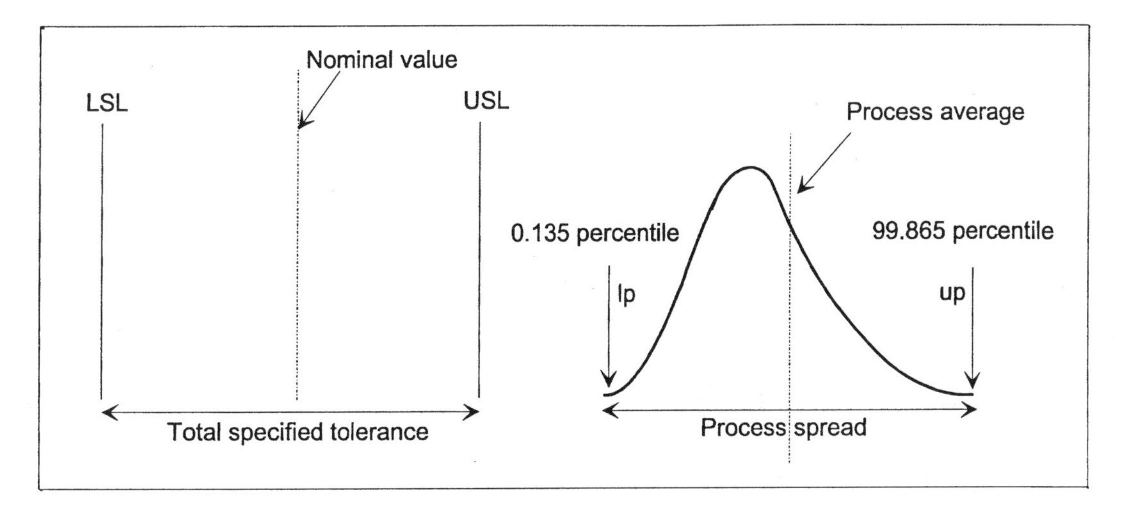

Figure 6.4–2. Process capability calculated with the percentile method, where up = upper percentile and lp = lower percentile.

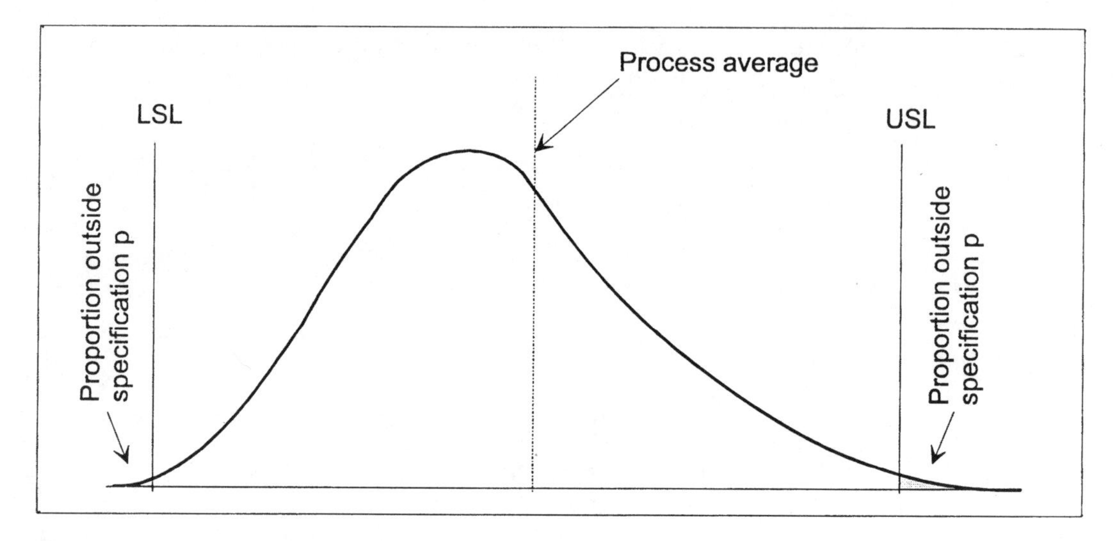

Figure 6.4–3. Process capability calculated from the proportion outside specification.

given identical capability indices, both distributions have the same proportion outside the critical specification limit. Note that it is not customary to calculate a process potential index on the basis of this method.

The capability index C_{pk} as calculated by this method is given by the following formula:

$$C_{pk} = \frac{u_{1-p}}{3}$$

The determination of u_{1-p} is as described in Chapter 2, Section 2.2.2.1.

6.5 Process Capability Indices: Overview

Figure 6.5–1 lists the various capability calculation methods available for different types of processes.

6.6 Index Calculation for the Two-Dimensional Normal Distribution

Similarly to the one-dimensional normal distribution, capability indices for the two-dimensional normal distribution are calculated on the basis of a variation region associated with a predetermined probability (e.g., $1 - \alpha = 99.73$ percent). Table 6.6–1 illustrates the connections between certain probabilities, normal distribution intervals (measured in standard deviations), and capability indices. Intermediate values can be calculated, based on the appropriate density function, both for one-dimensional and two-dimensional normal distributions.

For a given probability $1 - \alpha$, the two-dimensional case leads to a random variation region that may be shown on the x/y plot in the form of a probability ellipse. To

Process type	Method	Potential P_p, C_p	Capability P_{pk}, C_{pk}	Notes
normal distribution and $\mu_{(t)}$	Internal variation	$\frac{USL - LSL}{6\hat{\sigma}}$	$\frac{USL - \bar{\bar{x}}}{3\hat{\sigma}}$ $\frac{\bar{\bar{x}} - LSL}{3\hat{\sigma}}$	$\bar{\bar{x}}$ = Process average $\hat{\sigma} = \sqrt{\overline{s^2}}$ $= \bar{s} / a_n$ $= \bar{R} / d_n$
and $\sigma_{(t)}$ = const.	Total standard deviation	$\frac{USL - LSL}{6s_{tot}}$	$\frac{USL - \bar{\bar{x}}}{3s_{tot}}$ $\frac{\bar{\bar{x}} - LSL}{3s_{tot}}$	s_{tot} = Total standard deviation $s_{tot} \approx \hat{\sigma}$
nonnormal distribution and	Proportion outside specification	———	$\frac{u_{1-p}}{3}$	p = Proportion outside spec.
$\mu_{(t)}$, $\sigma_{(t)}$ = const.	Percentile method	$\frac{USL - LSL}{o_p - u_p}$	$\frac{USL - \bar{\bar{x}}}{o_p - x}$ $\frac{\bar{\bar{x}} - LSL}{\bar{\bar{x}} - u_p}$	o_p = upper percentile u_p = lower percentile
$\mu_{(t)}$ not const. and	Addition of variation	$\frac{USL - LSL}{6\hat{\sigma} + \bar{x}_{add}}$	$\frac{USL - \bar{\bar{x}}}{3\hat{\sigma} + \bar{x}_{add}/2}$ $\frac{\bar{\bar{x}} - LSL}{3\hat{\sigma} + \bar{x}_{add}/2}$	$\bar{x}_{add}$ = additional $\bar{x}$ variation
$\sigma_{(t)}$ = const.	Tolerance reduction	$\frac{USL - LSL - \bar{x}_{add}}{6\hat{\sigma}}$	$\frac{USL - \bar{\bar{x}} - \bar{x}_{add}/2}{3\hat{\sigma}}$ $\frac{\bar{\bar{x}} - LSL - \bar{x}_{add}/2}{3\hat{\sigma}}$	rarely used today!
	s_{total}	$\frac{USL - LSL}{6s_{total}}$	$\frac{USL - \bar{\bar{x}}}{3s_{total}}$ $\frac{\bar{\bar{x}} - LSL}{3s_{total}}$	$s_{total} \neq \hat{\sigma}$ Normal distribution applicable

Figure 6.5–1. Overview capability indices.

satisfy the predefined capability requirements, this ellipse must not pass beyond the tolerance ellipse.

Process Potential P_o

To determine process potential, one starts with an ellipse center positioned at the center of the tolerance region ($\mu_x = M_{px}$, $\mu_y = M_{py}$). Around this center point, an elliptical random variation region is formed that just touches the tolerance ellipse (Figure 6.6–1). The probability $1 - \alpha$ associated with this ellipse is used to determine the index of process potential P_o.

Table 6.6–1. Interdependence between random spread region, region of $\pm n\sigma$, and quality statistic, where σ = standard deviation and α = error probability.

Probability $1 - \alpha$ in %	Standard deviation regions in a Normal distribution	Capability index P_p or P_{pk} C_p or C_{pk}
99.73	$\pm 3\sigma$	1.0
99.9937	$\pm 4\sigma$	1.33
99.99994	$\pm 5\sigma$	1.67
99.9999998	$\pm 6\sigma$	2.0

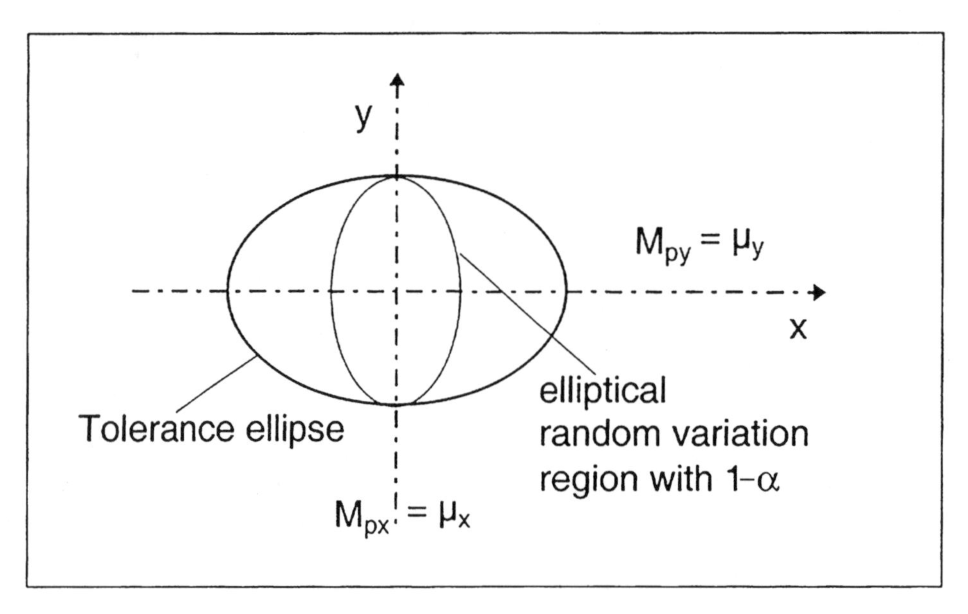

Figure 6.6–1. Determination of the capability potential P_o.

Process Capability P_{ok}

To determine the process capability index P_{ok}, an elliptical variation region that just touches the tolerance ellipse is determined (Figure 6.6–2). The estimated values for μ_x and μ_y of the two-dimensional normal distribution determine the location of the center of the ellipse. The probability $1 - \alpha$ associated with this ellipse is used to determine the index of process capability P_{ok}. If the point given by μ_x and μ_y is outside the tolerance ellipse, no capability index P_{ok} is given.

Example: Position Tolerance

Table 6.6–2 shows the x and y coordinates of bore centers. Determine the capability indices P_o and P_{ok} for these values.

The pairs of values are plotted in Figure 6.6–3. In addition, the probability ellipses for determination of P_o and P_{ok} are shown as well as the tolerance ellipse and the specification limits.

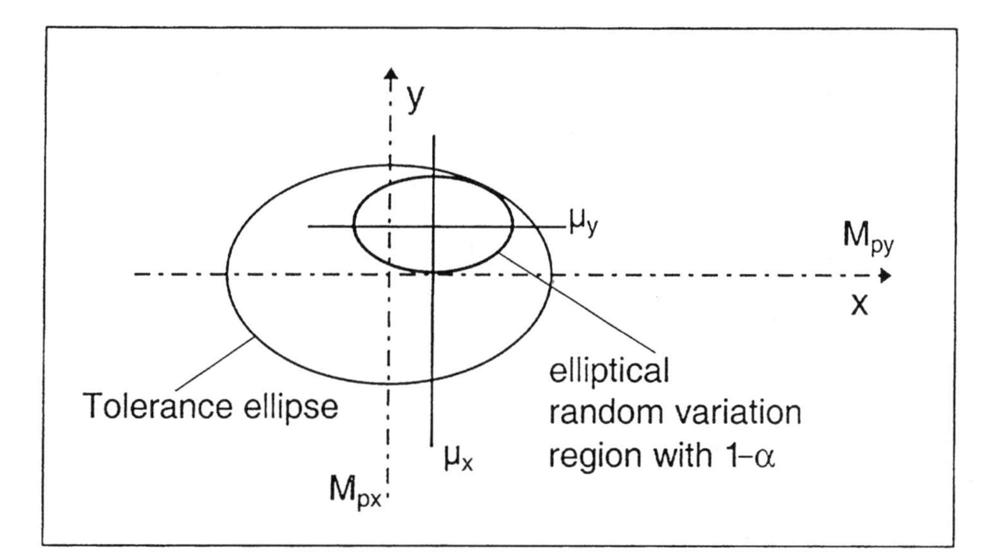

Figure 6.6–2. Determination of the capability indices P_{ok}.

Table 6.6–2. Bore centers.

Value no.	x position	y position	Value no.	x position	y position
1	–0.035	–0.010	21	0.045	–0.035
2	0.030	0.015	22	0.050	0.050
3	–0.020	–0.035	23	–0.015	–0.010
4	0.050	0.050	24	0.010	0.020
5	0.025	0.000	25	–0.020	0.010
6	0.030	0.030	26	0.050	0.030
7	0.050	–0.020	27	0.030	–0.015
8	0.025	0.010	28	0.010	–0.035
9	–0.015	–0.020	29	0.030	–0.025
10	0.050	0.050	30	0.020	0.030
11	–0.010	–0.010	31	–0.015	0.050
12	0.045	0.000	32	0.050	0.020
13	0.040	–0.025	33	–0.020	0.010
14	0.025	0.010	34	0.010	0.050
15	0.050	–0.010	35	0.000	–0.010
16	0.015	0.015	36	0.025	0.040
17	–0.050	0.050	37	0.050	–0.010
18	0.015	0.025	38	0.025	0.050
19	0.050	–0.030	39	–0.010	0.020
20	0.025	0.050	40	0.010	0.030

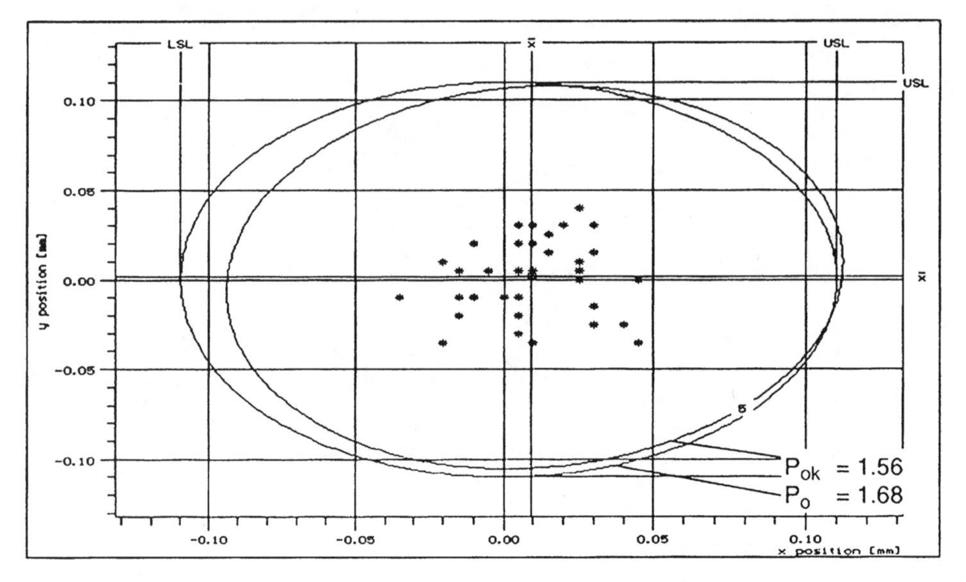

Figure 6.6–3. x/y plot.

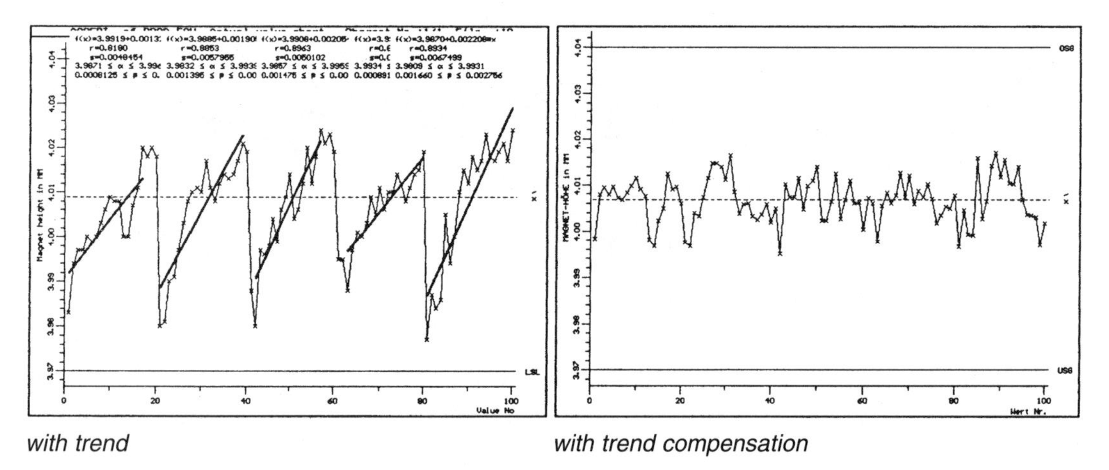

with trend *with trend compensation*

Figure 6.7–1. Trend compensation.

6.7 Index Calculation with Compensation for Additional $\bar{x}$ Variation

Frequently, especially in assessment of production equipment, the question arises: Would it make sense to compensate for the additional $\bar{x}$ variation, given that it is not caused by the equipment itself? A typical example is process model C, where the trends over time are caused by tool wear and not by the machine as such. Other examples are fluctuations in material or the use of differing setup methods by operators.

Example: Trend Process

Consider a process with USL = 4.04 and LSL = 3.97. The process mean is $\bar{x} = 4.0066$. The data used in this example are taken from Table 3.9–4. We want to find the capability indices for this process. By means of a regression line, the trend observed in the series

*"Wert Nr." (referred to in Figure 6.7–1(right)) means *value number.*

Table 6.7–1. Quality statistics with and without trend compensation.

	With trend	With trend compensation
Process Capability	$\bar{\bar{x}} = 4.0066$ $\hat{\sigma} = \bar{s}/a_n = 0.005284$ $\bar{x}_{add} = 0.01725$ Formula from Process model C (Section 6.3)	$\bar{\bar{x}} = 4.0066$ $s_{total} = 0.0049346$ Formula from Process model A (Section 6.2)
Process Potential	$C_p = \frac{USL - LSL}{6\hat{\sigma} + \bar{x}_{add}/2} = \frac{4.04 - 3.97}{6 \times 0.005284 + 0.01725}$ $= \frac{0.07}{0.04895} = 1.43$	$C_p = \frac{USL - LSL}{6s_{total}} = \frac{4.04 - 3.97}{6 \times 0.004934}$ $= \frac{0.07}{0.0296} = 2.36$
Estimator	$C_{pk} = \frac{USL - \bar{\bar{x}}}{3\hat{\sigma} + \bar{x}_{add}/2} = \frac{4.04 - 4.0066}{3 \times 0.005284 + 0.08628}$ $= \frac{0.00334}{0.02448} = 1.36$	$C_{pk} = \frac{USL - \bar{\bar{x}}}{6_{S_{total}}} = \frac{4.04 - 4.0066}{3 \times 0.004934}$ $= \frac{0.0334}{0.14802} = 2.26$

of measurements is determined and then compensated for (see Figure 6.7–1). After compensation, the plotted values only represent the differences between the measurement values and the appropriate regression line (Figure 6.7–1, right-hand side).

The resulting capability calculations are shown in Table 6.7–1.

The difference between the capability indices calculated with and without trend compensation would have to be ascribed to tool wear and not to the machine. This is a valid consideration from the point of view of the manufacturer of the machine. Hence, the purchase contract should clearly state which calculation methods will be used in capability assessment. When assessing the parts produced by the machine, however, the trend must of course always be taken into account in the calculation.

6.8 Special Case: Potential Smaller than Capability

There are many nonnormally distributed characteristics (shape, location, etc.) where it is only meaningful to specify a single specification limit. Hence, it would at first appear impossible to calculate process potential for such a characteristic. To solve this problem, we can use the natural limit of the characteristic as a second *specification limit* in the percentile formula for process potential and calculate a process potential index in the normal way. But, as process capability will always have to be calculated with regard to the specified limit (and never with regard to any natural limit), this index of *process potential* may turn out to be smaller than the capability index.

Example: $C_p < C_{pk}$

The data are from the Ford Motor Company Test Example No. 6 (see Chapter 11). The specification limits are LSL = 0 and USL = 0.1. The Rayleigh distribution is suggested as a suitable model. The process average is $\bar{\bar{x}}$ = 0.02527, and the process

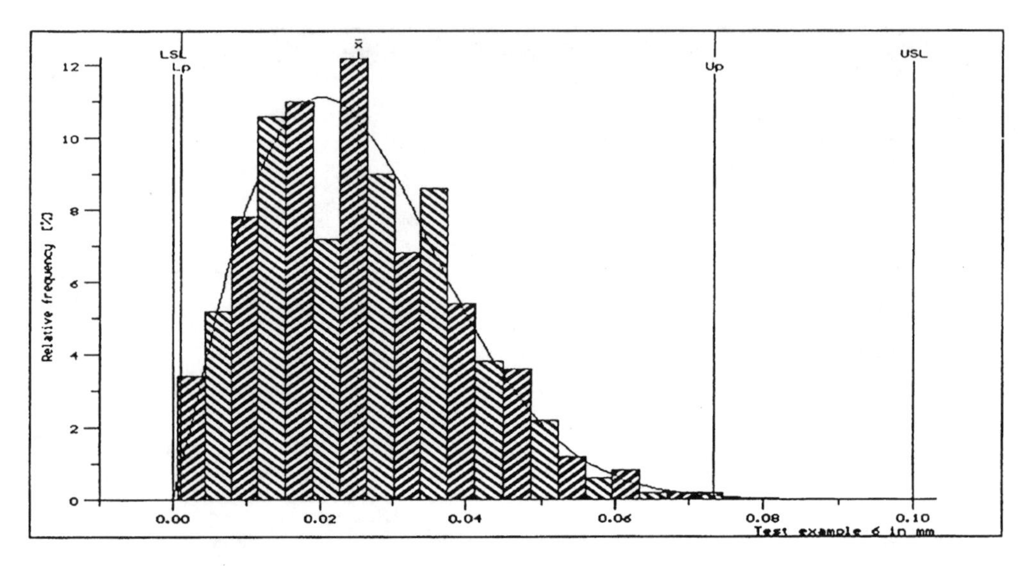

Figure 6.8–1. Rayleigh distribution.

spread is given by the two percentiles $p_u = 0.07483$ and $p_l = 0.00028$. Figure 6.8–1 illustrates this situation.

Using the percentile method (Section 6.4.1) we have

$$C_p = \frac{USL - LSL}{o_p - u_p} = \frac{0.1 - 0}{0.07483 - 0.0028} = 1.39$$

$$C_{up} = \frac{USL - \bar{\bar{x}}}{o_p - \bar{\bar{x}}} = \frac{0.1 - 0.02527}{0.07483 - 0.02527} = 1.51$$

$$C_{lp} = \frac{\bar{\bar{x}} - LSL}{o_p - u_p} = \frac{0.02527 - 0}{0.07483 - 0.0018} = 0.35$$

$$C_{pk} \stackrel{!}{=} C_{up} = 1.51 > C_p$$

6.9 Calculation of Quality Coefficients according to CNOMO

In the CNOMO norm [59], capability is termed the quality coefficient of the production machine and is abbreviated by CAM. This closely corresponds to the C_p value, since process variation is calculated from the subgroup statistics.

A further calculated value is the *location and variation coefficient of the production.* This is termed C_{pk}. Thus, the variation is determined from the total standard deviation, so this value is comparable to the C_{mk} factor.

Calculation of CAM

$$CAM = \frac{USL - LSL}{6 \cdot \hat{\sigma}_i}$$

The momentary standard deviation $\hat{\sigma}_i$ is estimated using d_5^*, which is calculated with a unilateral safety cutoff of 95 percent and weighted depending on the number of subgroups taken.

$$\hat{\sigma}_i = \frac{\overline{R}}{d_5^*}$$

$\overline{R}$ is the average of the ranges ($R = x_{max} - x_{min}$) from k subgroups.

The factor d^*_5 for subgroups of size n = 5 results from

$$d_5^* = d_5 - \alpha_{1-\alpha} \cdot \frac{\beta_5}{\sqrt{k}} \text{ where}$$

$d_5 = 2.326$ (see Table 10–7)

$\beta_5 = 0.864$ (see [58])

k = number of subgroups

$\alpha_{1-\alpha} = 1.645$ for the unilateral upper cutoff value of 95 percent (see Chapter 2, Section 2.2.2)

For six subgroups this means

$$d_5^* = 2.326 - 1.645 \cdot \frac{0.864}{\sqrt{6}} = 1.746$$

As an alternative, the d_5^* factor for the relative subgroup may be taken from the following table.

Number of subgroups	6	7	8	9	10	12	15	19	24	30	40
value d_5^* 95%	1.746	1.789	1.824	1.852	1.877	1.916	1.959	2.000	2.036	2.067	2.101

The calculated CAM value should not be smaller than 1.3. This analysis is allowed only if the homogeneity of the samples taken is insured. For control, the ranges R_i of the subgroups are compared to a limit. The following formula applies:

$$R_i < 0.643 \cdot \frac{USL - LSL}{1.3} \quad \text{where } i = 1, 2, \ldots, k$$

This corresponds closely to a stability examination through the upper control limits of a control chart.

C_{pk} Value Calculation

This coefficient is calculated only after the test for normal distribution. Then, this value results from

$$C_{pk} = \min\left\{\frac{USL - \overline{x}}{3 \cdot \hat{\sigma}_o}; \frac{\overline{x} - LSL}{3 \cdot \hat{\sigma}_o}\right\}$$

The production variation σ_o is estimated through the total standard deviation. This is weighted with the unilateral upper cutoff value of 95 percent and the subgroup size.

$$\hat{\sigma}_o = C \cdot \sqrt{\frac{1}{N-1} \cdot \sum_{i=1}^{N} (x_i - \bar{x})^2} \quad \text{where}$$

$$\bar{x} = \frac{1}{N} \cdot \sum_{i=1}^{N} x_i \qquad i = 1, 2, \ldots, N$$

$$C = \frac{1}{\sqrt{\frac{x^2_{\alpha;f}}{f}}} \qquad \text{where } f = N - 1$$

N = subgroup size

It is possible to calculate the C coefficients accordingly, or to take them from the following table:

Number of Subgroups	30	35	40	45	50	60	75	95	120	150	200
value C 95%	1.28	1.26	1.24	1.22	1.21	1.18	1.16	1.14	1.12	1.11	1.09

The calculated C_{pk} value must be larger than 1.0.

Note

Calculation of the CAM and C_{pk} value should be carried out only if homogeneity is insured and the values normally distributed. This requirement corresponds closely to process model A. Since real processes certainly conform only partially to this requirement, the calculation method described only rarely applies in practice.

Example CNOMO Norm

Table 6.9–1 contains 40 subgroups of size n = 5. Specification limits are USL = 21.0 mm and LSL = 20.0 mm.

The following statistical values result:

average $\bar{x} = 20.5904$
average of the ranges $\bar{R} = 0.216$
total standard deviation $s = 0.095082$

$$CAM = \frac{USL - LSL}{6 \cdot \hat{\sigma}_i} = \frac{21.0 - 20.0}{6 \cdot \frac{0.216}{2.101}} = 1.62$$

$$C_{pk} = \frac{USL - \bar{x}}{3 \cdot \sigma_o} = \frac{21.0 - 20.5904}{3 \cdot 0.109 \cdot 0.095082} = 1.32$$

Table 6.9–1. Characteristics for the determination of CAM and C_{pk} according to CNOMO.

1	2	3	4	5	6	7	8	9	10
20.83	20.64	20.68	20.70	20.75	20.56	20.66	20.80	20.55	20.77
20.47	20.69	20.57	20.56	20.56	20.58	20.44	20.52	20.54	20.59
20.64	20.67	20.57	20.42	20.48	20.64	20.35	20.53	20.57	20.71
20.64	20.70	20.67	20.69	20.67	20.67	20.53	20.79	20.57	20.55
20.59	20.29	20.65	20.63	20.58	20.73	20.55	20.70	20.57	20.55
11	**12**	**13**	**14**	**15**	**16**	**17**	**18**	**19**	**20**
20.62	20.62	20.50	20.56	20.63	20.50	20.65	20.63	20.55	20.57
20.41	20.67	20.62	20.71	20.66	20.69	20.53	20.68	20.52	20.46
20.78	20.45	20.54	20.55	20.49	20.40	20.69	20.70	20.65	20.58
20.67	20.53	20.62	20.63	20.58	20.78	20.76	20.60	20.51	20.50
20.64	20.65	20.64	20.64	20.49	20.55	20.58	20.48	20.52	20.53
21	**22**	**23**	**24**	**25**	**26**	**27**	**28**	**29**	**30**
20.44	20.53	20.75	20.68	20.56	20.72	20.60	20.49	20.64	20.65
20.59	20.67	20.59	20.55	20.67	20.47	20.59	20.55	20.58	20.64
20.61	20.59	20.58	20.50	20.66	20.50	20.62	20.56	20.44	20.54
20.51	20.50	20.45	20.77	20.57	20.71	20.56	20.43	20.41	20.53
20.54	20.58	20.55	20.60	20.48	20.51	20.50	20.62	20.38	20.44
31	**32**	**33**	**34**	**35**	**36**	**37**	**38**	**39**	**40**
20.72	20.55	20.45	20.46	20.57	20.53	20.60	20.68	20.67	20.66
20.79	20.60	20.65	20.58	20.69	20.60	20.57	20.55	20.60	20.59
20.58	20.68	20.55	20.47	20.69	20.55	20.62	20.68	20.69	20.71
20.52	20.69	20.54	20.50	20.48	20.64	20.56	20.69	20.61	20.52
20.45	20.53	20.74	20.44	20.70	20.61	20.46	20.59	20.76	20.64

7

Building a Qualification System

7.1 What Is Machine and Process Qualification?

The purpose of *machine and process assessment* is to demonstrate that the equipment itself or the process as a whole is able to perform *in statistical control,* or to give an indication of what should be done if this cannot be demonstrated. The action taken will be based on the estimated loss function for deviations from a given target value (see Chapter 1).

For most processes, it will not be possible to reduce variation to zero. To be able to compare processes independently of the various quality characteristics involved, certain terms such as *stable, in control,* and *capable* have been defined, as well as numerical indicators such as capability indices (see Chapters 3 and 6). The validity of the definition is subject to certain prerequisites, which can be verified on the basis of statistical methods using graphical analyses. These are summarized in the Assessment Report.

Contents of the Assessment Report

The assessment report documents the studies that were carried out, using numerical statistics as well as graphical representations. The report should

- Describe the methods used
- Clearly present all relevant statistics
- Provide a clear description of the state of affairs
- Show varying degrees of detail
- Include decision aids

Purpose of the Assessment Report

The purpose of the assessment report is

- To enable a clear decision as to whether the performance of a manufacturing process is in line with requirements
- To trigger improvement actions (systematic process assessment)
- To provide guidelines for continuous process monitoring (type of control

chart, sample size, and sampling frequency)
- To provide documented evidence to the manufacturing department that there may be a need for improved quality planning
- To serve as an aid in drawing up sorting instructions for processes that have failed to achieve performance targets

When Is an Assessment Report Required?

When should an assessment report be created? Basically, an assessment report is required for new equipment, after significant changes, and for basic assessment studies. The most important reasons are

- New manufacturing equipment
- New manufacturing process
- Initial assessment of critical characteristics
- Unstable process
- New materials, facilities or tools
- Ongoing verification of established processes
- Process improvement directed at specific concerns
- Establishing decision-making rules
- Etc.

The aim, once the process has been assessed and found to be satisfactory, is to keep process conditions stable. To achieve this aim, it is necessary to collect meaningful process data (on product characteristics and process parameters) that provide an accurate description of process behavior.

7.2 Why Do We Need Machine and Process Assessment?

No manufacturing enterprise will be able to avoid the subject of *machine and process assessment* for a number of reasons. Some of these are listed, but it should be noted that the significance of each of these points may vary from company to company. For this reason, we have not ranked the points.

Certification according to DIN EN ISO 9000ff

The DIN EN ISO 9004 standard [29] states in the section on *planning of production operations* that such operations should proceed under *controlled conditions.* To ensure controlled process conditions, process capability studies should be conducted to determine the potential effectiveness of a process. Hence, the application of modern statistical methods is an important element at all stages of the quality control loop. Such methods are not only used for capability studies but also for predicting the durability of components, assessment of measuring systems, data analysis, and performance assessment.

Typical methods include

- Design of experiments
- Analysis of variance and regression analysis
- Risk analysis

- Tests of significance
- Control charting
- Etc.

However, the standard does not define to what extent a company should deploy and make use of these methods.

Customer-Specific System

Very often suppliers are encouraged to align elements of their quality management system with the system of their customer. A typical example is Ford Motor Company's QS-9000 Standard [3] or the joint guidelines developed by Chrysler, General Motors, and Ford. In the French automotive industry, the CNOMO Standard is of prime importance. Other large companies have similar guidelines, and there are differing views on whether DIN EN ISO 9000ff certification in itself is sufficient, or what additional demands should be made.

Demonstrating Performance

Once an assessment system for machines, production equipment, and processes is in place, the evidence it creates can be used to demonstrate satisfactory production performance to the customer. This promotes a positive customer-supplier relationship and can enhance market opportunities.

Product Liability

The safety aspects of product quality should be identified with the aim of enhancing product safety and minimizing product liability. Using qualified machines and processes and monitoring them on an ongoing basis, with the aid of statistical methods, limits product liability risks. Of course, whether or not a company is liable for compensation will depend on the individual circumstances of each case. But, at any rate, the company will not have been guilty of gross negligence.

Due to changes in legislation, the importance of this aspect has grown and will continue to grow.

Continuous Process Improvement

A clear understanding of product characteristics and process factors, as well as their relationships, is of vital importance for continuous process improvement. Statistical methods, combined with careful logging of environmental conditions and process actions taken, can be used to distill this information from the collected data to achieve and maintain stable, low-variation processes.

Decision Aids

In production control, questions arise time and again as to how much process intervention is appropriate during continuous production, when such intervention should take place, and which actions should be taken. These decisions must be based on process knowledge. Hence, the creation of decision aids is a substantial part of any assessment process.

7.3 Statistical Methods as a Means to Achieving Lower Costs

In many companies, the *introduction of statistical methods* was considered equivalent to the introduction of SPC. The expense involved was often justified on the basis of outside pressures, without critically analyzing the needs or identifying measures to enhance efficiency. Without optimized efficiency, costs increased.

Rising Quality Costs

Studies performed within various companies have shown that quality costs amount on average to 5 percent of the turnover. The costs of nonconformities represent the largest part of this sum. Quality costs have increased heavily over the last few years and will continue to do so, if no countermeasures are taken. The reasons are

- More stringent demands by clients
- Increasing product complexity
- Shorter product life cycles.

In some industries, inspection costs may now be higher than actual production costs. Hence, it must be the aim of every company to reduce inspection costs by using qualified machines and processes. Increased net production capacity is an additional benefit of this.

SPC: Just for Show?

The costs of an SPC system include

- Provision of hardware (measuring systems, computers, network, etc.)
- Software licenses and customization
- Personnel and training
- Introduction, data administration, system maintenance
- Etc.

At this point, the system has become an end in itself and is maintained on the basis of *This is what the customer wants.* This has actually caused SPC to fall into disrepute. Some people quite rightly talk of *SPC = Show Program for Customers.*

What follows is meant as an antidote to this kind of argument and will focus on the positive aspects of statistical methods. In addition, we shall attempt to show the potential for rationalization in an *SPC-controlled* product environment.

Potential for Rationalization

The main focus here is improved process knowledge. The following points can help to reduce costs:

- Known relationships between product characteristics and process parameters can be exploited to reduce the number of characteristics and parameters monitored.
- More robust processes mean lower rework and reject rates and allow reduced sampling frequencies with smaller sample sizes.
- Automatic analysis can highlight problems. Special emphasis can then be placed on addressing these points. This ensures a more focused and efficient approach.

- Calculation of summary statistics makes it easier to administrate the data and saves paper as well as disk space. Even so, vital information remains available for long-term analyses.
- Consistent use of methods throughout the entire company, combined with correct statistical analysis, increases transparency and makes results comparable.
- Operating an assessment system creates expert knowledge, which in the medium and long term can be used for process optimization (knowledge-based SPC).
- The process knowledge gained can be transferred to similar products or characteristics.

An appropriate realization of the foregoing concepts will reduce data collection and data administration overheads. It will reduce not only the level of investment required (measuring systems, computer, and network facilities), but also the time spent on collecting, maintaining, and evaluating data. In a given company, it will be unlikely that all of these points will offer equal potential for rationalization, but even a single one of these points may constitute sufficient grounds to investigate the subject of rationalization in more detail.

The Ideal Scenario

The ideal scenario may only be a vision, but a vision is essential for moving in the right direction.

- Data are automatically collected and transferred to a computer system.
- Data are analyzed correctly, using the right statistical methods.
- A chart gives a visual representation of the current situation.
- *Bad* and *good* processes are identified automatically.
- *Deviations* from requirements are flagged automatically and are thoroughly investigated, with the causes established and appropriate corrective action recommended.
- Statistics are displayed on a chart and regularly updated.
- Important data are summarized for long-term analyses.
- Relationships between characteristics and process parameters are identified. The number of characteristics monitored can be reduced.
- Clear graphics provide added transparency.
- The effects of *noise factors* (i.e., uncontrolled or uncontrollable parameters) on quality characteristics are known. Processes can be made more robust. Sample size and sampling frequency can be reduced.

In this scenario, any statistics or graphics required by yourself, your supervisor, or your customer can be retrieved within seconds, simply by pressing a button. Using this information, you *get a feel* for your processes and learn to understand their behavior. As a result, you are able to take immediate and effective action if problems arise.

This scenario certainly represents a vision. The aim must be to convert this vision into a realizable goal, and to clearly define individual steps that will bring you closer to achieving it. The steps defined must be reproducible and measurable.

7.4 Information Levels

To achieve the goal, clear action steps must be identified. To illustrate this, we shall look at a company model with case studies that describe various situations and possible solutions. These concepts can then be transferred to individual real-life situations.

Figure 7.4–1 shows a process flow with several production lines. Raw materials and components are delivered by various suppliers. These are then processed using

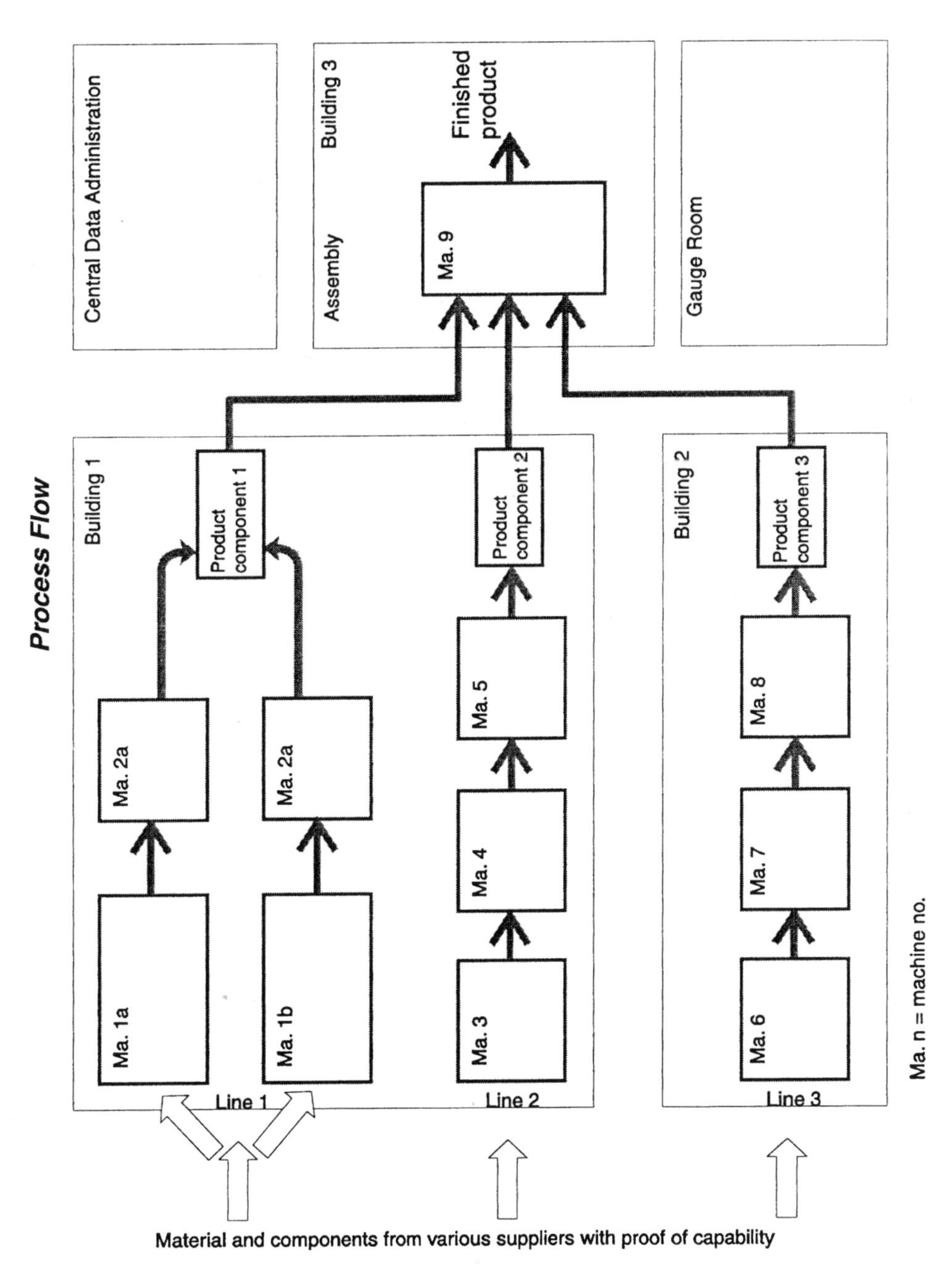

Figure 7.4–1. Process flow.

a variety of machines and tools. The result is a finished product component. To evaluate quality, measuring devices are positioned at appropriate points in the process flow. Some characteristics are checked in a central gauge room, whereas others are monitored by an automated SPC system or by means of manually plotted control charts (Figure 7.4–2). The manufacturing equipment can process different components, which are then combined to form various assemblies.

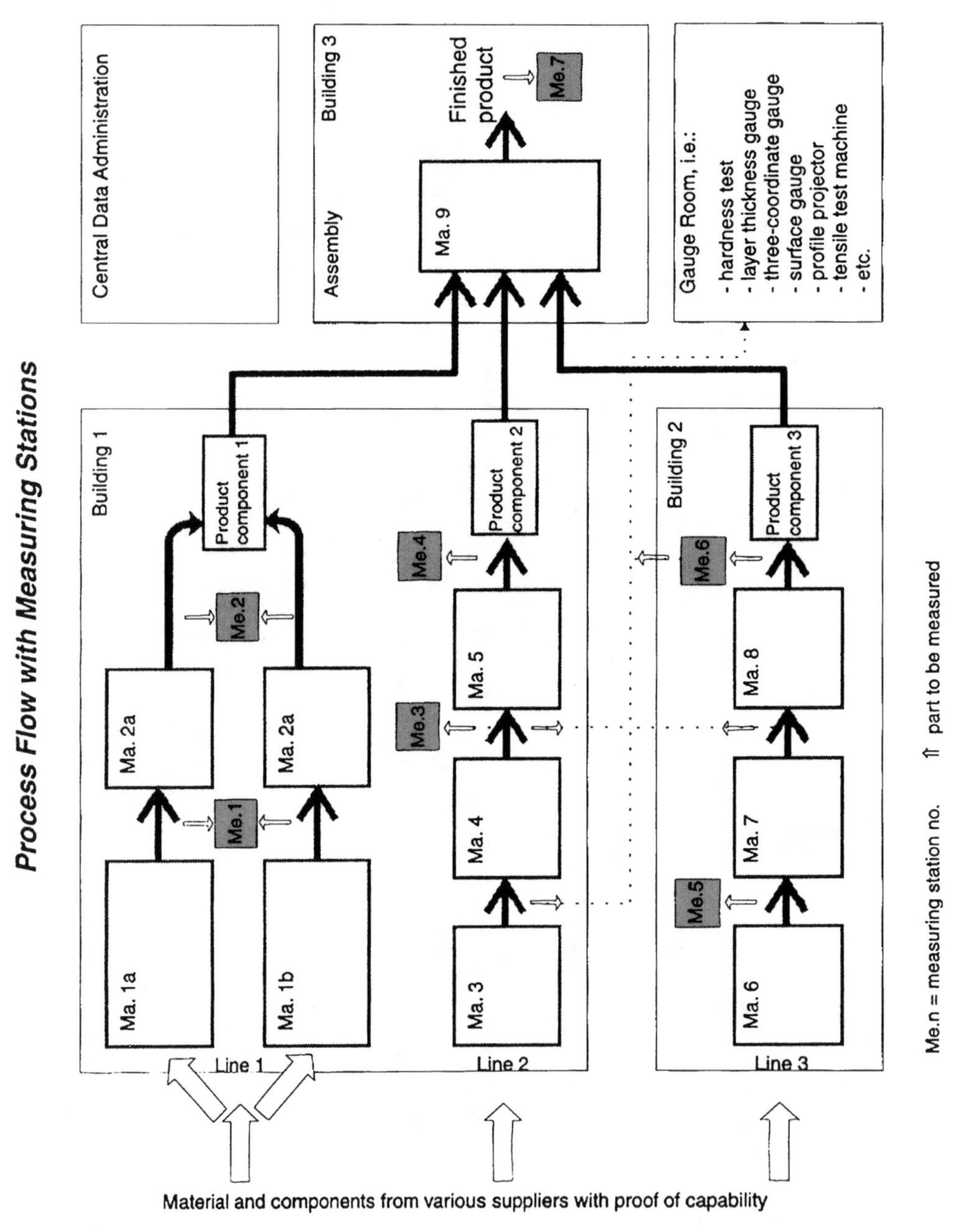

Figure 7.4–2. Process flow with measuring stations.

From the point of view of process assessment, there are three different levels of information. Each level can be considered individually. The three levels combined represent the process.

Individual Characteristics

On this level, characteristics suitable for SPC must be selected and monitored. This requires appropriate measuring systems. The sample size and sampling frequency must be optimized. The assessment may be based on statistics such as capability indices, estimates of process location and variation, and numerical descriptions of any additional variation of the mean or trend behavior.

If a central data administration station has access to the data collected by the various decentralized measurement systems, then a status display combining all these data would be useful to give an overview of current process behavior.

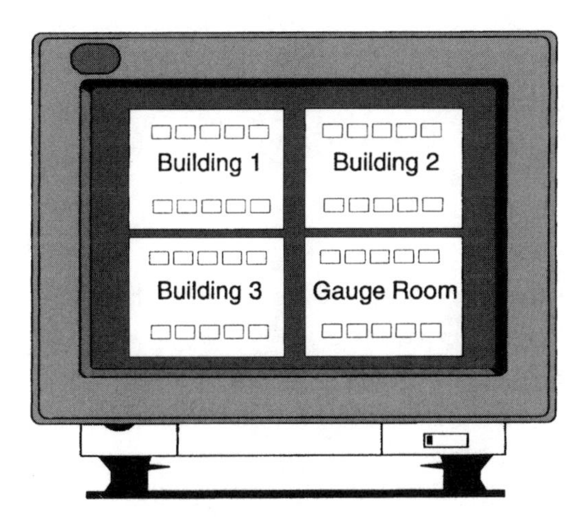

The status display indicates the status of each measurement system, e.g., by the following three colors:

- Red ⇒ process not O.K.
- Green ⇒ process O.K.
- Gray ⇒ system currently out of operation

Such a display can include additional details, depending on the screen area available and the number of status indicators required at any given time. Possible data are

- Data from the last 25 subgroups
- Statistics such as $\bar{x}$, $\tilde{x}$, s, R, etc.
- Information on trends or variation of the mean
- Descriptive data from the inspection instructions
- Distribution and process model with control chart
- Etc.

Such a status display enables the selection of problem areas for detailed analysis.

Manufacturing Line

This level consists of summarized information on units such as the following:

- Individual machines or equipment groups
- Tools
- Materials
- Suppliers
- Lots
- Operators/inspectors

This level can also relate to larger units such as individual buildings, departments, areas, or sections. Apart from capability indices, typical statistics include the proportion nonconforming or the rework rate, availability of equipment, information on process categories, and the proportion of stable or qualified processes. In this context, it is of vital importance that data are grouped and combined in a meaningful way.

Product Line

This level consists of statistics for the product and its components. In addition to the statistics mentioned previously, the nonconformity costs incurred for a given product line are of particular interest.

Routine Evaluation

Data analysis at these three levels may be prompted by external events or may be performed on a routine basis. Typical cycles are

- Daily
- Per shift
- Weekly, monthly, or annually
- At various intervals at the discretion of the operator
- Etc.

7.5 Global Approach

An automated procedure is a must for *user-friendly* process assessment. To this end, it is necessary to specify the type of analysis to be performed as well as the documentation format. Furthermore, it is necessary to determine which information should be retrieved and when. Figure 7.5–1 gives an outline of the process.

The Flow of Information

Measurement data collected by decentralized measuring systems are transferred to a central database for analysis. Various types of data management are possible (see Figure 7.6–3). The next step consists of selecting data from the pool, based on predefined or customized selection criteria. From these data, statistics are calculated and graphics created using the specified analysis methods. Various print templates are available for the presentation of results. The assessment report consists of the results transferred into the appropriate print template. The process is

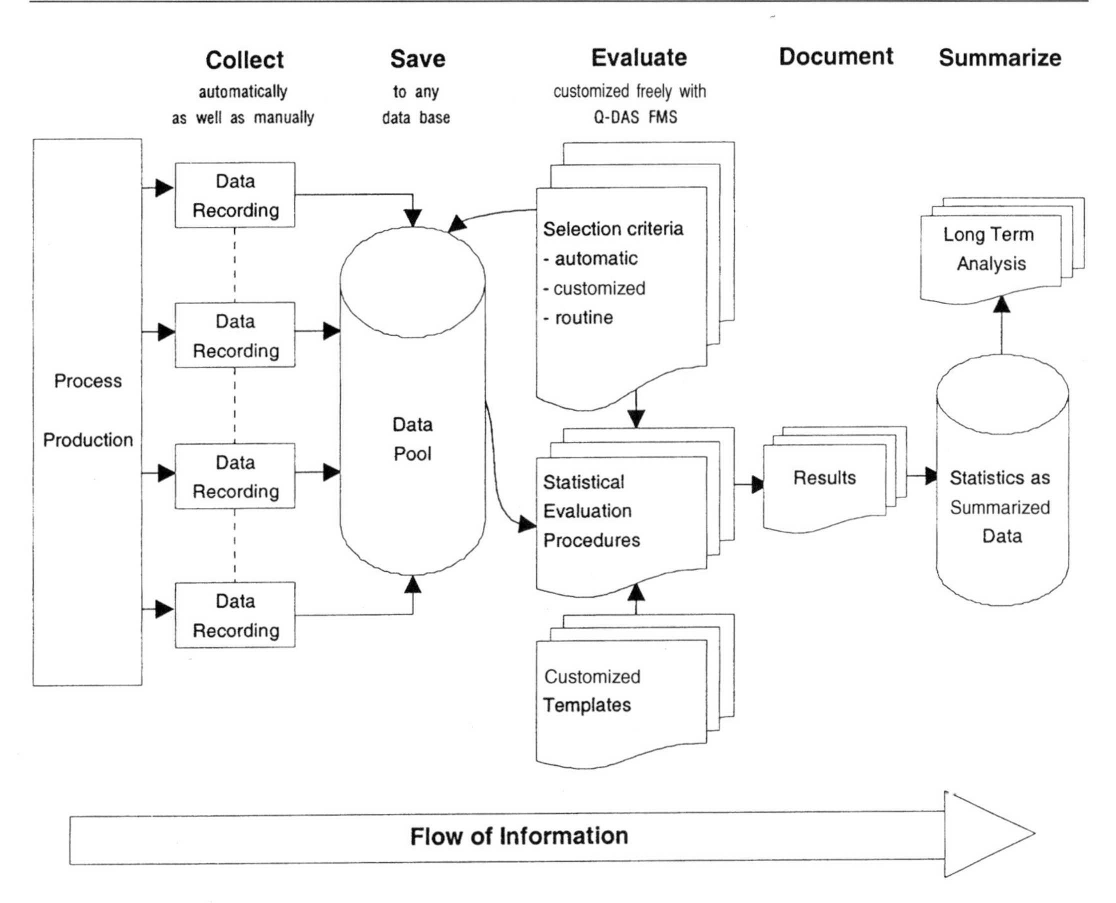

Figure 7.5–1. Flow of information with evaluation.

repeated at regular intervals, and records are maintained to provide data for long-term analyses.

Meaningful Documentation

Using predefined templates for data values, statistics, graphics, and textual information, various document types can be generated depending on the amount of information desired. Where possible, these documents should be generated automatically and at regular intervals without the need for personnel involvement. This ensures that there is a solid basis for medium-term and long-term analyses with a view to initiating improvement actions. In addition, users must be in a position to quickly generate these documents themselves as and when required. There are a number of software programs on the market to facilitate this.

Warning System

Individual analyses tend to be inefficient, especially where large quantities of data are involved. The solution is to collect the data centrally and to create a *warning system* based on *background* checks or on checks performed at regular intervals (e.g., during breaks or *off* shifts). Such a warning system includes statistical procedures to identify process changes or deviations from specified criteria. The permissible range of variation is also defined. If statistics fall outside their permitted range, the processes are flagged. This separates the wheat from the chaff. The results of the *good* processes are simply updated, but the *nonconforming* processes are investigated.

Statistical Methods Used in the Warning System

Statistical analyses, running in the background, are used to identify any problems. These analyses can be based on various methods:

- Numerical test procedures (normal distribution, randomness, outliers, trends, etc.)
- Statistics for process location or process variation, proportion nonconforming, capability indices
- Comparison with predefined limits
- Control chart stability

If there is no cause for concern, the results are simply collected and archived. A decision needs to be made as to the required retention period for certain statistics to enable long-term analyses. The statistics and graphics that will be printed out at regular intervals also need to be defined.

Statistical Analyses

Statistical analyses must be based on correct and consistent use of statistical methods. Guidelines must be defined and adhered to. This is important, because it ensures that the results can be used for long-term comparisons.

Results Format

Typical results formats are

- Statistics such as $\bar{x}$, s, capability indices, regression or correlation coefficients, proportion nonconforming
- Run chart comparisons for individuals, means, variances, c values
- Box plots
- Scatter plots
- Probability plots, histograms, control charts
- Multiple characteristics charts
- Number and type of process disturbances
- Out-of-control conditions and their frequency
- Etc.

Decision Rules

Based on the results and existing process knowledge, clear decision rules can be established to trigger predefined actions, e.g.,

- Assessment decision (process O.K./not O.K.)
- Process improvement actions
- Selection of the appropriate type of control chart

A decision table should be generated to provide a clear reference chart for rules and required actions.

Summarized Data

A decision needs to be made as to which statistics or descriptive data must be recorded in data summaries. The amount of information retained must be such that it is still possible to perform meaningful long-term analyses later on.

The records must contain

Process data
- Machine capability
- Process capability
- Proportion nonconforming
- Variability
- Trends
- Etc.

Work station data
- Machine number
- Material number
- Type of characteristic
- Tolerance limits
- Product, type
- Etc.

7.6 Ongoing Assessment

A process information system offering the possibility of generating assessment reports for machines and processes is of vital importance for ongoing product and process assessment and improvement. The global approach described earlier (see Figure 7.6–1) can be divided into the following individual steps.

Selection of Appropriate Characteristics

The first step consists of identifying suitable characteristics for process control based on correlations and external influences. Then, an appropriate measurement procedure must be selected along with a sampling frequency and sample size. Based on this, the type of monitoring, manual or automatic, must be determined. The inspection procedure must then become an institutionalized part of the process.

Measuring stations and inspection procedures can be entered into the general company model. Figure 7.6–2 illustrates this. Measuring stations (marked Me n) are created at appropriate points in the process flow. For special measuring requirements, process monitoring may involve samples of parts being taken to a gauge room for measuring and recording of results. The figure lists some typical types of measurements that might be performed in a gauge room.

Both quantitative and qualitative data are collected. Preference should be given to quantitative data, though, as they contain more objective information and are more easily integrated in automated systems.

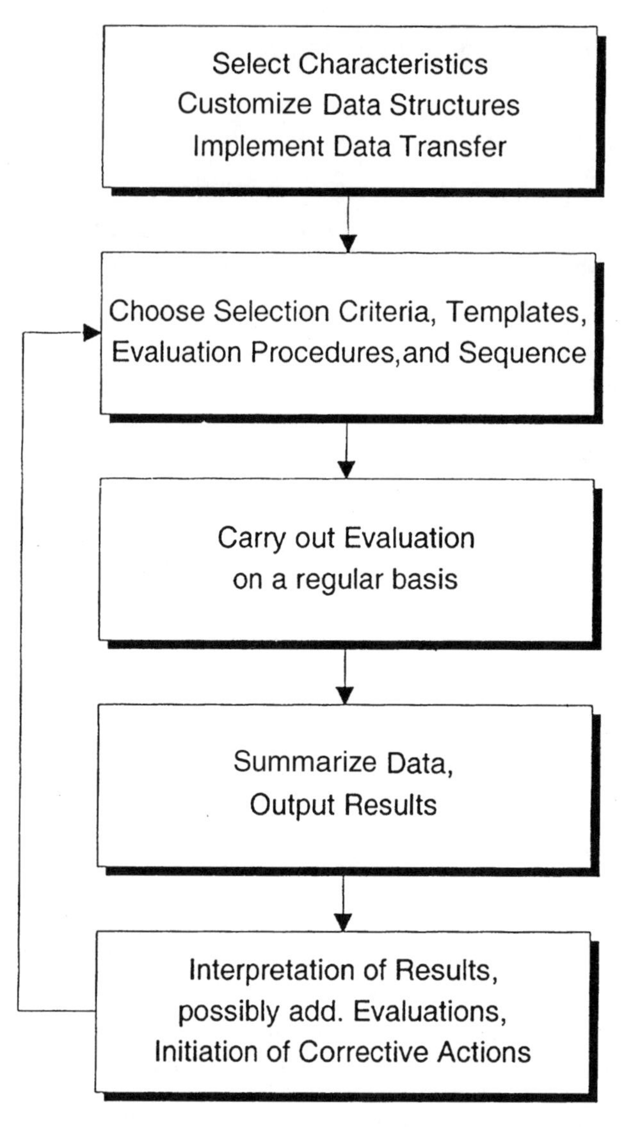

Figure 7.6–1. Continued qualification procedure.

Data Structures

Based on the individual requirements of the company, all necessary data must be available in the assessment system to ensure that users can properly administrate their *business processes* using the system. Hence, flexible customizable data structures are required. This is vital for user acceptance.

Data Transfer

Not all companies have automated and networked all their measuring stations, and this is not a requirement for a successful process assessment system. Cost should be the only criterion in deciding the appropriate degree of automation and network-

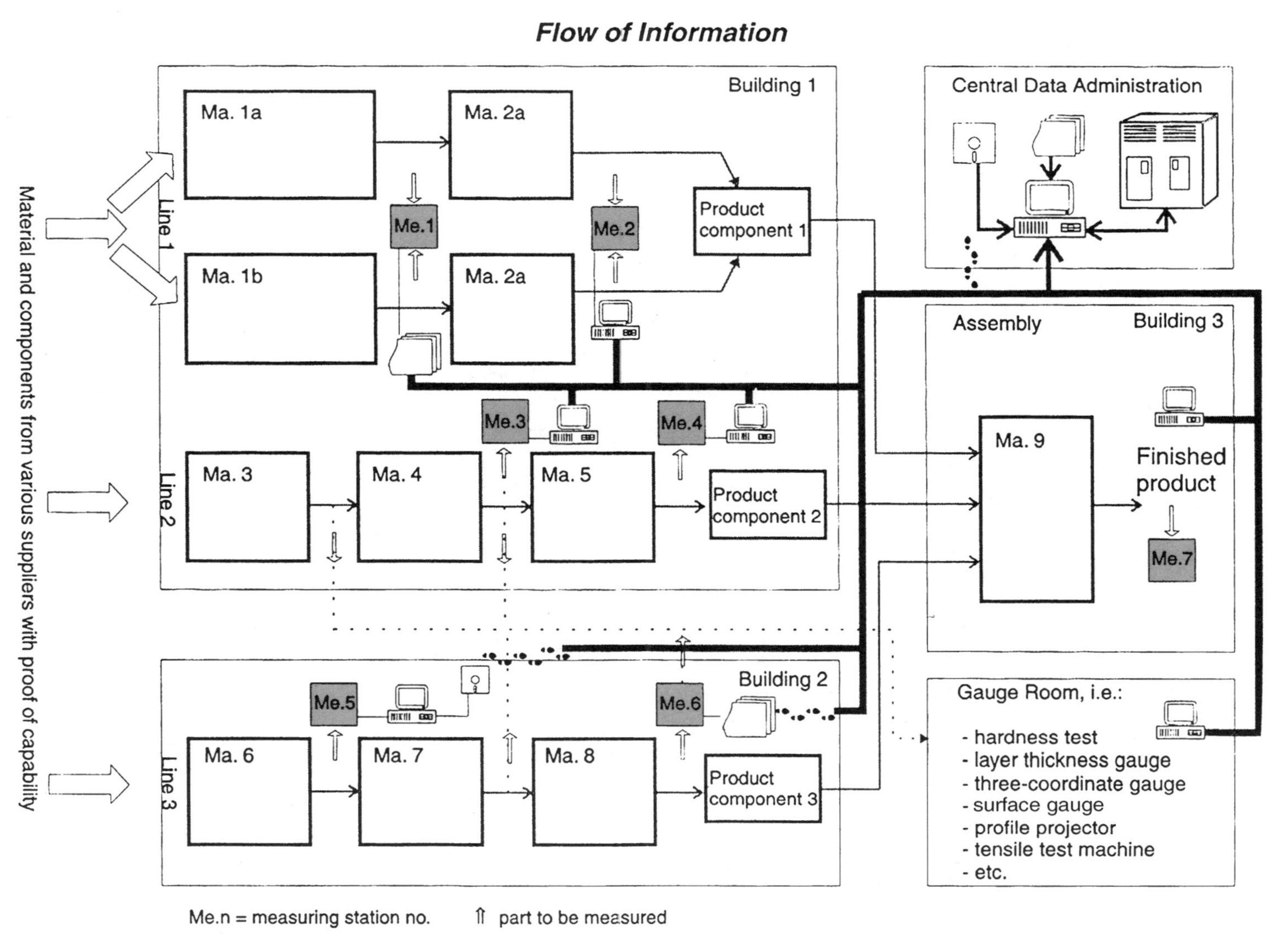

Figure 7.6–2. Flow of information.

ing. Decentralized data management is a viable alternative to network solutions. Here, data are transferred via diskette to a central computer. Decentralized analysis, combined with transfer of results for long-term records, saves administration effort and relieves the burden on the central data management site. Manually prepared control charts can be read in using special reading devices, evaluated, and the data transferred to the central database. For smaller amounts of data, manual entry of data into a computer system is a viable route. These possibilities are illustrated in Figure 7.6–3.

Where the data collected are stored and evaluated will depend on the available or planned computer network configuration. Figure 7.6–3 gives examples. Combinations of the options illustrated are possible as well. The precise approach in a given case will depend on

- Cost/benefit ratio
- Availability of equipment
- Data security considerations
- Available infrastructure
- Amount of data involved
- Requirements for up-to-the-minute data

Decentralized *monitoring* should, barring exceptional cases, always be a feature of the system.

Selection and Evaluation

The data collected according to the prescribed sampling frequency should be evaluated at predefined intervals. To ensure that this happens in the most efficient way possible, selection criteria and evaluation procedures should be predefined, and data analysis should be performed automatically. The results are entered in customized forms or templates. This regular analysis provides meaningful process documentation. Long-term evaluation of these data can reveal significant process changes and lead to the initiation of corrective action.

Summarized Data

Appropriate summary statistics need to be defined to restrict to reasonable proportions the amount of shelf space or hard disk space required. Once summary statistics have been calculated and the results recorded, the raw data can be deleted or retained for a specified period in a separate archive.

Accomplishing all the tasks just described is certainly not something that can be achieved in a matter of days. The important thing is to be aware of the possibilities, to establish goals, and to define measurable action steps. Meaningful statistical procedures can liberate a substantial potential for rationalization, especially for an existing SPC system. This can aid cost reduction and help justify new investment.

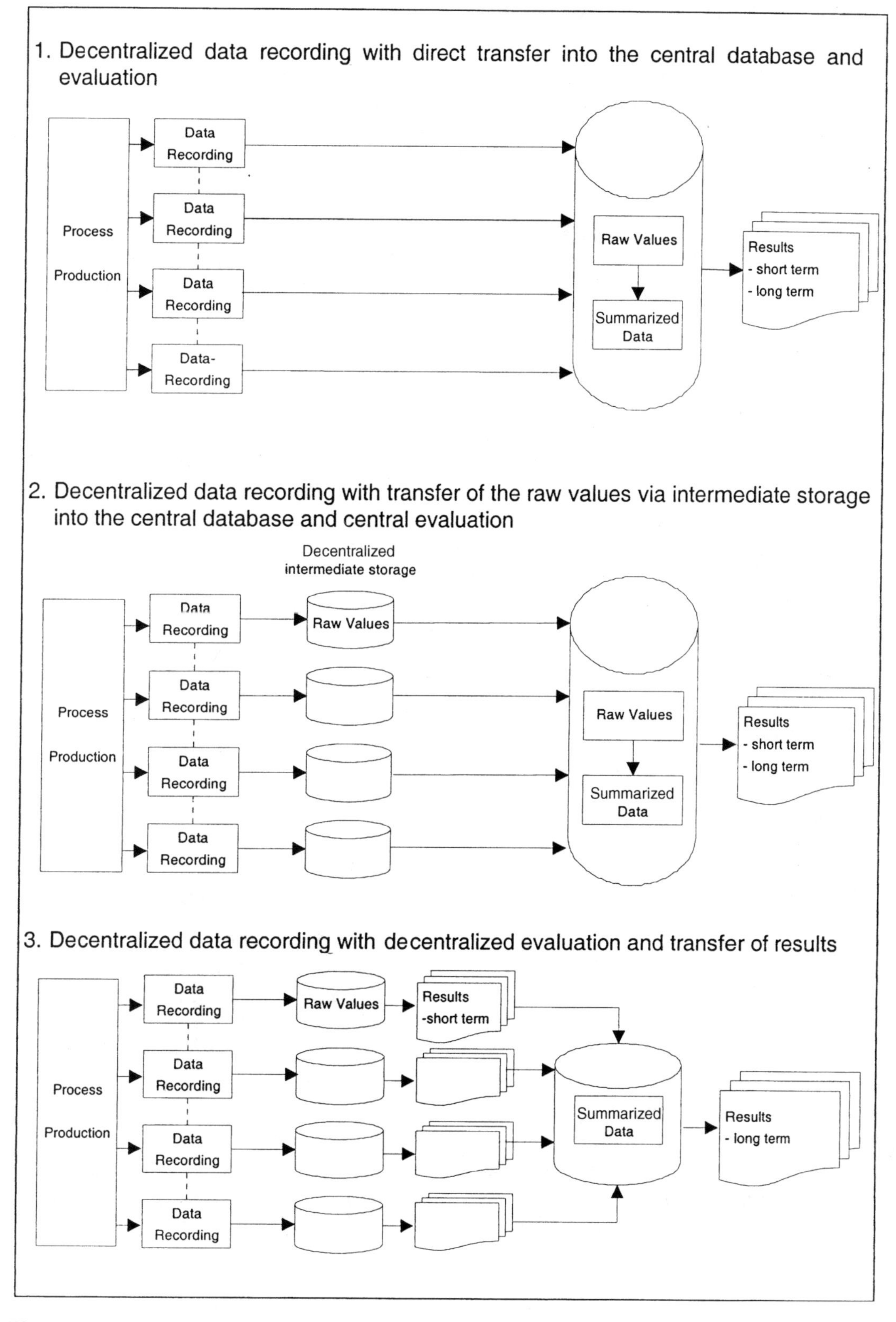

Figure 7.6–3. Possibilities of data storage and summary.

7.7 Case Study: Acceptance of a Machine Tool

Preliminary Remarks

The procedure for machine and process capability studies was mainly developed by the automobile manufacturers and their supplier community. The procedure begins with an initial run of a small number of parts to set up the equipment correctly. Then, a number of test pieces are manufactured and measured. The measurement data are subjected to a statistical analysis, and the machine or process is evaluated on the basis of this analysis. Machine or process capability indices are calculated, which quantify the ability of the equipment to consistently produce parts within specification. These indices are denoted by C_m, C_{mk} for machine capability, and P_p, P_{pk} or C_p, C_{pk} for process capability.

The automotive industry, in particular, developed their own guidelines, which are the basis for new acquisitions, i.e.,

Ford:	Production facilities -Guideline for performance rating -Capability rating with true position tolerances	[35]
Opel, Vauxhall, GM:	Machine runoff and acceptance specifications—LVQ1	[57]
PSA Peugeot, Citroën, Renault:	CNOMO norm E41.32.110.N	[58]
Volkswagen, Audi:	Production equipment guideline—BV 1.01	[68]

Information Gained

Direct investigation methods, such as geometry or positioning checks or static or thermal studies, focus on specific reproducible characteristics of the machine. In this way, assembly errors and appropriate design improvement action can be identified. Most studies of this type evaluate the machine under off-load conditions to exclude external influences and to make the study independent of varying operational conditions.

Capability studies, on the other hand, represent an indirect investigation method, because the machine is judged on the basis of the output it has produced. One disadvantage associated with this is the fact that the results are really only valid for the specific load conditions that prevailed at the time of the study, and that the multitude of external factors involved does not allow an independent assessment of the machine itself. While standardized machine tests, such as the production of simple test tools as per VDI 2851, are easily reproducible and thus allow a comparison of different machines, capability studies tell us if specific production requirements are fulfilled with a sufficient degree of accuracy and within the set time frame. Here the result is not just a function of machine quality and machine performance, but a function of all the factors affecting the process. The aim of a capability study, therefore, is not the classical assessment of machine accuracy, performance, or other desirable machine attributes, but only to assess the machine's ability to produce satisfactory product. In this sense, this type of study is not suitable for identifying weaknesses in machine design or identifying possible improvement actions to address any such design weaknesses.

External Factors

The main problem in evaluating a machine tool on the basis of a capability study consists in the numerous external factors that may affect the machine and reduce output quality, factors that are entirely beyond the machine manufacturer's control (Figure 7.7–1).

In addition to environmental conditions—such as fluctuations in environmental temperature—process parameters and tool wear have a direct impact on machine accuracy. Operating the machine with worn tools can also subject machine components to higher stresses, with the resulting deformation causing additional deviations. Differing allowances for unmachined parts or variations in clamping will have similar effects.

Another factor is the data evaluation process, which has an inherent degree of statistical uncertainty. The fact that the produced parts have to be measured also influences the results of a capability study since the indices are calculated from the standard deviation of the characteristic, which includes variation arising from the measurement process itself. Gauge capability studies often reveal that, for narrow tolerances, theoretical gauge performance requirements rise to levels that are almost impossible to achieve.

An important consideration in establishing the importance of a factor is the question of whether its effect on individual parts varies; i.e., does the factor affect the

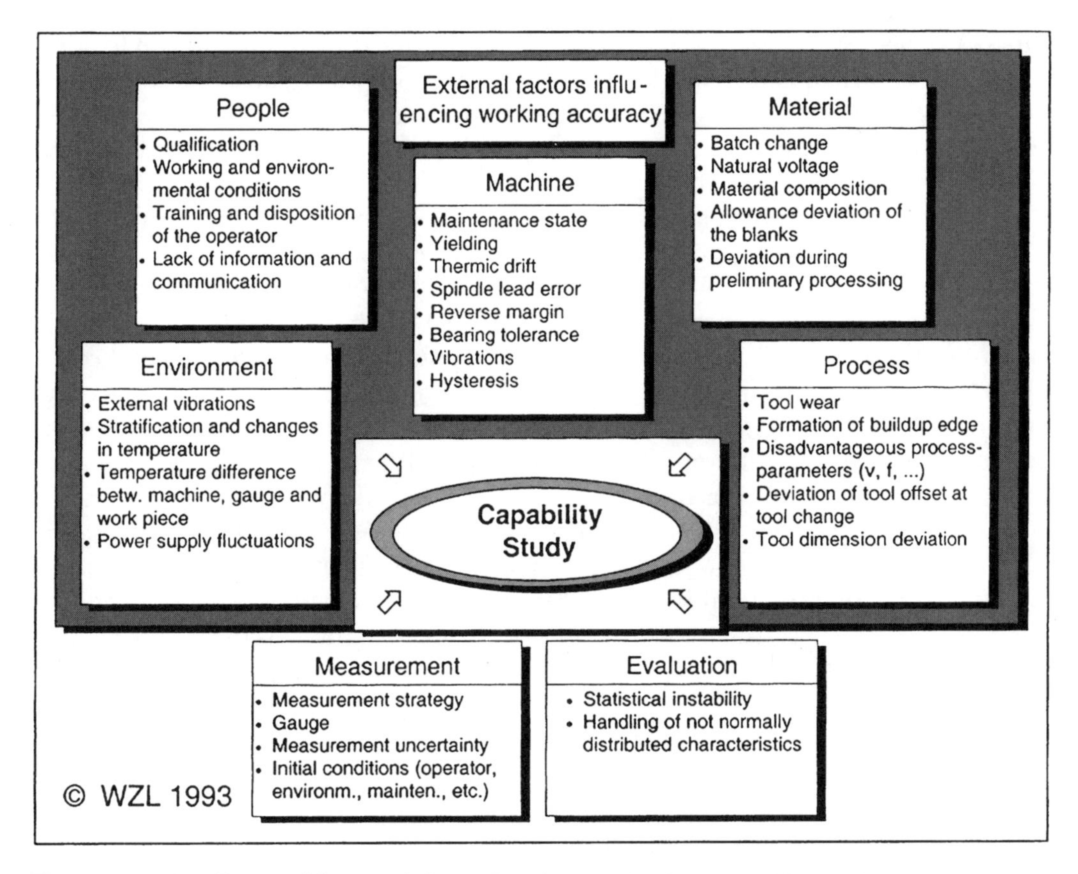

Figure 7.7–1. External factors influencing the result of a capability study.

accuracy of each part in the same way, or does its influence change over time? To give an example of the first category, consider the strain on a machine due to cutting forces. This is the same for each work piece, and hence it will only have the effect of shifting the average of the characteristic up or down. An example of the second category would be tool wear, which, through progressive dimensional change in the cutting tip and rising passive forces, will cause a corresponding progressive change in the measurements of the parts produced.

The main factors to be considered in establishing procedural guidelines and test conditions for capability studies are as follows:

- Process forces
- Tool wear
- Temperature factors
- Suitability of measurement processes

To ensure correct and comparable results, all statistical procedures and evaluation steps should be clearly defined.

Guidelines

The following represents a suggested general approach. It may need to be adapted to suit the specific manufacturing equipment and technology involved.

Figure 7.7–2 shows the basic process flow for a capability study. Once the process, the test conditions, and the evaluated variables have been agreed upon, the machine should be subjected to a warm-up period. This is followed by a start-up run to adjust the machine to target. Following this, a number of work pieces are manufactured in direct sequence, without any further adjustment or correction of the process, and taking all care to exclude any external factors such as vibration of the workshop floor or sudden changes in temperature.

Then the relevant characteristics are measured. This must be done using a measurement system whose capability for this specific measuring application has been demonstrated beforehand. The last step is the statistical evaluation of the results. In addition to identifying any trends due to tool wear or thermal drifts, this includes the identification of the distribution type. The capability indices determined on the basis of this distribution model indicate whether or not the machine is capable of fulfilling the acceptance criteria. In some cases, the calculation of capability indices may not be appropriate, and different decision criteria will have to be applied.

If the requirements are not met, one should first verify whether there were any abnormal occurrences during production. This can be established by checking the log, where such occurrences should be noted, and also from the statistical analysis, which will identify outliers. If it is found that there were such occurrences, production should be repeated. Where this would not appear to make sense, improvement actions must be initiated, or the agreed conditions or process parameters must be changed.

Agreements

Before the acceptance test proper can be performed, various points have to be agreed upon to ensure that

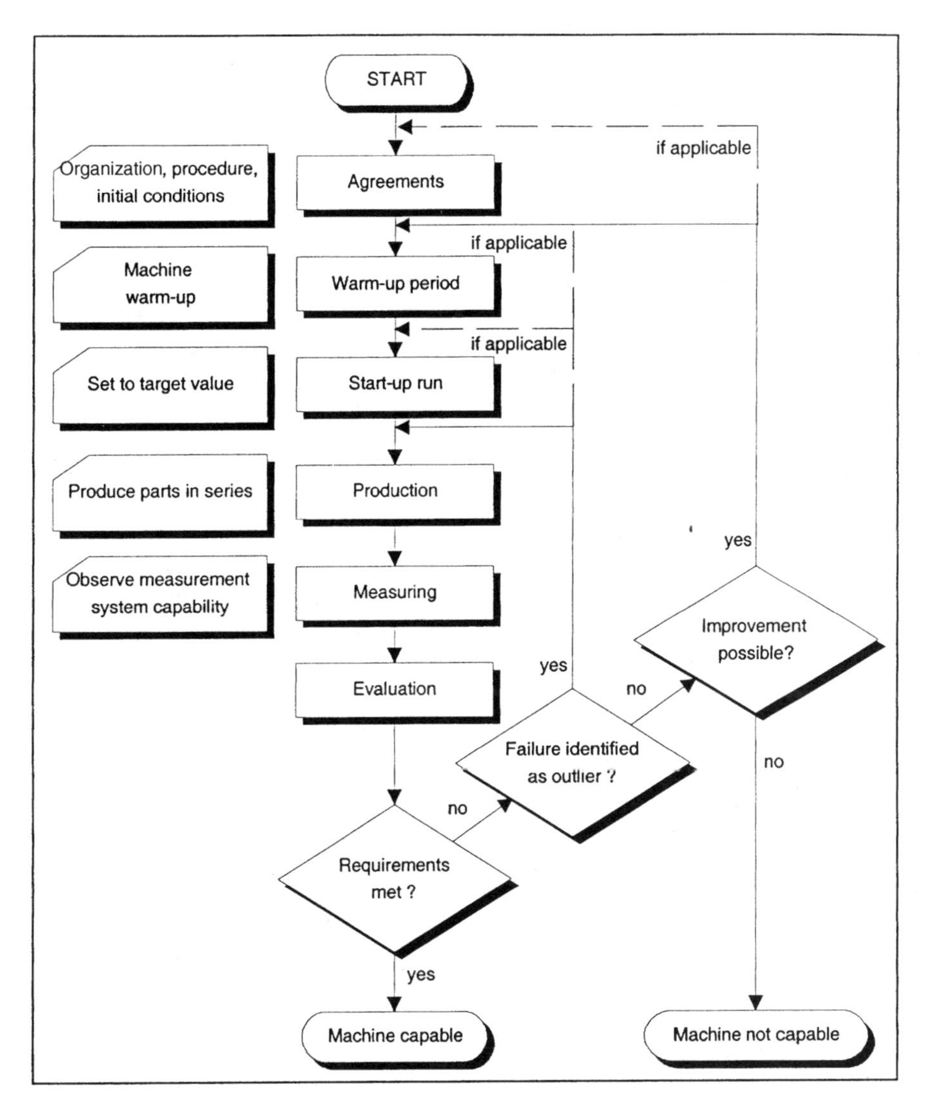

Figure 7.7–2. Basic procedure for a capability investigation.

- The machine is evaluated as objectively as possible, i.e., without external factors.
- Fulfillment of the requirements is technically feasible, allowing for the number of external factors, the uncertainty of measurement, and the uncertainty inherent in the statistical analysis.
- Both the manufacturer and the user of the machine are fully apprised of the procedure that will be followed and the information that will be gained by the test.
- Contractual agreements can be made that specify the scope, the procedure, and the assessment criteria of the acceptance test.

In addition to organizational issues, the test conditions have to be agreed upon. This includes, for example, the environmental temperature and its permissible range of variation during the testing period. The limits set will depend on the manufacturing application, which will also determine whether the machine is set up in the workshop or in a special air-conditioned room. For normal production applications, the following values represent a useful guideline:

Environmental temperature : ϑ_{env}, start $-3°C \leq \vartheta \leq \vartheta_{env}$, start $+3°C$

Temperature gradient : $\Delta\vartheta_{env} \pm 2°C/h$

A capable process requires capable input materials. Hence, it must be ensured that composition and material properties are not influenced by batch changes. Of particular importance is the agreement of over-allowances, to keep strain during roughing-down work within limits, and thus ensure that work pieces have a constant over-measure prior to finish-machining.

The number of work pieces depends on the time required to machine each piece. The production period should correspond to one shift, with samples of five taken during this period. The sampling frequency should be such that 50 parts are available for analysis. A total sample size of less than 25 parts is not acceptable for capability studies, as this would compromise the statistical significance of the results.

Furthermore, the technical details and duration of the warm-up phase must be agreed upon before the acceptance test is performed. The selection of a measurement system must be based on a clear understanding of the system's inherent uncertainty of measurement. This means that the capability of the measurement system must be demonstrated before it is used for measuring. Limits for trends due to thermal drift will depend on the production process, the machine size, and production and environmental conditions. For reasons that follow, this trend is often negligible for machines subjected to a capability study.

Warm-Up

As capability studies only make sense for machines used in mass production, and because such machines are usually subject to three-shift operation, actual production conditions are best simulated by including an appropriate warm-up period in the procedure. This can be accomplished by running an NC program in the night preceding the acceptance test. If, however, trends due to thermal drift are of particular interest to the user, or the duration of the warm-up stage cannot be extended sufficiently to achieve thermal equilibrium for the machine, then a permissible trend should be agreed upon prior to the study and allowed for during analysis.

Of greater importance for the thermal behavior of the machine are usually thermoelastic changes in shape due to disturbances such as breaks or mixed production. However, this tends to apply first and foremost to production of small-sized lots and should in any event be analyzed using other methods, such as a direct thermal study of the machine.

Start-Up Run

Unmachined work pieces conforming to the agreed-upon quality criteria must be made available and allowed to come to the temperature level of the production environment. New tools should not be used, as a high degree of wear is to be expected in the initial stages of operation and this causes, in addition to direct measurement deviation, a significant increase in passive forces. Hence, the tool should be allowed to perform some cutting work prior to the start-up run.

To allow for the trend caused by tool wear and to be able to separate this trend from any thermal drift, tool wear should be predicted, based on empirical knowledge or relevant literature, or measured using a brush analyzer or microscope. As the tool used is not completely new, it is sufficient to assume linear wear behavior and measure the tool before and after acceptance testing.

The purpose of the start-up run is to adjust production to target. For two-sided tolerances, this corresponds to the center of the tolerance interval, or as close as possible to zero for characteristics with a natural zero limit.

How accurately the process is adjusted to the target depends among other things on the amount of effort required to do this and the amount of importance attached to correct location of the mean. For instance, it may be that adjusting the process to the desired mean level requires a great deal of effort, but as such does not constitute a problem in practice. In such a case, it would make sense to only approximately adjust the process to the desired mean level and to define only the standard deviation or range as the assessment criterion.

In general, it should be verified, based on production of five pieces, that the critical distance Δx_{crit} of the average from the appropriate tolerance limit conforms to the following criterion:

$$\Delta x_{crit} \geq 0.45 \cdot T$$

Production

Floor vibrations caused by adjacent equipment and changes in environmental temperature must not exceed predefined limits. During production, the environmental temperature should, at regular intervals, be measured in the immediate vicinity of the machine and logged.

Parts should be manufactured in direct sequence, without interruptions. This must be a continuous process, since any changes in procedure, or in the amount of time used to machine individual parts, would affect the process and thus distort its normal behavior. Samples of five are taken at regular intervals throughout the production period. Process events and interventions should be logged to facilitate subsequent interpretation.

If the machine is capable of automatic adjustments, these should be allowed for in the capability study. The resulting distribution type of the characteristic must be accounted for in the analysis.

Measuring

The tolerance interval of the characteristic will result in certain requirements for the measuring device, the measuring location (air-conditioned gauge room, workshop), and the measuring strategy. All measurements must be performed by trained personnel. Care should be taken to ensure that the temperature of the measuring device and the measured work pieces does not differ from the environmental temperature.

The measuring device must have adequate resolution and an uncertainty of measurement that is as low as possible. It is desirable to have a measuring device that fulfills the following requirement:

$$\text{Scale interval} \leq \frac{\text{tolerance}}{20}$$

A less powerful resolution, i.e., a coarser scale graduation, is only acceptable if the standard deviation of the measuring device is very much smaller than the device's resolution, e.g., $24\ s_{gauge} \leq$ scale interval.

The measurement system's suitability for use in the process capability study must be demonstrated by performing a gauge capability study based on the tolerance or variation of the characteristic to be measured.

There are two types of gauge capability study, which are based, respectively, on

- Fifty repeat measurements of a standard under constant conditions, and calculation of the measurement system's standard deviation s_{gauge}
- Repeated measurement of ten work pieces by two or three operators with calculation of statistics for repeatability, reproducibility, and total gauge variation

A detailed description of these methods is given in Chapter 8.

8 Measurement System Capability

8.1 Introduction

8.1.1 Rationale

For quantitative characteristics, the acceptance of machinery and process equipment, process and product evaluation, or continuous process monitoring will be based on measurements of manufactured parts or process parameters. Special sensors, custom-made measurement systems, or commercially available standard devices are required to obtain these measurements.

To draw correct conclusions about current status from the measurement data obtained, measurements must be sufficiently accurate, with the required level of accuracy depending on the tolerance of the characteristic or on process performance. In the past, the suitability of measuring equipment would be examined mainly on the basis of minimum values given in standards or by verifying manufacturers' claims. When using this approach, checking procedures (see DIN 10012) and performance requirements for the measurement system (max. error range, repeatability, etc.) are device dependent. As a rule, only the measuring equipment itself is checked, and under ideal conditions, e.g., in a gauge room with trained personnel, with idealized production parts such as gauge masters or standards, and using standardized equipment. The associated methods and checking procedures in the form of inspection instructions are described with examples in VDI/VDE/DGQ guideline 2618. This approach is necessary for verifying manufacturers' claims for a new gauge, or when carrying out routine checks (monitoring of measurement systems) to detect changes or errors in gauge performance.

However, determining measuring equipment performance in this way does not provide an insight into how the equipment will perform under real operating conditions, for example,

- When used at the actual measuring location
- When used by several operators
- When used as part of a measuring device

- When used to measure real production parts
- When used under changing environmental conditions
- When used within a measuring sequence, where cumulative errors may become an issue, etc.

Thus, at best, the aforementioned procedures enable us to come to the theoretical conclusion that a gauge could, in principle, be suitable for measuring a characteristic at a given tolerance. However, to monitor whether the accuracy or variation of the gauge is adequate under real-life conditions for assessing a process that may only exhibit a small amount of variation, other methods and procedures are required.

In line with the philosophy of *never-ending improvement,* the objectives of process management are no longer centered on staying within the tolerance interval, but instead focus on achieving a continuous reduction in process variation. This implies a corresponding increase in the accuracy of the measurement systems and a reduction in their variation. If this is not the case, taking measurements almost amounts to a classification of the values, and the results are of little use to statistical analysis.

In the section entitled *Inspection, Measuring, and Test Equipment,* DIN EN ISO 9000 (et seq.) requires that measurement systems be assessed on the basis of so-called capability studies. However, it does not specify concrete procedures and calculation methods, and there are currently no corresponding standards. Thus, there is a dependency on company guidelines or recommendations of professional organizations. In recent years, these have become more and more widespread. One of the first companies to produce a guideline on this subject was General Motors [45] in 1987. In December 1989, Ford published a revised guideline [36] based on an earlier paper on this subject.

In Germany, the Robert Bosch Group [62] published a guideline in 1990 entitled *Ermittlung der Fähigkeit von Meßeinrichtungen unter Betriebsbedingungen* (Determination of Measurement System Capability under Actual Operating Conditions). Mercedes Benz published its own guide [56] in 1994.

In addition to ISO 9000, this whole subject has been given further impetus by VDA (German automotive industry association), volume 6.1, and especially by QS-9000 [3], generated by Ford, Chrysler, and General Motors. QS-9000 Element 4.11, *Control of Inspection, Measuring, and Test Equipment,* item 4.11.4, *Measurement System Analysis,* sets out the following requirement: *Appropriate statistical studies shall be conducted to analyze the variation present in the results of each type of measuring and test equipment system. The analytical methods and acceptance criteria used should conform to those in the* Measurement System Analysis *(MSA)[2] reference manual. Other analytical methods and acceptance criteria may be used if approved by the customer.*

The MSA reference manual does not describe any concrete procedures; rather, it is a guide describing fundamental principles, the importance of such studies, and the different methods of calculation. These can be used as a basis for generating procedures within companies. The authors of the MSA reference manual have themselves done this for their own companies (see Ford [36] or Opel, Vauxhall, GM [57]).

The popularity of these procedures will further increase, since there is no doubt about the importance of suitable measurement systems, especially for automatic data collection. Compliance with the procedures now often forms an integral part of contractual agreements. In addition, it is not only automotive manufacturers who check and assess the *methods used to ensure capable measurement systems and gauges* as part of their system audits. To illustrate recent thinking on the subject, the following is an extract from Ford Motor Company's Guidelines for *Measurement System and Equipment Capability*, published as Form EU 1880A in December 1989:

> *The analysis of a process can only be meaningful if the data on which the analysis is conducted is itself meaningful. Since the data will invariably be collected using a measurement system, it follows that the measurement system must produce meaningful data.*
>
> *To some degree,* all *measurement systems and equipment have some measurement error or uncertainty. It is important therefore to understand the magnitude of the variability this causes, and to hold such uncertainties within reasonable limits, consistent with the specific requirement of a particular measurement application.*
>
> *Since, due to measurement error, we cannot arrive at a completely* true value *of a measured quantity, we strive to approximate the truth by the best reasonable value that can be deducted from the measurements. To do this, we want to acknowledge the existence of* random *measurement variation that is natural, but we also want to recognize any* nonrandom *measurement variation.*
>
> *To determine the magnitude of variation, a* measurement system and equipment capability study *should be performed (*Gauge *Capability Study).*

8.1.2 Causes and Effects of Measurement Errors

The term *measurement system* refers not only to the measuring equipment itself but to all factors that influence the measurement process and all equipment components. A measurement system can thus be subdivided as shown in Figure 8.1–1.

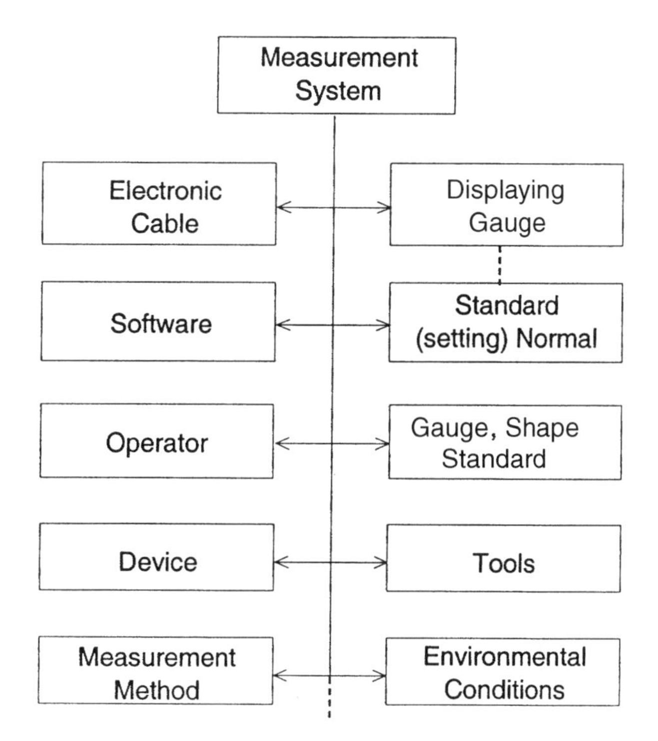

Figure 8.1–1. Definition of a measurement system.

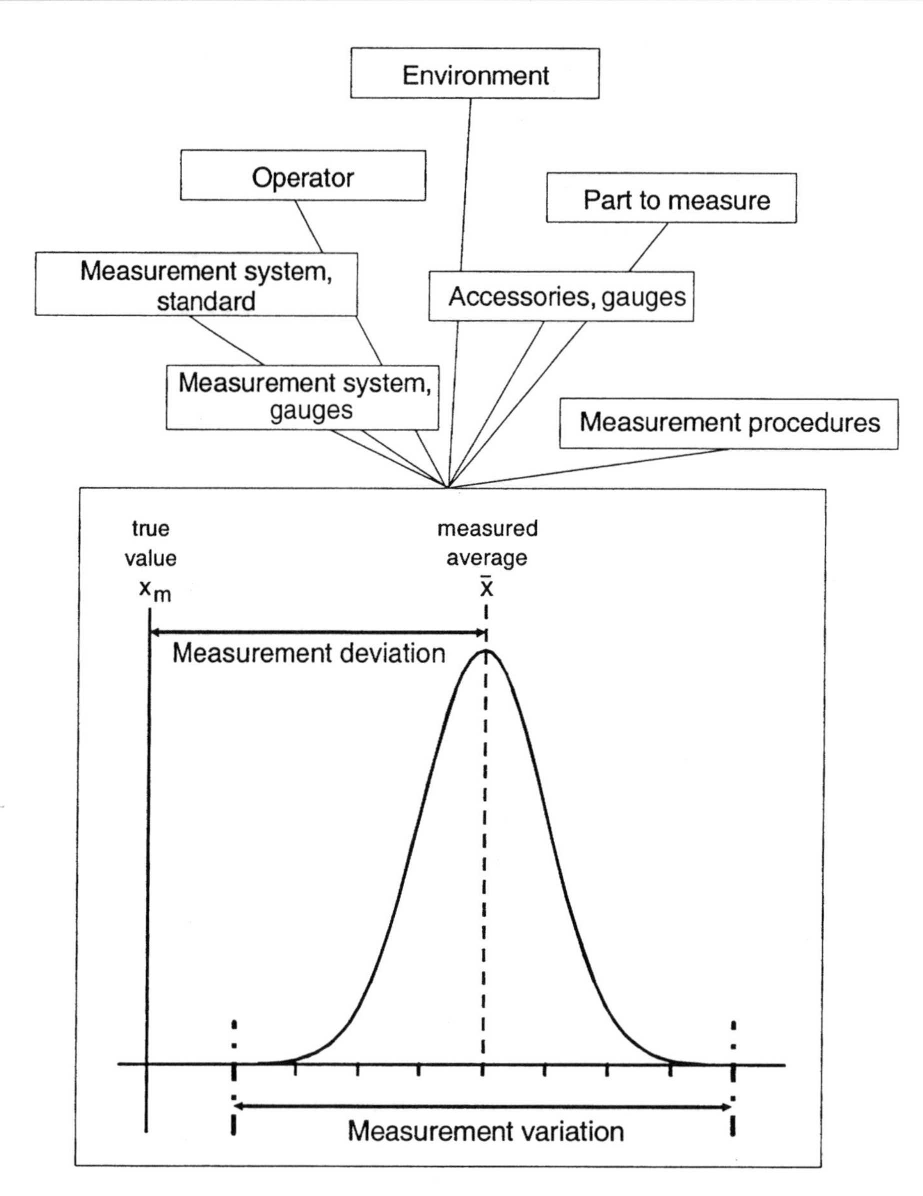

Figure 8.1–3. Causes of measurement system variation.

As a rule, it is not necessary to verify the suitability of the measurement equipment or measurement process if the measurement uncertainty resulting from the measurement process is small in relation to the characteristic variation to be measured.

Figure 8.1–3 shows the measurement value in terms of *location* and *variation* as a result of the different factors affecting the measurement result.

Figure 8.1–4 shows how systematic and random errors, due to unsuitable measurement procedures, can change the assessment of an ideal process.

In cases 2, 3, and 4, an *ideal process* will receive a *negative* evaluation simply because an unsuitable measurement procedure has been used.

Figure 8.1–5 illustrates the effect of differences between various appraisers (percentage deviation relative to the tolerance range or $\sigma_{process}$).

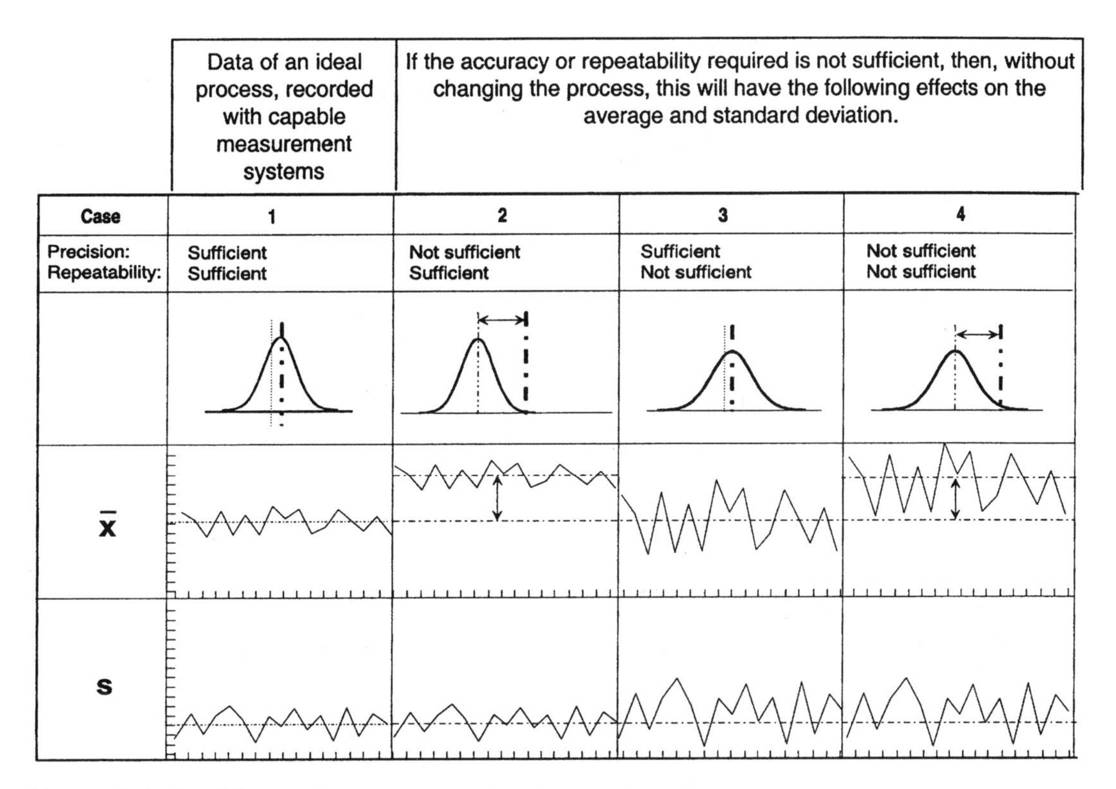

Figure 8.1–4. Effects of measurement system performance.

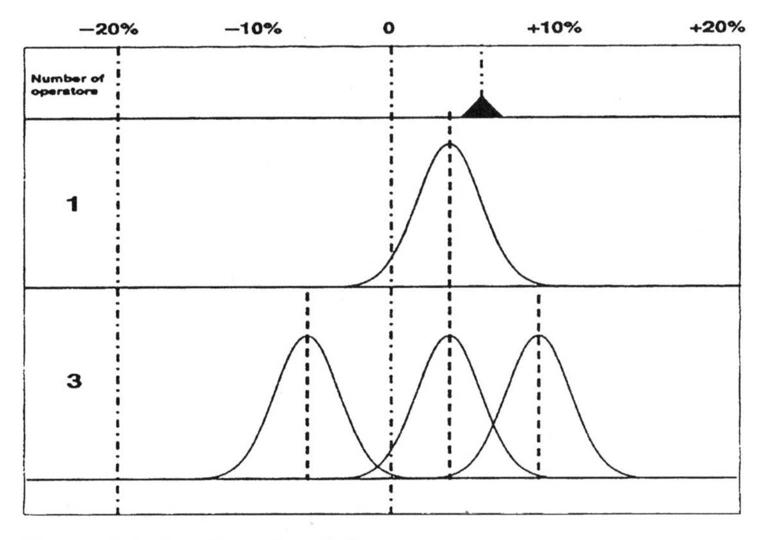

Figure 8.1–5. Appraiser influence on measurement systems.

8.1.3 Uncertainty of Measurement

The following is an overview of the different types of measurement error. Further information regarding the definitions of terms and the calculation or determination of the uncertainty can be found in the *Guide to the Expression of Uncertainty in Measurement,* GUM [31].

Systematic Error of Measurement

A constant deviation of a certain absolute value and sign (+ or –) may be observed when the same subject is measured several times under identical conditions (e.g., the same measuring equipment, the same measurement location, the same environmental conditions, etc.). A deviation of this kind may change in a predictable way if there is a change in conditions. The systematic error of measurement is also called *bias.*

The causes of bias may or may not be known. Bias tends to be caused by imperfections in the measuring equipment, the measurement procedure, or the subject, or by environmental factors.

Determinable Systematic Error of Measurement

If the systematic error is determinable (by calculation or by measurement), this information may be used for correction of the measured values. The measurement results will be incorrect if the determinable systematic error is not taken into account.

Indeterminable Systematic Error of Measurement

If the systematic error is not easily determinable, it may still be possible to estimate it. The bias will always have a certain sign (+ or –), but this will be unknown. Hence, a systematic error that can be estimated, but not precisely determined, is treated in the same way as random errors and is included with the sign ± in the calculation of the uncertainty of measurement.

Random Error of Measurement

Random errors are caused by indeterminable changes in standards, measurement equipment, measurement subjects, environment, and the operator taking the measurements. This type of error will, given the same conditions, not always be of the same order of magnitude. Individually, random errors are indeterminable; they introduce uncertainty into the measurement process and, hence, are one component of the uncertainty of measurement.

Systematic and random errors make up the so-called *uncertainty of measurement.*

Uncertainty of Measurement u

The result of a series of measurements is the mean value corrected for the known systematic error (bias) and an associated interval that is likely to contain the true value of the measured variable. The difference between the upper limit of the interval and the corrected mean value, or the difference between the corrected mean value and the lower limit, is known as the uncertainty of measurement.

Note: It is incorrect to describe the total width of the interval within which the true value is presumed to lie, i.e., the difference between its upper and lower limits, as the uncertainty of measurement.

The uncertainty of measurement u has two components. One component, u_r, represents the random errors, and the other component, u_s, represents the unknown systematic errors.

There are two methods for calculating u:

$$1.\ u = u_z + u_s$$

$$2.\ u = \sqrt{u_z^2 + u_s^2}$$

The first method is the simpler and, with respect to the risk of underestimating the uncertainty of measurement, the safer method. Its use is recommended whenever there is a large difference in the size of the two components. Where this is not the case, method 2 may be used.

If the unknown systematic errors cannot be estimated, u_r alone must be quoted as the uncertainty of measurement, and a corresponding note should be added.

Since only the random error is easily determinable by the user, it is not generally possible to give a precise figure for the uncertainty of measurement. As a general guide, it may be assumed that, with a good measuring device, the possible error corresponds to about one scale graduation in the case of an analog device. A measuring device with a $\frac{1}{10}$ mm vernier therefore has an error of about 0.1 mm, and a precision indicating gauge with a one µm scale graduation has an error of about one µm.

The DIN standards contain more detailed information about individual measuring devices. In most cases, these represent upper limits. In general, the equipment is better than indicated in the standard. Conversely, data from company documentation that do not refer to the standard are often overly optimistic.

Measuring equipment, e.g., caliper gauges, micrometers, etc., can often be tested using gauge blocks. They can, however, only be used to test precision indicating gauges at certain positions, and special test equipment is thus required for these. Go-no-go gauges can be tested using gauge blocks or appropriate measurement devices.

8.2 Testing of Manufacturers' Data in Accordance with DIN

The standards give procedures for checking and monitoring manufacturers' data and/or the basic suitability of standard measurement systems.

Table 8.2–1 on page 276 shows examples of maximum permissible device errors for standard measurement systems.

To determine the error range, measurement results taken at various points distributed over the whole measuring range of the gauge are compared with the true values. The differences are plotted in a graph against the corresponding points in the measuring range. The error range is given by the difference between the greatest and the smallest values. If the width of backlash also has to be taken into account, measurements need to be taken in both directions.

Examples

1. Display error range f_{max} for a micrometer is shown in Figure 8.2–1 on page 277.
2. Error range and width of backlash for a precision indicating gauge are shown in Figure 8.2–2 on page 277.

Table 8.2–1. Permissible measurement uncertainty for standard measurement systems.

	Permissible device error ± U			DIN
1. Calipers L meas. length in mm	(50 + 0.1 L) µm			862
2. Micrometer L′max. meas. Length of micrometer in mm (valid up to max. 500 mm) starting length set to 0	$(3 + \frac{L'}{50})$ µm (approximation)			863
3. Gauge blocks degree of accuracy 00 " 0 " 1 " 2	 (0.05 + 0.001 L) µm (0.10 + 0.002 L) µm (0.20 + 0.004 L) µm (0.40 + 0.008 L) µm			861
4. Dial gauges	ftot.	fw	U	878
display range 0.4 and 0.8 mm	9	3	6 µm	
" 3 mm	12	3	7 µm	
" 5 mm	14	3	8 µm	
" 10 mm	17	3	9 µm	
5. Lever gauges	13	3	8 µm	2271
6. Indicator				879
Scale value ≤ 1 µm	1.2	0.5	0.9 sv	
" > 1 µm	1.2	0.3	0.7 sv	
7. Indicator with electr. limit contacts				879
Scale value ≤ 1 µm	1.8	1	1.7 sv	
" > 1 µm	1.8	0.5	1.2 sv	
8. Pneumatic length measuring instruments				2271
without check	2.0	0.5	1.2 sv	
with check	1.5	0.5	1.0 sv	
9. Electrical length measurement for analog devices digital devices	 1% of final value* 0.5% of final value			no standard
10. Sinus guide bar	$U_{15°}$	$U_{60°}$		2273
Roller distance 100 mm	4″	11″		
200 mm	3″	10″		
300 mm	3″	9″		
400 mm and 500 mm	2″	9″		

**Final value is the maximum measurement value possible, starting from 0, in a set region.*

8.3 Advantages of Capability Studies

A gauge capability study should be carried out under actual working conditions. In this context, *actual working conditions* means that the study is carried out on-site, i.e., at the location where the gauge will actually be used; the measurement data used in the study will be taken by the future gauge operators themselves; and environmen-

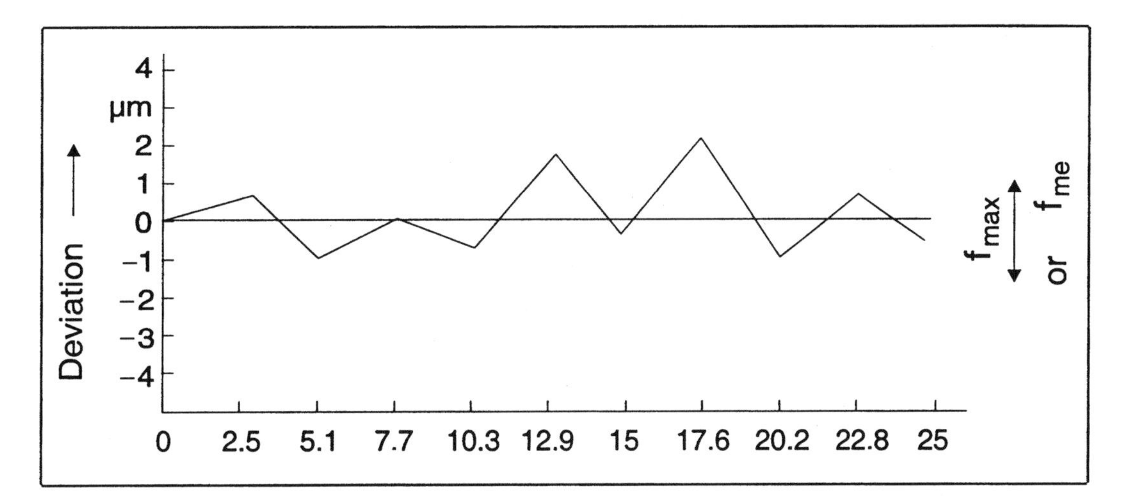

Figure 8.2–1. Error range of a micrometer.

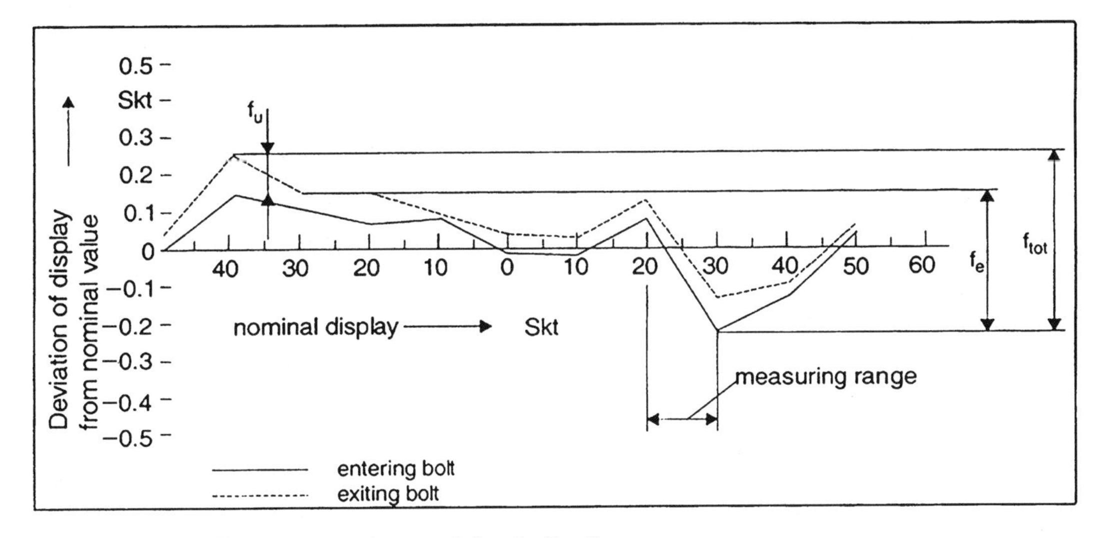

Figure 8.2–2. Error range of a precision indicating gauge.

tal conditions will be as close as possible to the actual conditions of service. This is necessary to ensure that the proof of gauge capability obtained in the study will retain its validity later on, when the gauge is actually used.

The main objectives of performing capability studies are

- To compare the performance of different gauges
- To confirm the suitability of gauges for use in production
- To provide an ongoing process control system for gauges
- To evaluate the performance of gauges
- To evaluate the reliability of measurement results
- To identify appraiser-induced or location-induced variation
- To determine failure causes in failure analyses

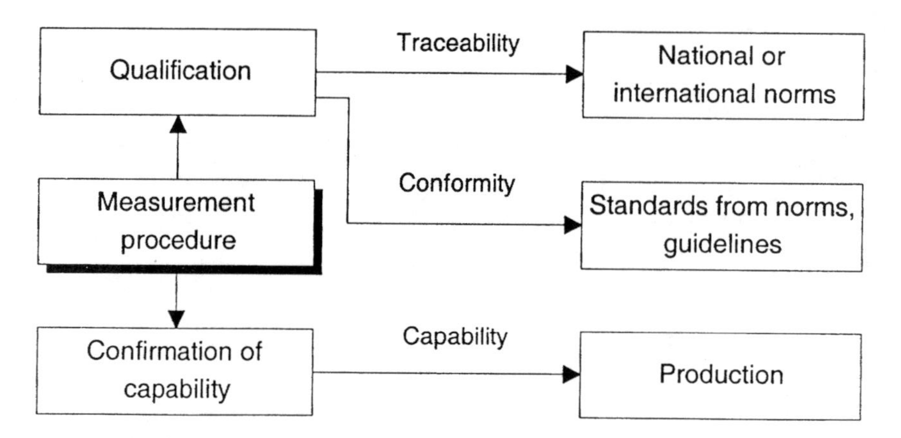

Figure 8.3–1. Qualification and confirmation of capability.

Capability studies make it possible to compare the performance of different gauges, to classify them, and to organize them according to groups. This provides a basis for accepting or rejecting new measurement systems.

Basically no measuring procedure should be used without evidence of capability. The purpose of a capability study is to verify that the procedure complies with the requirements set down in standards and guidelines, the manufacturers' data, etc., and to ensure tractability to national or international standards. Confirmation of capability testifies to the suitability of a procedure for *real-life,* on-site use in a specific measuring application (see Figure 8.3–1). At the same time, a link is established between the parts produced and the procedure used to measure them. Hence, when a measurement is taken, it can be traced back to a specified measurement procedure. This can be of great importance, especially in the context of product liability.

8.4 Definition of Concepts

SPC users have long been accustomed to the fact that capability definitions may differ from one company to the next. Hence, it is not surprising that, in the case of gauge capability, companies use a variety of different definitions, and requirements vary from one industry association to the next.

However, since the various procedures do not differ significantly, we shall start with a general overview and a definition of basic concepts. This will be followed by a description and evaluation of the differences between the procedures when making the calculations. Considerable differences occur at this point. Examples will be given to illustrate the various procedures.

In the evaluation of measuring equipment, it is generally necessary to distinguish between two situations:

Qualification

- Measuring equipment whose use is not limited to a particular application, such as general-purpose or handheld gauges, etc., should be evaluated on the basis of the relevant DIN-ISO standards and/or VDI/VDE/DGQ guidelines, or on the basis of manufacturers' data.

Verification of suitability

- Product-specific measuring equipment, which may comprise several measuring devices, requires additional evaluations to be performed (as described in the following sections). In particular, these must be performed for all *critical characteristics.*

The time and effort required to maintain an operational system for monitoring measurement equipment and qualification are enormous. In most companies, such systems must control several hundreds or thousands of gauges. This is very costly in terms of both time and money.

As we are here dealing exclusively with preventive measures whose economic efficiency is difficult to verify, a great deal of motivation is needed to both set up and maintain such a system. Management can play a key role in ensuring its success.

As in SPC, a further distinction should be made between an analysis phase, in which the basic suitability (capability) of the measurement procedure is assessed, and continuous on-site monitoring.

The definitions for the following terms apply regardless of the procedure adopted, and they are virtually the same for all guidelines:

- Accuracy
- Repeatability
- Reproducibility
- Stability
- Linearity

Unfortunately, not all of these definitions conform to norms (see DIN 55350, part 13 [24], or GUM [31]). However, they were integrated into general usage caused by the prevalence of the guidelines and the practical meaning. We recommend that the reader compare the definitions used with those of the norm when developing guidelines.

8.4.1 Accuracy

Accuracy is indicated by the difference between the average of a set of measurements, obtained by measuring the same gauge master or standard a number of times, and the true value of the master or standard, on which sufficiently precise information is available. The measurements should be taken by one appraiser, measuring one and the same gauge master or standard several times in the same location (see Figure 8.4–1).

8.4.2 Repeatability

One appraiser measures the same part (standard or normal production part) several times in succession, using the same measuring equipment in the same location in accordance with a specified measuring procedure (see Figure 8.4–2). The standard deviation of the measurements obtained will then be a measure of the repeatability.

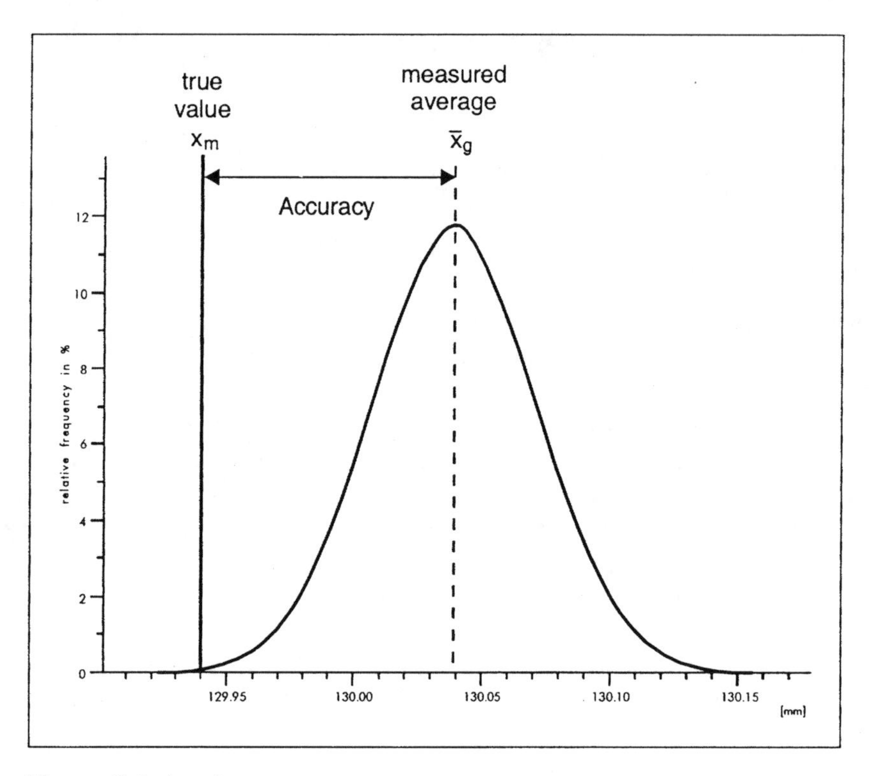

Figure 8.4–1. Accuracy.

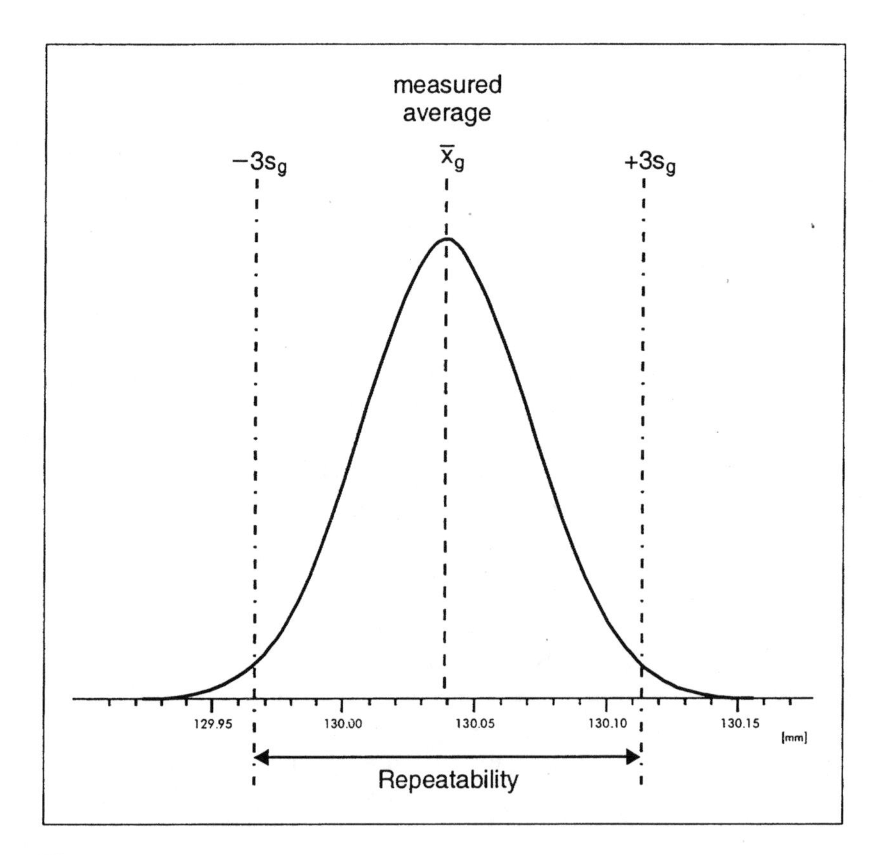

Figure 8.4–2. Repeatability.

8.4.3 Reproducibility

Following a specified measuring procedure, one part (normal production part) is measured several times. The measurements are taken in one of several ways:

- By different appraisers
- In different locations
- Using different gauges

Usually, two or three appraisers measure the same part, or one appraiser takes repeat measurements in different locations or using different gauges (see Figure 8.4–3).

In the course of a reproducibility study, only *one* of these three variables must be changed at a time. If this requirement is not fulfilled, it will no longer be possible to determine the effect of the individual factors, and the assessment will be of no value. The overall variation with respect to the overall average will then be a measure of the reproducibility.

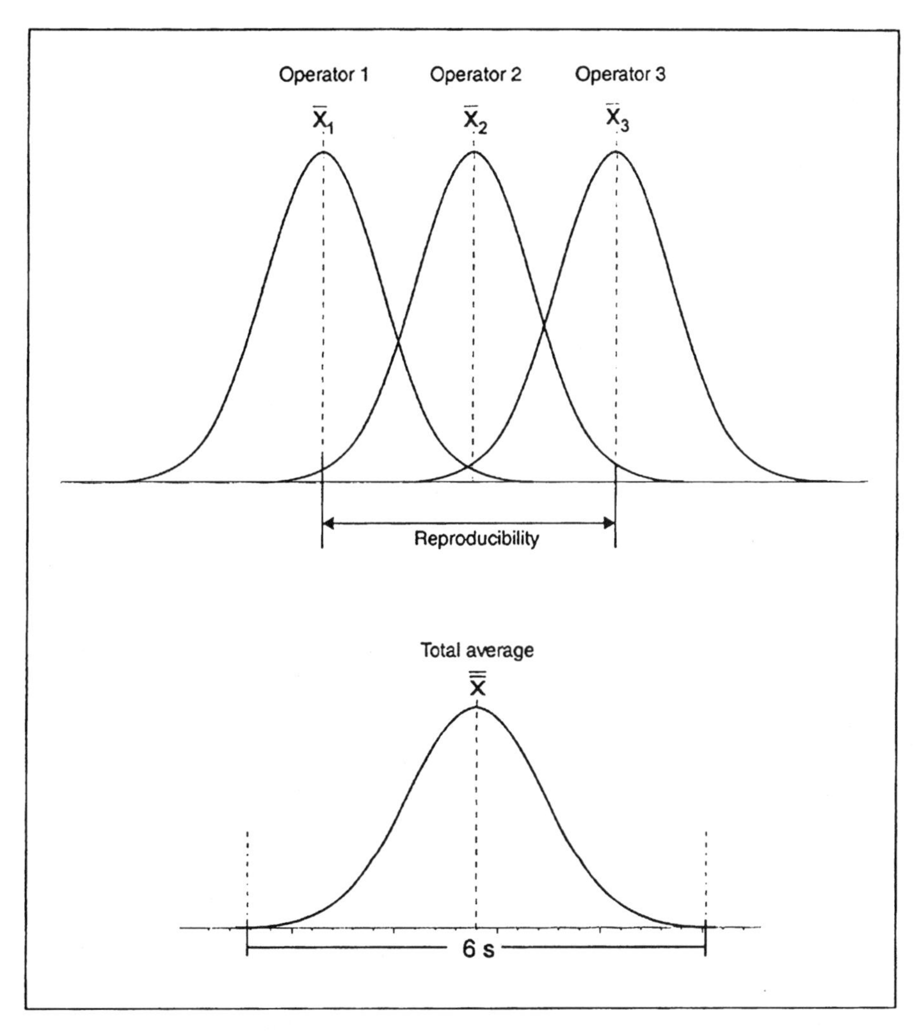

Figure 8.4–3. Reproducibility.

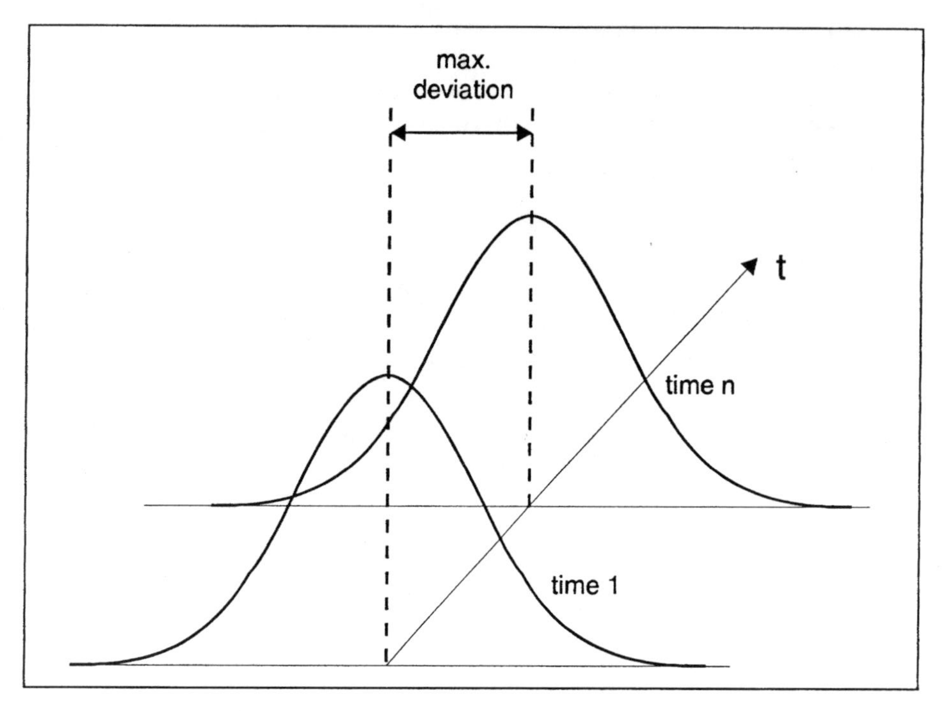

Figure 8.4–4. Stability.

8.4.4 Stability

At specified intervals and following a specified measuring procedure, one appraiser repeatedly measures the same part, using the same measuring equipment in the same location. The average measurements obtained at different times are then compared with each other (see Figure 8.4–4). The maximum difference between any two average measurements will then be a measure of the stability.

The variation in the measurements will include both random and systematic errors due to the appraiser, the gauge calibration, gauge wear, and changes in the measuring environment.

8.4.5 Linearity

A specified number of measurements is taken by one appraiser in the same location in accordance with a specified measuring procedure on standards that should cover the entire measuring range of the measuring equipment. The average of each set of measurements is compared against the true value of that standard. The varying differences between the average measurements and the corresponding true values will indicate the linearity. To obtain a line graph of the linearity-related measurement error, these differences can be plotted in a chart (see Figure 8.4–5).

8.5 Performing a Gauge Capability Study

As the term suggests, a gauge capability study bears certain similarities to a process capability study, both in concept and practical implementation. The following types of studies can be performed:

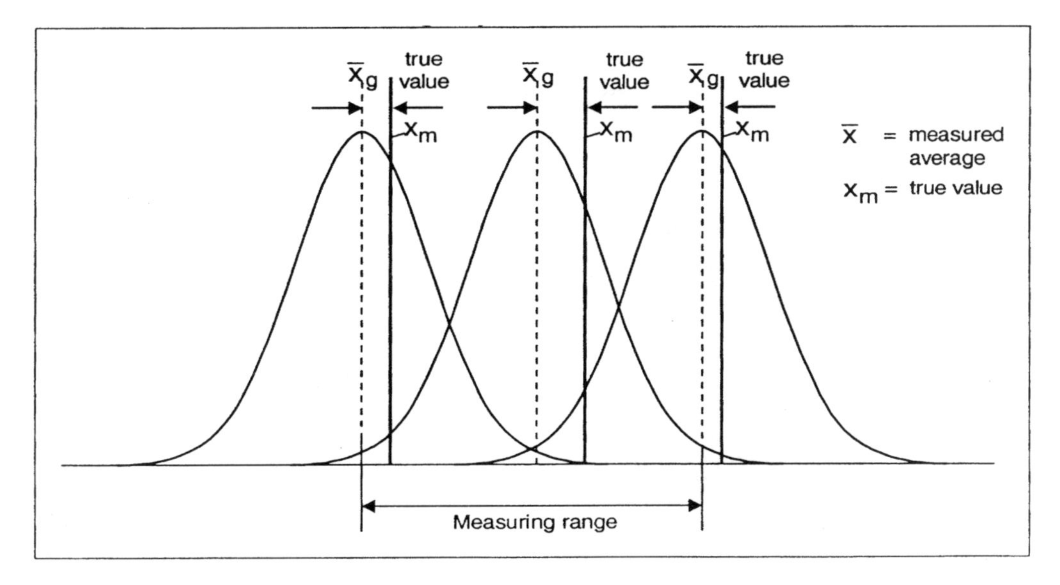

Figure 8.4–5. Linearity.

Analytical Phase

- Type 1 study: accuracy and repeatability (capability indices C_g, C_{gk})
- Type 2 study: repeatability and reproducibility and overall measurement error (R&R study with appraiser influence)
- Type 3 study: repeatability and overall measurement error (no appraiser influence)

Note on type 3. There is clearly no appraiser influence if the part is automatically inserted into and removed from the measurement system. However, the term *appraiser influence* can and should be defined more broadly. Design features such as limit stops, positioning devices, etc., can be used to minimize appraiser influence. The reproducibility is reduced, and the graphic representations do not reveal any significant differences between the appraisers.

Figure 8.5–1 explains the areas of application of the study types described. A type 1 study is mainly used by manufacturers for the construction of measurement equipment. It can be used to prove how suitable and capable the equipment is in principle. Only when this study has been completed successfully is it worthwhile conducting a type 2 or type 3 study for evaluation in the actual area of application. This procedure, the calculation method, and minimum requirements (e.g., capability index larger than 1.33) should form an integral part of any contractual agreement.

Furthermore, a type 1 study can be used to determine and monitor linearity and stability. This involves carrying out the capability study at several measuring points or at different times. If the values lie within the permissible range, such studies should be repeated at suitable time intervals.

Quotation from Ford guideline EU 1880A [37]: A type 1 study would primarily be used in evaluating new equipment at the supplier base prior to final shipment and installation. A type 2 study would primarily be aimed at evaluating new and existing equipment prior to final acceptance for use in production. This method can also be used for routine audits and recertification purposes.

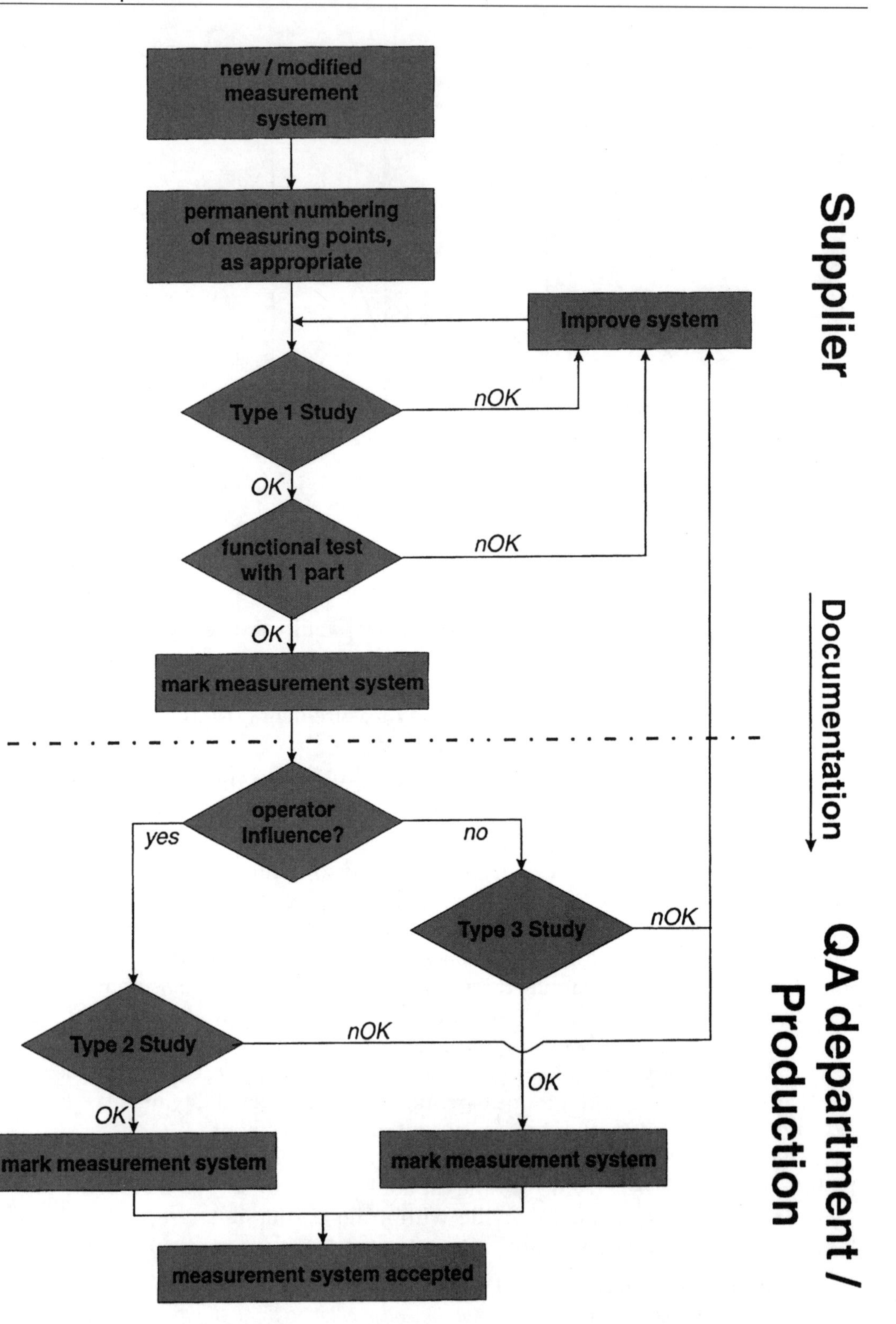

Figure 8.5–1. Process flow and relationships between the various types of studies.

Using a Control Chart for Ongoing Control

When deploying measurement systems, it is extremely important to maintain records verifying that the qualification or capability initially established is maintained over time. Information on qualification intervals is provided in the Appendix to DIN 10012. Stability studies (see Section 8.6) will verify whether the measurement system is still suitable for practical use. The time intervals should be determined depending on the application. Initial studies should be carried out after short time intervals. If adequate experience is available, the study interval can be extended accordingly. However the study interval is assessed, appraisers must receive appropriate training and instruction with regard to the incalculable consequences that could result from changes to the measurement system. Thus, it is essential that a stability study be performed in such situations.

Further possible differences when calculating the parameters are as follows:

- Calculation based on range or standard deviation or analysis of variance
- Confidence level (99 percent or 99.73 percent)
- Calculation based on: tolerance, process variation, or part variation
- Minimum requirements for evaluating suitability
- Whether there is compensation for gauge variation
- Whether within-part variation is taken into account

The various analysis methods differ in that the measure for the variation of the recorded values is estimated from the range, from an analysis of variance (ANOVA), or from the standard deviation. The measure of variation thus obtained is then compared either with the specified tolerance range T, the process variation $\sigma_{process}$, or the part variation. This produces the combinations shown in Figure 8.5–2.

The method of calculation used will depend primarily on the company. Where the process variation is not known, or if the part variation does not adequately reflect process behavior, the only available option is to compare the variation of the measurements with the tolerance.

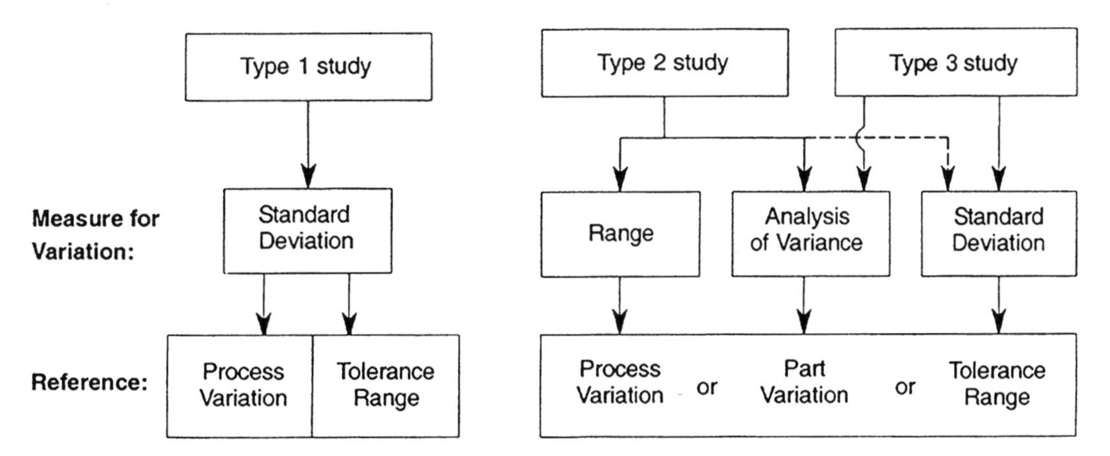

Figure 8.5–2. Different analysis methods.

8.5.1 Preparation

For any type of gauge capability study, there are certain general points concerning the selection of the measurement procedure and measuring equipment that need to be addressed prior to commencing the study:

- Define characteristic to be measured.
- Define measurement procedure.
- Define measuring range of equipment.
- Determine resolution of measuring equipment (two percent of characteristic tolerance is recommended; as a minimum requirement, the resolution should be five percent or better).
- Consider maximum speed of measurement.
- Consider accuracy of measurement (recommended error limits ten percent of tolerance, review manufacturer's accuracy data, specify error ranges).
- Assess environmental factors at measuring location.
- Ensure suitability of equipment for measuring location.

8.5.2 Type 1 Study

The type 1 study examines the inherent measurement variation of a gauge at the manufacturer's plant or on-site, using a gauge master or appropriate standard. The capability indices C_g and C_{gk} are calculated. This study evaluates accuracy and repeatability. The purpose of the type 1 study is to confirm compliance with manufacturer's specifications, especially for new or modified equipment and measurement systems, and to verify suitability of the measurement method.

An appropriate gauge master or standard should first be selected. The nominal value of this master or standard should correspond to the center of the specified tolerance interval. Fifty measurements (25 are the absolute minimum) are then taken. The master or standard must be put down and picked up after each measurement. The measurement results are recorded in a table, and a corresponding graph is plotted.

Preparation

- The gauge master or standard should have the same alignment (orientation) throughout the study.
- Whenever possible, the study should be carried out in an environment that matches the normal production environment.

Performing a Type 1 Study

- The gauge master or standard should be measured 50 times. Only in exceptional cases should the number of measurements be allowed to lie below 50, and it must never drop below 25. The master or standard should be removed from the fixture after each measurement. The measurement results should be entered in the table (Table 8.5–1) and a corresponding graph plotted.
- Calculate the mean value $\bar{x}_g$ and the standard deviation s_g of the set of measurements.

Table 8.5–1. Worksheet for recording and analyzing the values.

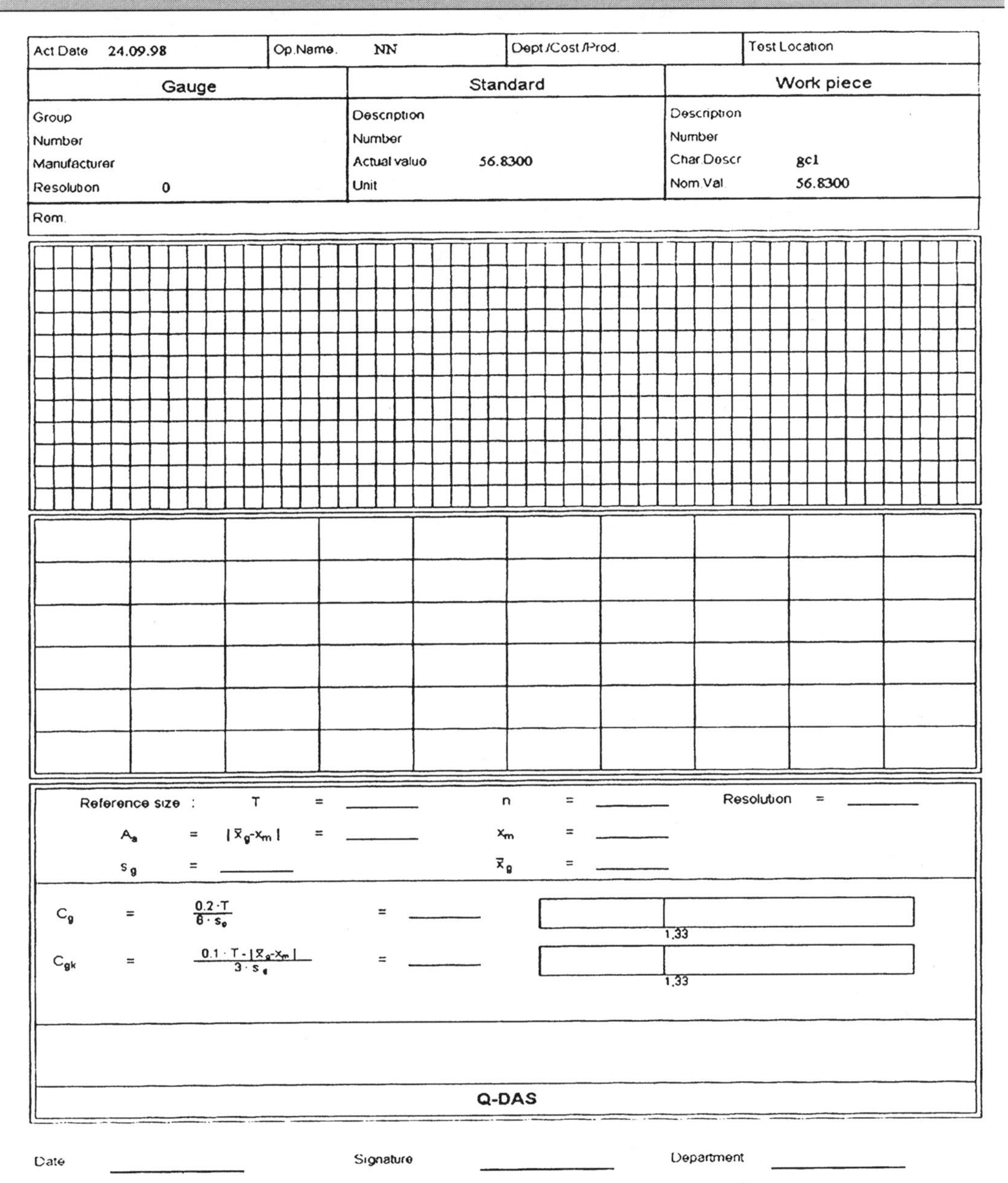

Act Date 24.09.98	Op.Name. NN	Dept./Cost./Prod.	Test Location

Gauge	Standard	Work piece
Group	Description	Description
Number	Number	Number
Manufacturer	Actual value 56.8300	Char.Descr. gc1
Resolution 0	Unit	Nom.Val. 56.8300

Rem.

Reference size : T = ______ n = ______ Resolution = ______

A_a = $|\bar{x}_g - x_m|$ = ______ x_m = ______

s_g = ______ $\bar{x}_g$ = ______

$C_g = \frac{0.2 \cdot T}{6 \cdot s_g}$ = ______ 1.33

$C_{gk} = \frac{0.1 \cdot T - |\bar{x}_g - x_m|}{3 \cdot s_g}$ = ______ 1.33

Q-DAS

Date ______ Signature ______ Department ______

- Calculate the process capability indices C_g and C_{gk} from the formulas given in the worksheet (Table 8.5–1 and Section 8.5.3).
- Compare the result with the specified minimum requirement.

Figure 8.5–3 on the next page shows the process flow for a type 1 study.

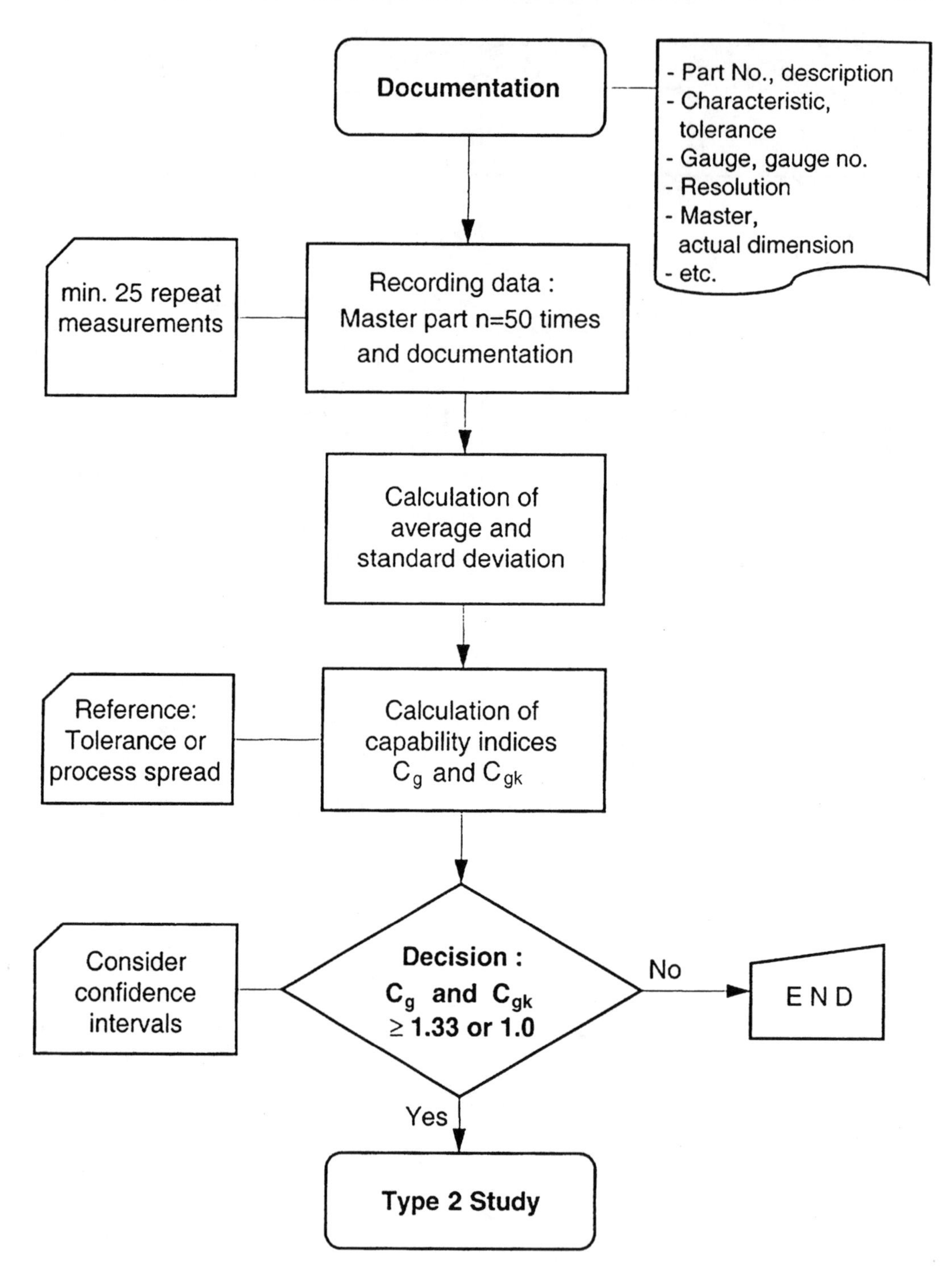

Figure 8.5–3. Type 1 study.

8.5.3 Using the Type 1 Study to Calculate Capability Indices

In analogy to process capability indices, the indices C_g and C_{gk} have been defined for measurement equipment. Figure 8.5–4 illustrates the mathematical basis of these indices.

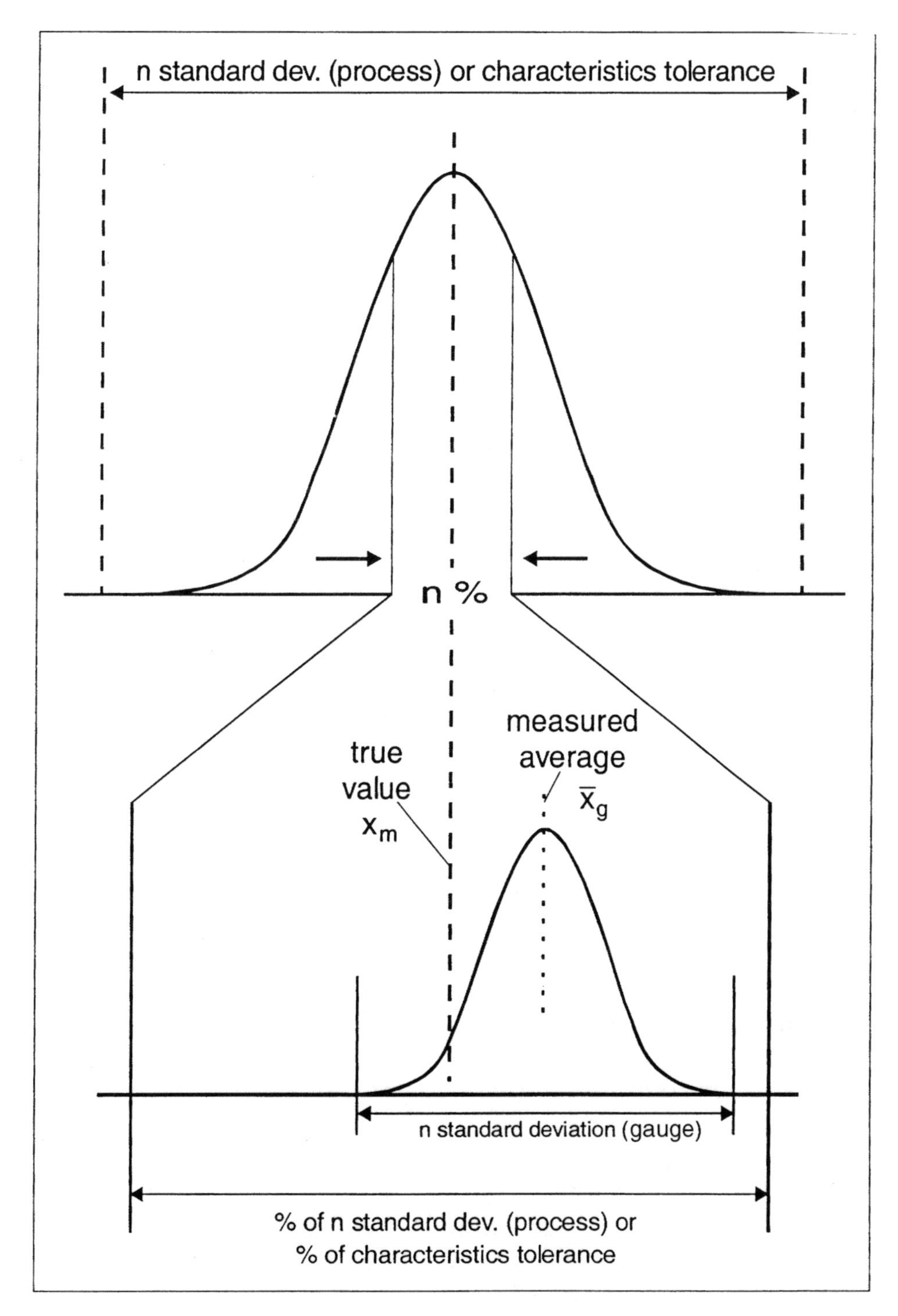

Figure 8.5–4. Calculation of capability indices.

Calculation of Capability Indices

according to [62] tolerance based

$$C_g = \frac{0.2 \cdot (USL - LSL)}{6s_g}$$

according to [36/37] process based

$$C_g = \frac{0.15 \cdot (6 \cdot \hat{\sigma}_{process})}{6 \cdot s_g} = \frac{0.15 \cdot \hat{\sigma}_{process}}{s_g}$$

for $\bar{x}_g > x_m$:

$$C_{gk} = \frac{(x_m + 0.1 \cdot T) - \bar{x}_g}{3 \cdot s_g}$$

$$C_{gk} = \frac{(x_m + 0.5 \cdot 0.15 \cdot 6 \cdot \hat{\sigma}_{process}) - \bar{x}_g}{3 \cdot g} = \frac{(x_m + 0.45 \cdot \hat{\sigma}_{process}) - \bar{x}_g}{3 \cdot s_g}$$

for $\bar{x}_g < x_m$:

$$C_{gk} = \frac{\bar{x}_g - (x_m - 0.1 \cdot T)}{3 \cdot s_g}$$

$$C_{gk} = \frac{\bar{x}_g - (x_m - 0.45 \cdot \sigma_{process})}{3 \cdot s_g}$$

Here x_m denotes the true value of the master or standard. $\bar{x}_g$ is the arithmetic mean, and s_g is the standard deviation of the measurement values:

$$\bar{x}_g = \frac{1}{n} \cdot \sum_{i=1}^{n} x_i \quad \text{and} \quad s_g = \sqrt{\frac{1}{n-1} \sum_{i=1}^{n} (x_i - \bar{x}_g)^2} \quad \text{where } i = 1, \ldots, n$$

The tolerance-based and the process-based evaluation produce the same result if $T = 6 \cdot \hat{\sigma}_{process}$. The difference between $\bar{x} < x_m$ and $\bar{x} > x_m$ (see also Figure 8.5–8) can be eliminated in the calculation of C_{gk} by using the following formula:

$$C_{gk} = \frac{0.1 \cdot T - |\bar{x}_g - x_m|}{3 \cdot s_g}$$

where $|\bar{x}_g - x_m|$ = Absolute value of difference between $\bar{x}_g$ and x_m (The result is always positive!)

x_m = true value of gauge master or standard

T = $USL - LSL$ (tolerance range)

With this method, the *accuracy* $|\bar{x}_g - x_m|$ is subtracted from the characteristic tolerance. Another method could be to increase the system variation by this amount (see [56]). The formula is then

$$C_{gk} = \frac{0.1 \cdot T}{3 \cdot s_g + |\bar{x}_g - x_m|}$$

Depending on the company standard (or current guidelines), the calculation is based on n percent of the specified tolerance range or six or eight $\sigma_{process}$. The mini-

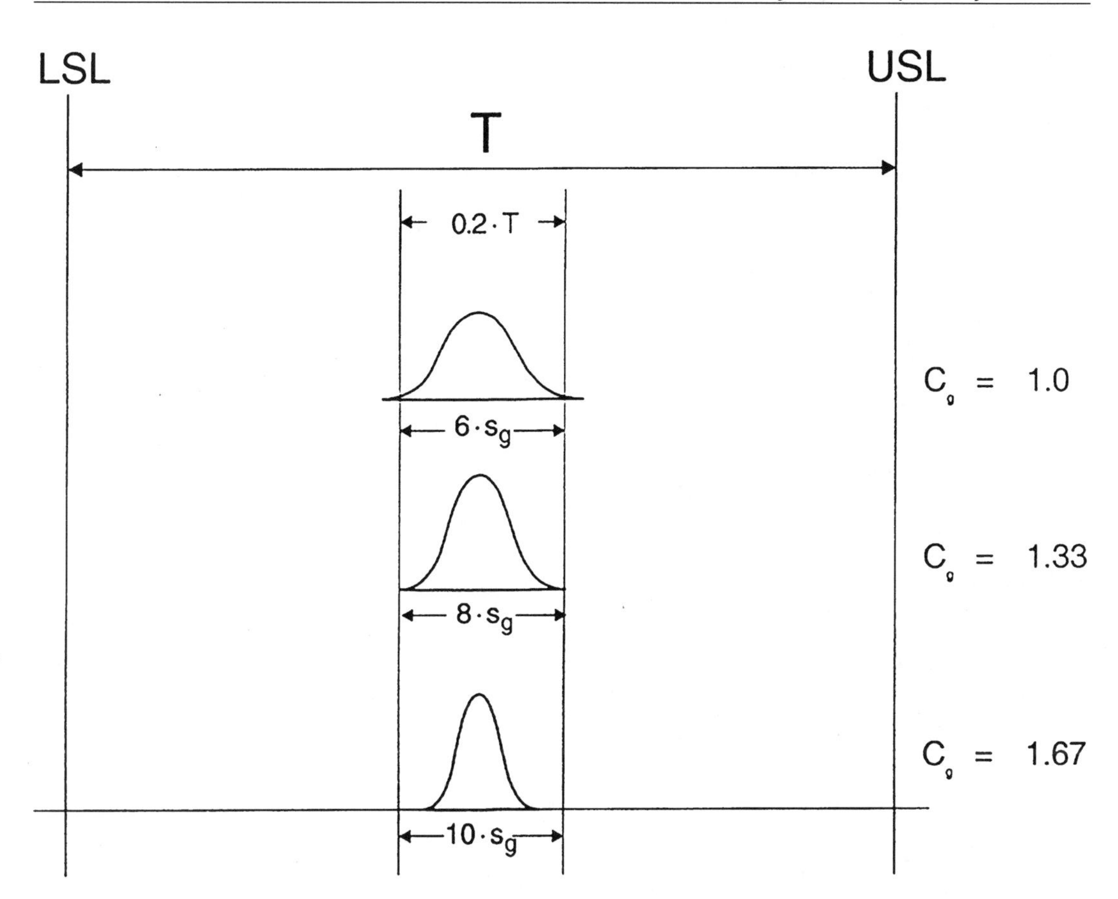

Figure 8.5–5. Relationship between s_g and C_g with tolerance-based calculation.

mum requirement for the capability indices C_g and C_{gk} can also vary (e.g., 1.0 in [36] and 1.33 in [62]).

Figure 8.5–5 shows the relation between the variation of the measuring equipment and the resulting capability index.

Notes

A calculation based on the range R would not be appropriate for the type 1 study due to the large sample size.

As an alternative to using the gauge capability indices C_g and C_{gk}, gauge performance can also be expressed in terms of a *percentage variability* statistic, which is given by:

$$\text{Percentage Variability} = \frac{15}{C_g} \text{ or } \frac{20}{C_g} \quad \text{for repeatability}$$

$$= \frac{15}{C_{gk}} \text{ or } \frac{20}{C_{gk}} \quad \text{for repeatability and accuracy}$$

Example of Type 1 Study: With Tolerance-Based Calculation

A standard with a true value of $x_m = 20.3$ mm was measured 50 times using a specified procedure. The following results were obtained:

20.303	20.311	20.311	20.313	20.306
20.301	20.297	20.309	20.303	20.296
20.304	20.295	20.308	20.308	20.306
20.303	20.302	20.304	20.298	20.299
20.306	20.304	20.298	20.306	20.300
20.296	20.298	20.308	20.303	20.302
20.301	20.295	20.302	20.310	20.303
20.300	20.301	20.294	20.304	20.307
20.307	20.307	20.302	20.309	20.303
20.305	20.312	20.304	20.305	20.305

Determine the capability indices, based on a tolerance range of USL = 20.45 and LSL = 20.15.

Number of values obtained	50
Smallest value	20.294
Largest value	20.313
Arithmetic mean	20.3035
Standard deviation s	0.00466

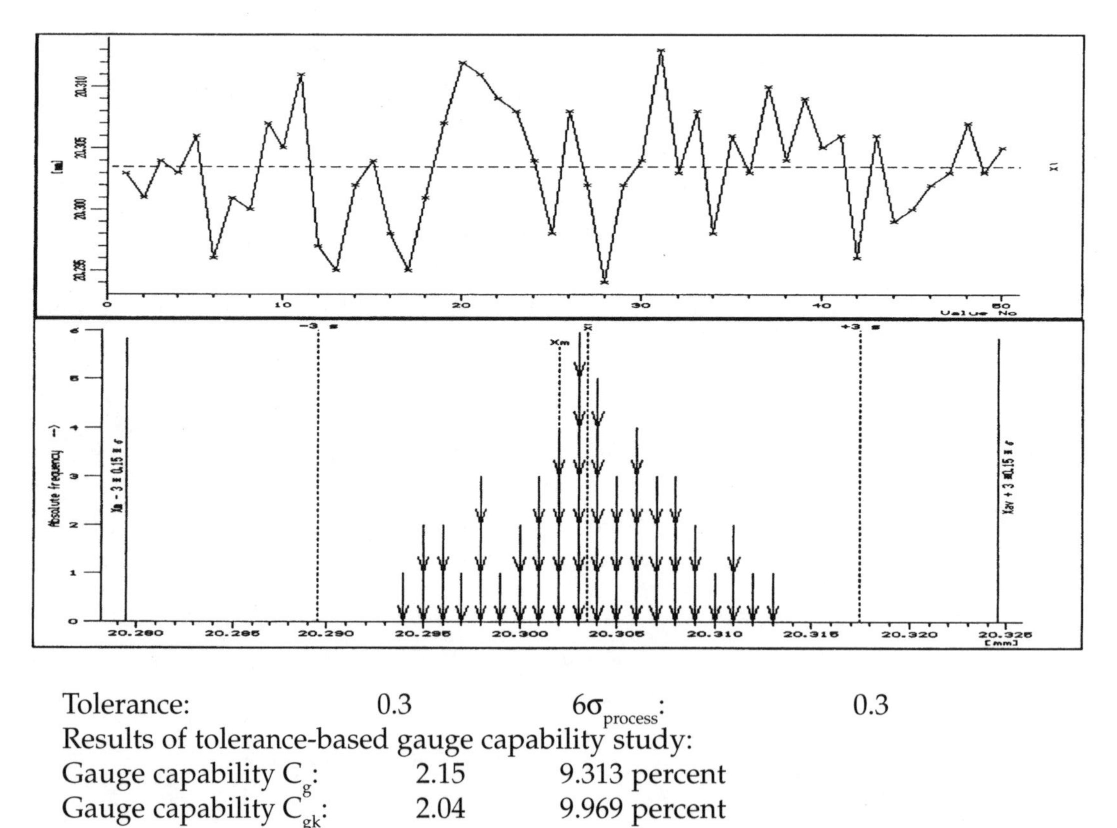

Tolerance: 0.3 $6\sigma_{process}$: 0.3

Results of tolerance-based gauge capability study:

Gauge capability C_g:	2.15	9.313 percent
Gauge capability C_{gk}:	2.04	9.969 percent

Note

A process-based evaluation produces the same results since, in this case, $T = 6 \cdot \hat{\sigma}_{process}$.

Example of Type 1 Study: With Process-Based Calculation

A standard with a true value of $x_m = 17.05$ mm was measured 50 times using a specified procedure. The following results were obtained:

17.050	17.050	17.050	17.030	17.040
17.060	17.050	17.040	17.050	17.030
17.040	17.070	17.050	17.050	17.050
17.060	17.050	17.030	17.060	17.060
17.040	17.050	17.050	17.050	17.050
17.050	17.040	17.040	17.040	17.050
17.040	17.040	17.040	17.060	17.070
17.060	17.060	17.060	17.050	17.060
17.050	17.050	17.040	17.070	17.060
17.040	17.060	17.040	17.040	17.050

Determine the capability indices, based on $\sigma_{process} = 0.07$.

Number of values obtained	50		
Smallest value	17.03	Largest value	17.07
Arithmetic mean	17.049	Standard deviation s	0.00998

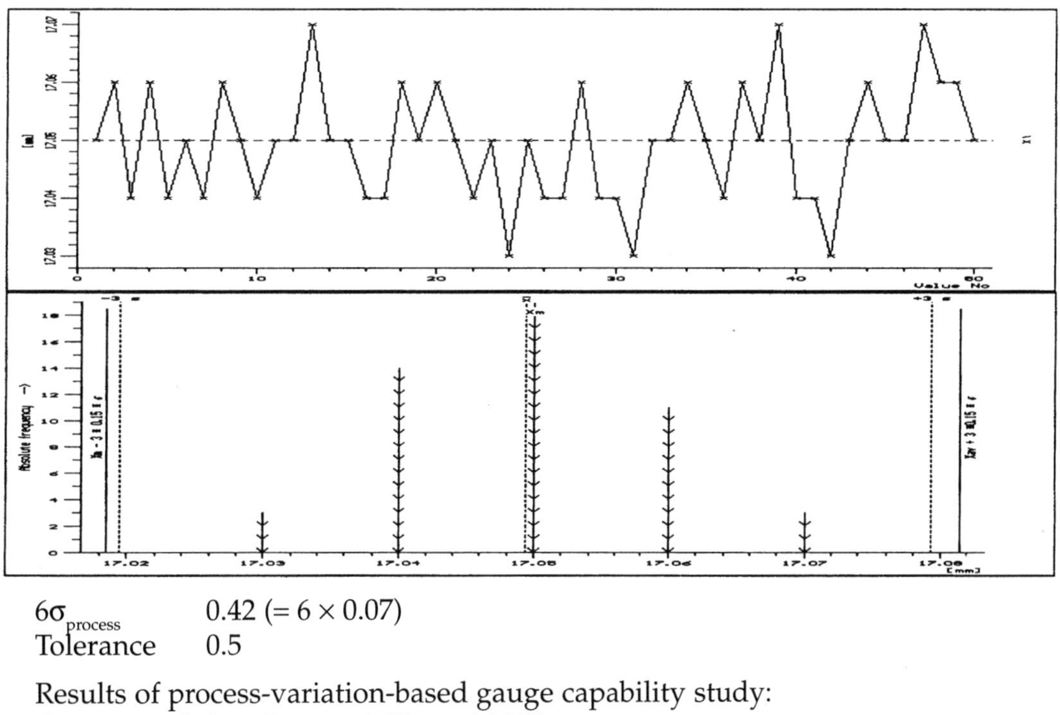

$6\sigma_{process}$	0.42 (= 6 × 0.07)
Tolerance	0.5

Results of process-variation-based gauge capability study:

Gauge capability C_g:	1.05	14.259 percent
Gauge capability C_{gk}:	1.03	14.536 percent

Compare these with the results of the tolerance-based evaluation:

Gauge capability C_g:	1.25	12.000 percent
Gauge capability C_{gk}:	1.23	12.195 percent

The issue of whether gauge assessment should be based on tolerance or process variation is illustrated in Figure 8.5–6.

The chart shown in Figure 8.5–6 shows that the process exhibits only very little variation within subgroups (see left-hand side of figure). A tolerance-based capability assessment would permit too great a gauge resolution (see right-hand side of figure). The measurements could no longer be determined with the required degree of accuracy. A statistical assessment of the process based on these data would be distorted. In such cases, the calculation method based on process variation is thus appropriate.

There are, however, two problems associated with this method.

1. The process variation is often not known (e.g., with new processes) and thus cannot be used as a reference figure unless special studies are carried out using extremely accurate measurement systems. These are, as a rule, either very time-consuming or they are not possible due to lack of suitable measurement systems.
2. If the process variation changes, this also has to be taken into account with the capability evaluation of the measurement system. Thus, comparisons of measurement system performance over time are not or only partially possible.

It is, therefore, recommended that the specified tolerance range be used as the basis of calculation. As a rule, this remains constant over longer periods of time, it is documented in the parts drawing, and it is an integral part of contractual agreements.

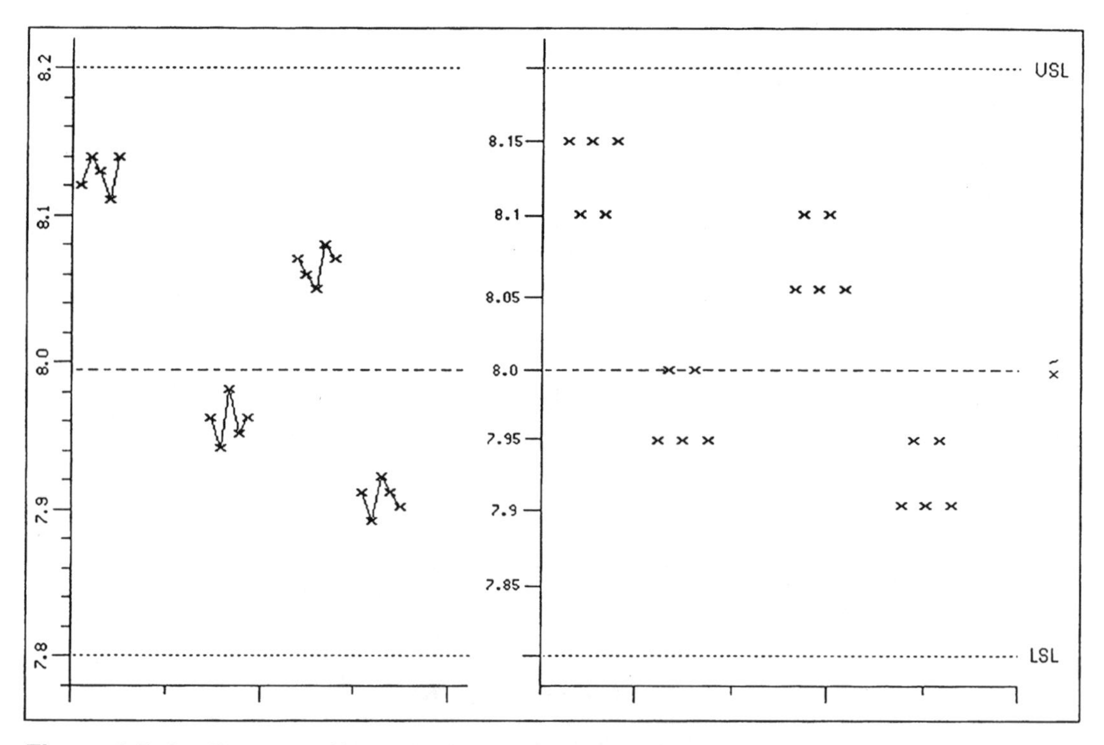

Figure 8.5–6. Process with small within-subgroup variation.

To prevent an inaccurate evaluation, the measurement system should have a two percent resolution (at least five percent) of the total specified tolerance range.

8.5.4 Unilateral Tolerances

The calculation of capability is based on the assumption that the measurement values are normally distributed. This also applies to characteristics that are not normally distributed (see Figure 8.5–7). To eliminate other distribution types as much as possible, for asymmetrical tolerances the gauge master or standard should again be placed as close as possible to the middle of the tolerance range.

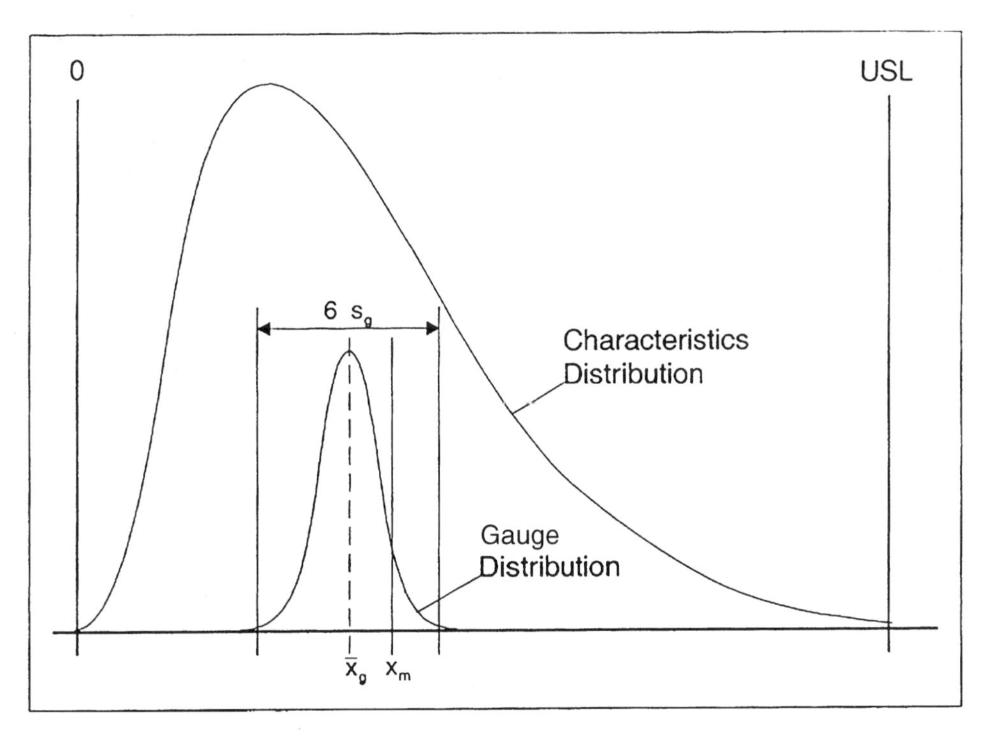

Figure 8.5–7. Measurement equipment variation for nonnormally distributed characteristics.

Here, we have to distinguish two situations. In one scenario, upper and lower limits are known (e.g., zero-limited characteristic); in the other, one limit is missing.

In the first case, it is only sensible to calculate C_{gk} by examining the tolerated limit.

$$C_{gk} = \frac{x_m + 0.1 \cdot T - \bar{x}_g}{3 \cdot s_g} \quad \text{or} \quad C_{gk} = \frac{\bar{x}_g - (x_m - 0.1 \cdot T)}{3 \cdot s_g}$$

Which one of these two formulas is used depends on the location of the average $\bar{x}_g$. This may be larger or smaller than x_m*. Contrary to the determination of process or machine capability indices, the capability index depends on this condition only and not on the distance of the average to the specification limits.

Figure 8.5–9 illustrates the situation for a zero-limited characteristic. The situation is similar for a characteristic that has a natural upper limit (see Figure 8.5–9).

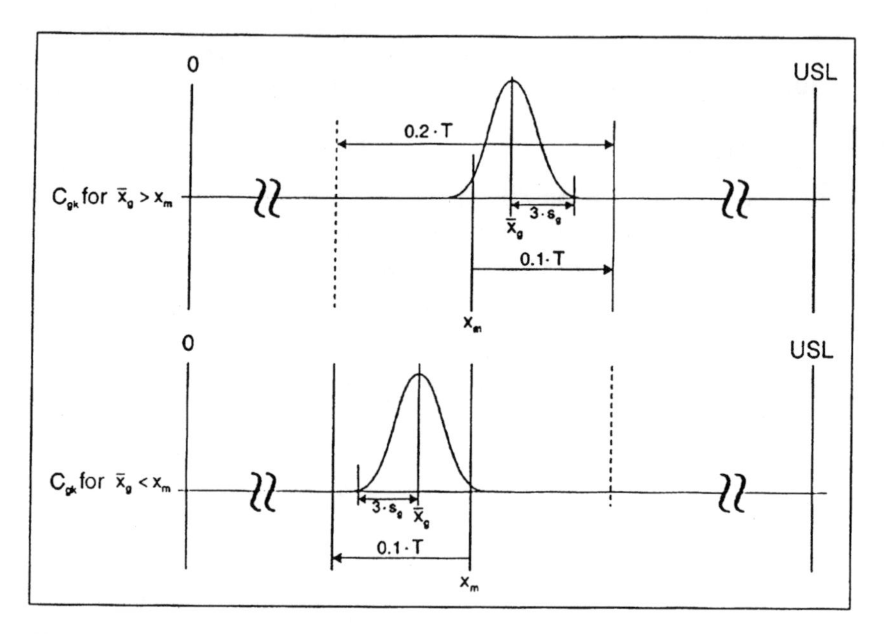

Figure 8.5–8. Determination of C_{gk} for unilateral characteristic.

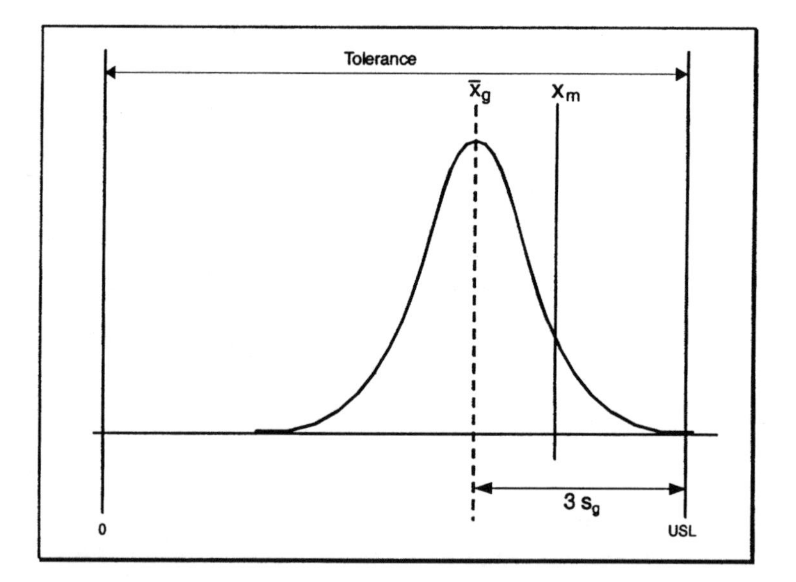

Figure 8.5–9. Capability study with zero-limited characteristic.

If there is only one limit (e.g., the lower specification limit, that is, the measured values have to be greater than a minimum value), it is not possible to determine C_g or C_{gk} based on the *tolerance interval.* To ensure that no values lying above USL or below LSL are accepted, the specification limit should be decreased or increased by $3s_g$ (corresponding to $C_g = 1.0$) or $4s_g$ (corresponding to $C_g = 1.33$), as appropriate.

Figure 8.5–10 illustrates the situation for $C_g = 1.0$ and an upper specification limit.

Note: If process spread is used as the basis of calculation, capability indices can be calculated in the normal way.

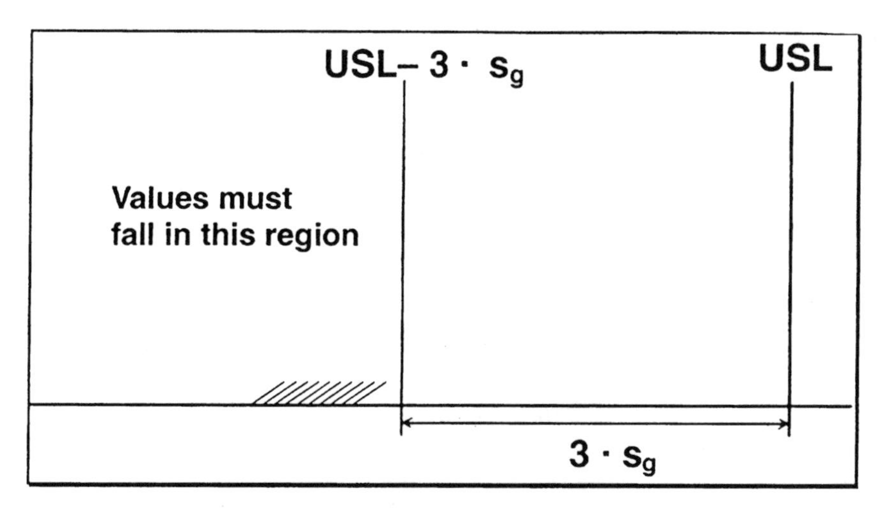

Figure 8.5–10. Adjusting the specification limit.

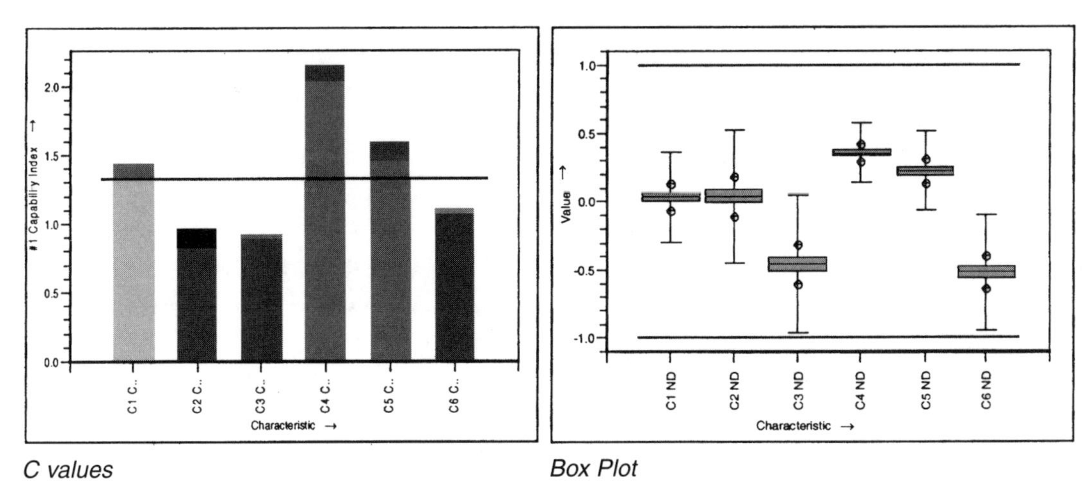

C values *Box Plot*

Figure 8.5–11. Overview of results for multiple characteristics.

8.5.5 Multiple Characteristics

If several characteristics of a part are studied at the same time, it is worthwhile summarizing all capability indices in a general overview (Figure 8.5–11). This is particularly useful for measurement systems comprising several components to ensure that details of those components that are functioning correctly are immediately accessible. The box plot provides similar information in a standardized format.

8.5.6 Type 2 Study (R&R Study)

It is only appropriate to perform a type 2 study if the capability of the measurement procedure has previously been confirmed in a type 1 study. As additional factors are now included in the study, it is likely that measurement variation will only increase as a result.

In a type 2 study, a number of normal production parts (usually ten) are measured on-site, using the same gauge, by three different appraisers (or, alternatively, one appraiser takes the measurement at different measuring locations or using different gauges). Each appraiser should measure each part two or three times. As far as possible, appraisers should not know the measurement values they or other appraisers previously obtained for this part. A statistic called *total gauge variation* is calculated and used to evaluate the performance of the measurement system. The following results are possible: The gauge

- Is suitable
- Must be improved (limited suitability)
- Is not acceptable (action required)

The type 2 study assesses repeatability and reproducibility (hence, *R&R study*); it is used to evaluate new or existing equipment prior to final acceptance for use in production. This study can also be used for routine audits and recertification purposes.

Preparing a Type 2 Study

Ten parts, ideally covering the full range of the process spread, are taken from the process and numbered one to ten. Three appraisers take two or three independent sets of measurements. All results are recorded on a worksheet (Table 8.5–2). Appraisers should not be able to see the results of the other appraisers or their own results obtained in the preceding trial.

Performing a Type 2 Study

- Identify the appraisers carrying out the study as A, B, and C.
- Select ten parts and identify them one to ten, or use some other identification system, ensuring that the identifications are not visible.
- If necessary, calibrate the gauge before taking the first measurement.
- Let appraiser A measure the ten parts and record the results in the corresponding column of the worksheet.
- Let appraiser B measure the ten parts in the same order—appraiser B should not be able to see the results obtained by appraiser A—and record the results in the corresponding column of the worksheet.
- Let appraiser C measure the ten parts in the same order—appraiser C should not be able to see the results obtained by appraisers A and B—and record the results in the corresponding column of the worksheet.
- Repeat the study one or two times until each appraiser has measured the ten parts two or three times.

Note

It may be worthwhile changing the order of the parts when repeating the measurements. The order must, however, be the same each time for all appraisers, and it must be guaranteed that measurement results are entered against the appropriate part number on the worksheet. Changing the order of the parts is a much more complicated procedure and is thus not often implemented in practice.

- Perform the calculations listed in the worksheet (see Table 8.5–2 and Section 8.5.7).

Table 8.5–2. Worksheet for R&R calculation.

Act.Date 24.09.98	Op.Name. NN	Dept./Cost./Prod.	Test Location

Gauge	Results Procedure 1	Work piece
Description Number Resolution 0 Tst Reas.	calculated C_g value --- calculated C_{gk} value ---	Wrk.pc.Descr. Wrk.pc.No. Char.No. 1 Nom.Val. 56.8300

Rem.

	Operator A				Operator B				Operator C			
Part	1	2	3	R	1	2	3	R	1	2	3	R
1												
2												
3												
4												
5												
6												
7												
8												
9												
10												

$\bar{R}_A$ = ______ $\bar{R}_B$ = ______ $\bar{R}_C$ = ______

$\bar{x}_A$ = $\bar{x}_B$ = $\bar{x}_C$ =

Repeatability (EV)	$K_1 \cdot \bar{\bar{R}}$	=	______
% Repeatability (%EV)	$\frac{100 \cdot EV}{RF}$ %	=	______%
Reproducibility (AV)	$K_2 \cdot \bar{x}_{Diff}$	=	______
% Reproducibility (%AV)	$\frac{100 \cdot AV}{RF}$ %	=	______%
Measurement System (R&R)	$\sqrt{EV^2+AV^2}$	=	______
%Measurement System (%R&R)	$\frac{100 \cdot R\&R}{RF}$ %	=	______%
Reference size (RF)	______	=	______
Upper control limit OEG $_R$	$D_4 \cdot \bar{R}$	=	______

10 30

D_4 = 2.57 K_1 = 3.04 K_2 = 2.70

Q-DAS ARM-Method

Date ______ Signature ______ Department ______

24 09 98 qs-STAT V3.09d IQPRO gc_2vl_n ZSPC project-version 2.beta C:\DATA\FORD_2.DFQ

Note

Ten parts are used to reduce the likelihood of the appraiser being able to remember previous measurement results for certain parts. The average range $\overline{R}$ is used to evaluate the repeatability, while the reproducibility is given by the difference between the greatest and the smallest arithmetic means ($\overline{x}_{Diff}$). The total variation of the gauge is used as acceptance criterion. As a general guide, the following table may be used:

	Total gauge variation	
Evaluation	**Existing measurement procedures**	**New measurement procedures**
Good	0% ... ≤ 20%	0% ... ≤ 10%
Marginal (limited suitability)	20% ≤ ... ≤ 30%	10% ≤ ... ≤ 30%
Not acceptable	30% ≤ ...	30% ≤ ...

Table 8.5–2 is an example of a worksheet used for recording and evaluating the sets of measurements and for reviewing the results.

Figure 8.5–12 shows the process flow for a type 2 study.

8.5.7 Calculating Measures of Variation in a Type 2 Study

The type 2 study is particularly suitable for assessing repeatability and reproducibility. The calculation results will indicate whether a gauge is suitable, of limited suitability, or unsuitable for measuring the variation of a measurement system. Four methods are described for evaluating a measurement system:

1. Range Method (RM)
2. Average and Range Method (ARM)
3. Average and standard deviation method
4. Analysis of Variance (ANOVA)

The procedure for the measurement process is very similar for all four methods. In each case, several appraisers (two or three) measure a number of parts (5–10). With the exception of the range method, the series of measurements should be repeated two or three times, depending on the specification.

Range Method

The range method is used for quickly determining the approximate variation of the measurement system. Two appraisers measure five parts. For each part, the range between the values measured by appraiser 1 and appraiser 2 is determined. The statistic in this method is the relation between the gauge variation and the process variation or tolerance.

$$\%R\&R = \frac{\overline{R} \cdot 4.33}{\text{Process variation}} \cdot 100\% \text{ or } \frac{\overline{R} \cdot 4.33}{\text{Tolerance}} \cdot 100\%$$

where $\overline{R}$ = average range taken from ranges of the two sets of measurements

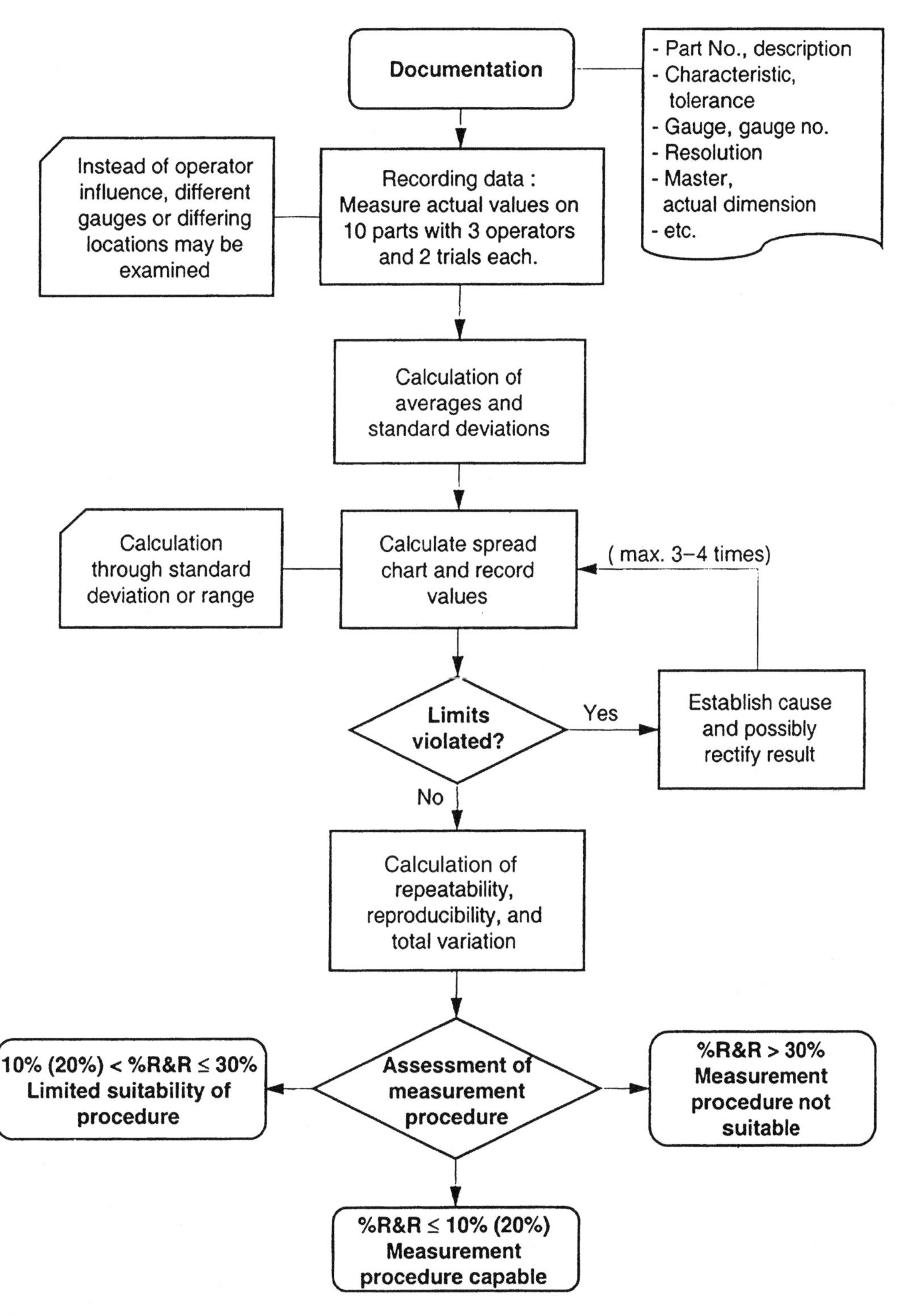

Figure 8.5–12. Type 2 study.

**This evaluation can also be done using an error chart.*

The factor percent R&R must be smaller than 20 percent. With this method, it is not possible to distinguish between repeatability and reproducibility.

Average and Range Method

With this method, two or three appraisers measure ten parts two or three times. The ten parts should cover the entire process spread. The measurement results of the individual appraisers can be displayed in the form of a run chart. The various results obtained by an appraiser for a part are examined together. In this way, it is possible to recognize significant variation between the series of measured values or the respective appraisers.

The statistics for assessing the measurement system are

EV = Equipment Variation/Repeatability

$$EV = \overline{R} \cdot K_1 \qquad K_1 \text{ (see Table 10.8)}$$

AV = Appraiser Variation/Reproducibility

$$AV = \sqrt{(\overline{x}_{Diff} \cdot K_2)^2 - \left[\frac{EV^2}{n \cdot r}\right]} \qquad K_2 \text{ (see Table 10.8)}$$

If

$$(\overline{x}_{Diff} \cdot K_2)^2 >> \left[\frac{EV^2}{n \cdot r}\right]$$

for reasons of simplicity, it is not necessary to carry out this correction. Then

$$AV = \overline{x}_{Diff} \cdot K_2$$

If

$$(K_2 \cdot \overline{x}_{Diff})^2 \leq \left[\frac{EV^2}{n \cdot r}\right]$$

then the result below the root is zero or negative, and reproducibility is small in comparison with repeatability and can be omitted.

R&R = Repeatability & Reproducibility

$$R\,\&\,R = \sqrt{EV^2 + AV^2}$$

PV = Part Variation

$$PV = R_p \cdot K_3 \qquad K_3 \text{ (see Table 10.8)}$$

where R_p = Range from $\overline{x}_{pi}$ where $i = 1 \ldots n$

$\overline{x}_{pi}$ = Mean of measurement values for part number i

The factors K_1, K_2, and K_3 are calculated from

$u_{1-\alpha/2} \cdot \frac{1}{d_2^*}$ with d_2^*

d_2^* = Values for the distribution of mean ranges (see Table 10.8)

$u_{1-\alpha/2} = 5.15$ for 99 percent

$u_{1-\alpha/2} = 6.0$ for 99.73 percent

K_1 depends on the number of trials, and the number of parts multiplied by the number of appraisers, K_2 depends on the number of appraisers, and K_3 depends on the number of parts. If the numbers of parts and appraisers are the same, then $K_2 = K_3$.

Average and Standard Deviation Method

The procedure for calculating the measurement data is the same as for the average and range method. However, only two measurements may be made of each part (see form in Table 8.5–2).

The parameters are calculated from

$\Delta = x_{1i} - x_{2i}$ where $i = 1, \ldots, 10$

Δ = Difference between sets of measurements 1 and 2 taken by one appraiser

For each appraiser (A, B, C), the standard deviation s can be determined from the differences.

$\bar{s} = \bar{s}_\Delta / \sqrt{2}$ where $\bar{s}_\Delta = \frac{1}{3}(s_{\Delta A} + s_{\Delta B} + s_{\Delta C})$

$$s_v = \sqrt{\frac{1}{2}\left[\left(\bar{x}_A - \bar{\bar{x}}\right) + \left(\bar{x}_B - \bar{\bar{x}}\right) + \left(\bar{x}_C - \bar{\bar{x}}\right)\right]}$$

where $\bar{x}_A, \bar{x}_B, \bar{x}_C$ = Average result of the respective appraiser

$R\&R = \sqrt{\bar{s}^2 + s_v^2}$ where $\bar{\bar{x}}$ = Average $\bar{x}_A, \bar{x}_B, \bar{x}_C$

Note: This method may only be used for two measurements per appraiser!

ANOVA Method

As far as the collection of measurement values is concerned, the ANOVA method (analysis of variance) is similar to the ARM. Evaluation, however, is considerably more complex and cannot be conducted manually. Where suitable programs (e.g., qs-STAT® by Q-DAS GmbH, Birkenau, Germany) are available, it is recommended that the ANOVA method be used. From a mathematical point of view, this is the more accurate method, and it can separate the variation components into the following groups:

- Measurement system repeatability
- Appraiser reproducibility
- Interaction
- Percent total variation (R&R)
- Part variation

The separation of the individual variation components means that it is easier to trace the causes of errors. It is thus easier to establish measures for improvement. The interpretation of the results is comparable with the ARM method. The *mathematical* aspects are described in the Appendix, Section 9.2.2.

Reviewing Statistics

The statistics just defined are the results of calculations obtained from the measurement values recorded. To assess suitability for a concrete application, the results can be reviewed in relation to a reference figure RF. The reference figures are as follows:

- The total variation (TV) with part variation
- The total variation (TV) from the process variation
- The tolerance T given in the part drawing
- The process variation

When the process variation is *unknown,* the total variation can be determined from

$$TV = \sqrt{R\&R^2 + PV^2}$$

If the process variation is *known,* the total variation is obtained from

$$TV = 5.15\ \frac{\text{process variation}}{6} \quad \text{for the 99 percent confidence level}$$

The statistics EV, AV, R&R, and PV are examined in relation to the reference figure and given as a percentage value. The *percentage components* are thus

Device variation: $\%EV = \frac{EV}{RF} \cdot 100\%$

Appraiser variation: $\%AV = \frac{AV}{RF} \cdot 100\%$

Part variation: $\%PV = \frac{PV}{RF} \cdot 100\%$

Measurement system variation: $\%R\&R = \frac{R\&R}{RF} \cdot 100\%$

The result is compared with predefined limits that are independent of the reference figure. The following limits are commonly used:

$0 \le \%R\&R \le 10\%\ (20\%)$	Gauge suitable
$10\ (20\%) < \%R\&R \le 30\%$	Gauge of limited suitability. Whether or not it is acceptable depends on the importance of the measurement task and the costs of the measurement system.
$30\% < \%R\&R$	Gauge not suitable. Errors should be analyzed.

Special features: Two factors, the appraiser and part irregularities, can considerably distort these measurements.

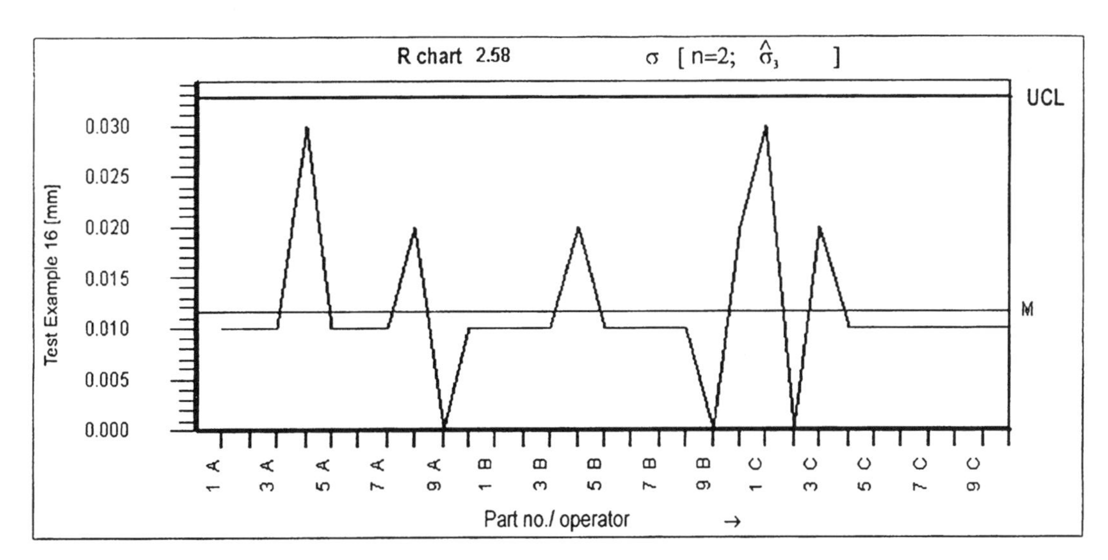

Figure 8.5–13. Quality control chart.

Appraiser

Different appraisers often handle the measurement system in different ways. To establish these differences, both the mean values and the ranges (difference between the sets of measurements) can be recorded on a control chart and checked to see whether they exceed the control limits. The control limits are calculated as on the $\overline{x}/R$ Shewhart chart (see [1]). When the system is not stable, the reasons for this have to be established and appropriate actions taken (operator training, change the fixture, etc.). As a rule, it is only necessary to complete a variation chart (R or s chart, see Figure 8.5–13). The control chart sometimes has the disadvantage of being too sensitive so that it never shows stability for some measuring procedures.

An error chart can be used as an alternative to or in addition to the control chart to assess this aspect. For each part, the deviation of each appraiser from the average for this part is recorded in the error chart (Figure 8.5–14a). The average measurement thus always corresponds to zero. The disadvantage of this graph is that the individual appraisers' results for all the various parts are difficult to compare. Therefore, the authors recommend that the results be presented separately for each appraiser (Figure 8.5–14b). At the same time, for comparative purposes, the diagram should contain the two sets of values taken by all the appraisers. The mean of the two sets of values is indicated by the bold line.

If significant differences are observed, corrective action should be taken. This could include instructions on how to handle the equipment or an improvement of the measuring device. Alternatively, the results could be recorded on a control chart.

Part Irregularities (Within-Part Variation)

Normal production parts are used for determining system variation, as already described. These may become deformed during the manufacturing process. When a

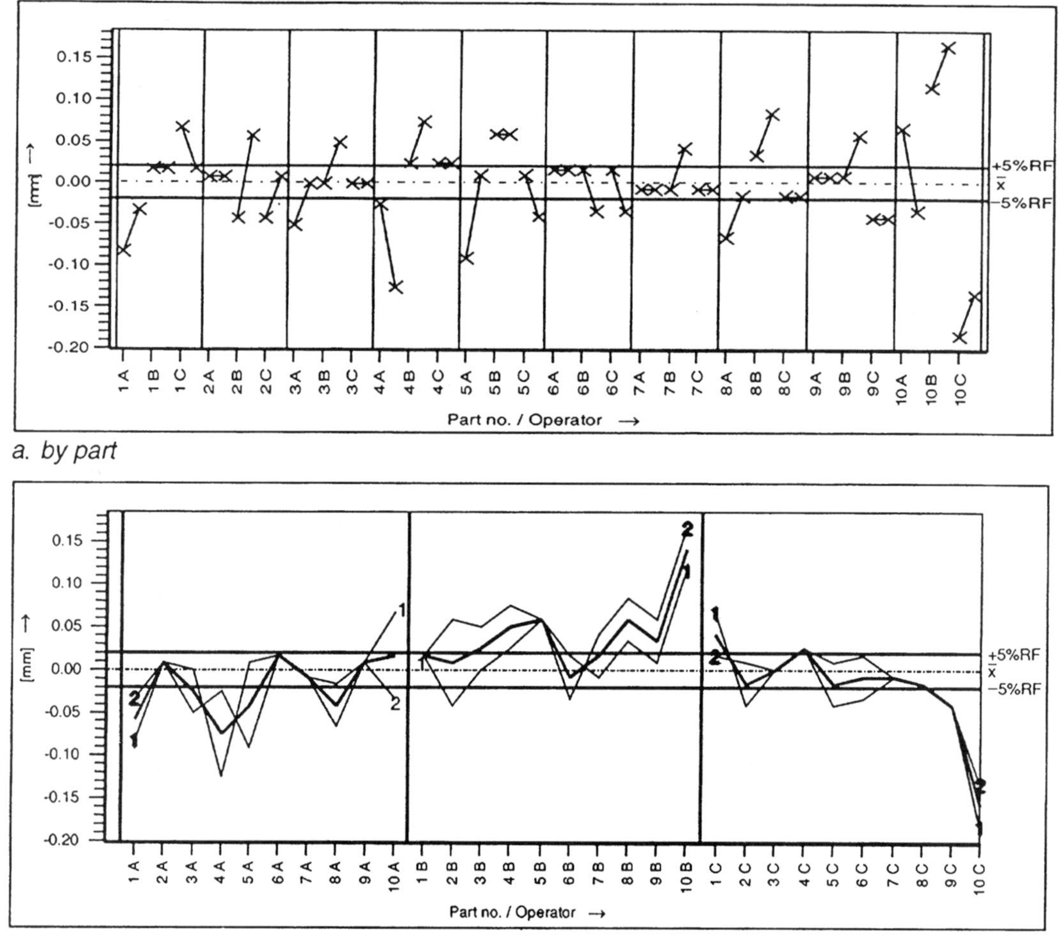

b. by appraiser

Figure 8.5–14. Error chart.

characteristic of a given part is measured several times, differences in the measurement position could then result in a further component of variation that is not caused by the gauge itself and for which adequate compensation must be made. There is a recommended procedure and calculation method for this situation that gives more accurate results but is much more complicated in terms of the measurement and analysis effort required. Thus, the need to apply this method should be assessed on a case-by-case basis. To eliminate this factor, the measuring position can be marked on the part. Repeat measurements can then be taken from exactly the same position. There is a very sound theoretical argument in favor of this procedure. It is, however, complex and thus often deemed impractical.

Example of a Type 2 Study: Calculating Percent R&R Based on Process Variation

The 60 measurement values given in the following table represent ten parts measured twice by three different appraisers. The process spread is given by $\hat{\sigma}_{process} = 0.06$ mm. The example was taken from the Ford test examples [40].

Appraiser 1		Appraiser 2		Appraiser 3	
Trial 1	Trial 2	Trial 1	Trial 2	Trial 1	Trial 2
100.01	100.00	100.00	99.99	100.01	99.98
100.01	100.02	99.99	100.00	100.00	100.00
100.01	100.00	100.01	100.00	100.01	99.99
100.02	99.99	100.01	99.99	100.01	100.00
100.00	100.01	100.00	99.99	100.00	100.01
100.01	100.00	100.00	99.99	100.00	99.99
99.99	100.00	100.00	99.99	99.99	100.00
100.01	99.99	100.01	100.00	100.00	99.99
100.01	100.01	100.00	100.00	100.00	99.99
100.01	100.00	99.99	100.01	99.99	100.00

$\overline{R}$ as mean value of $\overline{R}_1$, $\overline{R}_2$, $\overline{R}_3$	=	0.011667	mm
Upper control limit for R	=	0.038151	mm
$\overline{x}_{diff}$ as range for $\overline{x}_1$, $\overline{x}_2$, $\overline{x}_3$	=	0.007	mm

EV	= 0.06206	%EV	= 17.24	%
AV	= 0.02198	%AV	= 6.11	%
R&R	**= 0.06583**	**%R&R**	**= 18.29**	**%**

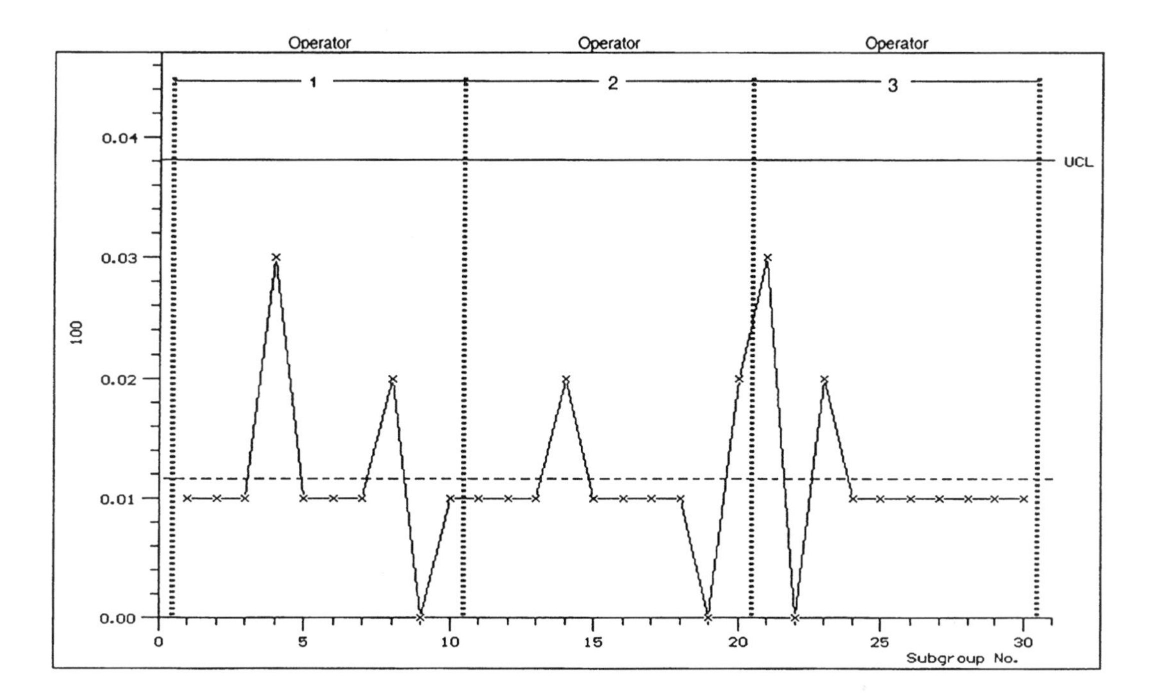

Example of a Type 2 Study: Calculating Percent R&R Based on Tolerance Spread

The 60 measurement values given in the following table represent ten parts measured twice by three different appraisers. The reference figure is the tolerance. The example was taken from the MSA Reference Manual [2].

Appraiser 1		Appraiser 2		Appraiser 3	
Trial 1	**Trial 2**	**Trial 1**	**Trial 2**	**Trial 1**	**Trial 2**
0.65	0.60	0.55	0.55	0.50	0.55
1.00	1.00	1.05	0.95	1.05	1.00
0.85	0.80	0.80	0.75	0.80	0.80
0.85	0.95	0.80	0.75	0.55	0.80
0.55	0.45	0.40	0.40	0.45	0.50
1.00	1.00	1.00	1.05	1.00	1.05
0.95	0.95	0.95	0.90	0.95	0.95
0.85	0.80	0.75	0.70	0.80	0.80
1.00	1.00	1.00	0.95	1.05	1.05
0.60	0.70	0.55	0.50	0.85	0.80

$\bar{\bar{R}}$ as mean value of $\bar{R}_1$, $\bar{R}_2$, $\bar{R}_3$		=	0.04	mm
Upper control limit for R		=	0.10753	mm
$\bar{x}_{diff}$ as range for $\bar{x}_1$, $\bar{x}_2$, $\bar{x}_3$		=	0.06	mm
EV	= 0.175	%EV	= 43.77	%
AV	= 0.162	%AV	= 40.46	%
R&R	**= 0.238**	**%R&R**	**= 59.60**	**%**

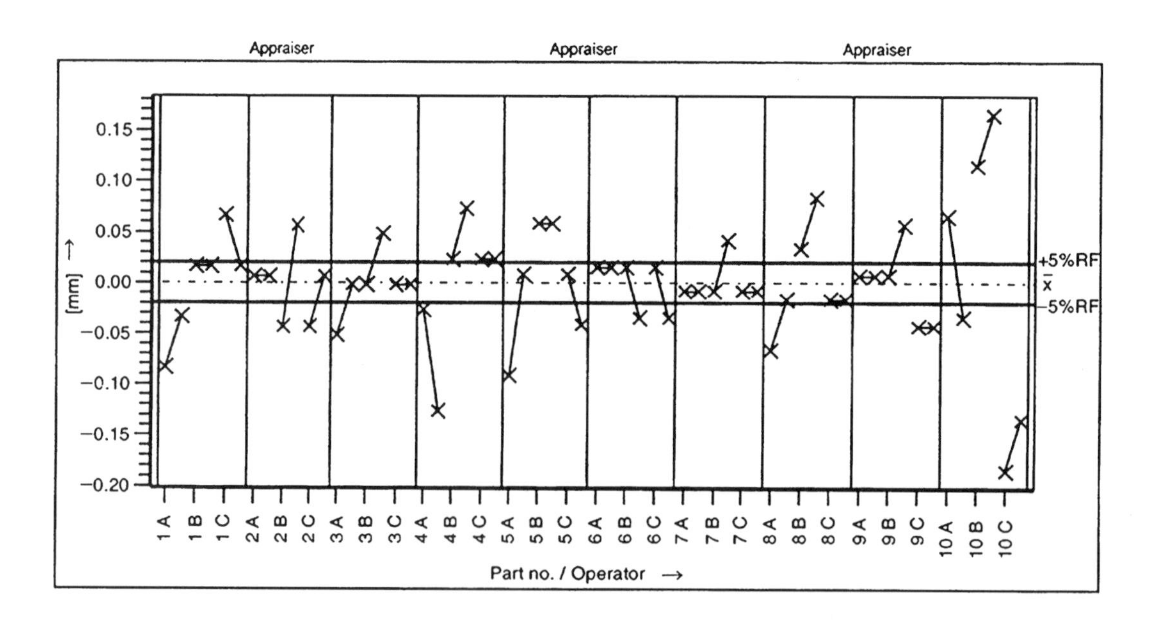

Comparison with the Reference Figure

In the ARM method described earlier, the so-called total variation TV is primarily taken as reference figure when evaluating suitability. This is made up of repeatability and reproducibility (R&R) and part variation PV. The part variation itself is thus determined by part selection. The parts have to cover the whole range of possible product variation as it occurs in production. If this is not possible, or if the parts are not selected correctly, the parts do not represent that actual process spread. This distorts the result. The result will not represent the quality of the measurement system since the selection of the parts has a strong influence on the result.

To avoid this issue, it is more appropriate to use the characteristic tolerance taken from the part drawing or, if known, the process spread as the reference figure.

The crucial issue is thus which reference figure to select. For clarification, Table 8.5–3 shows the results for two different sets of data, based on the three possible reference figures described.

The result clearly shows the effects of using different reference figures. Depending on the choice of reference figure, a measurement system is either suitable or not suitable. What causes the discrepancy in this example? The variation of the data in Test_16 is small compared to the tolerance ($PV/T = 0.023/0.4 = 0.0575$), but large for the data from the MSA reference manual ($PV/T = 0.9/0.4 = 2.25$).

The authors thus recommend that the characteristic tolerance be used as reference figure. One advantage of this is that it ensures the comparability of capability results determined at different times (for the same parts).

From all the different types of methods, a company must select the one that is most suitable for its own requirements and ensure that it is implemented throughout the company. This is the only way of ensuring that results can be meaningfully compared. The choice of method should also take practical and economic considerations into account. As there are so many measurement systems in use in any given company, implementing these studies is not a simple matter.

Table 8.5–3. Statistics depending on reference figure.

Statistics	% R&R / total variation	% R&R / process spread	% R&R / tolerance	Example
Repeatability Reproducibility Measurement system R&R	93.67 25.75 97.14	17.24 4.74 17.88	13.32 3.66 13.81	Test_16 from [40] Tolerance = 0.4 Process variation = 0.06 Part variation = 0.023
Result	not suitable	suitable	suitable	
Repeatability Reproducibility Measurement system R&R	18.7 16.8 25.2	35.0 31.4 47.0	43.8 39.2 58.8	MSA example from [2] Tolerance = 0.4 Process variation = 0.5 Part variation = 0.9
Result	limited suitability	not suitable	not suitable	

8.5.8 Type 3 Study

The type 3 study is a special case of the type 2 study, where the appraiser cannot exercise any influence on the measuring device (automatic handling). In this case, the study uses a larger number of parts (usually 25), each of which is measured twice by the automatic measuring equipment. Repeat runs with different appraisers are not necessary in this case. Apart from this, the procedure is the same as for the type 2 study. The worksheet shown in Table 8.5–4 is used for recording the values.

Table 8.5–4. Worksheet for type 3 study.

Act.Date 24.09.98	Op.Name. NN	Dept./Cost./Prod.	Test Location

Gauge		Results Procedure 1		Work piece	
Description		calculated C_g value	---	Wrk.pc.Descr.	
Number		calculated C_{gk} value	---	Wrk.pc.No.	
Resolution	0			Char.No.	
Tst.Reas.				Nom.Val.	56.8300
Rem.					

Part	Trial 1	Trial 2	Difference Δ	Evaluation
				Standard dev. the Variations s_Δ:
				s_Δ = ______
				Measurement System $\hat{\sigma}$:
				$\hat{\sigma} = s_\Delta / \sqrt{2}$ = ______
				Repeatability
				EV= ______ $\hat{\sigma}$ = ______
				% Repeatability
				%EV = $\frac{100 \cdot EV}{\_\_\_\_}$ = ______
				10 30

8.5.9 Calculating Variation in a Type 3 Study

With this method, appraiser influence is regarded as negligible. Thus, the total variation (R&R) is calculated from repeatability EV alone. This can be calculated from the standard deviation or the average range obtained from the sets of measurements.

*Standard deviation:**

$$EV = 5{,}15 \cdot \hat{\sigma} \quad \text{for 99 percent confidence level and}$$

$$EV = 6 \cdot \hat{\sigma} \quad \text{for 99.73 percent confidence level}$$

$$\hat{\sigma} = s_{\Delta}/\sqrt{2} \quad \text{and} \quad s_{\Delta} = \sqrt{\frac{1}{n-1}\sum_{i=1}^{k}(\overline{\Delta} - \Delta_i)}$$

$$\Delta_i = x_{1i} - x_{2i} \quad \text{difference from data sets 1 and 2}$$

$$\overline{\Delta} = \frac{1}{n}\sum_{i=1}^{n}\Delta_i \quad i = 1, 2, \ldots, n \quad n = \text{no. of parts}$$

Range:

$$EV = K_1 \cdot \overline{R} \quad \text{with } K_1 \text{ from Table 10.8}$$

$$\overline{R} = \frac{1}{n}\sum_{i=1}^{n}R_i \quad i = 1, 2, \ldots, n \quad n = \text{no. of parts}$$

$$R_i = x_{imax} - x_{imin} \quad \text{range from measurements of a part}$$

This results in *Measurement system variation:*

$$\%R\&R = \%EV = \frac{EV}{RF} \cdot 100\% \text{ where RF = reference figure}$$

The reference figure is

RF = T (characteristic tolerance) or
RF = $6 \cdot \sigma_{process}$ (process spread)

The result should be compared with the same limits that were used in the type 2 study. These are

$0 \leq$ %R&R $\leq$ 10% (20%)	Gauge suitable
10 (20%) $\leq$ %R&R $\leq$ 30%	Gauge of limited suitability
30 < %R&R	Gauge not suitable

As with the type 2 study, it is possible to calculate the measurement system variation using the balanced and simple ANOVA method (see Appendix, Section 9.2.3). With this method, it is possible to take more than two measurements of each part. Furthermore, unlike the range calculation method, the variation between the parts is indicated separately. As described for the ARM method, this variation can then be used to calculate the reference figure.

**This method can only be used for two measurements per part.*

Example of a Type 3 Study: Calculating Total Variation (Percent R&R) Based on Standard Deviation

Twenty-five parts are each measured twice. The tolerance (6.00 ± 0.03, T = 0.06 mm) is the reference figure. This example was taken from the Bosch guideline [62].

Part	Measurement set 1	Measurement set 2	Part	Measurement set 1	Measurement set 2
1	6.029	6.030	14	5.985	5.986
2	6.019	6.020	15	6.014	6.014
3	6.004	6.003	16	5.973	5.972
4	5.982	5.982	17	5.997	5.996
5	6.009	6.009	18	6.019	6.015
6	5.971	5.972	19	5.987	5.986
7	5.995	5.997	20	6.029	6.025
8	6.014	6.018	21	6.017	6.019
9	5.985	5.987	22	6.003	6.001
10	6.024	6.028	23	6.009	6.012
11	6.033	6.032	24	5.987	5.987
12	6.020	6.019	25	6.006	6.003
13	6.007	6.007			

The measurements produce the following results:

Standard deviation s_Δ	=	0.00212	mm
Measurement system variation $\hat{\sigma}$	=	0.0015	mm
Upper control limit for R	=	0.004488	mm
EV = 0.00772	%EV = %R&R = 12.87%	at 99% confidence level	
EV = 0.00899	%EV = %R&R = 14.99%	at 99.73% confidence level	

Figure 8.5–15 shows Δ_i (difference between sets 1 and 2) and a range that represents ± 5 percent of the reference figure (in this case, tolerance).

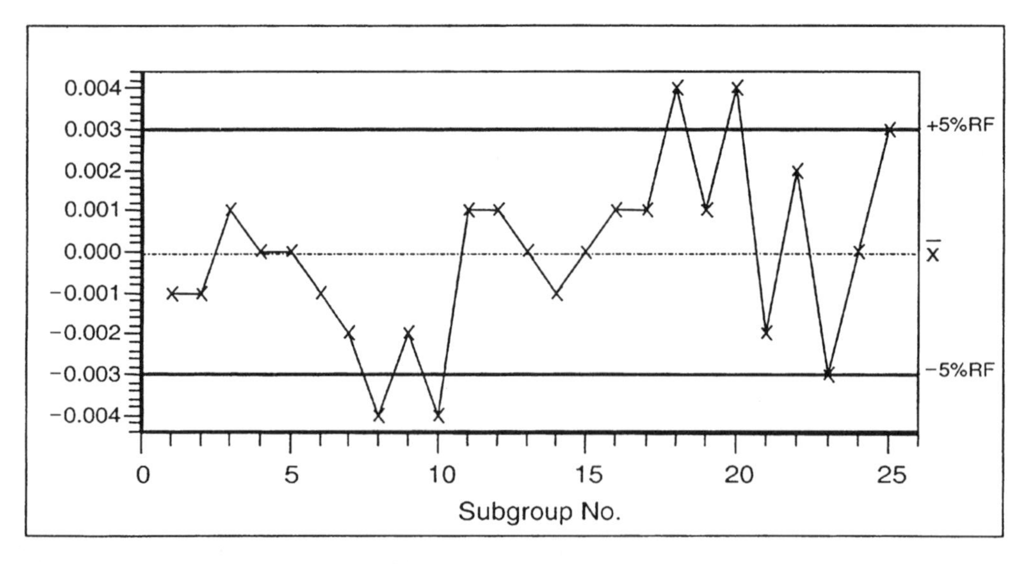

Figure 8.5–15. Run chart of differences between sets of measurements.

Example of a Type 3 Study: Calculating Percent Total Variation (R&R) Based on Range

Twenty-five parts were each measured twice. The tolerance is the reference value (56.830 ± 0.01 mm). The example was taken from the Ford guideline [37].

Part	Measurement set 1	Measurement set 2	Part	Measurement set 1	Measurement set 2
1	56.8338	56.8321	14	56.8273	56.8282
2	56.8310	56.8320	15	56.8235	56.8247
3	56.8359	56.8368	16	56.8177	56.8186
4	56.8291	56.8257	17	56.8332	56.8339
5	56.8294	56.8290	18	56.8305	56.8321
6	56.8206	56.8202	19	56.8228	56.8233
7	56.8410	56.8410	20	56.8310	56.8300
8	56.8305	56.8295	21	56.8318	56.8312
9	56.8269	56.8274	22	56.8344	56.8358
10	56.8386	56.8386	23	56.8327	56.8329
11	56.8321	56.8323	24	56.8213	56.8221
12	56.8331	56.8334	25	56.8205	56.8215
13	56.8382	56.8400			

The measured values produce the following results:

Average range $\overline{R}$	=	0.00932	mm
Measurement system variation $\hat{\sigma}$	=	0.000826	mm
Upper control limit for R	=	0.00305	mm
EV = 0.004254	%EV = %R&R = 21.27%	at 99% confidence level	
EV = 0.004956	%EV = %R&R = 24.78%	at 99.73% confidence level	

Using a control chart (see Figure 8.5–16 on page 314, in this case, an R chart with 99 percent control limits), it is possible to examine whether the measurement system is subject to excessive fluctuations in spread. The upper control limit is determined using the usual Shewhart formula for R charts.

8.6 Test for Stability to Determine an Appropriate Calibration and Adjustment Interval

General Points

The following procedure can be used for repeated assessment of measurement results taken under normal operating conditions. By analyzing the variation pattern of the measurement values over time, it is possible to determine an appropriate calibration and adjustment interval for the measurement system.

The test for stability should, where possible, be performed over a period of at least one day. During this time, 25 sample readings should be taken at regular intervals. The gauge master or standard that was used in the type 1 study (C_{gk} study) may also be used for the stability test.

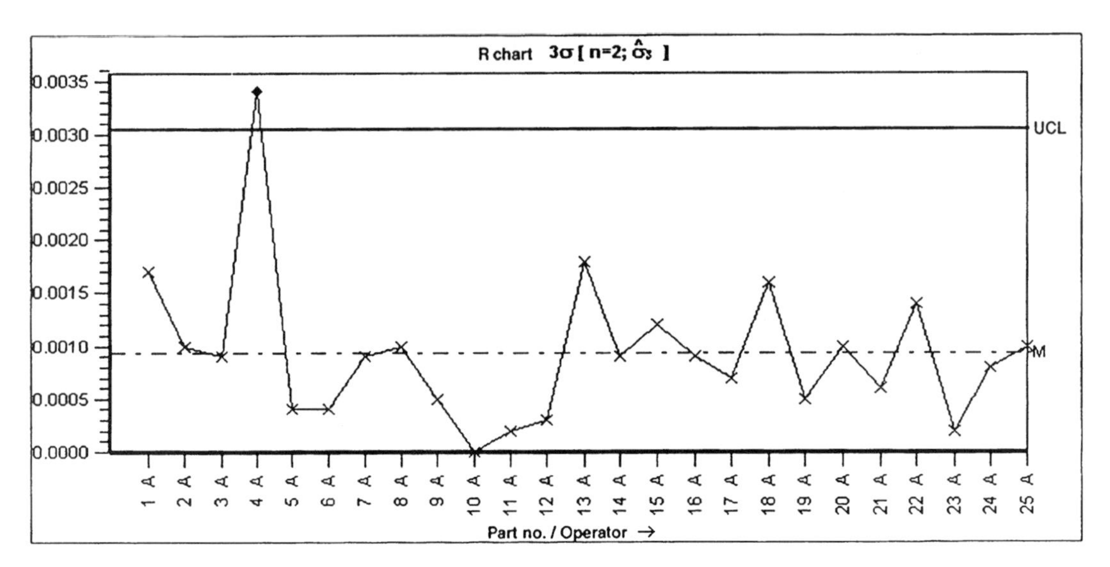

Figure 8.5–16. Range chart.

Note: If the first few sample measurements show chaotic fluctuations or a definite trend, the study should be stopped immediately and the causes corrected. The study should then be restarted, possibly with a shorter overall duration and shorter measurement intervals.

Procedure

- The true value of the master or standard should correspond to the center of the tolerance interval.
- Immediately prior to the study, calibrate the measurement system using the master or standard that is going to be measured.
- Do not make any adjustments during the study.
- At regular intervals, enter the measured values (usually individual values) in the worksheet.
- An evaluation based on moving averages and standard deviations with n = 3 is recommended.
- Plot the $\bar{x}$ values and connect them with lines; this results in a smoothed chart trace.

Note: In many cases, the smoothed chart trace approximates a straight line. In nonlinear cases, sections of the trace may be approximated by straight lines.

Analysis

Plot a tolerance band corresponding to ±5 percent of the specified tolerance on the average chart. Based on this, the following situations may be distinguished:

Case 1: If all the measurement values are within the tolerance band (e.g., ±5 percent), then one calibration per shift, performed at the beginning of the shift, is sufficient.

Case 2: If a trend leads to values outside the tolerance band, the interval between calibrations should be shortened to ensure that the ±5 percent limits are not exceeded.

Case 3: If, despite optimization of the calibration interval, there are still frequent violations of the upper or lower limit, then the measurement system must be improved. *Note:* Any instability of the measuring device should also become apparent during the type 1 study (C_{gk} study).

Case 4: Very small tolerances may require a calibration before each measurement. In this case, the test for stability does not apply.

8.7 Evaluation Based on the CNOMO Guideline

In France, evaluation based on the CNOMO E41.36.110.N measurement system capability standard [59] is widespread. This method combines the type 1 and 2 studies that have been described. With this method, the measurement system is analyzed in three stages.

Preparatory Stage

Stage 1: In measurement laboratory with gauge master or standard (as type 1 study)

Stage 2: In measurement laboratory with production part (as type 2 study)

Acceptance by the Manufacturer/in the Plant

Stage 3: In measurement laboratory/production area with several production parts and five repeat measurements per part (as type 2)

The results obtained are analyzed and compared with the performance requirements to evaluate suitability. Unlike the type 2 study, only one appraiser measures the parts.

Determination of Statistics

The following statistics are calculated at each stage:

Stage 1: Repeatability of measurement equipment with gauge master or standard

The gauge master or standard is measured five times, and the following statistics are determined from the data:

Average $\bar{y}_e$

Variance V_e

Uncertainty $I_e = 2\sqrt{V_e}$

I_e must meet the requirements set down in Table 8.7–1 on the next page.

Stage 2: Repeatability of measurement equipment with production part

The part is measured five times, and the following statistics are determined from the data:

Average $\bar{y}_r$

Variance V_r

Uncertainty $I_r = 2\sqrt{V_r}$

I_r must meet the requirements set down in Table 8.7–1.

Table 8.7–1. Performance requirements for measurement equipment.

Statistic	Requirement for T > 16µm and Q > 5	Requirement for T = 16µm or Q = 5
Resolution	$= \leq T/20$	$= \leq T/10$
$\pm I_e$	$= \leq \pm T/20$	$= \leq \pm T/10$
$\pm I_r$	$= \leq \pm T/8$	$= \leq \pm T/4$
$\pm I_{precision\ lab}$	$= \leq \pm T/16$	$= \leq \pm T/8$
$\pm I_g$	$= \leq \pm T/8$	$= \leq \pm T/4$
CMC	$= \geq 4$	$= \geq 2$

Stage 3: Five or more parts are measured at least five times (y_{ij}).

Precision measurements (x_i) of the same parts are then taken in a suitable laboratory. After the nth measurement in the laboratory, the mean is calculated for each part and, at the same time, the uncertainty of measurement in the laboratory is divided by $\sqrt{n}$.

The mean value x and the variance V_g are then calculated for the differences ($y_{ij} - x_i$):

$$I_g = \text{Uncertainty of gauge}$$

$$= |x| + 2\sqrt{(V_g + V_e)}$$

The uncertainty of measurement in the laboratory where the precision measurements are taken, the resolution of the measuring equipment I_g, and the calculated capability index CMC $= \frac{T}{(2 I_g)}$ have to meet the requirements set down in Table 8.7–1.

The reference figure is, in all cases, the tolerance T. The requirements (apart from special gauge specifications) can be taken from Table 8.7–1. Q stands for quality index.

Example of a CNOMO Study

The following example is taken from the aforementioned CNOMO standard E41.36.110.N (enclosure 2). The specification limits are USL = 10.025 mm and LSL = 9.975 mm. For the gauge master (stage 1), the following values are obtained: 2, 1, 2, 1, 1. This gives $I_e = 1.0954$.

The measurement equipment produced the following values for the production part (stage 2): 50, 49, 48, 50, 49, 48, 50, 47, 49, 50. This gives $I_r = 2.1082$.

The form in Figure 8.7–1 contains the measurement values and results of stage 1 and stage 2.

Comparison of the results with the performance limits given in Table 8.7–1 shows that the measurement system can be regarded as suitable.

8.8 Measurement System Capability: Special Cases

Not all gauges or inspection devices can be dealt with according to the methods described. There are no appropriate methods for attribute gauges such as plug gauges, templates, snap, or functional gauges. It is also not possible to evaluate mea-

	Efficiency tolerance of gauges Specific gauges			
Acceptance		Operator		Date
Plant	Producer		Equipment	Order no.
Work part	Characteristic			
Nominal Dimension 10.0000	max 10.0250		min 9.9750	
IT:		Unit used		

Preparatory phases:

Repeatability of measurement on the center of the setting standard

Measurement	1	2	3	4	5
Value $y_{\bullet j}$	10.0020	10.0010	10.0020	10.0010	10.0010

Average: $\bar{y}_\bullet = \frac{1}{5}\Sigma y_{\bullet j} = 10.0014$

Repeatability Variation: $V_\bullet = \frac{1}{4}\Sigma (y_{\bullet j}-\bar{y}_\bullet)^2 = 0.0000003$

Insecurity: $I_\bullet = 2\sqrt{V_\bullet} = 0.0010954$

Decision:

IT/20 = 0.0025

Measurement system capable

Repeatability of measurement on the center of the work part

Measurement	1	2	3	4	5
Value y_j	10.0500	10.0490	10.0480	10.0500	10.0490
Measurement	6	7	8	9	10
Value y_j	10.0480	10.0500	10.0470	10 0490	10.0500

Average: $\bar{y} = \frac{1}{10}\Sigma y_j = 10.049$

Repeatability Variation: $V_r = \frac{1}{9}\Sigma (y_j-\bar{y})^2 = 0.0000011111$

Insecurity: $I_r = 2\sqrt{V_r} = 0.0021082$

Decision:

IT/8 = 0.00625

Measurement system capable

E41.36.110.N

Figure 8.7–1. CNOMO form.

surement systems where measurement values of a part characteristic are taken, but the measurements cannot really be repeated, for example, hardness, surface, and profile testers.

In the case of hardness testers, the method can be used for comparative studies as a basis for acceptance of new hardness testers. It can also be used to illustrate and compare existing equipment to evaluate ongoing stability and capability.

The measurement systems and equipment mentioned have to be checked in accordance with current standards and guidelines such as ISO, British Standard, DIN, ASME, VDI, manufacturer's guidelines, etc. Each case should be examined on an individual basis to ascertain whether it is still necessary to carry out a measurement system capability study.

Experience tells us that at low tolerances (e.g., smaller than 15 μm, or at equivalent orders of magnitude for other types of characteristics), the technical feasibility of a measurement system capability study is pushed to its limits. Consequently, the results of the study may not be acceptable. In this case, a type 1 study can be performed as follows, without statistical analysis:

- Take 50 measurements of the gauge master or standard. All measurements must be repeatable and lie within ten percent of the tolerance range. This range represents ± 5 percent of the tolerance range relative to the master. It is thus possible to calculate an upper and lower capability limit in accordance with the calculation of C_{gk}. A tolerance range of ± 0.0075 mm and a gauge master or standard of $x_m = 56.830$ mm produce the following results:

Up. Cap. Lim.: $x_m + 0.5 \times 0.1 \times \text{tolerance} = 56.830 + 0.5 \times 0.1 \times 0.015 \approx 56.831$ μm
Lo. Cap. Lim.: $x_m - 0.5 \times 0.1 \times \text{tolerance} = 56.830 - 0.5 \times 0.1 \times 0.015 \approx 56.829$ μm

- Now take 25 measurements of a production part. The range of the measured values must be less than 15 percent of the tolerance. This test corresponds in a way to the evaluation based on C_g. For a specified tolerance of 0.015 μm, this means that the range R of the set of values should be $\leq 0.015 \times 0.015 = 0.00225$ μm.

Further exceptions are

- Dynamic measurement systems with static calibration
- Measurement systems without gauge masters
- Measurement systems whose gauge masters are of a different shape from the production parts

To be able to evaluate these measurement methods, a type 2 study is used in a modified form. This involves measuring five production parts, each five times. Before each measurement is taken, the part has to be removed from the fixture and then reinserted. The measurement results should be recorded in the following table:

Meas. no.	1	2	3	4	5
1					
2					
3					
4					
5					
$\bar{x}_i$					
R_i					

If only one production part is available, the same number of measurements is taken of the one part. This involves inserting the part and measuring it five times. It is then removed from the fixture, reinserted and again measured five times. The above table can be used for recording the results.

The following calculations can be made from the results obtained:

Repeatability $EV = \bar{R} \cdot K_1$ where $K_1 = 2.58$ for $n = 5$

Reproducibility $AV = \bar{x}_{Diff} \cdot K_2$ where $K_2 = 2.42$ for $n = 5$

Total gauge variation $R\&R = \sqrt{EV^2 + AV^2}$

The constants K_1 and K_2 can be taken from Table 10.8. For n = 5, the values are as follows:

K_1 = 2.21 for 99 percent confidence level
K_1 = 2.58 for 99.73 percent confidence level
K_2 = 2.08 for 99 percent confidence level
K_2 = 2.42 for 99.73 percent confidence level

The result should be divided by the specified reference figure and then assessed against the usual acceptance criteria.

Example: Measurement System without Gauge Master

The diameter of a connecting rod was measured with an electronic measuring device. The specification of the characteristic is 56.830 mm ± 0.01 mm. The connecting rod was inserted into a measuring fixture and measured five times without being removed. The connecting rod was then removed from the fixture and reinserted four times. The same measurement was repeated. The following values were measured:

	Repeat measurements with insertion and removal of part				
Meas. no.	**1**	**2**	**3**	**4**	**5**
1	56.8340	56.8330	56.8340	56.8343	56.8341
2	56.8337	56.8333	56.8339	56.8346	56.8339
3	56.8336	56.8335	56.8337	56.8352	56.8344
4	56.8343	56.8335	56.8342	56.8346	56.8341
5	56.8338	56.8328	56.8337	56.8348	56.8343
$\bar{x}_i$	56.83388	56.83322	56.83390	56.83470	56.83416
R_i	0.00070	0.00070	0.00050	0.00090	0.00050

Taking the tolerance as the reference figure, the analysis of these values produces the following results:

Average range $\bar{R}$		=		0.00066	mm
Difference of mean values $\bar{x}_{Diff}$		=		0.0015	mm
EV =	0.001703	%EV	=	8.52%	
AV =	0.03582	%AV	=	17.91%	
R&R =	0.03966	%R&R	=	19.83%	

8.9 Frequently Asked Questions about Measurement System Capability

Studies to verify the suitability of measurement systems are generally very time-consuming, and where a large number of measurement systems are used, it may not always be possible to carry them out for all systems. Thus, in practical application, the following questions arise:

- When should capability studies be carried out?
- How are new measurement systems assessed?
- How should similar measurement systems be assessed?
- Can groups of measurement devices be evaluated together?
- How often should capability studies be carried out?
- Should capability studies be performed for standard measurement systems?
- Is it possible to carry out a capability study for destructive testing?
- What are the critical acceptance criteria?
- Which acceptance criteria cannot be fulfilled?
- Has there been full compliance with the relevant customer's guideline?
- What action should be taken with *incapable* measurement systems?

There is no standard response to these questions. The practical and technical context should be taken into account in each case. Keeping this in mind, the following statements can be made:

- Capability studies should, as a rule, be carried out for all characteristics that are classified as critical or require documentation. This is particularly important where the specification limits are narrow. If the measurement method is changed or if new measurement systems are developed, their suitability also has to be verified. For tolerances smaller than 15 µm, special procedures should be agreed upon. Where new measurement systems are concerned, there is often a lack of suitable measurement procedures or reference figures for correct evaluation. In these cases, the manufacturer must work out suitable assessment methods and make these readily available. In doing so, the manufacturer should involve qualified personnel and, if necessary, consult official bodies.
- If several different measurement systems or their components are used for the same measurement application, their reproducibility should be verified on the basis of a small number of sample systems. A type 2 study can be used for this purpose. Instead of having different appraisers, different measurement systems are used by one appraiser. If the total variation of the measurement systems is sufficiently small, it can be assumed that any measurement system can be selected from the relevant group for the respective measurement task. Such a study should be performed on a sample rather than a 100 percent basis and documented accordingly. Regular capability studies ensure that the individual measurement devices do not change significantly.
- The frequency of capability studies depends on the results of the study and on the stability of the measurement system. A type 1 study can be performed to verify stability. You will find information regarding the frequency of stability

studies in Section 8.6. As a rule, frequent studies will also be necessary whenever the result of a capability study lies close to or above the limit value.

- A capability study may not be necessary for standard measurement equipment if the expected uncertainty of measurement is sufficiently small in relation to the characteristic tolerance. As a rule, there is a large number of such measurement devices within a company. For economic reasons alone, it is not possible to examine all devices for each application. The suitability of other standard measuring equipment for the measurement tasks can be verified by means of individual studies or comparative measurements (see the previous paragraph on measurement equipment groups).
- For destructive testing, a capability study is not generally possible. If no gauge master or standard is available, it is not possible to implement a type 1 study. A type 2 study can only be conducted if it can be assumed that the material is homogeneous so that measurements of the original parts can be repeated. In this way, for example, two sections can be removed from a piece of wire regarded as homogeneous and used for the study. Whether or not these assumptions are fulfilled must be decided on a case-by-case basis. The same applies to hardness testers.
- With many existing measurement systems, it is not possible to conform to the QS-9000 [3] requirement for capable measurement systems with percent R&R ≤ 10 percent. Thus, it is recommended that this requirement only be applied to new measurement systems since the replacement of existing measurement systems is not practical for financial reasons.

  If the requirement of percent R&R ≤ 30 percent is not fulfilled, the capability studies should be performed at shorter intervals, stability studies should be performed more frequently, and, if necessary, systems should be recalibrated before each measurement.
- As discussed, acceptance criteria are set down in company guidelines. As a rule, these guidelines also state that other methods and performance limits are acceptable, provided that the customer is in agreement. For this reason, especially where critical measurement methods are concerned, it is important for criteria to be agreed upon with the customer on a case-by-case basis. Auditors often overlook this possibility when certifying quality systems!
- The previous point enables a company to develop and implement its own instructions on how to evaluate measurement systems, provided it uses customary standards as a point of reference. A supplier of several customers cannot implement a multitude of procedures for economic reasons alone.
- In the case of measurement systems that, in accordance with the methods and acceptance criteria described, are *not capable,* the feasibility of using a more accurate system is the first aspect to consider. If there are no alternatives, check whether the causes of incapability can be remedied. If none of the attempts at improvement are successful, it may be possible to accept the measurement system without statistical analysis. In this case, the repeatability is checked by measuring a production part at least 25 times, but no more than 50 times. If the maximum range of the measured values does not exceed 15 percent of the

tolerance, the measurement system should be accepted. In any case, it should be monitored by means of regular stability studies.

- If there are no suitable measurement systems, the following temporary measures can be taken for maintaining production until a permanent solution has been found:

  A general method that can be implemented on a daily basis is to monitor the same part several times. Random effects can be reduced if the measurements are carried out a number of times for the same production part and if the mean value of the measurements is then calculated. The measurement system should be calibrated before each measurement of the production part.

  Additional and more exact measuring devices can be used to complement the measurement system. In this way, for example, it is possible to monitor shape tolerances such as roundness, straightness, parallelism, etc., with additional measuring equipment that is installed in air-conditioned rooms.

The measures taken should always be documented in the relevant inspection instructions.

9 Appendix

9.1 Q-Das® Products

Figure 9.1–1 on the following page portrays the conceptual formulation of the modules, or the Q-DAS® products for statistical analysis and their uses.

The examples presented in this volume may be treated using the modules

- Sample analysis
- Process analysis
- Measurement system capability

You can get a demo version from Q-DAS® GmbH, Balzenbacher Str. 57, 69488 Birkenau, according to the current price list. The data and sample files are included! Furthermore, Q-DAS® offers a CBT program that guides the user in solving the examples.

9.2 Analysis of Variance Models

9.2.1 Process Evaluation

The *analysis of variance* (ANOVA) is a general procedure of inductive statistics for the examination of the influence of one or more factors on the target size. The variation of the measurement values is broken up into single components relative to the factors regarded in the model.

Generally, it is presumed that distribution of random influences is normal. It is possible to test statistical hypotheses for effects of individual factors or interactions between factors by applying the F-test, and to estimate the variation components according to the method of the least square.

Hereafter, we restrict ourselves to the one-way analysis of variance; i.e., examine the effect of a factor present on m levels on the target size Y.

A factor is described as *not random* (fixed, systematic) if the main interest is fixed on certain, set levels of the factor. A factor is called *random* if every level included in the examination was selected randomly from one (finite or infinite) population of factor levels.

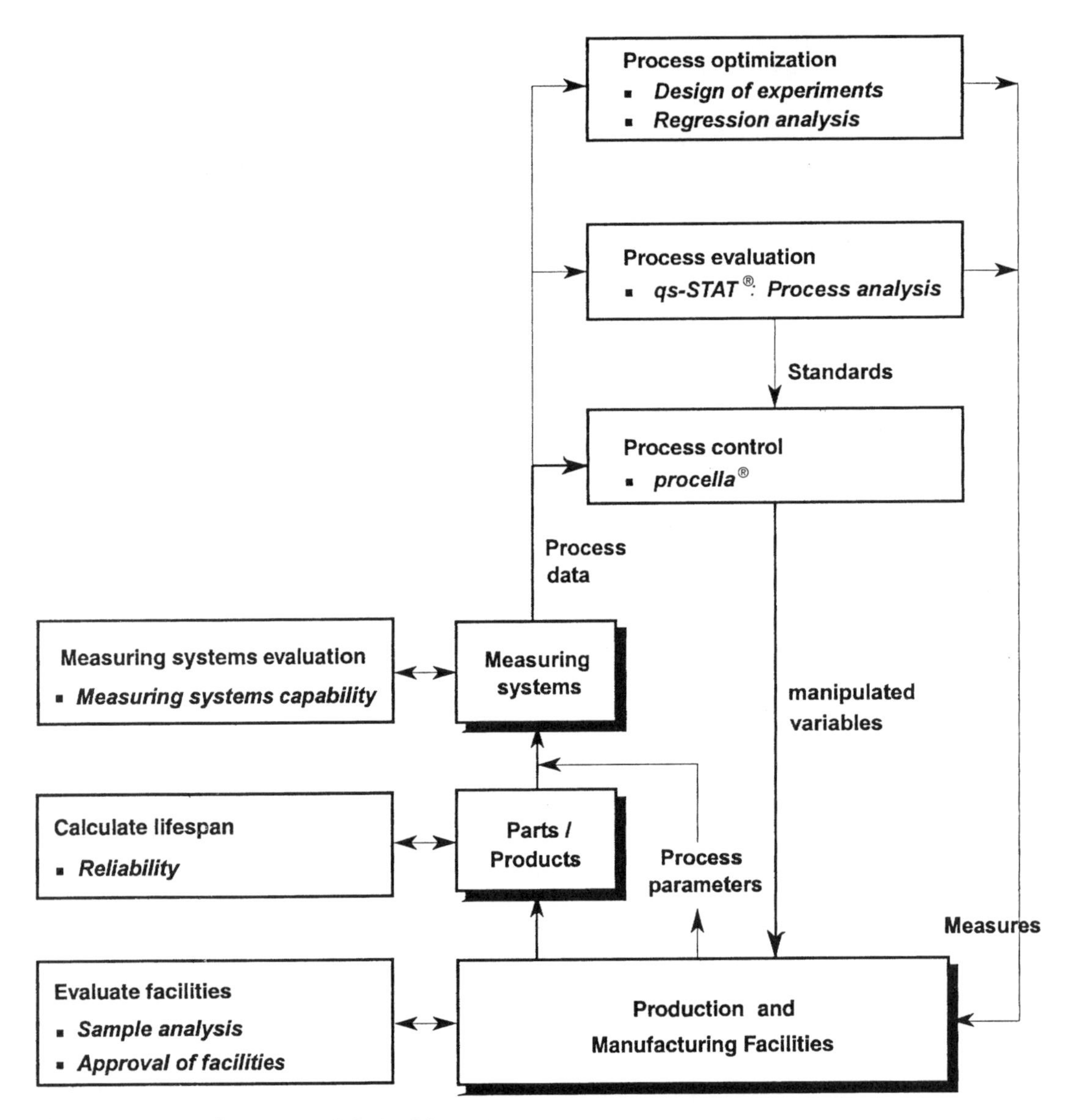

Figure 9.1–1. Overview of Q-DAS® products

Example

Model I: Investigation into the influence of m diverse types of feed on the milk output of cows

Model II: Comparison of m machines producing the same product, selecting m machines randomly out of the available M machines

In the case of a process with fluctuating averages, model II must be applied. Different questions are treated in both models; however, identical descriptive statistics and test statistics are needed.

Individual values exist for every one of the m levels of factor A, which makes a total of $N = m \cdot n$ measurement values $y_{11}, \ldots, y_{mn}$. The results $y_{11}, \ldots, y_{mn}$ are realizations of the observed random variables $y_{11}, \ldots, y_{mn}$.

It is presumed that

$$Y_{ij} = \mu + \alpha_i + \varepsilon_{ij} \quad \text{where} \quad i = 1,\ldots, m, j = 1,\ldots n,$$

where μ = grand average

α_i = incremental effect of level i of the observed factor A

ε_{ij} = additive random proportion (error)

ε_{ij} is a stochastic independently $N(0, \sigma_\varepsilon^2)$ distributed random variable. The two models differ with regard to α_i.

Model I. It is assumed that only the examined m levels of factor A exist or are of interest; e., α_i is a constant parameter, where

$$\sum_{i=1}^{m} \alpha_i = 0$$

The parameters μ, α_i, and σ_ε^2 are unknown in model I.

Model II. The examined m levels of factor A were selected randomly from M existing levels (M >> m). α_i is stochastic independently $N(0, \sigma_A^2)$ distributed random variables. The parameters μ, σ_A^2, and σ_ε^2 are unknown.

Typical questions for every statistical model are to estimate (as close as possible) the unknown parameter or to test hypotheses on these parameters. The simple null hypotheses are:

$$H_0 : \alpha_1 = \alpha_2 = \ldots = \alpha_m = 0$$

in model I, and

$$H_0 : \sigma_A^2 = 0$$

in model II.

The interim results required for the solution are traditionally recorded into an ANOVA table. This standard procedure is common practice, especially if no closed computer supported evaluation system is available.

Analysis of Variance Table (ANOVA Table)

Source of variation	SQ	Degrees of freedom (f)	MQ = SQ/f	F	Model I E(MQ)	Model II E(MQ)
Factor A	$\sum_{i=1}^{m} n(\bar{y}_{i\bullet} - \bar{y}_{i\bullet\bullet})^2$	m – 1	MQ_A	$\frac{MQ_A}{MQ_E}$	$\sigma_\varepsilon^2 + \frac{n}{m-1}\sum_{i=1}^{m} \alpha_i^2$	$\sigma_\varepsilon^2 + n\sigma_A^2$
Error E	$\sum_{i=1}^{m}\sum_{j=1}^{n} (y_{ij} - \bar{y}_{i\bullet})^2$	m(n – 1)	MQ_E		σ_ε^2	σ_ε^2
Total	$\sum_{i=1}^{m}\sum_{j=1}^{n} (y_{ij} - \bar{y}_{\bullet\bullet})^2$	mn – 1				

where $\bar{y}_{i\bullet} = \frac{1}{n}\sum_{j=1}^{n} y_{ij}$ and $\bar{y}_{\bullet\bullet} = \frac{1}{m \cdot n}\sum_{i=1}^{m}\sum_{j=1}^{n} y_{ij}$

MQ = Mean square total
SQ = sum of squares
f = degrees of freedom
F = test statistic of the F-test

The test statistic of the basic F-test is the quotient $\frac{MQ_A}{MQ_E}$.

The null hypothesis H_0 is rejected if

$$\frac{MQ_A}{MQ_E} > F_{(m-1,m(n-1),(1-\alpha))}$$

where $F_{(m-1,m(n-1),(1-a))}$ α-fractile of the F distribution with (m – 1) and m(n – 1) degrees of freedom.

Estimation of the Unknown Parameters

Model I

$$\mu := \bar{y}_{\bullet\bullet}$$

$$\hat{\alpha}_i := \bar{y}_{i\bullet} - \bar{y}_{\bullet\bullet}$$

$$\hat{\sigma}_\varepsilon^2 := \frac{1}{m(n-1)}\sum_{i=1}^{m}\sum_{j=1}^{n}(y_{ij} - \bar{y}_{i\bullet})^2 \quad (= MQ_E)$$

are equally best absolutely unbiased (UMVU) estimators and the least squares estimators for the respective model parameter.

Model II

$$\hat{\mu} := \bar{y}_{\bullet\bullet}$$

$$\hat{\sigma}_\varepsilon^2 := \frac{1}{m(n-1)}\sum_{i=1}^{m}\sum_{j=1}^{n}(y_{ij} - \bar{y}_{i\bullet})^2 \quad (= MQ_E)$$

$$\hat{\sigma}_A^2 := \frac{1}{n}\left(\frac{1}{m-1}\sum_{i=1}^{m} n(\bar{y}_{i\bullet} - \bar{y}_{\bullet\bullet})^2 - \hat{\sigma}_\varepsilon^2\right) \left(= \frac{1}{n}(MQ_A - MQ_E)\right)$$

are UMVU estimators for the respective model parameters. A negative value for $\hat{\sigma}_A^2$ may be the result when calculating the estimator. In this case, $\hat{\sigma}_A^2$ is set equal to zero: $\hat{\sigma}_A^2 := 0$.

A further estimating problem may be expressed in model II. The observed random variables Y_{ij} are not independent stochastically; however, the same normal distribution applies for them, which is $N(\mu, \sigma_A^2 + \sigma_\varepsilon^2)$. The uniformly minimum variance unbiased (UMVU) estimator for the variance $\sigma_{ges}^2 := \sigma_A^2 + \sigma_\varepsilon^2$ of this normal distribution is

$$\sigma_{tot}^2 := \sigma_A^2 + \hat{\sigma}_\varepsilon^2$$

It must be noted that this question is sensible *only* with model II and *not* with model I, and σ_{tot}^2 is *not* the subgroup variance of the total sample $y_{11}, \ldots, y_{mn}$.

Only in connection with the respective null hypothesis is

$$\frac{1}{mn-1}\sum_{i=1}^{m}\sum_{j=1}^{n}(y_{ij}-\bar{y}_{\bullet\bullet})^2$$

the unbiased estimator for σ^2_{tot}. If the null hypothesis is true, that is, a single subgroup problem exists, this variance estimation will be unbiased.

9.2.2 Measurement System Analysis: Type 2 Study

P operators use a gauge to measure *T* parts with *W* repetitions each.

It is assumed that every measurement value is composed from the total average of the measurement values, operator influence, part influence, the influence of interactions between operator and part, as well as residual deviation (gauge influence), i.e.,

measurement value by operator on part in repetition =
Total average + Operator influence
+ Part influence
+ Influence of (operator measures part)
+ Residual deviation

To be able to evaluate the influences separately, first of all, the sum of the square deviations over all measurement values is divided into partial sums, and from these the variances are calculated. For calculation,

The average of *Operator p measures part t* over the repetitions $:X_{pt\cdot}$
The average over the measurement values of operator *p* $:X_{p\cdot\cdot}$
The average over the measurement values of part *t* $:X_{\cdot t\cdot}$
The total average $:X_{\cdots}$

Sum of square deviations between the *p* operators:

$$\Sigma P := tw\,[(X_{1}..-X...)^2+(X_{2}..-X...)^2+(X_{3}..-X...)^2\ldots(X_{p}..-X...)^2]$$

with degree of freedom

$$f_{IV} := p-1;$$

Sum of square deviations between the *t* parts:

$$\Sigma T := pw\,[(X._{1}.-X...)^2+(X._{2}.-X...)^2+(X._{3}.-X...)^2\ldots(X._{t}.-X...)^2]$$

with degree of freedom

$$f_{III} := t-1$$

Sum of square deviations through interaction (*p* measures *t*):

$$\Sigma PT := w\sum\nolimits_{i=1..p}\sum\nolimits_{j=1}(X_{ij}.-X_{i}..-X._{j}.-X...)^2$$

with degree of freedom

$$f_{II} := (p-1)(t-1)$$

Sum of square deviations within repetitions of operator *p* measures part *t:*

$$\Sigma E := \sum\nolimits_{i=1..p}\sum\nolimits_{j=1..t}\sum\nolimits_{k=1..w}(X_{ijk}-X_{ij}.)^2$$

with degree of freedom

$$f_I := pt\,(w-1)$$

The sum of square deviations over all measurement values is thus

$$\Sigma P + \Sigma T + \Sigma PT + \Sigma E$$

The following variances are calculated for measurement system analysis. Here, it is differentiated whether the influence of the interaction is significant or not. (F-test, test value s^2_{PT}/s^2_E, crit. value $F_{fII,\, fI,\, 1-\alpha}$).

The variances are calculated accordingly from the quotient of the sum of square deviations divided by the corresponding degree of freedom:

Variance operator influence	$s^2_P := \Sigma P/f_{IV}$
Variance part influence	$s^2_T := \Sigma T/f_{III}$

In case of significant interaction

Variance interaction	$s^2_{PT} := \Sigma PT/f_{II}$
Variance gauge influence	$s^2_E := \Sigma E/f_I$

In case of nonsignificant interaction

Variance ADDinteraction/gauge	$s^2_{add} := (\Sigma E + \Sigma PT)/(f_I + f_{II})$

1. Significant interaction influence:

Confidence intervals for level 1–α are calculated from

$$\frac{f_I}{\chi^2_{fI,1-\alpha/2}} s_E^2 \le \sigma_E^2 \le \frac{f_I}{\chi^2_{fI,\alpha/2}} s_E^2$$

$$\frac{s_E^2}{w}\left(\frac{s_{PT}^2/s_E^2}{F_{fII,fI,1-\alpha/2}}-1\right) \le \sigma_{PT}^2 \le \frac{s_E^2}{w}\left(\frac{s_{PT}^2/s_E^2}{F_{fII,fI,\alpha/2}}-1\right)$$

$$\frac{s_{PT}^2}{tw}\left(\frac{s_P^2/s_{PT}^2}{F_{fIII,fII,1-\alpha/2}}-1\right) \le \sigma_P^2 \le \frac{s_{PT}^2}{tw}\left(\frac{s_P^2/s_{PT}^2}{F_{fIII,fII,\alpha/2}}-1\right)$$

$$\frac{s_{PT}^2}{pw}\left(\frac{s_T^2/s_{PT}^2}{F_{fIV,fII,1-\alpha/2}}-1\right) \le \sigma_T^2 \le \frac{s_{PT}^2}{pw}\left(\frac{s_T^2/s_{PT}^2}{F_{fIV,fII,\alpha/2}}-1\right)$$

$$\frac{1}{tw}\left(\frac{f_{III}}{\chi^2_{fIII,1-\alpha/2}} s_P^2 + t(w-1)s_E^2 + (t-1)s_{PT}^2\right) \le \sigma_P^2 + \sigma_T^2 + \sigma_{PT}^2 \le \frac{1}{tw}\left(\frac{f_{III}}{\chi^2_{fIII,\alpha/2}} s_P^2 + t(w-1)s_E^2 + (t-1)s_{PT}^2\right)$$

Using the calculated values

Gauge:	VE	:=	s^2_E
Interaction (operator measures part):	VW	:=	$(s^2_{PT} - s^2_E)/w$
Operator:	VP	:=	$(s^2_P - s^2_{PT})/tw$
Part:	VT	:=	$(s^2_T - s^2_{PT})/pw$

it is possible to draw conclusions on the influence of the individual components (the product 5.15 * s equals a proportion of 99 percent of the values in case of a normally distributed population):

EV (**e**quipment **v**ariation)	:	$5.15\sqrt{VE}$
AV (**a**ppraiser **v**ariation)	:	$5.15\sqrt{VP}$
IA (**i**nter**a**ction)	:	$5.15\sqrt{VW}$
PV (**p**art **v**ariation)	:	$5.15\sqrt{VT}$
R&R (**r**epeatability & **r**eproducibility)	:	$\sqrt{EV^2 + AV^2 + IA^2}$

2. Nonsignificant interaction influence:

The confidence intervals for level $1 - \alpha$ are calculated from

$$\frac{f_I + f_{II}}{\chi^2_{fI+fII,1-\alpha/2}} s^2_{add} \le \sigma^2_{add} \le \frac{f_I + f_{II}}{\chi^2_{fI+fII,\alpha/2}} s^2_{add}$$

$$\frac{s^2_{add}}{tw}\left(\frac{s^2_P / s^2_{add}}{F_{fIII,fI+fII,1-\alpha/2}} - 1\right) \le \sigma^2_P \le \frac{s^2_{add}}{tw}\left(\frac{s^2_P / s^2_{add}}{F_{fIII,fI+fII,\alpha/2}} - 1\right)$$

$$\frac{s^2_{add}}{pw}\left(\frac{s^2_T / s^2_{add}}{F_{fIV,fI+fII,1-\alpha/2}} - 1\right) \le \sigma^2_T \le \frac{s^2_{add}}{pw}\left(\frac{s^2_T / s^2_{add}}{F_{fIV,fIfII,\alpha/2}} - 1\right)$$

$$\frac{1}{tw}\left(\frac{f_{III}}{\chi^2_{fIII,1-\alpha/2}} s^2_P + (tw-1)s^2_{add}\right) \le \sigma^2_P + \sigma^2_T \le \frac{1}{tw}\left(\frac{f_{III}}{\chi^2_{fIII,1-\alpha/2}} s^2_P + (tw-1)s^2_{add}\right)$$

Using the calculated values

Gauge:	VE	:=	s^2_{add}
Operator:	VP	:=	$(s^2_P - s^2_{add})/tw$
Part:	VT	:=	$(s^2_T - s^2_{add})/pw$

it is possible to draw conclusions on the influence of the individual components (the product 5.15 * s equals a proportion of 99 percent of the values in case of a normally distributed population):

EV (**e**quipment **v**ariation)	:	$5.15\sqrt{VE}$
AV (**a**ppraiser **v**ariation)	:	$5.15\sqrt{VP}$
PV (**p**art **v**ariation)	:	$5.15\sqrt{VT}$
R& R (**r**epeatability & **r**eproducibility)	:	$\sqrt{EV^2 + AV^2}$

Example

	Operator 1		**Operator 2**	
	Rep. 1	**Rep 2.**	**Rep. 1**	**Rep. 2**
Part 1	2	1	1	1
Part 2	1	1	1	2
Part 3	2	1	1	1
Part 4	3	2	1	2
Part 5	1	3	1	1

For manual calculation of the individual averages, sums of square deviations, and variances, the ANOVA table (see [46]) is used:

	Operator 1		**Operator 2**		Σ	Σ	(Σ)²	Σ()²
	W1+W2	W1²+W2²	W1+W2	W1²+W2²				
T1	3		2		a1=5		c1=25	e1=13
T1		5		2		b1=7		
T2	2		3		a2=5		c2=25	e2=13
T2		2		5		b2=7		
T3	3		2		a3=5		c3=25	e3=13
T3		5		2		b3=7		
T4	5		3		a4=8		c4=64	e4=34
T4		13		5		b4=18		
T5	4		2		a5=6		c5=36	e5=20
T5		10		2		b5=12		
Σ	A1=17		A2=12		A=29		C=175	
Σ		B1=35		B2=16		B=51		
(Σ)²	D1=289		D2=144		D=433			
Σ()²	E1=63		E2=30					E=93

This results in the following statistics:

$X_{pt\cdot}$ = Sum W1 + W2 of operator p, part t divided by no. of repetitions:

$X_{11\cdot} = 3/2 = 1.5$ $\quad X_{12\cdot} = 2/2 = 1$ $\quad X_{13\cdot} = 3/2 = 1.5$

$X_{14\cdot} = 5/2 = 2.5$ $\quad X_{15\cdot} = 4/2 = 2$

$X_{21\cdot} = 2/2 = 1$ $\quad X_{22\cdot} = 3/2 = 1.5$ $\quad X_{33\cdot} = 2/2 = 1$

$X_{24\cdot} = 3/2 = 1.5$ $\quad X_{25\cdot} = 2/2 = 1$

$X_{p\cdot\cdot} = A_p$ divided by parts *rep.:

$X_{1\cdot\cdot} = 17/10 = 1.7$ $\quad X_{2\cdot\cdot} = 12/10 = 1.2$

$X_{\cdot t\cdot} = a_t$ divided by operator *rep.:

$X_{\cdot 1\cdot} = 5/4 = 1.25$ $\quad X_{\cdot 2\cdot} = 5/4 = 1.25$ $\quad X_{\cdot 3\cdot} = 5/4 = 1.25$

$X_{\cdot 4\cdot} = 8/4 = 2$ $\quad X_{\cdot 5\cdot} = 6/4 = 1.5$

$X... = A$ divided by operator *parts *rep. = 29/20 = 1.45.

$$\begin{aligned}
\Sigma P &= D/(tw) - A^2/(ptw)\\
&= 433/10 - 841/20\\
&= 1.25\\
s^2_P &= 1.25/1 = 1.25\\
\Sigma T &= C/(pw) - A^2/(ptw)\\
&= 175/4 - 841/20\\
&= 1.7\\
s^2_T &= 1.7/4 = 0.425\\
\Sigma PT &= E/w - C/(pw) - D/(tw) + A^2/(ptw)\\
&= 93/2 - 175/4 - 433/10 + 841/20\\
&= 1.5\\
s^2_{PT} &= 1.5/4 = 0.375\\
\Sigma E &= B - E/w\\
&= 51 - 93/2\\
&= 4.5\\
s^2_E &= 4.5/10 = 0.45
\end{aligned}$$

Test value F-test:

$$s^2_{PT}/s^2_E = 0.375/0.45 = 0.8334 < 3.48 = F_{10,4,1-95\%}$$

thus interaction is not significant, i.e.,

$$\begin{aligned}
s^2_{add} &= (\Sigma E + \Sigma PT)/ptw - p - t + 1\\
&= 6/14 = 0.4285
\end{aligned}$$

Gauge:	VE	=	0.429
Operator:	VP	=	0.0821
Part:	VT	=	0(da < 0)

EV	(equipment variation)	: $5.15\sqrt{VE}$	=	3.373,
AV	(appraiser variation)	: $5.15\sqrt{VP}$	=	1.476,
PV	(part variation)	: $5.15\sqrt{VT}$	=	0
R&R	(repeatability & reproducibility)	: $\sqrt{EV^2 + AV^2}$	=	3.682.

The result R&R must be put into relation to a given reference figure (RF):

$$\%R\&R = \frac{R\&R}{RF} \cdot 100\%$$

This result is to be compared to the stipulated acceptance criteria.

9.2.3 Measurement System Analysis: Type 3 Study

For the evaluation of an automatic measurement system, the model of the balanced simple analysis of variance with random components offers itself. Contrary to the standard deviation method described in Chapter 8, Section 8.5.9, this method may be used more than twice for testing. It is presumed that every measurement value is composed from

total average + part influence + gauge influence

To evaluate gauge influence only, part influence must be kept low through proper measures. This may be realized by marking of the measurement spots.

The sum of square deviations of repetitions (= measurements per part):

$$\sum E = \sum_{i=j}^{n} \sum_{j=1}^{k} \left(X_{ij} - X_{i\bullet}\right)^2 \quad \text{where } X_{i\bullet} = \text{average of measurements per part}$$

$i = 1, 2, \ldots, n$ = number of parts
$j = 1, 2, \ldots, k$ = number of measurements per part

This is used for calculating *gauge variation:*

$$s_E^2 = \frac{1}{f} \sum E \quad \text{with degree of freedom} \quad f = n \cdot (k-1).$$

$EV = 5.15 \cdot s_E$ for confidence interval 99 percent
$EV = 6 \cdot s_E$ for confidence interval 99.73 percent

For calculation of the total gauge variation percent R&R, EV is set in relation to a given reference figure (RF):

$$\%R\&R = \%EV = \frac{EV}{RF} \cdot 100\%$$

This calculated value must be compared to the stipulated acceptance criteria. Typical reference figures are tolerance, 6 * process variation, or part variation (= the variation between the different parts, PV = part variation). This may be determined from the square deviation between the parts:

$$\sum T = k \sum_{i=1}^{n} (x_{i\bullet} - x_{\bullet\bullet})^2 \quad \text{where } i = 1, 2, \ldots, n = \text{number of parts}$$

k = number of measurements per part
$x_{\bullet\bullet}$ = total average

$$s_T^2 = \frac{\sum T}{f_T} \quad \text{and} \quad VT = \frac{\left(s_T^2 - s_E^2\right)}{k} \quad \text{with degree of freedom } f_T = n - 1$$

$PV = 5.15 \cdot VT$ for confidence interval 99 percent
$PV = 6 \cdot VT$ for conficence interval 99.73 percent

If no significant part variation exists, then $VT < 0$ is possible. In this case, part variation is not allowed to be taken as reference figure.

Example

Ten parts are measured twice. Characteristics tolerance is 0.06 mm.

i	x_{1i}	x_{2i}	$x_{i\bullet}$	$(x_{1i}-x_{i\bullet})^2$	$(x_{2i}-x_{i\bullet})^2$	E	T
1	6.029	6.030	6.0295	0.00000025	0.00000025	0.0000005	0.00065536
2	6.019	6.020	6.0195	0.00000025	0.00000025	0.0000005	0.00024336
3	6.004	6.003	6.0035	0.00000025	0.00000025	0.0000005	0.00000016
4	5.982	5.982	5.9820	0.00000000	0.00000000	0.0000000	0.00047961
5	6.009	6.009	6.0090	0.00000000	0.00000000	0.0000000	0.00002601
6	5.971	5.972	5.9715	0.00000025	0.00000025	0.0000005	0.00104976
7	5.995	5.997	5.9960	0.00000100	0.00000100	0.0000020	0.00006241
8	6.014	6.018	6.0160	0.00000400	0.00000400	0.0000080	0.00014641
9	5.985	5.987	5.9860	0.00000100	0.00000100	0.0000020	0.00032041
10	6.024	6.028	6.0260	0.00000400	0.00000400	0.0000080	0.00048841
	$x_{\bullet\bullet} = 6.0039$					$\Sigma E = 0.0000220$	$\Sigma T = 0.00347190$

For Measurement System Analysis

$$s_E^2 = \frac{0.000022}{10(2-1)} = 0.0000022$$

$$EV = 6 \cdot \sqrt{0.0000022} = 0.0089 \quad \text{for confidence interval 99.73 percent}$$

$$\%R\&R = \%EV = \frac{0.0089}{0.06} \cdot 100\% = 14.8\% \quad \text{with tolerance as reference figure}$$

For Variation between the Parts

$$s_T^2 = \frac{0.0034719}{9} = 0.000386$$

$$VT = \frac{(0.000386 - 0.000022)}{2} = 0.0001819$$

$$PV = 6 \cdot \sqrt{0.0001819} = 0.0809 \quad \text{for confidence interval 99.73 percent}$$

9.3 Formulas

Normal Distribution

$$g(x) = \frac{1}{\sqrt{2\pi}\sigma} \exp\left\{-\frac{1}{2}\left(\frac{x-\mu}{\sigma}\right)^2\right\} \qquad -\infty < x < \infty$$

$$G(x) = \frac{1}{\sqrt{2\pi}\sigma} \int_{-\infty}^{x} \exp\left\{-\frac{1}{2}\left(\frac{t-\mu}{\sigma}\right)^2\right\} dt$$

Lognormal Distribution

$$g(x) = \frac{1}{\sqrt{2\pi}\sigma} \frac{1}{x-a} \exp\left\{-\frac{1}{2}\left(\frac{\text{In}(x-a)-\mu}{\sigma}\right)^2\right\} \qquad a < x < \infty$$

$$G(x) = \int_a^x g(t)dt$$

Weibull Distribution

$$g(x) = \frac{\beta}{\alpha}\left(\frac{x-a}{\alpha}\right)^{\beta-1} \exp\left\{-\left(\frac{x-a}{\alpha}\right)^{\beta}\right\} \qquad a \le x < \infty$$

$$G(x) = 1 - \exp\left\{-\left(\frac{x-a}{\alpha}\right)^{\beta}\right\}$$

Folded Normal Distribution

$$g(x) = \frac{2}{\sqrt{2\pi}\sigma} \exp\left\{-\frac{1}{2}\left(\frac{|x-\mu|}{\sigma}\right)^2\right\} \qquad 0 \le |x-a| < \infty$$

$$G(x) = \int_0^{|x-a|} g(t)dt$$

Rayleigh Distribution

$$g(x) = \frac{2(x-a)}{\alpha^2} \exp\left\{-\left(\frac{x-a}{\alpha}\right)^2\right\} \qquad a \le x < \infty$$

$$G(x) = 1 - \exp\left\{-\left(\frac{x-a}{\alpha}\right)^2\right\}$$

Two-Dimensional Normal Distribution

$$g_{(x,y)} = \frac{1}{2\pi \times \sigma_x \times \sigma_y \times \sqrt{1-\rho^2}} \exp\left[-\frac{1}{2(1-\rho^2)}(u^2 - 2\rho uv + v^2)\right] \qquad u = \frac{x-\mu_x}{\sigma_x}; \quad v = \frac{y-\mu_y}{\sigma_y}; \quad -\infty < u, v < +\infty$$

10

Tables

Factors for estimating the standard deviation in the case of individual and median values distribution are shown in Table 10–1.

Table 10–1. Factors for estimating the standard deviation.

n	a_n	d_n	c_n
2	0.798	1.128	1.000
3	0.886	1.693	1.160
4	0.921	2.059	1.092
5	0.940	2.326	1.197
6	0.952	2.534	1.135
7	0.959	2.704	1.214
8	0.965	2.847	1.160
9	0.969	2.970	1.223
10	0.973	3.078	1.176
11	0.975	3.173	1.228
12	0.978	3.258	1.187
13	0.979	3.336	1.232
14	0.981	3.407	1.196
15	0.982	3.472	1.235
16	0.983	3.532	1.202
17	0.985	3.588	1.237
18	0.985	3.640	1.207
19	0.986	3.689	1.239
20	0.987	3.735	1.212
21	0.988	3.778	1.240
22	0.988	3.819	1.216
23	0.989	3.858	1.241
24	0.989	3.859	1.218
25	0.990	3.930	1.242

Factors for calculating the warning ($1 - \alpha = 95\%$) and control limits ($1 - \alpha = 99\%$) in *average charts* without regarding limit values are shown in Table 10–2.

Table 10–2. Factors for calculating the limits of average charts.

n	A_W	A_E
2	1.386	1.821
3	1.132	1.487
4	0.980	1.288
5	0.877	1.152
6	0.800	1.052
7	0.741	0.974
8	0.693	0.911
9	0.653	0.859
10	0.620	0.815
11	0.591	0.777
12	0.566	0.744
13	0.544	0.714
14	0.524	0.688
15	0.506	0.665
16	0.490	0.644
17	0.475	0.625
18	0.462	0.607
19	0.450	0.591
20	0.438	0.576
21	0.428	0.562
22	0.418	0.549
23	0.409	0.537
24	0.400	0.526
25	0.392	0.515

Factors for calculating the warning ($1 - \alpha = 95\%$) and control limits ($1 - \alpha = 99\%$) of *median charts* without regarding limit values are shown in Table 10–3.

Table 10–3. Factors for calculating the limits of median charts.

n	C_W	C_E
2	1.386	1.821
3	1.313	1.725
4	1.070	1.406
5	1.049	1.379
6	0.908	1.194
7	0.899	1.182
8	0.804	1.056
9	0.799	1.050
10	0.729	0.958
11	0.726	0.954
12	0.672	0.883
13	0.670	0.880
14	0.626	0.823
15	0.625	0.821
16	0.589	0.774
17	0.588	0.773
18	0.558	0.733
19	0.557	0.732
20	0.531	0.698
21	0.530	0.697
22	0.508	0.668
23	0.507	0.667
24	0.487	0.640
25	0.487	0.640

Factors for calculating the warning ($1 - \alpha = 95\%$) and control limits ($1 - \alpha = 99\%$) of *raw value charts* without regarding limit values are shown in Table 10–4.

Table 10–4. Factors for calculating the limits of raw value charts.

n	E_W	E_E
2	2.236	2.806
3	2.388	2.934
4	2.491	3.022
5	2.569	3.089
6	2.631	3.143
7	2.683	3.188
8	2.727	3.226
9	2.766	3.260
10	2.800	3.289
11	2.830	3.316
12	2.858	3.340
13	2.883	3.362
14	2.906	3.383
15	2.928	3.402
16	2.948	3.419
17	2.966	3.436
18	2.984	3.451
19	3.000	3.466
20	3.016	3.480
21	3.031	3.493
22	3.045	3.505
23	3.058	3.517
24	3.071	3.528
25	3.083	3.539

Factors for calculating the warning ($1 - \alpha = 95\%$) and control limits ($1 - \alpha = 99\%$) of *standard deviation charts* are shown in Table 10–5.

Table 10–5. Factors for calculating the limits of standard deviation charts.

n	B_{Wlo}	B_{Wup}	B_{Elo}	B_{Eup}
2	0.031	2.241	0.006	2.807
3	0.159	1.921	0.071	2.302
4	0.268	1.765	0.155	2.069
5	0.348	1.669	0.227	1.927
6	0.408	1.602	0.287	1.830
7	0.454	1.552	0.336	1.758
8	0.491	1.512	0.376	1.702
9	0.522	1.480	0.410	1.657
10	0.548	1.454	0.439	1.619
11	0.570	1.431	0.464	1.587
12	0.589	1.412	0.486	1.560
13	0.606	1.395	0.506	1.536
14	0.621	1.379	0.524	1.515
15	0.634	1.366	0.540	1.496
16	0.646	1.354	0.554	1.479
17	0.657	1.343	0.567	1.463
18	0.667	1.333	0.579	1.450
19	0.676	1.323	0.590	1.437
20	0.685	1.315	0.600	1.425
21	0.692	1.307	0.610	1.414
22	0.700	1.300	0.619	1.404
23	0.707	1.293	0.627	1.395
24	0.713	1.287	0.635	1.386
25	0.719	1.281	0.642	1.378

Factors for calculating the warning ($1 - \alpha = 95\%$) and control limits ($1 - \alpha = 99\%$) *of range charts* are shown in Table 10–6.

Table 10-6. Factors for calculating the limits of range charts.

n	D_{Wlo}	D_{Wup}	D_{Elo}	D_{Eup}
2	0.044	3.170	0.009	3.970
3	0.303	3.682	0.135	4.424
4	0.595	3.984	0.343	4.694
5	0.850	4.197	0.555	4.886
6	1.066	4.361	0.749	5.033
7	1.251	4.494	0.922	5.154
8	1.410	4.605	1.075	5.255
9	1.550	4.700	1.212	5.341
10	1.674	4.784	1.335	5.418
11	1.784	4.858	1.446	5.485
12	1.884	4.925	1.547	5.546
13	1.976	4.985	1.639	5.602
14	2.059	5.041	1.724	5.652
15	2.136	5.092	1.803	5.699
16	2.207	5.139	1.876	5.742
17	2.274	5.183	1.944	5.783
18	2.336	5.224	2.008	5.820
19	2.394	5.262	2.068	5.856
20	2.449	5.299	2.125	5.889
21	2.500	5.333	2.178	5.921
22	2.549	5.365	2.229	5.951
23	2.596	5.396	2.277	5.979
24	2.640	5.425	2.323	6.006
25	2.682	5.453	2.366	6.032

The factors for calculation of the control limits are based on a confidence level of 99.73 percent (see Table 10–7).

Table 10-7. Factors from [38].

n	A_2	d_2	D_3	D_4	A_3	c_4	B_3	B_4	$\tilde{A}_2$
2	1.880	1.128	—	3.267	2.659	0.7979	—	3.267	1.880
3	1.023	1.693	—	2.574	1.954	0.8862	—	2.568	1.187
4	0.729	2.059	—	2.282	1.628	0.9213	—	2.266	0.796
5	0.577	2.326	—	2.114	1.427	0.9400	—	2.089	0.691
6	0.483	2.534	—	2.004	1.287	0.9515	0.030	1.970	0.548
7	0.419	2.704	0.076	1.924	1.182	0.9594	0.118	1.882	0.508
8	0.373	2.847	0.136	1.864	1.099	0.9650	0.185	1.815	0.433
9	0.337	2.970	0.184	1.816	1.032	0.9693	0.239	1.761	0.412
10	0.308	3.078	0.223	1.777	0.975	0.9727	0.284	1.716	0.362
11	0.285	3.173	0.256	1.744	0.927	0.9754	0.321	1.679	—
12	0.266	3.258	0.283	1.717	0.886	0.9776	0.354	1.646	—
13	0.249	3.336	0.307	1.693	0.850	0.9794	0.382	1.618	—
14	0.235	3.407	0.328	1.672	0.817	0.9810	0.406	1.594	—
15	0.223	3.472	0.347	1.653	0.789	0.9823	0.428	1.572	—
16	0.212	3.532	0.363	1.637	0.763	0.9835	0.448	1.552	—
17	0.203	3.588	0.378	1.622	0.739	0.9845	0.466	1.534	—
18	0.194	3.640	0.391	1.608	0.718	0.9854	0.482	1.518	—
19	0.187	3.689	0.403	1.597	0.698	0.9862	0.497	1.503	—
20	0.180	3.735	0.415	1.585	0.680	0.9869	0.510	1.490	—
21	0.173	3.778	0.425	1.575	0.663	0.9876	0.523	1.477	—
22	0.167	3.819	0.434	1.566	0.647	0.9882	0.534	1.466	—
23	0.162	3.858	0.443	1.557	0.633	0.9887	0.545	1.455	—
24	0.157	3.895	0.451	1.548	0.619	0.9892	0.555	1.445	—
25	0.153	3.931	0.459	1.541	0.606	0.9896	0.565	1.435	—

n	2	3	4	5	6	7	8	9	10
E_2	2.66	1.77	1.46	1.29	1.18	1.11	1.05	1.01	.98

Table 10–8. d_2^* values for K factors.

Subgroup size: Number of operators (k) * number of pieces (n)	Subgroup size: number of trials (r) for K_1 or number of operators (k) for K_2													
d_2^*	2	3	4	5	6	7	8	9	10	11	12	13	14	15
1	1.41	1.91	2.24	2.48	2.67	2.83	2.96	3.08	3.18	3.27	3.35	3.42	3.49	3.55
2	1.28	1.81	2.15	2.40	2.60	2.77	2.91	3.02	3.13	3.22	3.30	3.38	3.45	3.51
3	1.23	1.77	2.12	2.38	2.58	2.75	2.89	3.01	3.11	3.21	3.29	3.37	3.43	3.50
4	1.21	1.75	2.11	2.37	2.57	2.74	2.88	3.00	3.10	3.20	3.28	3.36	3.43	3.49
5	1.19	1.74	2.10	2.36	2.56	2.73	2.87	2.99	3.10	3.19	3.28	3.35	3.42	3.49
6	1.18	1.73	2.09	2.35	2.56	2.73	2.87	2.99	3.10	3.19	3.27	3.35	3.42	3.49
7	1.17	1.73	2.08	2.35	2.55	2.72	2.87	2.99	3.10	3.19	3.27	3.35	3.42	3.48
8	1.17	1.72	2.08	2.35	2.55	2.72	2.87	2.98	3.09	3.19	3.27	3.35	3.42	3.48
9	1.16	1.72	2.08	2.34	2.55	2.72	2.86	2.98	3.09	3.18	3.27	3.35	3.42	3.48
10	1.16	1.72	2.08	2.34	2.55	2.72	2.86	2.98	3.09	3.18	3.27	3.34	3.42	3.48
11	1.16	1.71	2.08	2.34	2.55	2.72	2.86	2.98	3.09	3.18	3.27	3.34	3.41	3.48
12	1.15	1.71	2.07	2.34	2.55	2.72	2.85	2.98	3.09	3.18	3.27	3.34	3.41	3.48
13	1.15	1.71	2.07	2.34	2.55	2.71	2.85	2.98	3.09	3.18	3.27	3.34	3.41	3.48
14	1.15	1.71	2.07	2.34	2.54	2.71	2.85	2.98	3.08	3.18	3.27	3.34	3.41	3.48
15	1.15	1.71	2.07	2.34	2.54	2.71	2.85	2.98	3.08	3.18	3.26	3.34	3.41	3.48
>15	1.128	1.693	2.059	2.326	2.534	2.704	2.847	2.970	3.078	3.173	3.258	3.336	3.407	3.472

The K factors K_1 and K_2 are calculated from $\frac{5.152}{d_2^*}$ to represent 99 percent of the ND (see Table 10–8).

K_1 is dependent on the number of trials r and the number of pieces n times the number of operators k.

K_2 is dependent on the number of operators. Since there is only one range calculation, only line one is applicable.

Examples

1. 2 trials (r = 2), 3 operators (k = 3), 10 pieces (n = 10)
   $k \cdot n = 3 \cdot 10 = 30$ so the >15 line is applicable
   $d_2^* = 1.128$
   $$K_1 = \frac{5.152}{1.128} = 4.567$$
2. 10 trials (r = 10), 1 operator (k = 1), 5 parts (n = 5)
   $k \cdot n = 1 \cdot 5 = 5$ $d_2^* = 3.10$
   $$K_1 = \frac{5.152}{3.1} = 1.66$$
3. 3 operators (k = 3)
   $d_2^* = 1.91$
   $$K_2 = \frac{5.152}{1.91} = 2.697$$

11
Ford Test Examples

Ford Test Examples

Verification of SPC Software

Amendments were made by the authors for this edition. These are printed in italics.

Example number	Data file name	Decimal points	Total sample size	Subgroup sample size	Purpose of test
1	TEST_1	3	500	5	correct indication of out-of-control conditions ($\frac{\bar{x}}{s}$ chart)
2	TEST_2	4	100	5	accuracy of calculation
3	TEST_3	2	875	7	$\frac{\bar{x}}{s}$ chart, capability indices for normal distribution
4	TEST_4	1	200	5	$\frac{\bar{x}}{s}$ chart, capability indices for lognormal distribution
5	TEST_5	0	200	5	$\frac{\bar{x}}{s}$ chart, capability indices for Weibull distribution
6	TEST_6	3	500	5	$\frac{\bar{x}}{s}$ chart, capability indices for Rayleigh distribution
7	TEST_7	3	600	3	$\frac{\bar{x}}{s}$ chart, capability indices for folded normal distribution
8	TEST_8	2	500	5	$\frac{\bar{x}}{s}$ chart, capability indices for fix tooling (tool change)
9	TEST_9	2	600	6	$\frac{\bar{x}}{s}$ chart, capability indices for trend production (tool wear)
10	TEST_10	1	102	$\frac{3}{3}$	moving averages and moving s chart (moving averages and R chart)
11	TEST_11	1	150	$\frac{1}{3}$	individual and moving s chart (individual and moving R chart)
12	TEST_12	2	500	5	correct indication of out-of-control conditions, $\frac{\bar{x}}{R}$ chart
13	TEST_13	3	500	5	$\frac{\bar{x}}{R}$ chart, capability indices for normal distribution
14	TEST_14	3	50	50	measurement capability type 1 study based on tolerance
15	TEST_15	2	50	50	measurement capability type 1 study based on process performance
16	TEST_16	2	60	10	measurement capability type 2 study
17	*TEST_17*	*3*	*50*	*2*	*measurement capability type 3 study (standard deviation)*
18	*TEST_18*	*4*	*50*	*2*	*measurement capability type 3 study (range)*

Test 1

Purpose

This set of test data checks whether, in connection with the use of $\bar{x}/s$ control charts, all out-of-control conditions are indicated in conformance with Ford requirements.

Procedure

Input of 500 measurements in subgroups of five to $\bar{x}/s$ control chart with known control limits calculated based on previous process performance.

$$\text{Specification} = 20 \pm 0.04 \text{ mm}$$

$\bar{x}$ chart			s chart		
UCL	=	20.01646 mm	UCL	=	0.017452 mm
LCL	=	19.99261 mm			

Capability indices based on previous performance when this process was operating under stable conditions:

$$\mathbf{C_p = 1.50} \qquad \mathbf{C_{pk} = 1.33}$$

Calculation using the mixed distribution results based on the percentile method in the capability indices $T_p = 0.8$ and $T_{pk} = 1.33$, which are termed T_p or T_{pk} for better identification because of instability.

Results

Histogram with all input data in comparison to specification to indicate number of parts and percentage out of tolerance.

$$\bar{x} = 20.00453 \text{ mm}$$

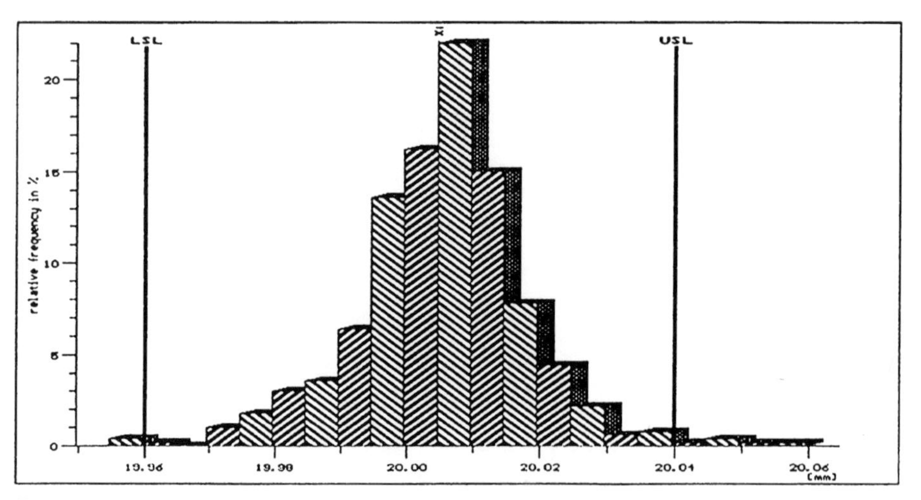

Class interval: 0.005 mm; Class with highest frequency: 20.0045 mm–20.0095 mm.

Displayed histogram of all data in comparison to specification indicates seven measuring results (1.4 percent) out of tolerance (five above = one percent; two below = 0.4 percent).

$\bar{x}$ control chart indicating all out-of-control conditions

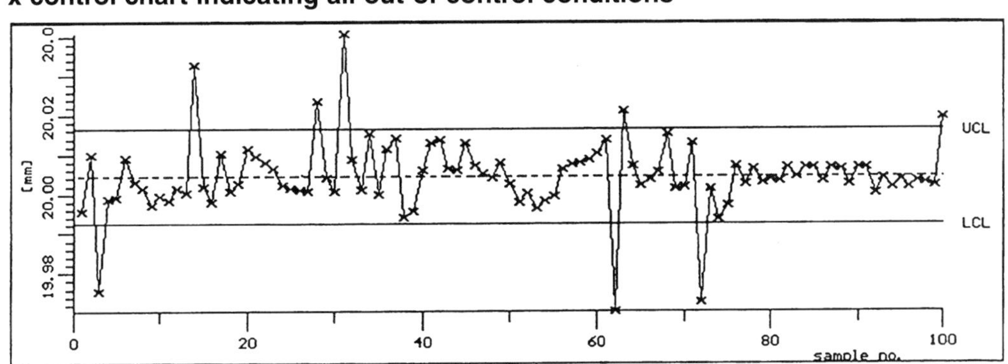

Subgroup	Result/violation	Subgroup	Result/violation
1	$\bar{x}_1 = 19.9958$ mm	32 to 56	middle third 40%
3	violation of LCL	53 to 61	run upward
7 to 13	run below centerline	62	violation of LCL
14	violation of UCL	63	violation of UCL
1 to 25	middle third 40%	72	violation of LCL
20 to 27	run downward	51 to 75	middle third 40%
28	violation of UCL	75 to 99	middle third 92%
31	violation of UCL	92 to 99	run below centerline
40 to 47	run above centerline	76 to 100	middle third 92%
30 to 54	middle third 40%	100	violation of UCL
31 to 55	middle third 36%		$\bar{x}_{100} = 20.0196$ mm

s control chart indicating all out-of-control conditions

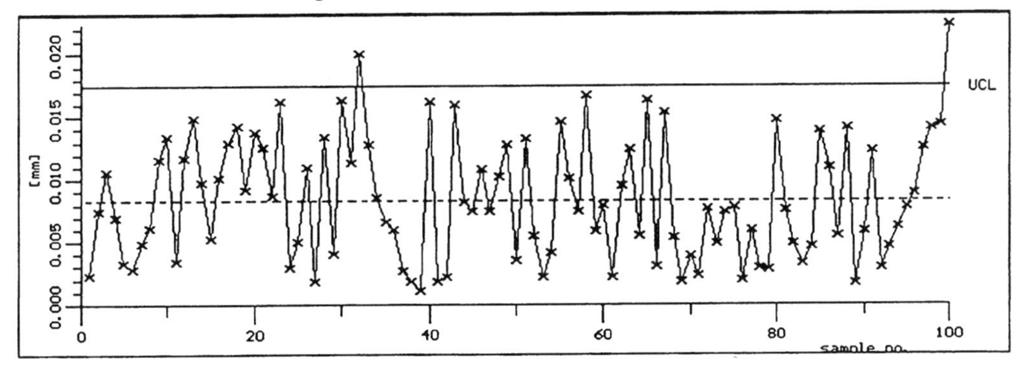

Subgroup	Result/violation	Subgroup	Result/violation
1	$s_1 = 0.00228$ mm	68 to 79	run below centerline
16 to 23	run above centerline	92 to 100	run upward
32	violation of UCL	100	violation of UCL
32 to 39	run downward		$s_{100} = 0.02230$ mm

As indicated by the control chart, this process is out of control. If capability indices based on $\frac{\bar{x}}{s}$ control chart data are calculated, even though the process is not stable, it would result in meaningless numbers. How misleading those numbers are is indicated by the confirmation of the previously valid values of 1.50 and 1.33. Should those results still be acknowledged as capability indices, the conclusion will be drawn that the process is still operating well within specification. But that is totally incorrect as proven by 1.4 percent parts out of tolerance.

Test 2

Purpose

Main purpose of this set of normally distributed test data is to verify accuracy of calculation.

Procedure

Input of 100 measurements in subgroups of five from the start of a new production under normal production conditions.

Specification = 14.070 + 0.005/–0.010 USL = 14.075 mm
LSL = 14.060 mm

Results

$\bar{x}$ of the first subgroup of 5 = 14.068620 mm
s of the first subgroup of 5 = 0.0008167 mm

$\bar{x}$ of all 20 subgroups = 14.067923 mm
$\hat{\sigma}$ calculated from $\bar{s}$ of the chart = 0.0012414 mm

$\bar{x}$ chart			s chart		
UCL	=	14.069589 mm	UCL	=	0.0024378 mm
LCL	=	14.066257 mm			

Histogram with all input data in comparison to specification indicates all parts are within specification.

Capability indices:

$P_p = 2.01$ $P_{pk} = 1.90$

Within this example, capability indices will result in the same numerical values regardless of whether calculation is based on control chart data or total population.

$\overline{x}$/s - control chart indicating process performance

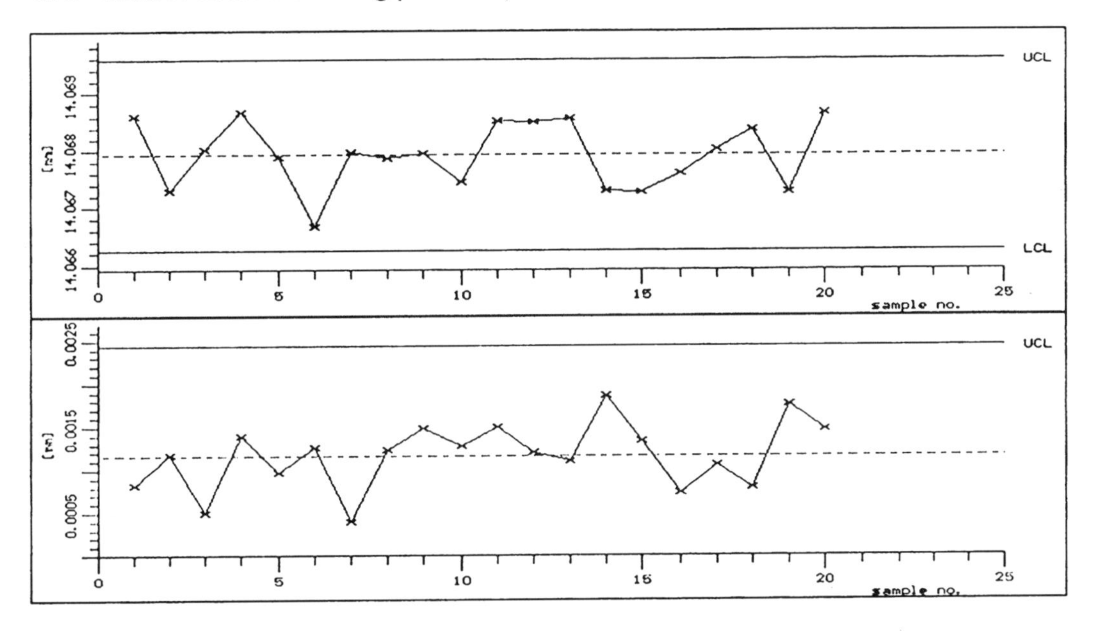

Histogram with all input data in comparison to specification

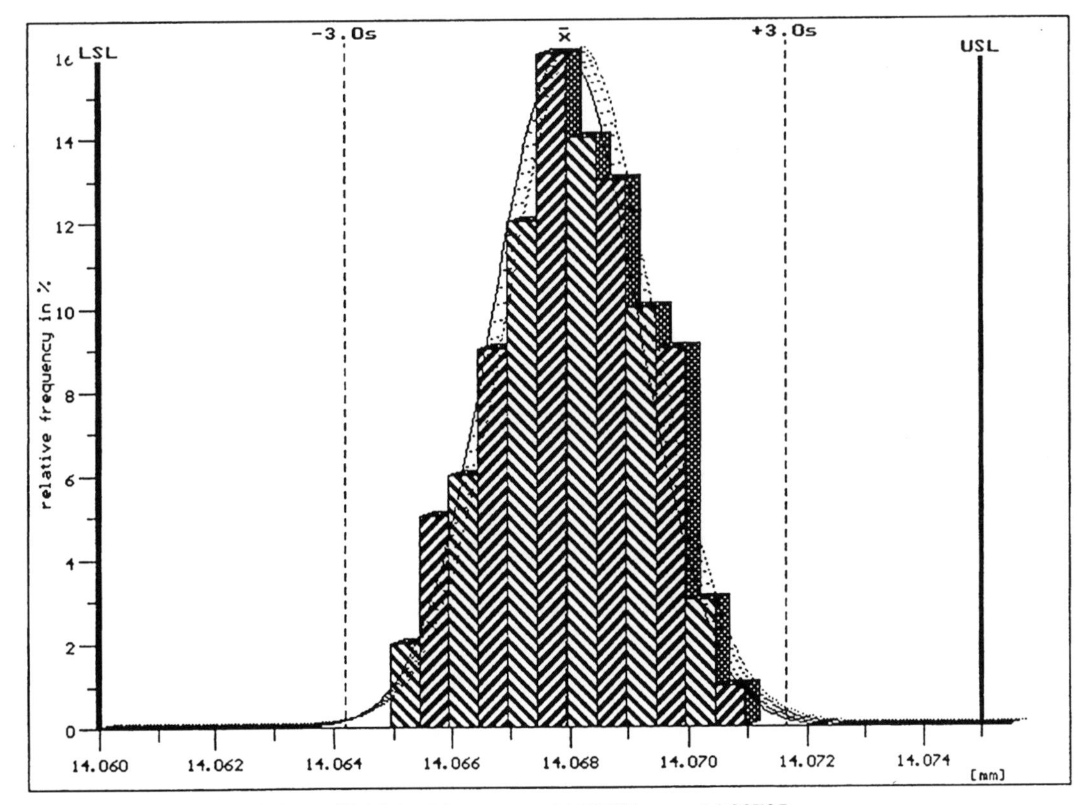

class interval: 0.005mm / class with highest frequency: 14.06745 mm - 14.06795 mm

Test 3

Purpose

This set of test data checks whether calculations required in connection with the use of $\bar{x}/s$ control charts are performed correctly and capability indices are calculated in conformance with Ford requirements. Data supplied fit best to a normal distribution and serve the purpose of checking this calculation mode.

Procedure

Input of 875 measurements in subgroups of seven representing a normal production run over more than 20 production days.

Specification = $130.0 +{}^{0.25}_{-0.10}$ USL = 130.25 mm
LSL = 129.90 mm

Results

$\bar{x}$ of the first subgroup of 7 = 130.0214 mm
s of the first subgroup of 7 = 0.03485 mm

$\bar{x}$ of all 125 subgroups = 130.0392 mm
$\hat{\sigma}$ calculated from $\bar{s}$ of the chart = 0.03260 mm

$\bar{x}$ chart			s chart		
UCL	=	130.0761 mm	UCL	=	0.05884 mm
LCL	=	130.0022 mm	LCL	=	0.00369 mm

Stability assessment: Stability confirmed.
Borderline out-of-control conditions that follow should not be indicated as out-of-control.

$\bar{x}$ chart		s chart	
subgroup	**results**	**subgroup**	**results**
71–77	6 intervals increasing	8–13	6 points below centerline
99–104	6 points below centerline	90–95	6 points above centerline
110–116	6 intervals decreasing	98–104	6 intervals decreasing

Histogram with all input data in comparison to specification indicates all parts are within specification.
Normal distribution confirmed by straight-line comparison on probability plot displayed.
Capability indices:

$C_p = 1.79$ $C_{pk} = 1.42$

Within this example, capability indices will result in the same numerical values regardless of whether calculation is based on control chart data or total population.

$\overline{x}$/s - control chart indicating process performance

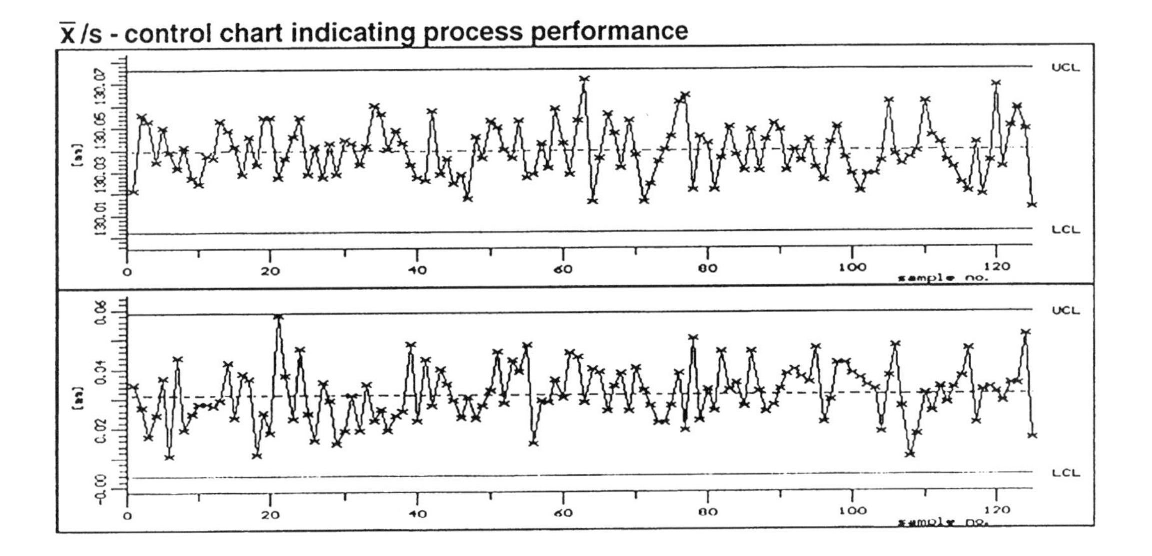

Histogram with all input data in comparison to specification

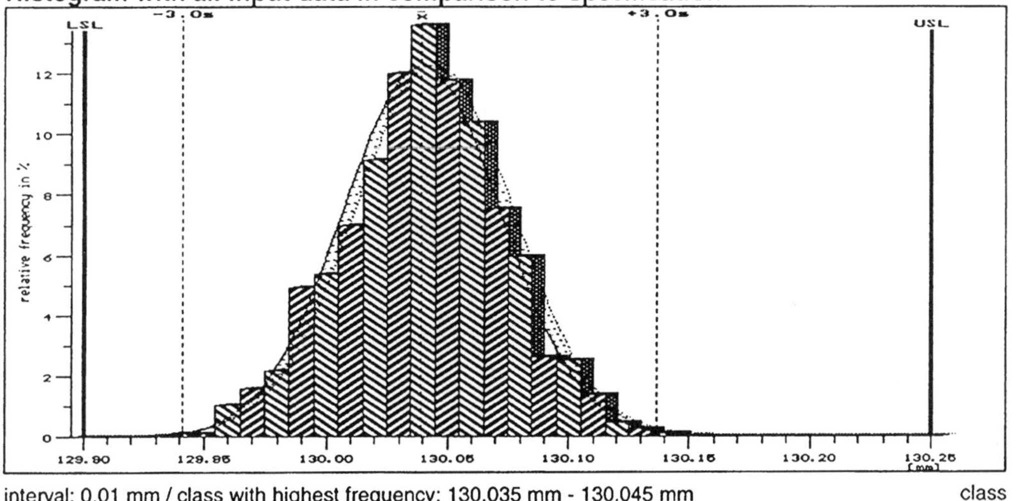

interval: 0.01 mm / class with highest frequency: 130.035 mm - 130.045 mm

class

Straight line comparison on probability plot for normal distribution

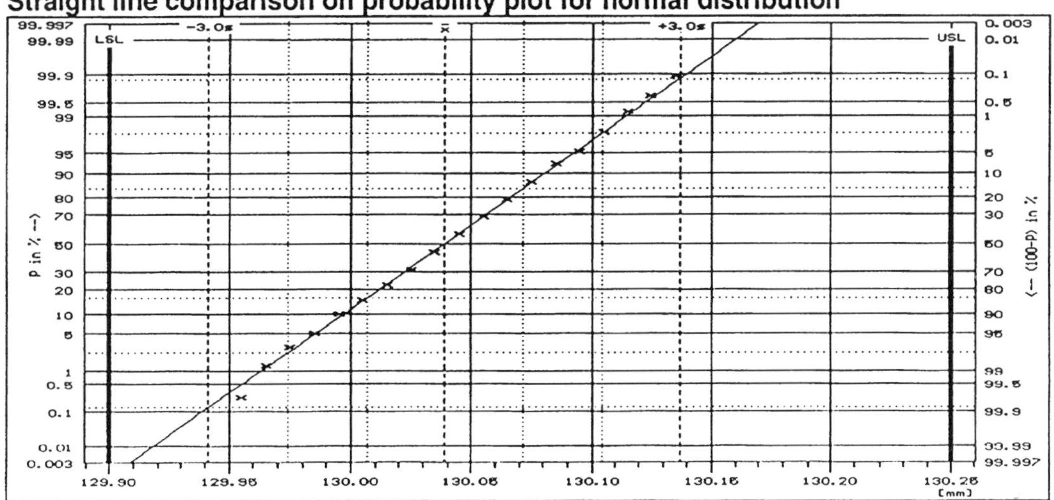

Test 4

Purpose

This set of test data checks whether calculations required in connection with the use of $\bar{x}/s$ control charts are performed correctly and capability indices are calculated in conformance with Ford requirements. Data supplied fit best to a lognormal distribution and serve the purpose of checking this calculation mode.

Procedure

Input of 200 measurements in subgroups of five from the start of a new production under normal production conditions.

Specification = max. 5.0 USL = 5.0 mm
LSL = none

Results

$\bar{x}$ of all 40 subgroups = 0.504 mm
$\hat{\sigma}$ calculated from $\bar{s}$ of the chart = 0.3530 mm

$\bar{x}$ chart			s chart		
UCL	=	0.978 mm	UCL	=	0.6933 mm
LCL	=	0.030 mm			

Stability assessment: Stability confirmed.

Histogram with all input data in comparison to specification indicates all parts are within specification.

Lognormal distribution confirmed by straight-line comparison on probability plot displayed.

Capability indices based on 200 measurements, percentile method, and lognormal distribution.

P_p not calculated as max. 5.0 is as unilateral tolerance, $\mathbf{P_{pk} = 1.72.}$

As, due to the nature of this process, results below zero are not possible, zero can be used as lower limit to calculate P_p. It makes no sense, however, to consider this bounded limit as *lower limit* for P_{pk} calculation.

When zero is regarded as lower specification limit, P_p can be calculated and will result in $\mathbf{P_p = 1.63.}$

When calculating capability indices for a lognormal distribution, a better fit may be found using an offset factor determined by regression analysis.

Results for this example with offset factor 0.03 are $\mathbf{P_p = 1.78}$ and $\mathbf{P_{pk} = 1.92.}$

It would be totally incorrect to calculate capability indices based on $\bar{\bar{x}}$ and $\bar{s}$ of control chart assuming a normal distribution. Because data will not fit a normal distribution, calculation based on control chart data only ($P_{pk} = 4.24$ for max. limit) will not correctly express process performance.

$\overline{x}$/s - control chart indicating process performance

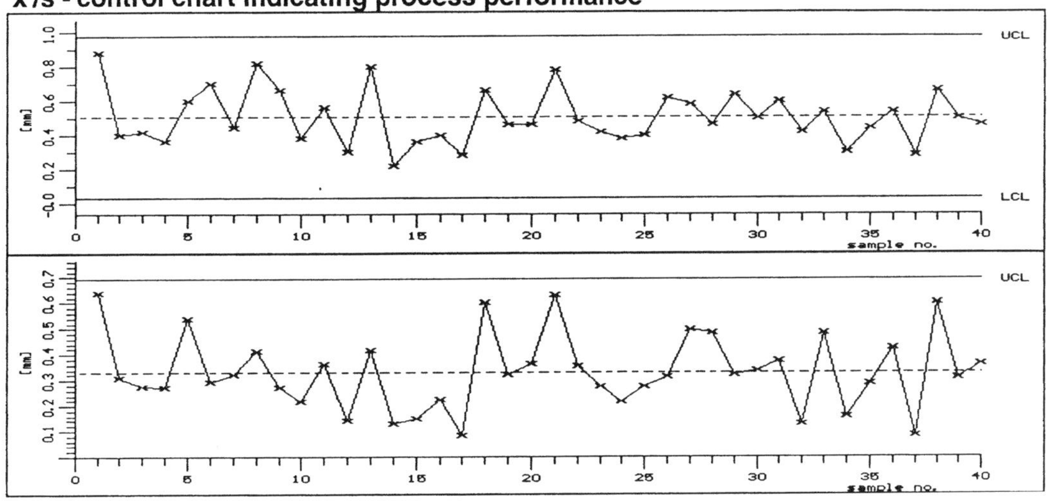

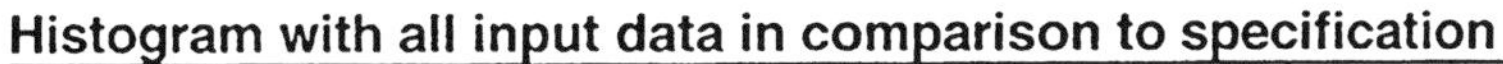

Histogram with all input data in comparison to specification

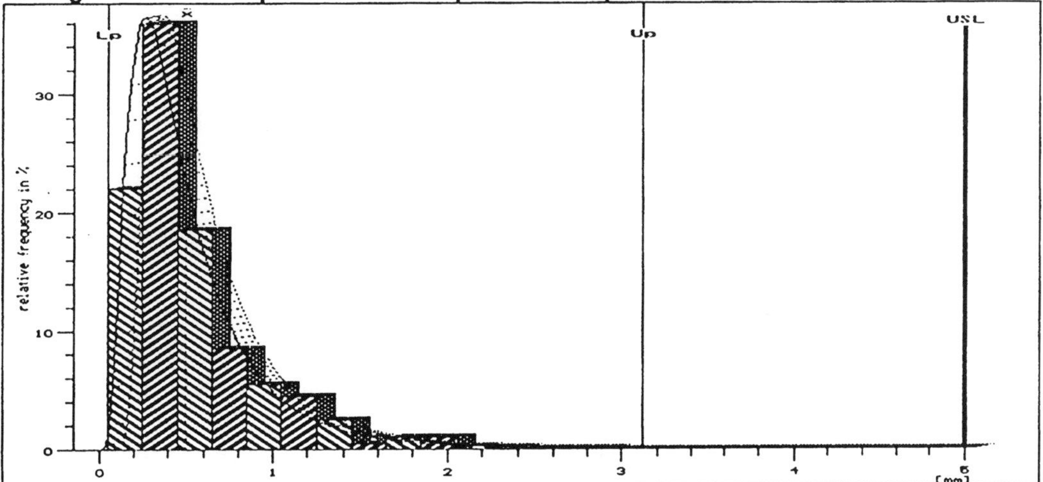

class interval: 0.2 mm / class with highest frequency: 0.25 mm - 0.45 mm

Probability plot for log. normal distribution (based on actual values)

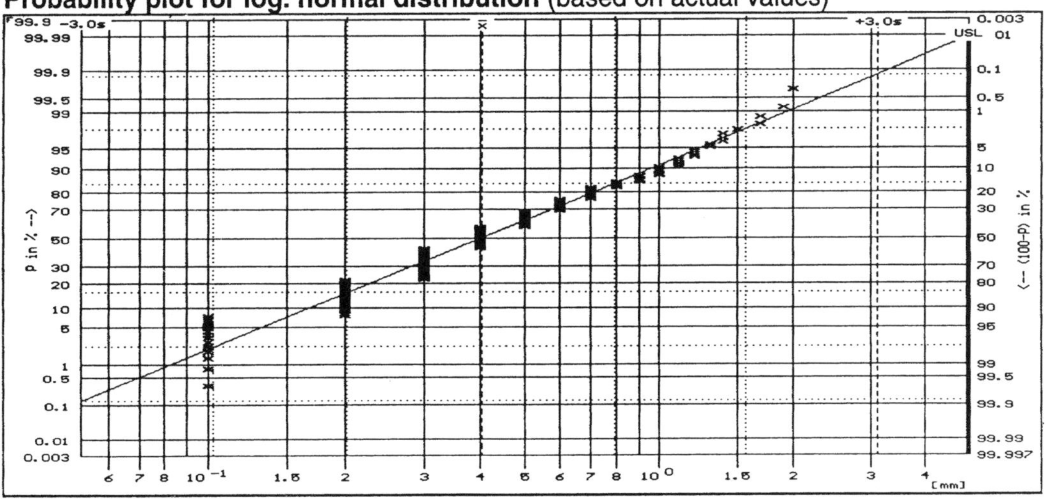

Test 5

Purpose

This set of test data checks whether calculations required in connection with the use of $\bar{x}/s$ control charts are performed correctly and capability indices are calculated in conformance with Ford requirements. Data supplied fit best to a Weibull distribution and serve the purpose of checking the calculation mode for Weibull distribution.

Procedure

Input of 200 measurements in subgroups of five from the start of a new production under normal production conditions.

Specification = $770\,^{+150}_{-270}$ USL = 920 Nm
LSL = 500 Nm

Results

$\bar{\bar{x}}$ of all 40 subgroups = 718.30 Nm
$\hat{\sigma}$ calculated from $\bar{s}$ of the chart = 61.212 Nm

$\bar{x}$ chart			s chart		
UCL	=	800.42 Nm	UCL	=	120.20 Nm
LCL	=	636.18 Nm			

Stability assessment: Stability confirmed.

Histogram with all input data in comparison to specification indicates all parts within specification.

Weibull distribution confirmed by straight-line comparison on probability plot displayed.

Capability indices based on 200 measurements, percentile method, and Weibull distribution:

$\mathbf{P_p = 1.14}$ $\mathbf{P_{pk} = 0.90}$

It should be noted, however, that different methods for evaluation of Weibull distribution are available and the application of different regression models will slightly influence capability indices.

Maximum likelihood for parameter α and β:

without offset $\mathbf{P_p = 1.14}$ $\mathbf{P_{pk} = 0.90}$
with offset* $\mathbf{P_p = 1.09}$ $\mathbf{P_{pk} = 0.84}$

Regression analysis:

without offset $\mathbf{P_p = 1.08}$ $\mathbf{P_{pk} = 0.86}$
with offset* $\mathbf{P_p = 1.05}$ $\mathbf{P_{pk} = 0.81}$

**Offset for this example is 450.*

$\overline{X}$/s - control chart indicating process performance

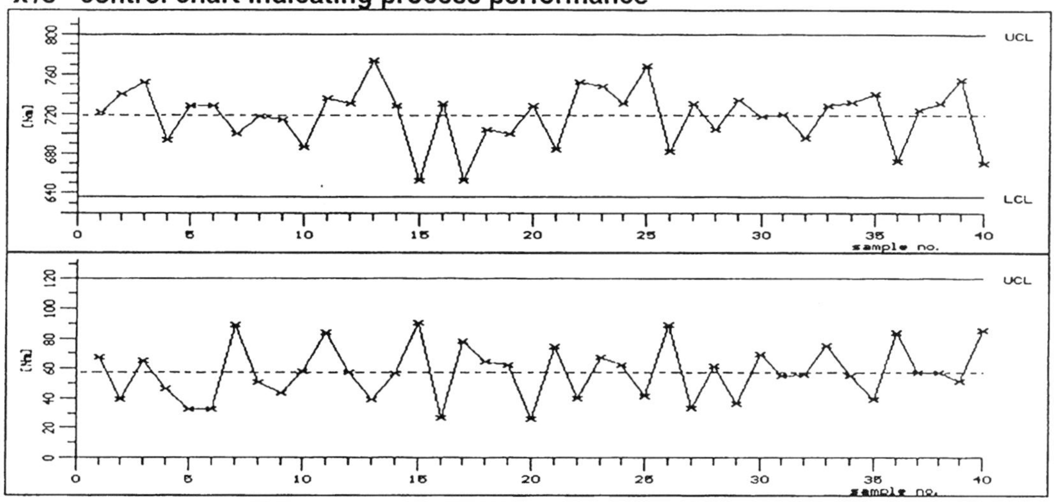

Histogram with all input data in comparison to specification

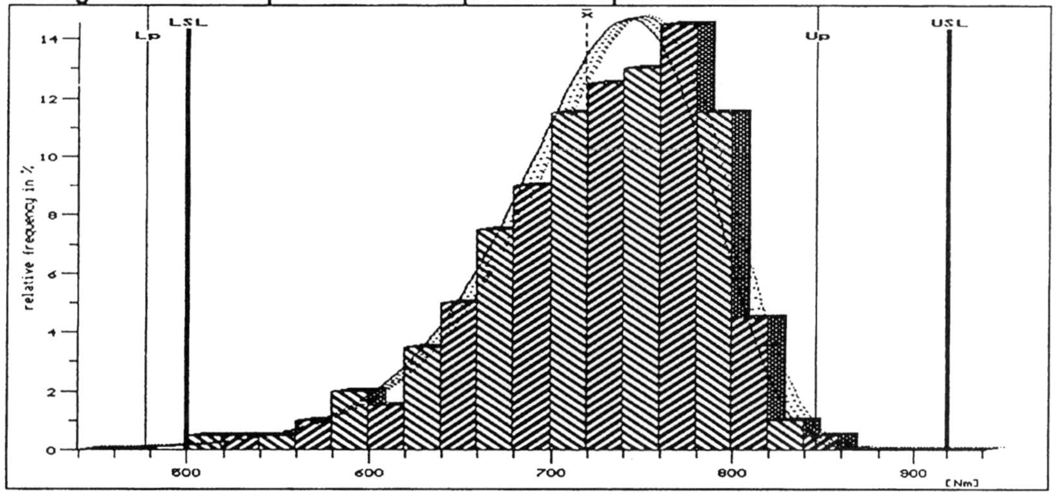

class interval: 20 Nm / class with the highest frequency: 759.5 Nm - 779.5 Nm

Straight line comparison on probability plot for Weibull distribution

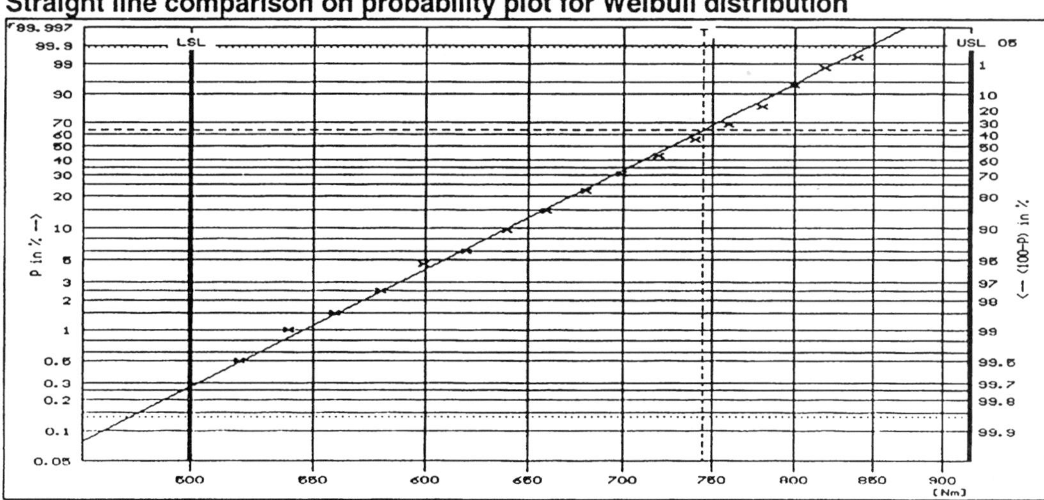

Test 6

Purpose

This set of test data checks whether calculations required in connection with the use of $\bar{x}$/s control charts are performed correctly and capability indices are calculated in conformance with Ford requirements. Data supplied fit best to a Rayleigh distribution and serve the purpose of checking this calculation mode.

Procedure

Input of 500 measurements in subgroups of five representing a normal production run over more than 20 production days.

Specification = max. 0.10 USL = 0.10 mm
LSL = none

Results

$\bar{\bar{x}}$ of all 40 subgroups = 0.02527 mm
$\hat{\sigma}$ calculated from $\bar{s}$ of the chart = 0.013510 mm

$\bar{x}$ chart			s chart		
UCL LCL	= =	0.04340 mm 0.00715 mm	UCL	=	0.026529 mm

Stability assessment: Stability confirmed.

Histogram with all input data in comparison to specification indicates all parts within specification.

Rayleigh distribution confirmed by straight-line comparison on probability plot displayed.

Capability indices based on 500 measurements, percentile method, and Rayleigh distribution.

C_p = not calculated as max. 0.10 is a unilateral tolerance. $\mathbf{C_{pk} = 1.56}$

As, due to the nature of this process, results below zero are not possible, zero can be used as lower limit to calculate C_p. It makes no sense, however, to consider this bounded limit as *lower limit* for C_{pk} calculation.

When zero is regarded as lower specification limit, C_p can be calculated and will result in

$$\mathbf{C_p = 1.38}$$

When capability indices for this Rayleigh distribution are calculated using an offset factor of 0.0008 found by regression, results are as follows:

$\mathbf{C_p = 1.34}$ $\mathbf{C_{pk} = 1.51}$

Capability indices calculated assuming other types of distribution:

Folded normal distribution $C_{pk} = 0.98$
Folded normal distribution with offset (–0.001) $C_{pk} = 1.02$
Lognormal distribution $C_{pk} = 0.56$
Lognormal distribution offset (0.031) $C_{pk} = 1.34$
Normal distribution $C_{pk} = 1.84$

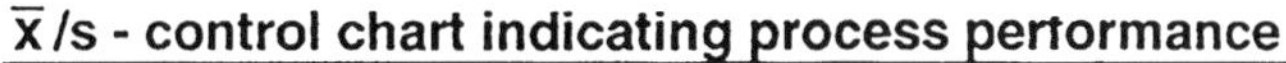

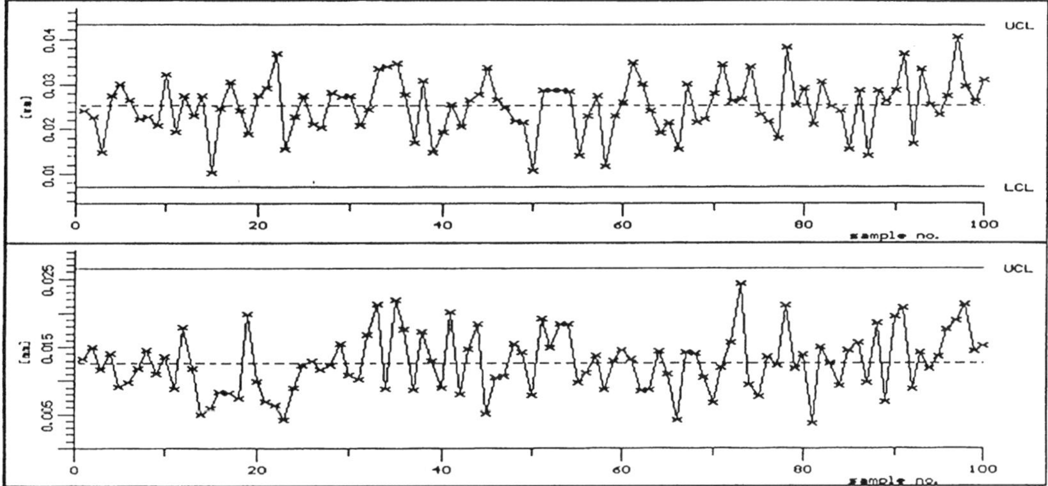

Histogram with all input data in comparison to specification

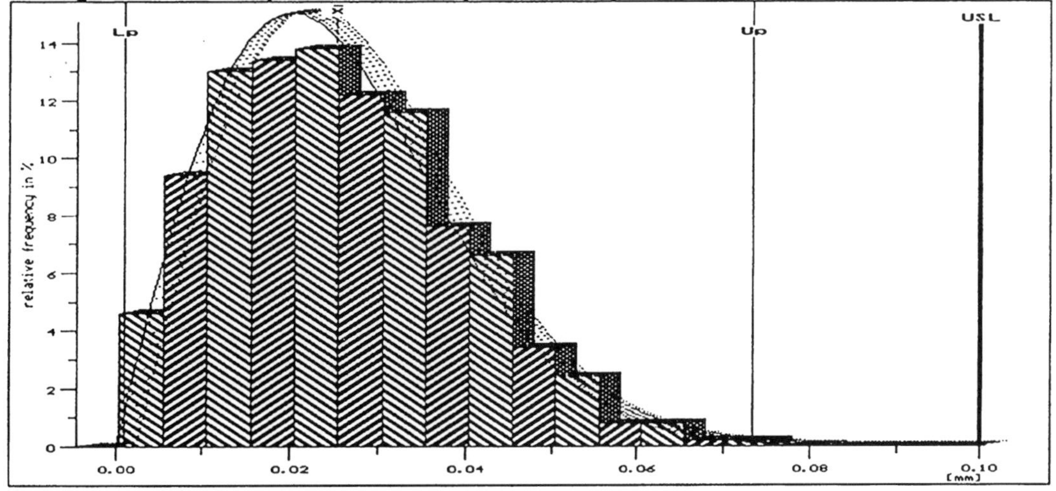

class interval: 0.005 mm / class with the highest frequency: 0.0205 mm - 0.0255 mm

Straight line comparison on probability plot for Rayleigh Distribution

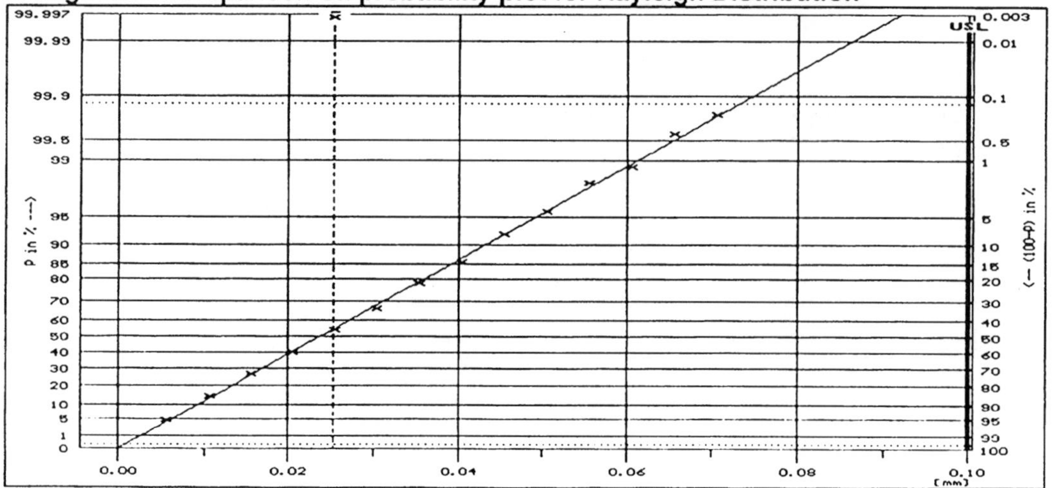

Test 7

Purpose

This set of test data checks whether calculations required in connection with the use of $\bar{x}/s$ control charts are performed correctly and capability indices are calculated in conformance with Ford requirements. Data supplied fit best to a folded normal distribution and serve the purpose of checking the calculation mode for a folded normal distribution.

Procedure

Input of 600 measurements in subgroups of three representing a normal production run over more than 20 production days.

Specification = max. 0.04 USL = 0.04 mm
LSL = none

Results

$\bar{x}$ of all 200 subgroups = 0.00829 mm
$\hat{\sigma}$ calculated from s of the chart = 0.004871 mm

$\bar{x}$ chart			s chart		
UCL	=	0.01672 mm	UCL	=	0.011083 mm
LCL	=	–0.00015 mm*			

**As actual $\bar{x}$ values below zero are not possible, calculated LCL should be ignored.*

Stability assessment: Stability confirmed.

Histogram with all input data in comparison to specification indicates all parts within specification.

Folded normal distribution confirmed by straight-line comparison on probability plot displayed.

Capability indices based on 600 measurements, percentile method, and folded normal distribution.

C_p = not calculated as max. 0.04 mm is a unilateral tolerance. $\mathbf{C_{pk} = 1.67}$

As, due to the nature of this process, results below zero are not possible, zero can be used as lower limit to calculate C_p. It makes no sense, however, to consider this bounded limit as *lower limit* for C_{pk} calculation.

When zero is regarded as lower specification limit, C_p can be calculated and would result in

$$\mathbf{C_p = 1.59}$$

The calculation of capability indices for a folded distribution should not be carried out without the use of an offset factor, since results obtained will not express process performance correctly. Calculating capability indices without offset factor for this example (offset factor is –0.002) would result in

$$C_p = 1.20 \text{ and } C_{pk} = 1.27$$

$\overline{x}$/s - control chart indicating process performance

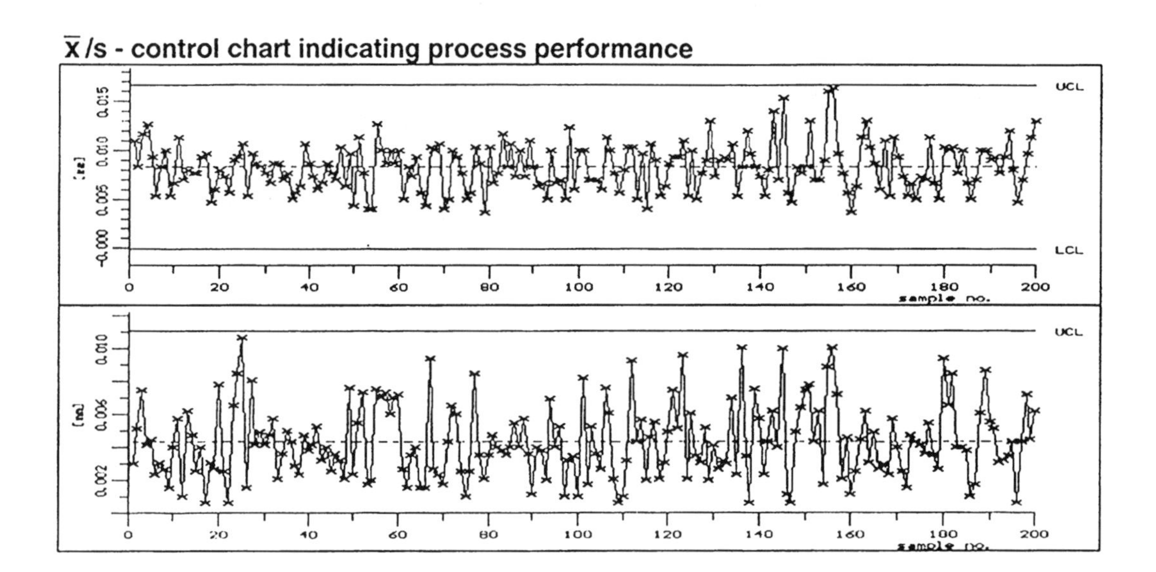

Histogram with all input data in comparison to specification

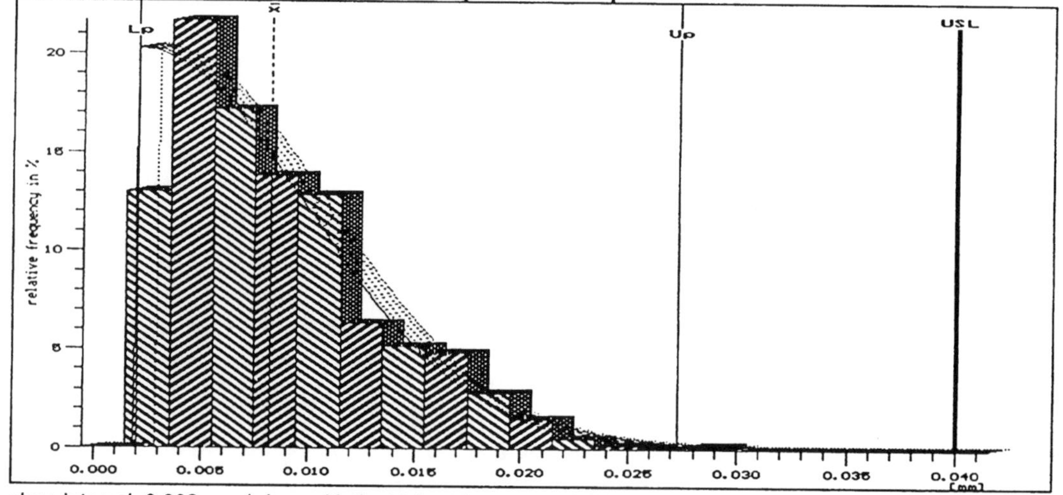

class interval: 0.002 mm / class with the highest frequency: 0.0035 mm - 0.0055 mm

Probability plot for folded normal distribution (based on actual values)

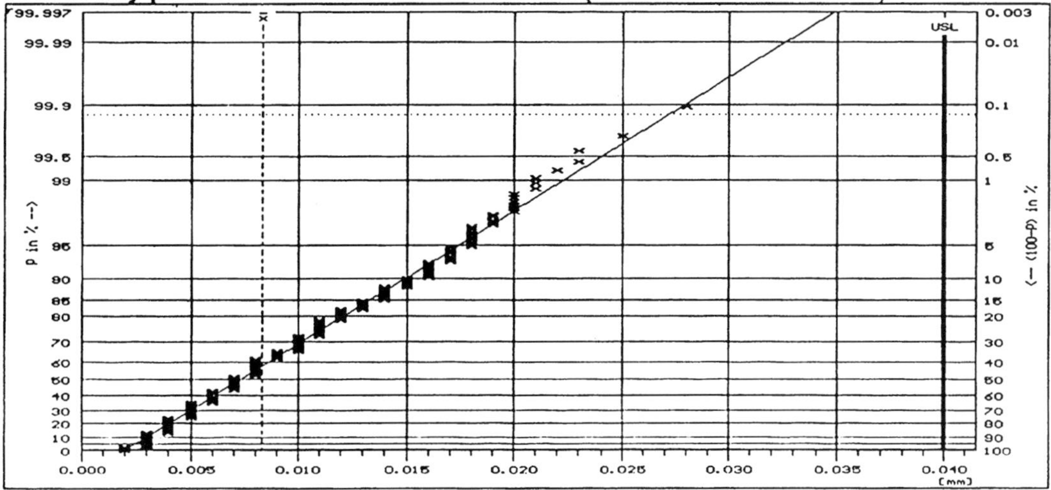

Test 8

Purpose

This set of test data checks whether calculations required in connection with the use of $\bar{x}/s$ control charts are performed correctly and capability indices are calculated in conformance with Ford requirements. Data supplied represent a fixed tooling situation and serve for purpose of checking this calculation mode.

Procedure

Input of 500 measurements in subgroups of five representing a normal production run over more than 20 production days.

Specification = 30.0 ± 0.13 USL = 30.130 mm
LSL = 29.870 mm

Control limits calculated for normal Ford $\bar{x}/s$ control chart:

UCL $\bar{x}$ chart = 30.0203 mm
LCL $\bar{x}$ chart = 29.9930 mm
UCL s chart = 0.01996 mm

Stability is confirmed for s but not confirmed for $\bar{x}$.

As confirmed by investigation, additional variation of $\bar{x}$ is caused by different tools. Based on the decision to accept this additional $\bar{x}$ variation as an inherent part of this process, it has to be treated as common cause variation. To acknowledge this fact, control limits have to be adjusted.

Accepted additional $\bar{x}$ variation = 0.11 mm

Results

$\bar{\bar{x}}$ of all 100 subgroups = 30.0067 mm
$\hat{\sigma}$ calculated from $\bar{s}$ of the chart = 0.01017 mm

Control limits for special situation control chart: When a control chart for special process situation is justified, only a point beyond control limit remains to be indicated as an out-of-control condition.

$\bar{x}$ chart			s chart		
UCL	=	30.0753 mm	UCL	=	0.01996 mm*
LCL	=	29.9381 mm			

**Control limit for the s chart is unchanged compared to the normal Ford $\bar{x}/s$ control chart.*

Capability indices:

$C_p = 1.52$ $C_{pk} = 1.44$

Histogram with all data in comparison to specification indicates all parts within specification.

Based on the mixed distribution, a capability index of $C_p = 1.58$ and $C_{pk} = 1.48$ results. The higher indices are caused by a better adaptation of the distribution model.

$\bar{x}$/s - control chart indicating process performance

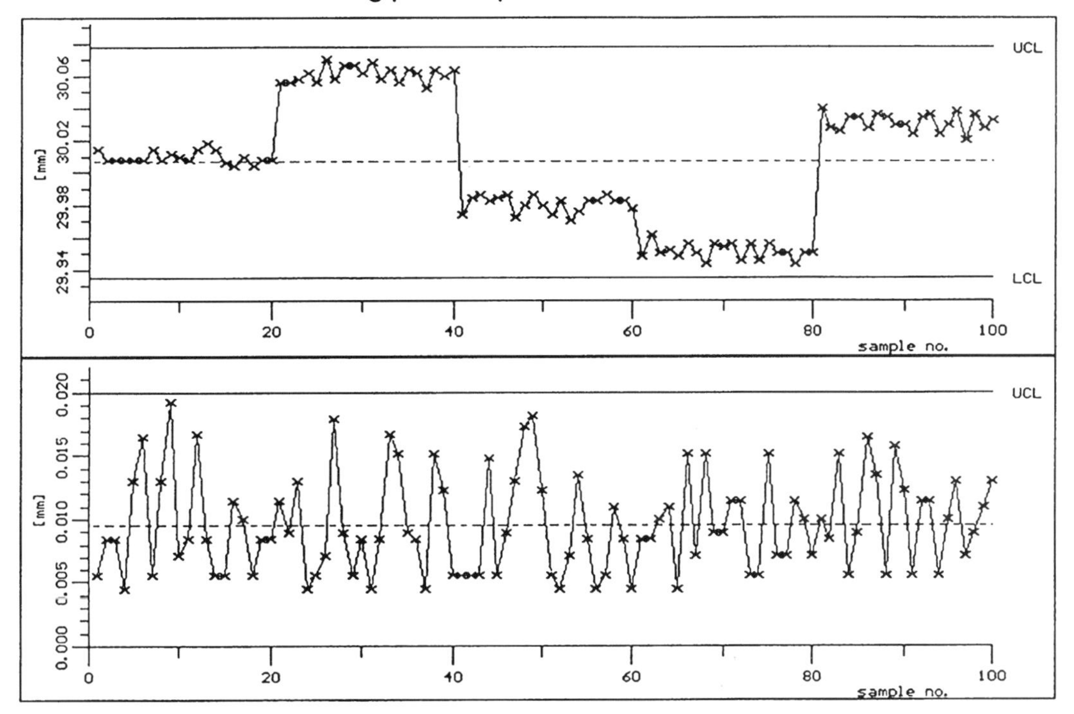

Histogram with all input data in comparison to specification

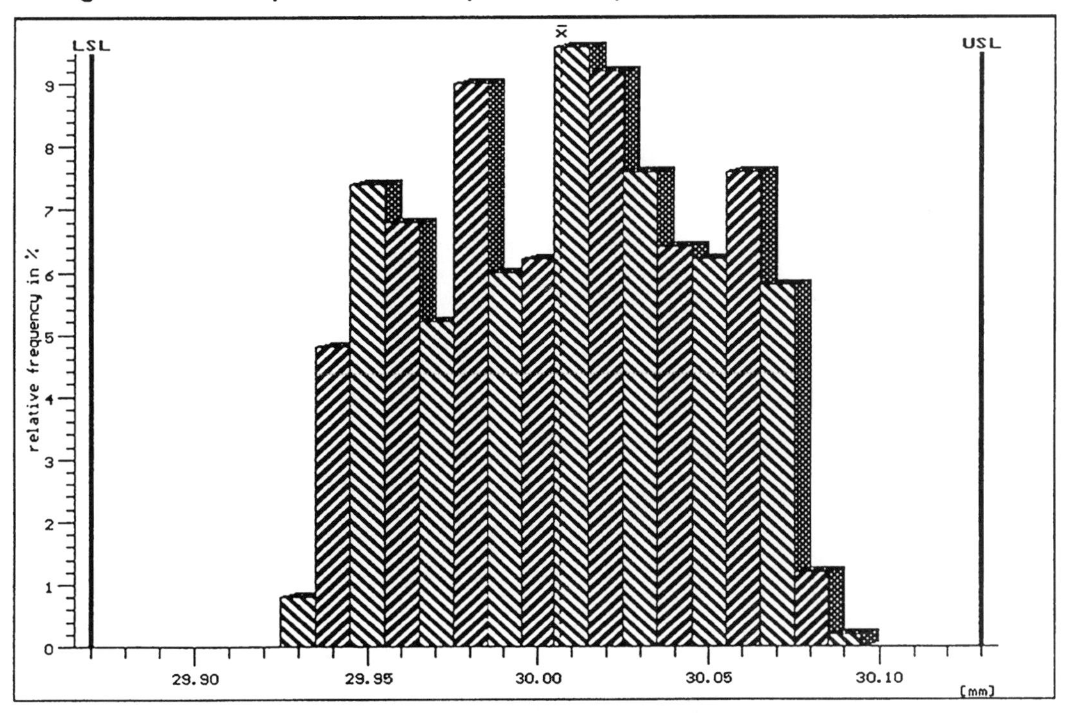

class interval: 0.01 mm / class with highest frequency: 30.005 mm - 30.015 mm

Test 9

Purpose

This set of test data checks whether calculations required in connection with the use of $\bar{x}/s$ control charts are performed correctly and capability indices are calculated in conformance with Ford requirements. Data supplied represent a tool wear situation and serve the purpose of checking this calculation mode.

Procedure

Input of 600 measurements in subgroups of six representing a normal production run over more than 20 production days.

Specification = 20.0 ± 0.30 USL = 20.300 mm
LSL = 19.700 mm

Control limits calculated for normal Ford $\bar{x}/s$ control chart:

UCL $\bar{x}$ chart = 20.0342 mm
LCL $\bar{x}$ chart = 19.9604 mm
UCL s chart = 0.05649 mm
LCL s chart = 0.00086 mm

Stability is confirmed for s but not confirmed for $\bar{x}$.

As confirmed by investigation, additional variation of $\bar{x}$ is caused by tool wear. Based on the decision to accept this additional $\bar{x}$ variation as an inherent part of this process, it has to be treated as common cause variation. To acknowledge this fact, control limits have to be adjusted.

Accepted additional $\bar{x}$ variation = 0.20 mm

Results

$\bar{\bar{x}}$ of all 100 subgroups = 19.9973 mm
$\hat{\sigma}$ calculated from $\bar{s}$ of the chart = 0.03012 mm

Control limits for special situation control chart: When a control chart for special process situation is justified, only a point beyond control limit remains to be indicated as an out-of-control condition.

$\bar{x}$ chart			s chart*		
UCL	=	20.1342 mm	UCL	=	0.05649 mm
LCL	=	19.8604 mm	LCL	=	0.00086 mm

**Control limits for s charts are unchanged compared to the normal Ford $\bar{x}/s$ control chart.*

Capability indices:

$C_p = 1.58$ $C_{pk} = 1.56$

Histogram with all data in comparison to specification indicates all parts within specification.

Based on the mixed distribution, a capability index of $C_p = 1.87$ and $C_{pk} = 1.69$ results. The higher indices are caused by a better adaptation of the distribution model.

$\bar{x}$/s - control chart indicating process performance

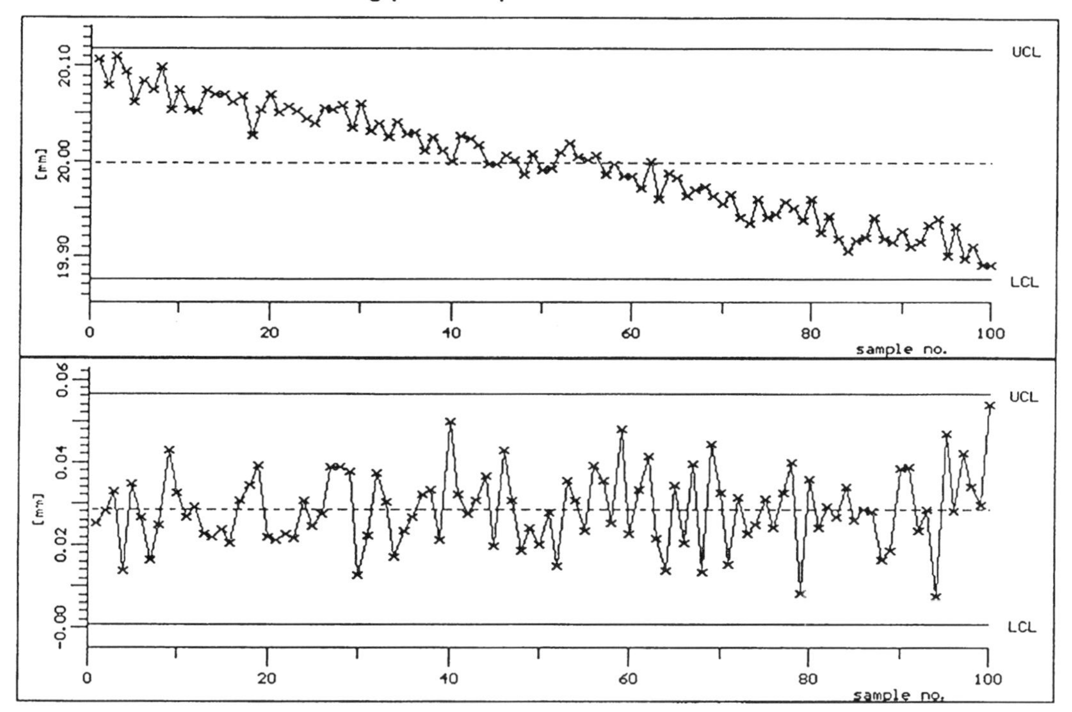

Histogram with all input data in comparison to specification

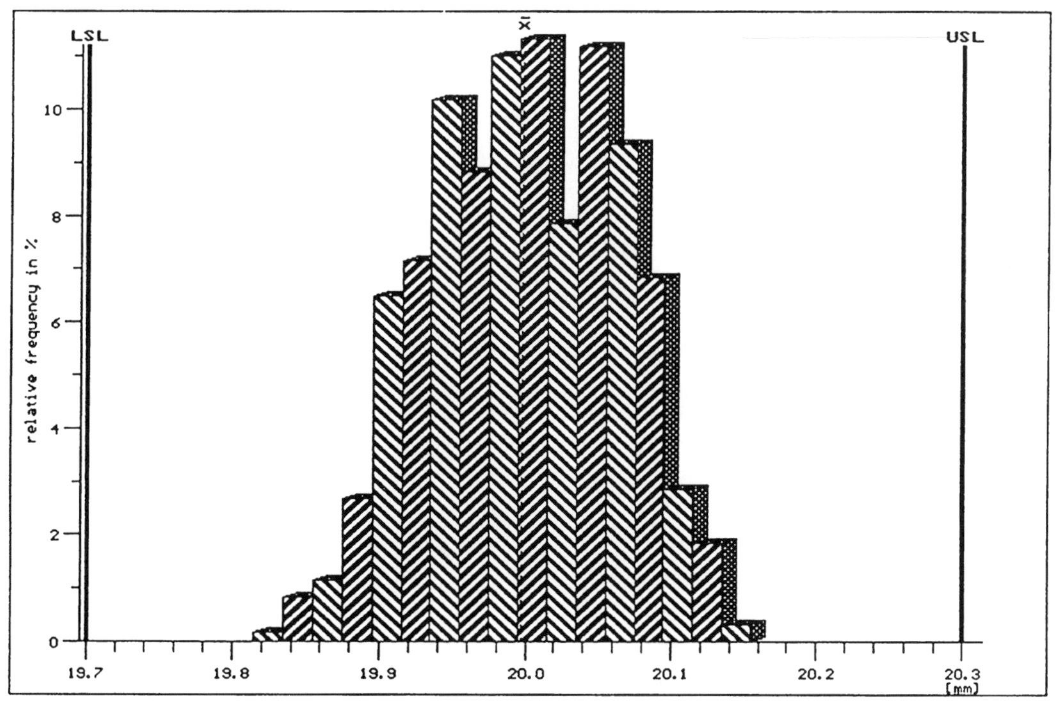

class interval: 0.02 mm / class with highest frequency: 19.995 mm - 20.015 mm

Test 10

Purpose

This set of normally distributed test data checks calculations required in connection with the use of moving average and moving s control charts are performed correctly and capability indices are calculated in conformance with Ford requirements.

Procedure

Input of 102 measurements to form moving subgroup samples of three for both $\bar{x}$ and s chart or $\bar{x}$ and R chart respectively. Data represent the start of a new production under normal production conditions.

Specification = 65 ± 5 USL = 70.00%
LSL = 60.00%

Results

$\bar{x}$ of first subgroup of 3 = 64.73% $\bar{x}$ of last subgroup of 3 = 64.13%
s of first subgroup of 3 = 0.777% s of last subgroup of 3 = 1.443%
R of first subgroup of 3 = 1.5% R of last subgroup of 3 = 2.5%

$\bar{\bar{x}}$ calculated from all subgroup $\bar{x}$ values = 64.921%
$\hat{\sigma}$ calculated from $\bar{s}$ of the control chart = 1.2350%
$\hat{\sigma}$ calculated from $\bar{R}$ of the control chart = 1.2274%

average chart			moving s chart		
UCL	=	67.060%	UCL	=	2.8099%
LCL	=	62.782%			

average chart			moving Range chart		
UCL	=	67.047%	UCL	=	5.3488%
LCL	=	62.795%			

Stability assessment: Stability confirmed.

Histogram of all data in comparison to specification indicates all parts within specification (not displayed).

Normal distribution confirmed by straight-line comparison on probability plot displayed.

Capability indices:

Based on $\bar{\bar{x}}$ and $\bar{s}$ of the control chart **P_p = 1.35** **P_{pk} = 1.33**
Based on $\bar{\bar{x}}$ and $\bar{R}$ of the control chart **P_p = 1.36** **P_{pk} = 1.34**
Based on total population **P_p = 1.50** **P_{pk} = 1.48**
(s = 1.10943, $\bar{x}$ = 64.916)

Moving average and moving s chart indicating process performance

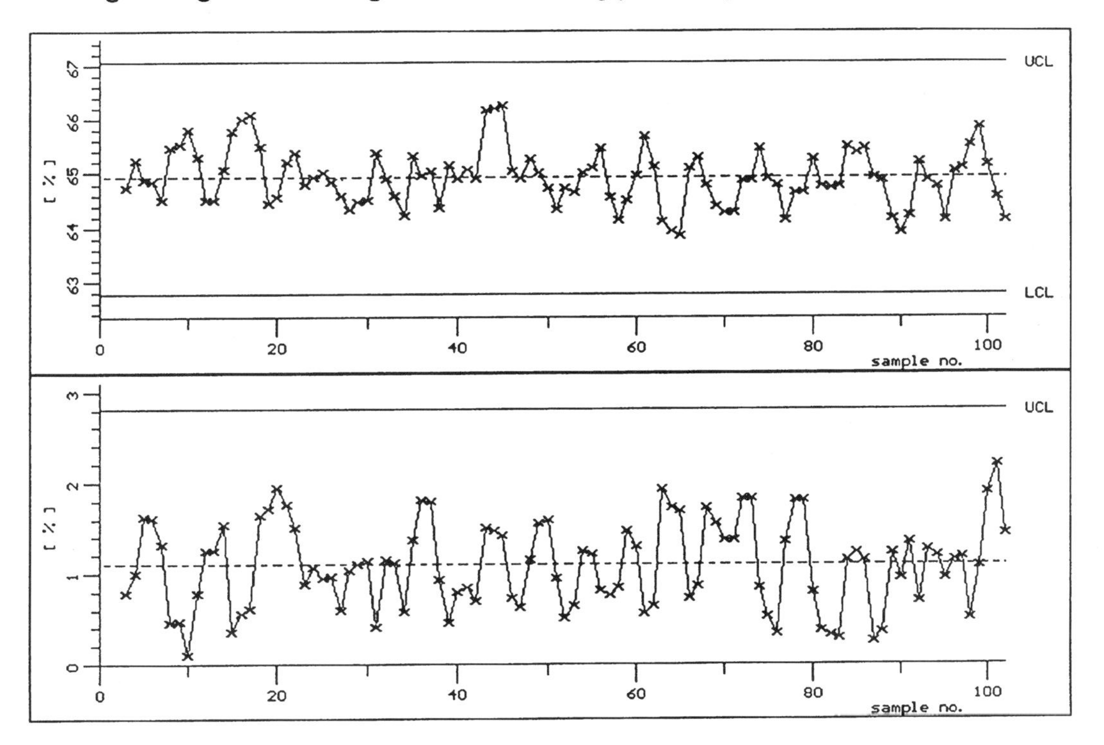

Straight line comparison on probability plot for normal distribution

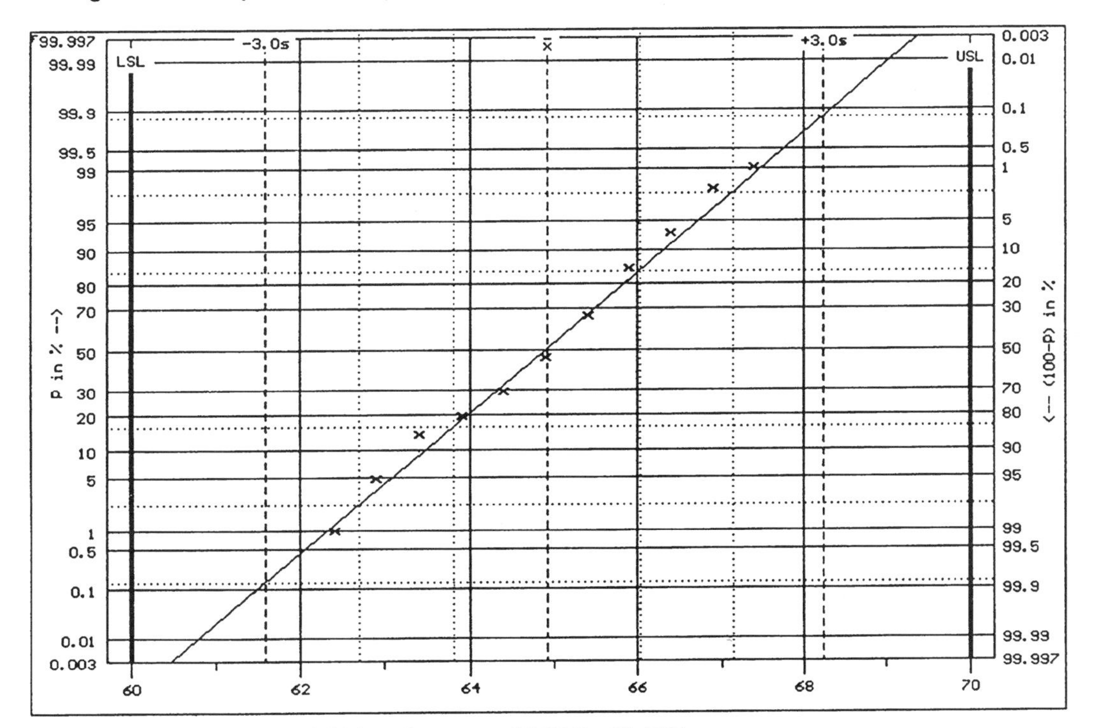

class interval: 0.5% / class with highest frequency: 64.95 % - 65.45 %

Test 11

Purpose

This set of normally distributed test data checks whether calculations required in connection with the use of individual and moving s control charts are performed correctly and capability indices are calculated in conformance with Ford requirements.

Procedure

Input of 150 measurements as individual results for $\bar{x}$ chart and to form moving subgroup samples of three for s chart or R chart respectively. Data represent a normal production run over more than 20 production days.

Specification = $5 + {}^{2}_{-3}$ USL = 7.00%
LSL = 2.00%

Results

$\bar{x}$ of all 150 individual x values = 4.491%
s of the first subgroup of 3 = 0.400%
R of the first subgroup of 3 = 0.8%
s of the first subgroup of 3 = 0.586%
R of the first subgroup of 3 = 1.1%
$\hat{\sigma}$ calculated from $\bar{s}$ of the control chart = 0.548%
$\hat{\sigma}$ calculated from $\bar{R}$ of the chart = 0.546%

individual chart			moving s chart		
UCL	=	6.137%	UCL	=	1.2479%
LCL	=	2.846%			

individual chart			moving Range chart		
UCL	=	6.128%	UCL	=	2.3775%
LCL	=	2.855%			

Stability assessment: Stability confirmed.

Histogram of all data in comparison to specification indicates all parts within specification (not displayed).

Normal distribution confirmed by straight-line comparison on probability plot displayed.

Capability indices:

Based on $\bar{x}$ and $\bar{s}$ of control chart	C_p = **1.52**	C_{pk} = **1.51**
Based on $\bar{x}$ and $\bar{R}$ of control chart	C_p = **1.53**	C_{pk} = **1.52**
Based on total population (s = 0.50805, $\bar{x}$ = 4.491)	C_p = **1.64**	C_{pk} = **1.63**

Individual and moving s chart indicating process performance

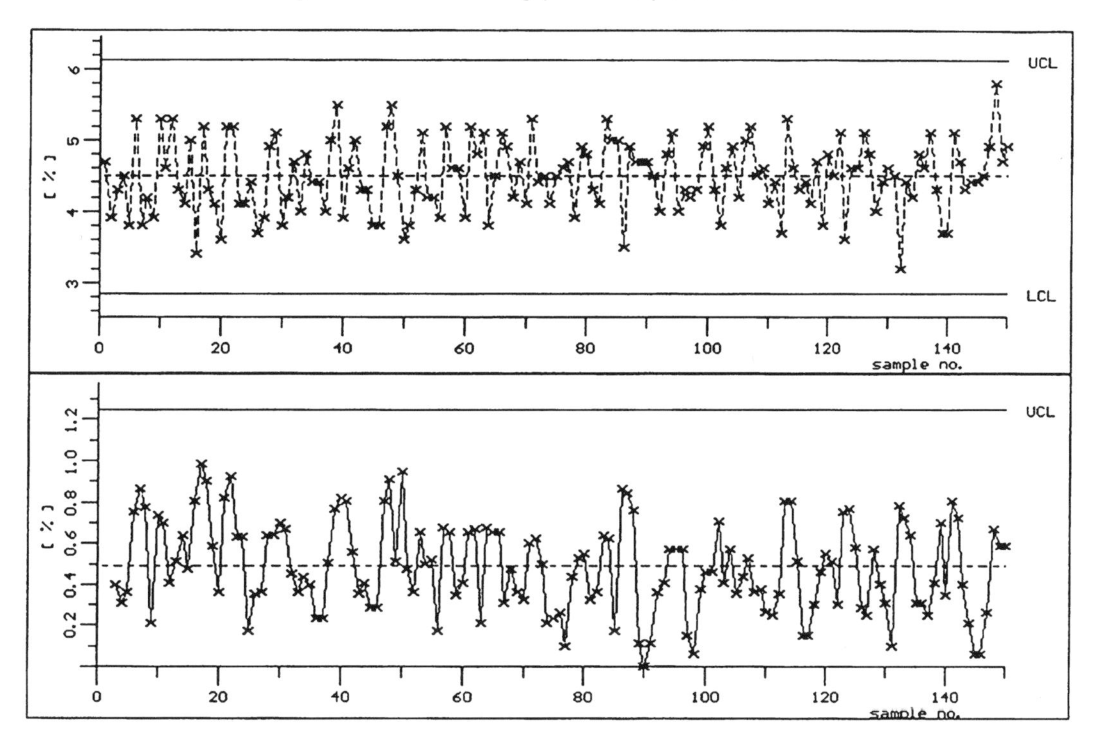

Straight line comparison on probability plot for normal distribution

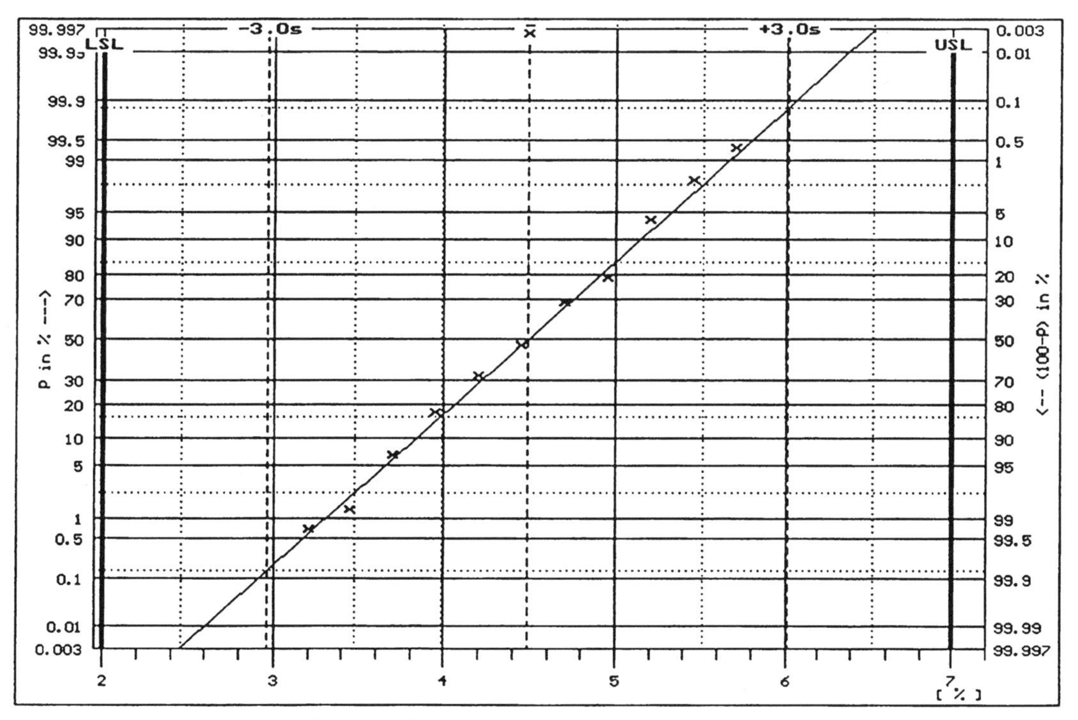

class interval: 0.25% / class with highest frequency: 4.45 % - 4.70 %

Test 12

Purpose

This set of test data checks whether, in connection with the use of $\bar{x}$/R control charts, all out-of-control conditions are indicated in conformance with Ford requirements.

Procedure

Input of 500 measurements in subgroups of five to $\bar{x}$/R control chart with known control limits calculated based on previous process performance.

Specification = 26.5 ± 0.5 USL = 27.0 mm
LSL = 26.0 mm

$\bar{x}$ chart			R chart		
UCL	=	26.6386 mm	UCL	=	0.51138 mm
LCL	=	26.3595 mm			

Capability indices based on previous performance when this process was operating under stable conditions:

$$\mathbf{C_p = 1.60 \qquad C_{pk} = 1.60}$$

Calculation using the mixed distribution results based on the percentile method in the capability indices $T_p = 0.86$ and $T_{pk} = 0.85$, which are termed T_p or T_{pk} for better identification because of instability.

Results

Histogram with all input data in comparison to specification to indicate number of parts and percentage out of tolerance.

$$\bar{x} = 26.4991 \text{ mm}$$

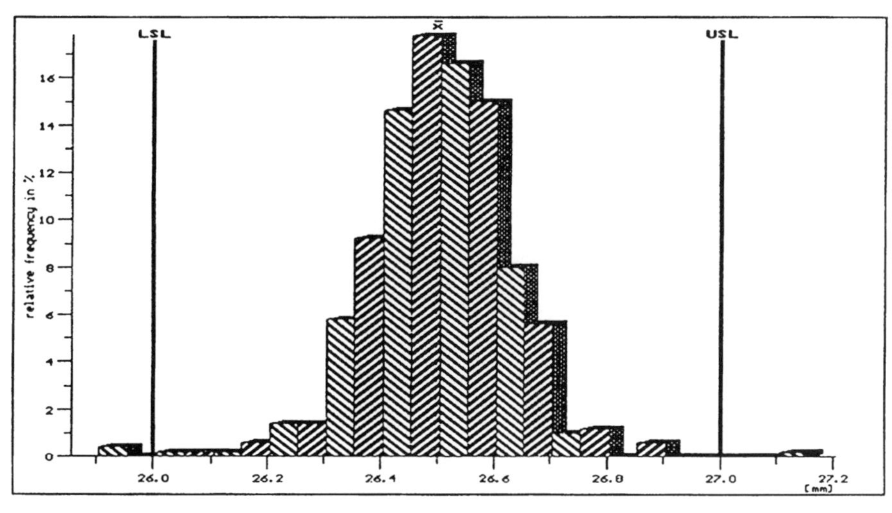

class interval: 0.05 mm / class with highest frequency: 26.455 mm - 26.505mm

Displayed histogram of all data in comparison to specification indicates three measuring results (0.6 percent) out of tolerance (one above = 0.2 percent; two below = 0.4 percent).

$\bar{x}$ - control chart indicating all out of control conditions

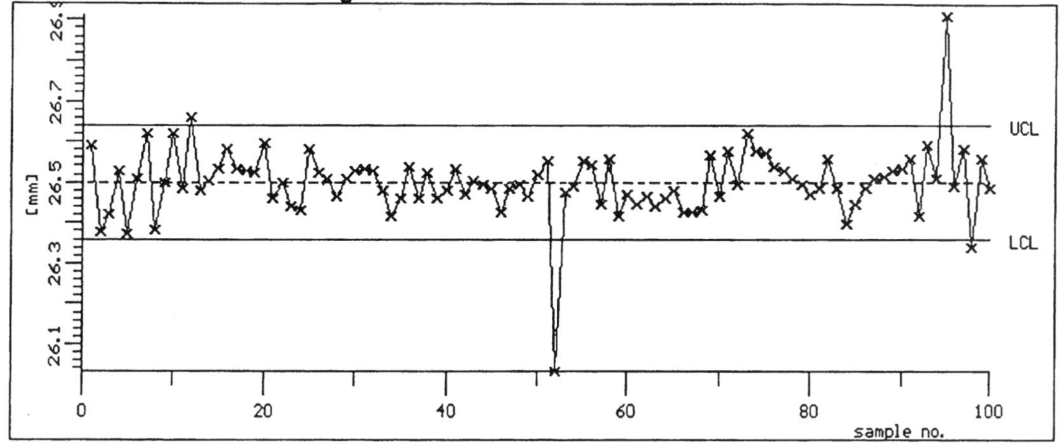

Subgroup	Result/violation	Subgroup	Result/violation
1	$\bar{x}_1 = 26.592$ mm	51 to 75	middle third 36%
12	violation of UCL	52 to 76	middle third 40%
14 to 20	run above centerline	73 to 80	run downward
26 to 50	middle third 92%	84 to 91	run upward
52	violation of LCL	95	violation of UCL
59 to 68	run below centerline	98	violation of LCL
50 to 74	middle third 40%	100	$\bar{x}_{100} = 26.482$ mm

R - control chart indicating all out of control conditions

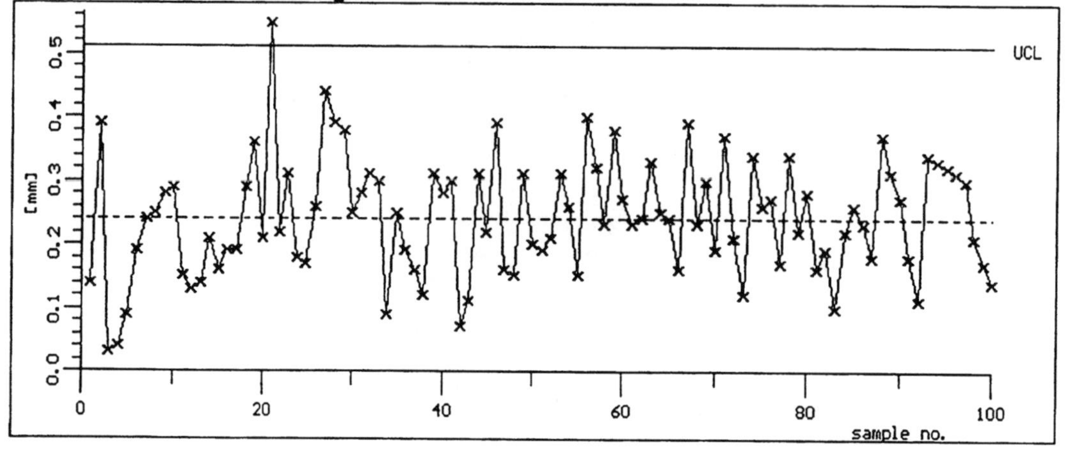

Subgroup	Result/violation	Subgroup	Result/violation
1	$R_1 = 0.14$ mm	21	violation of UCL
3 to 10	run upward	26 to 33	run above centerline
11 to 17	run below centerline	93 to 100	run downward
		100	$R_{100} = 0.14$ mm

$\overline{x}$ of all input data = 26.4991 mm $\qquad$ $\hat{\sigma}$ based on $\overline{R}$ of chart = 0.1040 mm

Do not calculate capability indices as process is not stable.

Previous capability indices are no longer valid as process is no longer in control (ref.: comment for test 1).

Test 13

Purpose

This set of test data checks whether calculations required in connection with the use of $\bar{x}/R$ control charts are performed correctly and capability indices are calculated in conformance with Ford requirements. Data supplied fit best to a normal distribution and serve the purpose of checking this calculation mode.

Procedure

Input of 500 measurements in subgroups of five representing a normal production run over more than 20 production days.

Specification = 28.50 ± 0.30 USL = 28.80 mm
LSL = 28.20 mm

Results

$\bar{x}$ of the first subgroup of 5 = 28.5174 mm
R of the first subgroup of 5 = 0.134 mm
$\bar{\bar{x}}$ of all 100 subgroups = 28.54925 mm
$\hat{\sigma}$ calculated from $\bar{R}$ of the chart = 0.049235 mm

$\bar{x}$ chart			R chart		
UCL	=	28.6153 mm	UCL	=	0.2421 mm
LCL	=	28.4832 mm			

Stability assessment: Stability confirmed.

Histogram with all input data in comparison to specification indicates all parts within specification.

Normal distribution confirmed by straight-line comparison on probability plot displayed.

Capability indices:

$\mathbf{C_p = 2.03} \qquad \mathbf{C_{pk} = 1.70}$

Within this example, capability indices will result in the same numerical values regardless of whether calculation is based on control chart data or total population.

$\overline{X}$/R - control chart indicating process performance

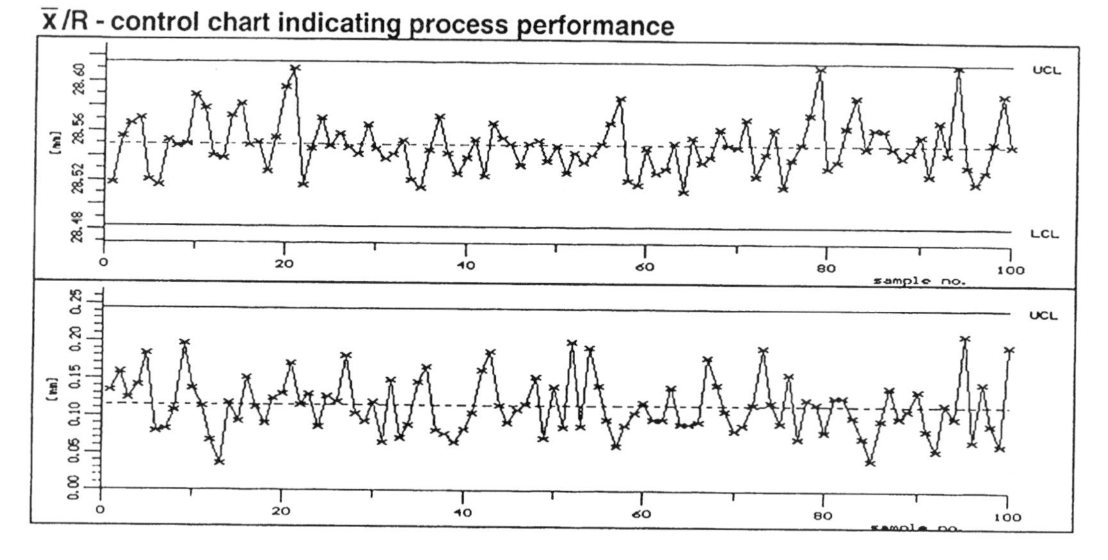

Histogram with all input data in comparison to specification

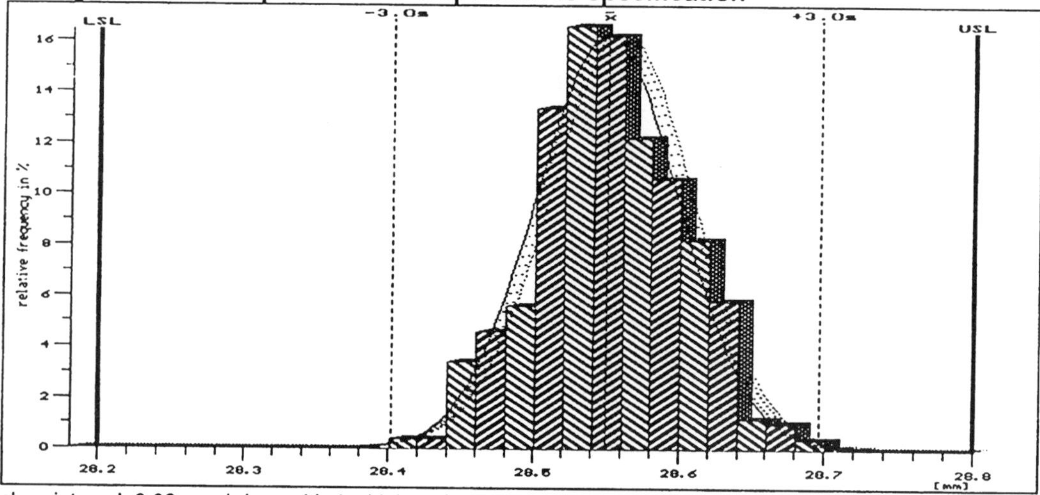

class interval: 0.02 mm / class with the highest frequency: 28.5205 mm - 28.5405 mm

Straight line comparison on probability plot for normal distribution

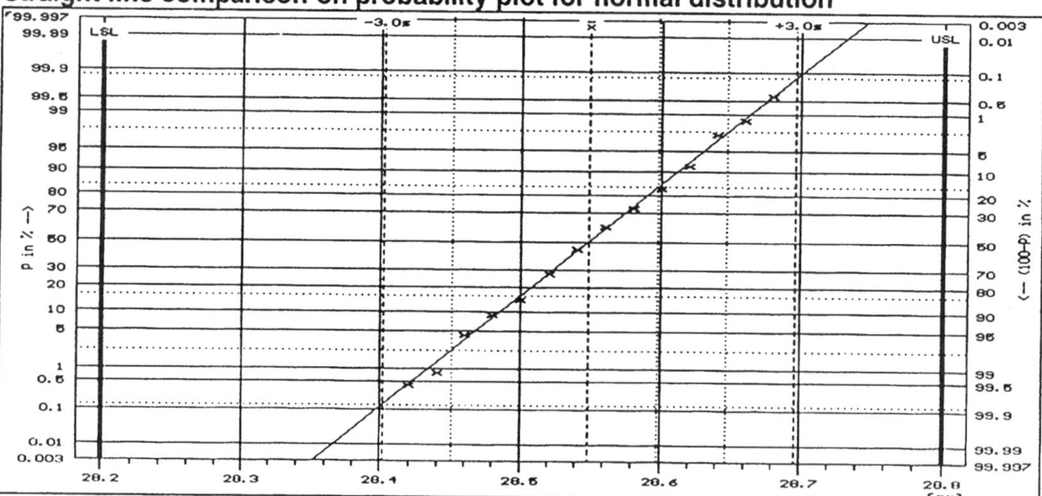

Test 14

Purpose

This set of test data checks whether calculations required in connection with a type 1 measurement system capability study based on tolerance are performed correctly.

Procedure

Input of 50 measurements as individual results.

Specification = 20.30 ± 0.15 mm   Actual value of master = 20.302 mm

Results

$\hat{\sigma}_{gauge}$	=	0.004657 mm	
C_g	=	1.61	
C_{gk}	=	1.50	
% variation(C_g)	=	9.31	%
% variation(C_{gk})	=	9.97	%

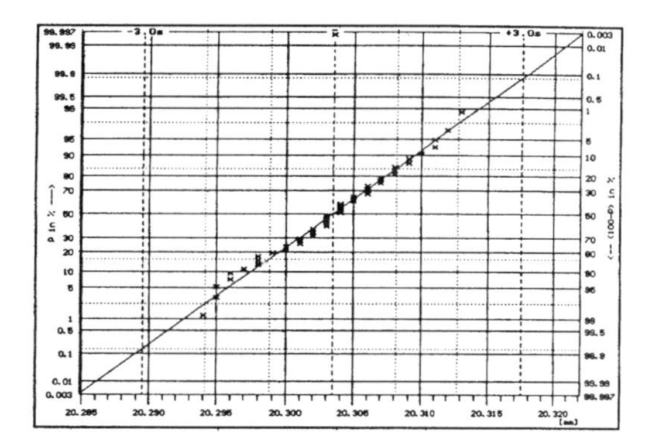

Test 15

Purpose

This set of data checks whether calculations required in connection with a type 1 measurement system capability study based on process performance are performed correctly.

Procedure

Input of 50 measurements as individual results.

Specification = 17.00 ± 0.25 mm Actual value of master = 17.05 mm

$$\hat{\sigma}_{process} = 0.07 \text{ mm}$$

Results

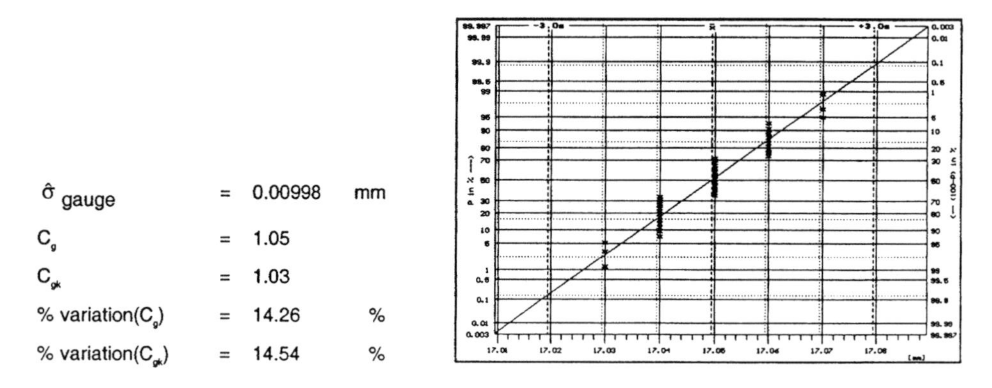

$\hat{\sigma}$ gauge	=	0.00998	mm
C_g	=	1.05	
C_{gk}	=	1.03	
% variation(C_g)	=	14.26	%
% variation(C_{gk})	=	14.54	%

Test 16

Purpose

This set of test data checks whether calculations required in connection with a type 2 measurement system capability study are performed correctly.

Procedure

Input of 60 measurements determined by three operators checking ten products twice. Results to be taken from the following table.

$$\hat{\sigma}_{process} = 0.06 \text{ mm}$$

Operator A		Operator B		Operator C	
1st trial	2nd trial	1st trial	2nd trial	1st trial	2nd trial
100.01	100.00	100.00	99.99	100.01	99.98
100.01	100.02	99.99	100.00	100.00	100.00
100.01	100.00	100.01	100.00	100.01	99.99
100.02	99.99	100.01	99.99	100.01	100.00
100.00	100.01	100.00	99.99	100.00	100.01
100.01	100.00	100.00	99.99	100.00	99.99
99.99	100.00	100.00	99.99	99.99	100.00
100.01	99.99	100.01	100.00	100.00	99.99
100.01	100.01	100.00	100.00	100.00	99.99
100.01	100.00	99.99	100.01	99.99	100.00

Results

$\bar{R}$ as average of R_A, R_B, R_C		=	0.011667 mm
Upper control limit R		=	0.038151 mm
$\bar{x}_{diff}$ as range between $\bar{x}_A$, $\bar{x}_B$, $\bar{x}_C$		=	0.007 mm
Repeatability factor = 0.06206	Repeatability	=	17.24%
Reproducibility factor = 0.02198	Reproducibility	=	6.11%
Total variation factor = 0.06583	Total variation	=	18.29%

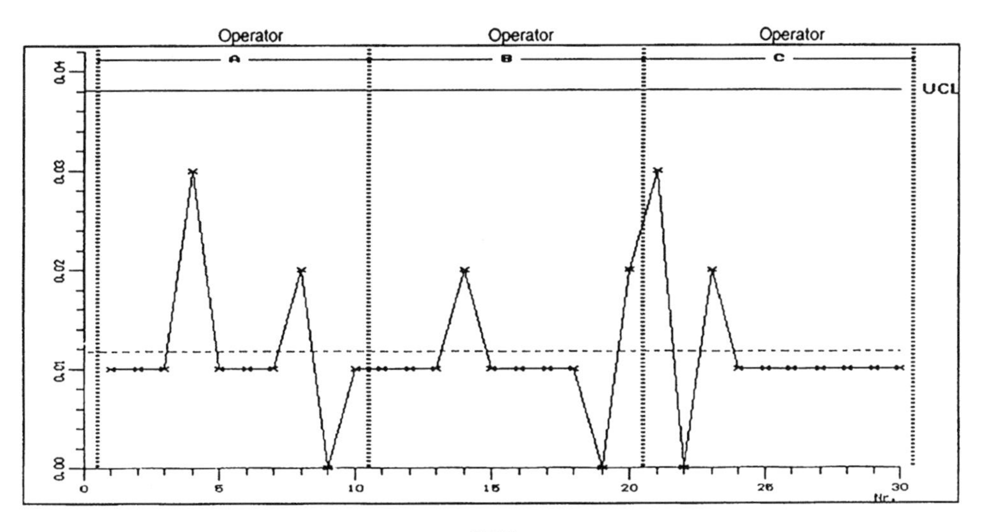

Test 17

Purpose

This set of test data checks whether calculations required in connection with a type 3 measurement system capability study are performed correctly for automatic measurement systems.

Procedure

Input of 50 measurements according to the following table as result of a study performed on 25 parts measured twice. Capability must be calculated through the standard deviation.

Specification: 6.00 ± 0.03, $T = 0.06$ mm

Part	1st trial	2nd trial	Part	1st trial	2nd trial
1	6.029	6.030	14	5.985	5.986
2	6.019	6.020	15	6.014	6.014
3	6.004	6.003	16	5.973	5.972
4	5.982	5.982	17	5.997	5.996
5	6.009	6.009	18	6.019	6.015
6	5.971	5.972	19	5.987	5.986
7	5.995	5.997	20	6.029	6.025
8	6.014	6.018	21	6.017	6.019
9	5.985	5.987	22	6.003	6.001
10	6.024	6.028	23	6.009	6.012
11	6.033	6.032	24	5.987	5.987
12	6.020	6.019	25	6.006	6.003
13	6.007	6.007			

Results

Standard deviation s_Δ	=	0.00212 mm
Gauge variation $\hat{\sigma}$	=	0.0015 mm
Upper control limit for R	=	0.004488 mm

EV = 0.00772 %EV = %R&R = 12.87% for confidence interval 99%
EV = 0.00899 %EV = %R&R = 14.99% for confidence interval 99.73%

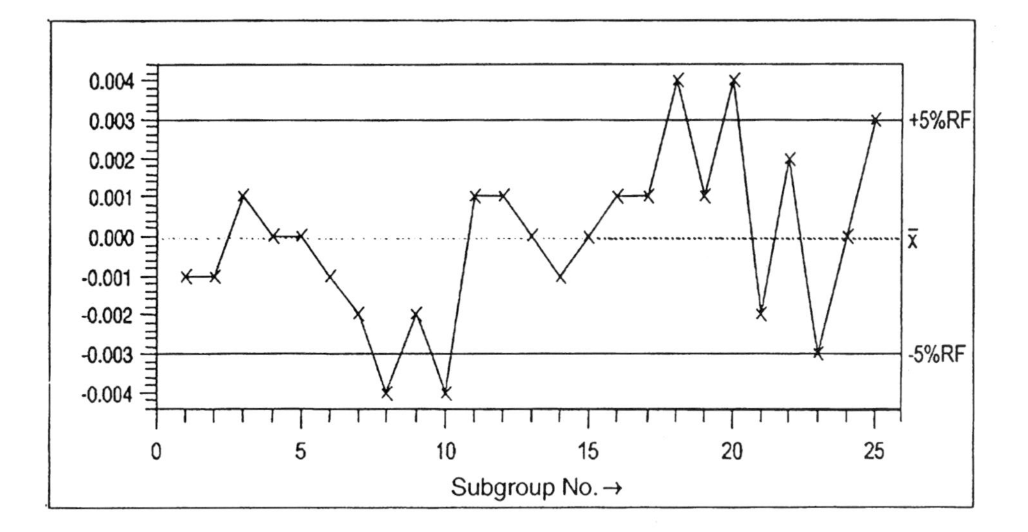
0.004
0.003
0.002
0.001
0.000
-0.001
-0.002
-0.003
-0.004
+5%RF
$\bar{X}$
-5%RF
0
5
10
15
20
25
Subgroup No. →

Test 18

Purpose

This set of test data checks whether calculations required in connection with a type 3 measurement system capability study are performed correctly for automatic measurement systems according to Ford guidelines.

Procedure

Input of 50 measurements according to the following table as result of a study performed on 25 parts measured twice. Capability must be calculated through the range.

Specification: 56.83 ± 0.01, T = 0.02 mm

Part	1st trial	2nd trial	Part	1st trial	2nd trial
1	56.8338	56.8321	14	56.8273	56.8282
2	56.8310	56.8320	15	56.8235	56.8247
3	56.8359	56.8368	16	56.8177	56.8186
4	56.8291	56.8257	17	56.8332	56.8339
5	56.8294	56.8290	18	56.8305	56.8321
6	56.8206	56.8202	19	56.8228	56.8233
7	56.8410	56.8401	20	56.8310	56.8300
8	56.8305	56.8295	21	56.8318	56.8312
9	56.8269	56.8274	22	56.8344	56.8358
10	56.8386	56.8386	23	56.8327	56.8329
11	56.8321	56.8323	24	56.8213	56.8221
12	56.8331	56.8334	25	56.8205	56.8215
13	56.8382	56.8400			

Results

Average range $\bar{R}$	=	0.00932 mm
Gauge variation $\hat{\sigma}$	=	0.000826 mm
Upper control limit for R	=	0.00305 mm

EV = 0.004254	%EV = %R&R = 21.27%	for confidence interval 99%
EV = 0.004956	%EV = %R&R = 24.78%	for confidence interval 99.73%

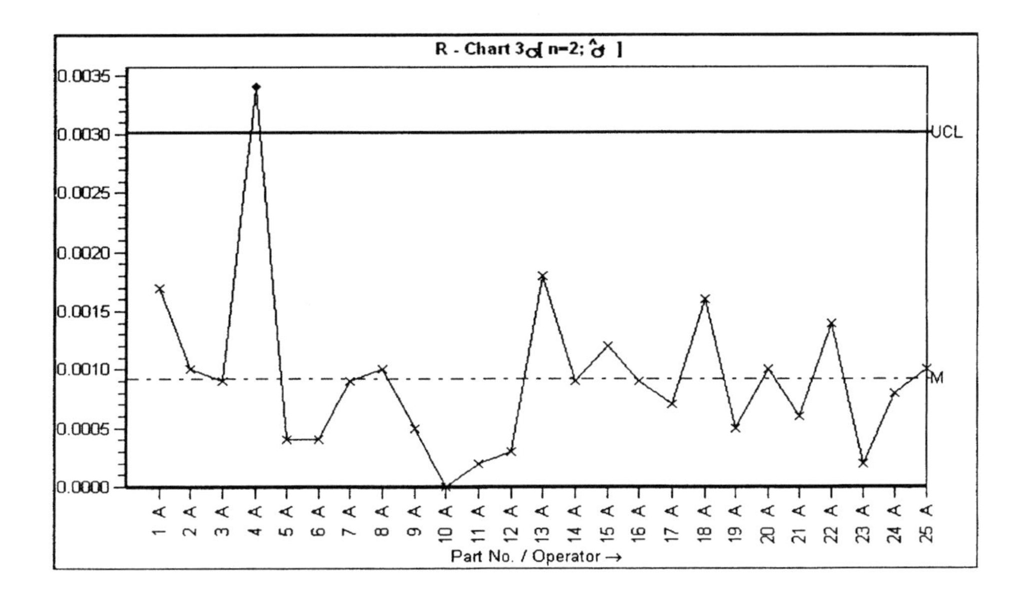
R - Chart 3σ[n=2; σ̂]
0.0035
0.0030
0.0025
0.0020
0.0015
0.0010
0.0005
0.0000
UCL
M
1 A
2 A
3 A
4 A
5 A
6 A
7 A
8 A
9 A
10 A
11 A
12 A
13 A
14 A
15 A
16 A
17 A
18 A
19 A
20 A
21 A
22 A
23 A
24 A
25 A
Part No. / Operator →

12 Reference

12.1 Bibliography

[1] A.I.A.G. 1991. *Fundamental Statistical Process Control, Reference Manual.* Michigan, USA.

[2] A.I.A.G. 1990. *Measurement Systems Analysis, Reference Manual.* Michigan, USA.

[3] Chrysler Corp., Ford Motor Co., General Motors Corp. 1994. *Quality System Requirements, QS-9000.*

[4] Abramowitz, M., and I.A. Stegun. 1972. *Handbook of Mathematical Functions.* New York: Dover Publications.

[5] Anghel, C., H. Hausberger, and W. Streinz. 1992; 1993. Unsymmetriegrößen erster und zweiter Art richtig auswerten. Teil 1: Unsymmetriegrößen erster Art. *QZ* 37: 12 (1992). Teil 2: Unsymmetriegrößen zweiter Art. *QZ* 38: 1 (1993). München: Carl Hanser Verlag.

[6] Anghel, C. 1993. Nomogramme für Betragsverteilungen erster und zweiter Art. *QZ* 38: Heft 9, S. 523 bis 524. München: Carl Hanser Verlag.

[7] Bläsing, J.P. 1986ff. *Praxishandbuch der Qualitätssicherung Band 1–4* München: gmft Gesellschaft für Management und Technologie.

[8] D'Agostino R. 1971. An omnibus test of normality for moderate and large size samples. *Biometrika* 58: S. 341–348.

[9] DataMyte Corporation. 1987. *DataMyte Handbook.* 3d ed. Minnetonka.

[10] DGQ–NTG Schrift Nr. 12-51: Software-Qualitätssicherung-Aufgaben, Möglichkeiten, Lösungen. VDE-Verlag GmbH, Offenbach.

[11] DGQ: Deutsche Gesellschaft für Qualität. 1993. Schrift Nr. 11-04: Begriffe im Bereich der Qualitätssicherung. Berlin: Beuth Verlag.

[12] DGQ: Deutsche Gesellschaft für Qualität. 1988. Lehrgangsunterlagen: SPC II (November). Frankfurt.

[13] DGQ: Deutsche Gesellschaft für Qualität. Lehrgangsunterlagen: QII. Frankfurt.

[14] DGQ: Deutsche Gesellschaft für Qualität. Arbeitsgruppe 136: *Prüfmittelüberwachung.* Frankfurt.

[15] Dietrich, E., and A. Schulze. 1991. *Fähige Meßverfahren—Die Basis der statistichen Prozßlenkung.* QZ 36: 3, S. 153–159. München: Carl Hanser Verlag.

[16] Dietrich, E., and A. Schulze. 1992. *Qualitätsregelkarten und Fähigkeitsindizes bilden eine Einheit. Berechnung der Fähigkeitsindizes im Vergleich.* QZ 37: 8, S. 466 bis 472. München: Carl Hanser Verlag.

[17] Dietrich, E., and A. Schulze. 1994. *Geeignete Verteilungsmodelle finden.* Birkenau: Q-Das® GmbH.

[18] Dietrich, E., and A. Schulze. 1994. *Out of control?* QZ 39: 4, S. 416–418. München: Carl Hanser Verlag.

[19] Dietrich, E., and A. Schulze. 1989; 1990. *Rechnergestützte Verfahren zur statistischen Prozeßlenkung.* Ein Modell–Vorschlag. Teil 1: QZ 34: 12, S. 677–681 (1989). Teil 2: QZ 35: 1, S. 49–53 (1990). München: Carl Hanser Verlag.

[20] Dietrich, E., and A. Schulze. 1994. *Maschinen-und Prozeßqualifikation.* Birkenau: Q-Das® GmbH.

[21] Dietrich, E., and A. Schulze. 1993. *Zuverlässigkeitsanalyse.* Birkenau: Q-Das® GmbH.

[22] DIN: Deutsches Institut für Normung. DIN 1319, Teil 1-4: Grundbegriffe der Meßtechnik, Behandlung von Unsicherheiten bei der Auswertung von Messungen. Berlin: Beuth Verlag.

[23] DIN: Deutsches Institut für Normung. 1987. Din 2257, Teil 1-2: Begriffe der Längenprüftechnik: Einheiten, Tätigkeiten, Prüfmittel, meßtechnische Begriffe. Berlin: Beuth Verlag.

[24] DIN: Deutsches Institut für Normung. 1989. DIN 55350, Teil 13: Begriffe der Qualitätssicherung und Statistik; Begriffe zur Genauigkeit von Ermittlungsverfahren und Ermittlungsergebnissen. Berlin: Beuth Verlag.

[25] DIN: Deutsches Institut für Normung. DIN ISO 10012, Teil 1: Forderungen an die Qualitätssicherung von Meßmitteln. Berlin: Beuth Verlag.

[26] DIN: Deutsches Institut für Normung. DIN ISO 5479: Test auf Normalverteilung. Berlin: Beuth Verlag.

[27] DIN: Deutsches Institut für Normung. DIN ISO 5725: Ermittlung der Wiederhol- und Vergleichspräzision von festgelegten Meßverfahren durch Ringversuche. Berlin: Beuth Verlag.

[28] DIN: Deutsches Institut für Normung. 1993. DIN Normenentwurf 55303, Teil 7 (Juli). Berlin: Beuth Verlag.

[29] DIN: Deutsches Institut für Normung. 1993. DIN ISO 9000ff: Qualitätsmanagement- und Qualitätssicherungsnormen. Berlin: Beuth Verlag.

[30] DIN: Deutsches Institut für Normung. 1986. DIN Taschenbuch 11: Längenprüftechnik 1. Berlin: Beuth Verlag.

[31] DIN: Deutsches Institut für Normung. 1995. Leitfaden zur Angabe der Unsicherheit beim Messen (GUM). Translation of *Guide to the Expression of Uncertainty in Measurement.* Berlin: Beuth Verlag.

[32] Duncan, A.J. 1974. *Quality Control and Industrial Statistics.* Irwin: Homewood.

[33] Elderton, W. P., and N.L. Johnson. 1969. *Systems of Frequency Curves.* Cambridge: University Press.

[34] Ford AG: EU 882. 1991. Richtlinie für Untersuchungen der vorläufigen und fortdauernden Prozeßfähigkeit. Köln.

[35] Ford AG: Fertigungseinrichtungen. 1992; 1995. Richtlinie zur Leistungsbeurteilung. Köln (Januar 1992). Auswertung von Positionstoleranzen. Köln (Februar 1995).

[36] Ford Motor Co.: EU 1880A. 1989. *Measurement System and Equipment Capability* (Dezember). London.

[37] Ford Motor Co.: EU 1880B. 1997. Richtlinie: Fähigkeit von Mess-Systemen und Messmitteln. Translation of EU 1880A. Köln (Oktober).

[38] Ford Motor Co.: EU 880. 1987. *SPC Instruction Guide.* London.

[39] Ford Motor Co.: EU 881. 1986. *SPC for Dimensionless Materials.* London.

[40] Ford Motor Co.: Q-DAS GmbH. 1991. Ford Testbeispiele, Beurteilung von SPC Software. Birkenau.

[41] Ford Motor Co.: Q101. 1990. *Worldwide Quality System Standard.*

[42] Franzkowski, R. 1981. Beurteilung der Normalität. *QZ Qualität und Zuverlässigkeit* 26: 4. München: Carl Hanser Verlag.

[43] Franzkowski, R. 1980. DGQ-Schrift Nr. 16-36: Multivariate Fertigungsüberwachung. Berlin: Beuth Verlag.

[44] Geiger, W. 1976. Gefaltete und Betragsverteilungen. *QZ* 21: 7, S. 156-160. Freiburg und Berlin: Rudolf Haufe Verlag.

[45] General Motors Co. 1987. Journal GM 1390 (Juni). Detroit.

[46] Graf, U., H.J. Henning, and K. Stange. *Formeln und Tebellen der mathematischen Statistik.* Berlin: Springer Verlag; New York: Heidelberg.

[47] Hartung, J. 1982. *Statistik-Handbuch der angewandten Statistik.* München, Wien: R. Oldenbourg Verlag.

[48] ISO/TC 69/S 6/WG 3. Measurement methods and results: *Measurement uncertainty.*

[49] ISO/TC 69/S 6/WG 4. Measurement methods and results: *Linear Calibration using reference material.*

[50] John, B. 1979. *Statistische Verfahren für Technische Meßreihen.* München, Wien: Carl Hanser Verlag.

[51] Kreyzig, E. 1977. *Statistische Methoden und Ihre Anwendung.* Göttingen: Vandenhoeck & Ruprecht.

[52] Lehmann, E.L. 1986. *Testing Statistical Hypothesis.* 2d ed. New York: Wiley & Sons.

[53] Lehmann, E.L. 1983. *Theory of Point Estimation.* New York: Wiley & Sons.

[54] Masing, W. 1994. Handbuch Qualitätsmanagement, 3, überarbeitete Auflage. München: Carl Hanser Verlag.

[55] Mercedes Benz AG. 1991. *Statistische Prozeßregelung (SPC). Leitfaden zur Anwendung.* Stuttgart.

[56] Mercedes Benz AG. 1994. *Meßgerätefähigkeitsuntersuchung. Leitfaden zur Anwendung.* Stuttgart.

[57] Opel, Vauxhall, General Motors. 1996. *Ergänzung der GM Richtlinie B-01* (November): *Abnahme von Meßmitteln für PT und Chassis-Werke; Qualitätsabnahme von Fertigungseinrichtungen LVQ-1.* Rüsselsheim.

[58] PSA Peugeot, Citroën, Renault. 1991. CNOMO Norm E41.32.110.N (Juli). Produktionsmittel, Zulassung der Funktionsfähigkeit von Produktionsmitteln zur Ausführung von Merkmalen entsprechend einem Normalgesetz.

[59] PSA Peugeot, Citroën, Renault. 1991. CNOMO Norm E41.36.110.N (Oktober). Produktionsmittel, Zulassung der Funktionsfähigkeit von Meβmitteln, Spezifische Prüfmittel.

[60] Rinne, H., and H.-J Mittag. 1993. *Statistical Methods of Quality Assurance.* London: Chapman & Hall.

[61] Robert Bosch GmbH. 1990. Schriftenreihe "Qualitätssicherung in der Bosch-Gruppe Nr. 9"; Technische Statistik Maschinen- und Prozeβfähigkeit. Stuttgart.

[62] Robert Bosch GmbH. 1990. Schriftenreihe "Qualitätssicherung in der Bosch-Gruppe Nr. 10"; Technische Statistik Fähigkeit von Meβeinrichtungen. Stuttgart.

[63] Shapiro, S. 1980. *How to test normality and other distributional assumptions.* Vol. 3. American Society for Quality Control.

[64] Seimens AG. 1993. *Maschinen- und Prozeßqualifikation.* München.

[65] Timischl, W. 1996. Qualitätssicherung–Statistische Methoden, 2, durchgesehene Auflage. München: Carl Hanser Verlag.

[66] VDA: Verband der Automobilindustrie. 1984. VDA-Schrift Nr. 3: Zuverlässigkeitssicherung bei Automobilherstellern und Lieferanten. Frankfurt: VDA.

[67] VDMA: Verband Deutscher Maschinen- und Anlagenbau e.V. 1995. VDMA 8669: Fähigkeitsuntersuchung zur Abnahme spanender Werzeugmaschinen. Berlin: Beuth-Verlag GmbH.

[68] Volkswagen AG; Audi AG. 1995. *BV 1.01; Betriebsmittel-Vorschriften* (Mai).

12.2 Abbreviations

$1 - \alpha$ . . . . . . confidence level
α . . . . probability for type 1 error = error probability
A_a . . . . accuracy
a_n . . . . parameter for determination of the variance estimator (total population) from the variance (sample)
ANOVA . . . analysis of variance
ARM . . . . . average and range method
AV . . . . reproducibility/appraiser variation
%AV . . . . . . reproducibility/appraiser variation in % relative to a reference figure
β . . . . probability for type 2 error
χ^2 . . . . critical value of the χ^2 distribution
C_g . . . . gauge capability potential
C_{gk} . . . . gauge capability index
c_{crit} . . . . distance of the average from the specification limits in standard deviation units
C_m . . . . machine capability potential
C_{mk} . . . . machine capability index
c_n . . . . parameter determined by the ratio of $\sigma_{\tilde{x}}$ to $\sigma_{\bar{x}}$
C_p . . . . continuing capability potential
C_{pk} . . . . continuing capability index
D_4 . . . . factor for the number of trials per operator for the determination of the control limits (R chart)
Δ_i . . . . difference between two measurement series
d_n . . . . parameter for determination of the variance estimator (total population) from the range (sample)
EV . . . . repeatability/equipment variation
%EV . . . . . . repeatability/equipment variation in % relative to a reference figure
f . . . . degree of freedom
F . . . . critical value of the F-distribution
g . . . . individual probability for discrete characteristics
G . . . . sum probability for discrete characteristics
g(u) . . . . . . standard density function of the normal distribution
G(u) . . . . . . standard cumulative function of the normal distribution
g_1 . . . . skewness
g_2 . . . . excess
H_0 . . . . null hypothesis
H_1 . . . . alternative hypothesis
IA . . . . interaction
k . . . . number of classes in the histogram
k . . . . number of samples
K_1 . . . . factor for number of trials per operator
K_2 . . . . factor for number of operators
K_3 . . . . factor for calculation of part variation
LCL . . . . . . lower control limit

LLV . . . . . . lower limit value
LSL. . . . . . . lower specification limit
LSL. . . . . . . lower warning limit
μ . . . . arithmetic mean; average of the total population
N . . . . batch size
n. . . . . subgroup size
UCL . . . . . . upper control limit
ULV . . . . . . upper limit value
USL . . . . . . upper specification limit
UWL . . . . . upper warning limit
P. . . . . general specification of the probability
$\hat{p}$. . . . . proportion of nonconforming units (sample)
p. . . . . proportion of nonconforming units (population)
P_a . . . . acceptance probability
P_p . . . . preliminary capability potential
P_{pk} . . . . preliminary capability index
PV . . . . part variation
%PV . . . . . . part variation in % relative to a reference figure
QCC. . . . . . quality control chart
r . . . . number of trials per operator
R. . . . . range
$\bar{\bar{R}}$. . . . . average range of variances of all operators (from $\bar{R}_{A,B,C}$)
$\bar{R}_{A,B,C}$ . . . . . average range of variances of the individual operators
r_{tot} . . . . regression coefficient of all the values
$r_{25\%}$ . . . . regression coefficient of 25% of the values
R&R . . . . . . repeatability & reproducibility
%R&R . . . . repeatability & reproducibility in % relative to a reference figure
RF. . . . . reference figure
s . . . . standard deviation of the subgroup
s^2. . . . . variance of the subgroup
$\bar{s}$ . . . . average sample standard deviations
s_Δ . . . . standard deviation of deviations between two trials
s_g . . . . standard gauge deviation
S_M . . . . total gauge variation
σ. . . . . population standard deviation
σ^2 . . . . population variance
$\hat{\sigma}$. . . . . estimator for the standard deviation of the total population
t . . . . critical value of the t distribution
T. . . . . total tolerance
TV . . . . total variation of a measurement system
u. . . . . standard size for $(x - \mu)/\sigma$ of the normal distribution
ω. . . . . class interval
x . . . . number of nonconformities with discrete characteristics
x_i. . . . . value of a continuous characteristic
x_{max} . . . . maximum value

x_{min} . . . . minimum value
$\bar{x}$ . . . . average of values in a subgroup
$\bar{x}_{A,B,C}$ . . . . . . average of measurement series (trials)
$\bar{x}_{Diff}$ . . . . maximum difference between averages from several trials (from $\bar{x}_{A,B,C}$)
$\bar{x}_g$ . . . . average gauge
$\bar{\bar{x}}$ . . . . average of subgroup averages; the measured process average
$\tilde{x}$ . . . . the median of values in a subgroup
x_m . . . . true value of standard, setting master

Index

Page numbers in *italic* indicate figures. Page numbers followed by "t" indicate tables.

D

E

Q